Quantum Processes, Systems, and Information

A new and exciting approach to the basics of quantum theory, this undergraduate textbook contains extensive discussions of conceptual puzzles and over 800 exercises and problems.

Beginning with three elementary "qubit" systems, the book develops the formalism of quantum theory, addresses questions of measurement and distinguishability, and explores the dynamics of quantum systems. In addition to the standard topics covered in other textbooks, it also covers communication and measurement, quantum entanglement, entropy and thermodynamics, and quantum information processing.

This textbook gives a broad view of quantum theory by emphasizing dynamical evolution, and exploring conceptual and foundational issues. It focuses on contemporary topics, including measurement, time evolution, open systems, quantum entanglement, and the role of information.

Benjamin Schumacher is Professor of Physics at Kenyon College. He coined the term "qubit" and invented quantum data compression, among other contributions to quantum information theory.

Michael D. Westmoreland is Professor of Mathematics at Denison University. Trained as an algebraist, for many years he has researched nonstandard logics, models of computation, and quantum information theory.

The authors are long-time research collaborators and have made numerous joint contributions to quantum channel capacity theorems and other aspects of quantum information science.

Quantum Processes, Systems, and Information

BENJAMIN SCHUMACHER

Kenyon College

MICHAEL D. WESTMORELAND

Denison University

CAMBRIDGE
UNIVERSITY PRESS

CAMBRIDGE
UNIVERSITY PRESS

University Printing House, Cambridge CB2 8BS, United Kingdom

Cambridge University Press is part of the University of Cambridge.

It furthers the University's mission by disseminating knowledge in the pursuit of education, learning and research at the highest international levels of excellence.

www.cambridge.org
Information on this title: www.cambridge.org/9780521875349

© B. Schumacher and M. Westmoreland 2010

First published 2010

A catalogue record for this publication is available from the British Library

Library of Congress Cataloguing in Publication data

Schumacher, Benjamin.
Quantum processes, systems, and information / Benjamin Schumacher, Michael Westmoreland.
p.   cm.
ISBN 978-0-521-87534-9   (Hardback)
1. Quantum theory–Textbooks.   I. Westmoreland, Michael D.   II. Title.
QC174.12.S385 2010
530.12–dc22

2009039353

ISBN 978-0-521-87534-9 Hardback

Contents

Preface

The last two decades have seen the development of the new field of quantum information science, which analyzes how quantum systems may be used to store, transmit, and process information. This field encompasses a growing body of new insights into the basic properties of quantum systems and processes and sheds new light on the conceptual foundations of quantum theory. It has also inspired a great deal of contemporary research in optical, atomic, molecular, and solid state physics. Yet quantum information has so far had little impact on the way that quantum mechanics is taught.

Quantum Processes, Systems, and Information is designed to be both an undergraduate textbook on quantum mechanics and an exploration of the physical meaning and significance of information. We do not regard these two aims as incompatible. In fact, we believe that attention to both subjects can lead to a deeper understanding of each. Therefore, the essential "story" of this book is very different from that found in most existing undergraduate textbooks.

Roughly speaking, the book is organized into five parts:

- Part I (Chapters 1–5) presents the basic outline of quantum theory, including a development of the essential ideas for simple "qubit" systems, a more general mathematical treatment, basic theorems about information and uncertainty, and an introduction to quantum dynamics.
- Part II (Chapters 6–9) extends the theory in several ways, discussing quantum entanglement, ideas of quantum information, density operators for mixed states, and dynamics and measurement on open systems.
- Part III (Chapters 10–14) uses the basic theory to discuss several specific quantum systems, including particles moving in one or more dimensions, systems with orbital or intrinsic angular momentum, harmonic oscillators and related systems, and systems containing many particles.
- Part IV (Chapters 15–17) deals with the stationary states of particles moving in 1-D and 3-D potentials, including variational and perturbation methods.
- Part V (Chapters 18–20) further develops the ideas of quantum information, examining quantum information processing, NMR systems, the meaning of classical and quantum entropy, and the idea of error correction.

These chapters are followed by Appendices on probability (Appendix A), Fourier series and Fourier transforms (Appendix B), Gaussian functions (Appendix C) and generalized quantum evolution (Appendix D).

Part I is the basis for all further work in the text. The remaining parts follow two quasi-independent tracks:

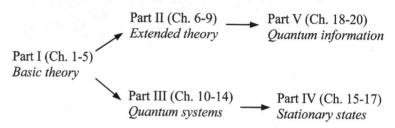

Part II (Ch. 6-9) → Part V (Ch. 18-20)
Extended theory   *Quantum information*

Part I (Ch. 1-5)
Basic theory

Part III (Ch. 10-14) → Part IV (Ch. 15-17)
Quantum systems   *Stationary states*

Thus, this book could be used as a text for either an upper-track or a lower-track style of course.[1]

We, however, strongly recommend including material from both tracks. This book is written from the conviction that a modern student of physics needs a broader set of concepts than conventional quantum mechanics textbooks now provide. Unitary time evolution, quantum entanglement, density operator methods, open systems, thermodynamics, concepts of communication, and information processing – all of these are at least as essential to the meaning of quantum theory as is solving the time-independent Schrödinger equation.

As we wrote this book, we had the benefit of useful and inspiring conversations with a great many colleagues and friends. Among these we wish particularly to express our gratitude to Charles Bennett, Herb Bernstein, Carl Caves, Chris Fuchs, Lucien Hardy, David Mermin, Michael Nielsen, and Bill Wootters. In a similar vein, we would also like to thank the other members of the (fondly remembered) Central Ohio Quantum Conspiracy: Michael Nathanson, Kat Christandl Gillen, and Lee Kennard. We have also received valuable input on the book from Matthew Neal and Ron Winters of Denison University and Ian Durham of St. Anselm College.

An early version of this book was used as an experimental textbook for a quantum mechanics course at Kenyon College, and the students in that course deserve their own thanks: Andrew Berger, Stephanie Hemmingson, John Hungerford, Lee Kennard, Joey Konieczny, Jeff Lanz, Max Lavrentovich, David Lenkner, Nikhil Nagendra, Alex Rantz, David Slochower, Jeremy Spater, Will Stanton, Adam Tassile, Chris Yorlano, and Matt Zaremsky.

Our faculty colleagues at both Kenyon College and Denison University have been wonderfully supportive throughout this project. One of us (MDW) is grateful to acknowledge a Robert C. Good Faculty Fellowship from Denison University. We also thank our editor at Cambridge University Press, Simon Capelin, for providing the initial impetus and for considerable patience and encouragement throughout.

---

[1] There are a few minor dependencies not indicated in this chart, but these can be easily accommodated in practice. The general discussion of composite systems in Section 6.1 is a useful preparation for work on many-particle systems in Chapter 14. The analysis of thermal states of a ladder system (Section 13.4) depends on the density operator formalism, but may be omitted if Chapter 8 has not been covered.

We are more grateful than we can readily express for the continuing love and support of our wives, Carol Schumacher and Bonnie Westmoreland. And finally, a word to our children, Barry, Patrick and Carolyn Westmoreland, and Sarah and Glynis Schumacher: This is what we have been so busy doing for the last few years. We hope you like it, because we are dedicating it to you.

Benjamin Schumacher
Department of Physics
Kenyon College

Michael D. Westmoreland
Department of Mathematics
and Computer Science
Denison University

Bits and quanta

1.1 Information and bits

On the evening of 18 April 1775, British troops garrisoned in Boston prepared to move west to the towns of Lexington and Concord to seize the weapons and capture the leaders of the rebellious American colonists. The colonists had anticipated such a move and prepared for it. However, there were two possible routes by which the British might leave the city: by land via Boston Neck, or directly across the water of Boston Harbor. The colonists had established a system of spies and couriers to carry the word ahead of the advancing troops, informing the colonial militias exactly when, and by what road, the British were coming.

The vital message was delivered first by signal lamps hung in the steeple of Christ Church in Boston and observed by watchers over the harbor in Charlestown. As Henry Wadsworth Longfellow later wrote,

> One if by land, and two if by sea;
> And I on the opposite shore will be,
> Ready to ride and spread the alarm
> Through every Middlesex village and farm . . .

Two lamps: the British were crossing the harbor. A silversmith named Paul Revere, who had helped to organize the communication network, was dispatched on horseback to carry the news to Lexington. He stopped at houses all along the way and called out the local militia. By dawn on 19 April, the militiamen were facing the British on Lexington Common. The first battle of the American Revolutionary War had begun.

In the United States, Paul Revere[1] is remembered as a hero of the Revolutionary War, not for his later military career but for his "midnight ride" in 1775. Revere is famous to this day as a carrier of information.

Information is one of our central themes, and over the course of this book we will formalize the concept, generalize it, and subject it to extensive analysis. The story of Paul Revere illustrates several key ideas. What, after all, is information? We can give a heuristic definition that, though we will later stretch and generalize it almost beyond recognition, will serve to guide our discussion.

Information is the ability to distinguish reliably between possible alternatives.

---

[1] William Dawes and Dr. Samuel Prescott, who accompanied Revere on the ride and helped spread the word, are almost forgotten – perhaps because they were never immortalized in verse by Longfellow.

Before Paul Revere's ride, the militiamen of Lexington could not tell whether the British were coming or not, or by what route. After Revere had reached them, they could distinguish which possibility was correct. They had gained *information*.

We can also distinguish between an abstract *message* and the physical *signal* that represents the message. The association between message and signal is called a *code*. For instance, here is the code used by the colonists for their church steeple signal.

Signal	Message
0 lamp	The British are not coming.
1 lamp	The British are coming by land.
2 lamps	The British are coming by sea.

Mathematically, the code is a function from a set of possible messages to a set of possible states of the physical system that will carry the message – in this case, the lamp configuration in the church tower.

There is no requirement that all possible signals are used in the code. (The Boston spies could have hung three lamps, or six, though their distant compatriots would have been rather perplexed.) On the other hand, the association between message and signal should be one-to-one, so that distinct messages are represented by distinct signals. Otherwise, it is not possible to deduce the message reliably from the signal.

Another very important point is that information can be *transformed* from one physical representation to another, so that the same message can be encoded into quite different physical signals. Paul Revere's message was not only represented by lamps in a church, but also by neural activity in his brain and then by patterns of sound waves as he cried, "The British are coming!"

Exercise 1.1 Identify at least seven distinct physical representations that this sentence has had from the time we wrote it to the time you read it.

The fact that the same message can be carried by very different signals is a fundamental truth about information.

This *transformability* of information allows us to simplify matters considerably, for we can always represent a message using signals of a standard type. The universal "currency" for information theory is the *bit*. The term *bit* is a generic term for a physical system with two possible distinguishable states. The states may be designated *no* and *yes*, *off* and *on*, or by the binary digits 0 and 1. These two states may be distinct voltage levels in an electrical device, two directions of magnetization in a small region of a computer disk, the presence or absence of a light pulse, two possible patterns of ink on a piece of paper, etc. All of these bits are *isomorphic*, in that information represented by one type of bit can be converted to another type of bit by a physical process.

A single bit has a limited "capacity" to represent information. We cannot store an entire book in one bit. The reason is that there are too many possible books (that is, possible messages) to be represented in a one-to-one manner by the two states 0 and 1. Somewhere in the code, we would inevitably have something like this:

Signal	**Message**
⋮	
0	*Alice in Wonderland* by Lewis Carroll
0	*The Guide for the Perplexed* by Moses Maimonides
⋮	

From a bit in the state 0, it would be impossible to choose the correct book in a reliable way.

To represent an entire book, we must use strings or sequences of many bits. If we have a string of n bits, then the number of distinct states available to us is

$$\text{\# of states} = \underbrace{2 \times \cdots \times 2}_{n \text{ times}} = 2^n. \tag{1.1}$$

Exercise 1.2 A *byte* is a string of eight bits. How many possible states are there for one byte?

Suppose that there are M possible messages. If the number of possible signals is at least as large as M, then we can find a code in which the message can be reliably inferred from the signal. We can thus determine whether the message can be represented by n bits.

- If $M > 2^n$, then n bits are not enough.
- If $M \le 2^n$, then n bits are enough.

The number n of bits necessary to represent a given message is a way of measuring "how much information" is in the message. This is a very practical sort of measure, since it tells us what resources (bits) are necessary to perform a particular task (represent the message faithfully).

We define the *entropy H* of our message to be

$$H = \log M, \tag{1.2}$$

where M is the number of possible messages. (From now on, unless we otherwise indicate, "log" will denote the logarithms with base 2: $\log \equiv \log_2$.) The entropy H is a measure of the information content of the message. From our discussion above, we see that if $n < H$, n bits will not be enough to represent the message faithfully. On the other hand, if $n \ge H$, then n bits will be enough. Thus, H measures the number of bits that the message requires.[2]

We can think of H as a measure of *uncertainty* – that is, of how much we do not know before we get the message. It is also a measure of how our uncertainty is reduced when we identify which message is the right one. In other words, before we receive and decode the signal, there are M possibilities and $H = \log M$. Afterward, we have uniquely identified the right message, and the entropy is now $H' = \log 1 = 0$.

We have called H the "entropy," which is the name used for H by Claude Shannon in his pioneering work on the mathematical theory of information. The name harkens back to

---

[2] Anticipating later developments, we should note here that our definition of H implicitly assumes that the M possible messages are all *equally likely*. To cope with more general situations, we will need a more general expression for the entropy. However, Eq. 1.2 will do for our present purposes.

thermodynamics, and for very good reason. Ludwig Boltzmann showed that if a macroscopic system has W possible microscopic states, then the thermodynamic entropy S_θ is

$$S_\theta = k_B \ln W, \qquad (1.3)$$

where $k_B = 1.38 \times 10^{-23}$ J/K, called *Boltzmann's constant*. This famous relation (which is inscribed on Boltzmann's tomb in Vienna) can be viewed as a fundamental link between information and thermodynamics. Up to an overall constant factor, the thermodynamic entropy S_θ is just a measure of our uncertainty of the microstate of the system.

Exercise 1.3 A liter of air under ordinary conditions has a thermodynamic entropy of about 5 J/K. How many bits would be necessary to represent the microstate of a liter of air?

We will have more to say about the connection between information and thermodynamics later on.

Suppose A and B are two messages, having M_A and M_B possible values respectively. The two messages taken together form a joint message that we denote AB. If A and B are independent of each other, every combination of A and B values is a possible joint message, and so $M_{AB} = M_A M_B$. In this case the entropy is additive:

$$H(AB) = H(A) + H(B). \qquad (1.4)$$

On the other hand, if the messages are not independent, it may be that some combinations of A and B are not allowed, so that the joint entropy $H(AB)$ may be less than $H(A) + H(B)$.

Exercise 1.4 Suppose A and B each have 16 possible values. What is the joint entropy $H(AB)$ (a) if the messages are independent, and (b) if B is known to be an exact copy of A?

How much information was contained in the message sent from the Christ Church steeple in 1775? There were three possible messages, and so $H = \log 3 \approx 1.58$. This means that one bit would not suffice to represent the message, but two bits would be more than enough. That much is clear; but can we give a more exact meaning to H? Does it make sense to say that a message contains 1.58 bits of information?

Suppose our message is a decimal digit, which can take on values 0 through 9. There are ten possible values for this message, so the entropy is $H = \log 10 \approx 3.32$. We shall need at least four bits to represent the digit. But imagine that our task is to encode, not just a single digit, but a whole sequence of independent digits. We could simply set aside four bits per digit, but we can do better by considering groups (or *blocks*) of three digits. Each group has $10^3 = 1000$ possible values, and so has an entropy of $3 \log 10 = 9.97$. Therefore we can encode three digits in ten bits, using (on average) $10/3 \approx 3.33$ bits per digit. This is more efficient, and is very close to using $\log 10$ bits per digit.

Exercise 1.5 Devise a binary code for triples of digits (as described above) and use your code to represent the first dozen digits of π.

This motivates the following argument. Consider a message having an entropy H. If we have a long sequence of independent messages of this type, we can group them into blocks and encode the blocks into bits. If the blocks have n messages, each block has an entropy nH. Let N be the minimum number of bits needed to represent a message block. This will be the smallest integer that is at least as big as nH, and so

$$nH \leq N < nH + 1. \tag{1.5}$$

Calculating the number of bits required on a "per message" basis, we are using $K = N/n$ bits per message, and

$$H \leq K < H + \frac{1}{n}. \tag{1.6}$$

If we consider very large message blocks, $n \gg 1$ and so $1/n$ is very small. The two ends of the inequality chain squeeze together, and for large blocks we will use almost exactly H bits per message to represent the information. Therefore, if we encode our messages "wholesale," the entropy H precisely measures the number of bits per message that we need.

Exercise 1.6 Consider a type of message that has three possible values (like the message of the colonial spies in Boston). Calculate the minimum number of bits required to encode blocks of 2, 3, 5, 10, or 100 such messages. In each case, also calculate the number of bits used per message.

Things become more complicated in the presence of noise. *Noise* is a general term for any process that prevents a signal from being transferred and read unambiguously. For example, imagine that there had been fog on Boston Harbor on that April night in 1775. In a heavy fog, the church steeple might not have been visible at all from Charlestown, and no information would have been conveyed. In a lighter mist, the observers might have been able to see that there were lamps in the steeple, but not been able to count them. They would then have known that the British troops were on the move, but not which way they were going. A part of the information would have been transmitted successfully, but not all.

It is possible to formalize this notion of partial information. Before any communication takes place, there are M possible messages and the entropy is $H = \log M$. Afterward, we have reduced the number of possible messages from M to M', but because of noise $M' > 1$. The amount of information conveyed in this process is defined to be

$$H - H' = \log \frac{M}{M'}. \tag{1.7}$$

Exercise 1.7 A friend is thinking of a number between 1 and 20 (inclusive). She tells you that the number is prime. How much information has she given you?

The concept of information is fundamental in scientific fields ranging from molecular biology to economics, not to mention computer science, statistics, and various branches of engineering. It is also, as we will see, an important unifying idea in physics.

1.2 Wave–particle duality

Since the 17th Century, there have been two basic theories about the physical nature of light. Isaac Newton believed that light is composed of huge numbers of particle-like "corpuscles." Christiaan Huygens favored the idea that light is a wave phenomenon, a moving periodic disturbance analogous to sound. Both theories explain the obvious facts about light, though in different ways. For example, we observe that two beams of light can pass through one another without affecting each other. In the Newtonian corpuscle theory, this simply means that the light particles do not interact with each other. In the Huygensian wave theory, it implies that light waves obey the *principle of superposition*: the total light wave is simply the sum of the waves of the two individual beams.

To take another example, we observe that the shadows of solid objects have sharp edges. This is easily explained by the Newtonian theory, since the light particles move in straight lines through empty space. On the other hand, this observation seems at first to be a fatal blow to the wave theory, because waves moving past an obstacle should spread out in the space beyond. However, if the wavelength of light were very short, then this spreading might be too small to notice. For over a hundred years, the known experimental facts about light were not sufficient to settle whether light was a particle phenomenon or a wave phenomenon, and both theories had many adherents.

Then, in 1801, Thomas Young performed a crucial experiment in which Huygens's wave theory was decisively vindicated. This was the famous two-slit experiment.

Suppose that a beam of monochromatic light shines on a barrier with a single narrow opening, or "slit." The light that passes through the slit falls on a screen some distance away. We observe that the light makes a small smudge on the screen. (For thin slits, this smudge of light actually gets wider when the slit is made narrower, and on either side of the main smudge there are several much dimmer smudges. These facts are already difficult to explain without the wave theory, but we will skip this point for now.)

Light passing through another slit elsewhere in the barrier will make a similar smudge centered on a different point. But suppose two nearby slits are both open at once. If we imagine that light is simply a stream of non-interacting Newtonian corpuscles, we would expect to see a somewhat broader and brighter smudge of light, the result of the two corpuscle-showers from the individual slits.

But what happens in fact (as Young observed) is that the region of overlap of the two smudges shows a pattern of light and dark bands called *interference fringes*, see Fig. 1.1.

This is really strange. Consider a point on the screen in the middle of one of the dark fringes. When either one of the slits is open, some light does fall on this point. But when both slits are open, the spot is dark. In other words, we can *decrease* the intensity of light at some points by *increasing* the amount of light that passes through the barrier.

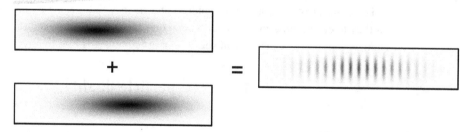

Fig. 1.1   The light patterns from two single slits combine to form a pattern of interference fringes. (For clarity on the printed page, the negative of the pattern is shown; more ink means higher intensity.)

The situation is no less peculiar for the bright fringes. Take a point in the middle of one of these. When either slit is opened, the intensity of light at the point has some value I. But with both slits open, instead of an intensity $2I$ (as we might have expected), we see an intensity of $4I$! The *average* of the intensity over the light and dark fringes is indeed $2I$, but the pattern of light on the screen is less uniform than a particle theory of light would suggest.

Young realized that this curious behavior could easily be explained by the wave theory of light. Waves emerge from each of the two slits, and the combined wave at the screen is just the sum of the two disturbances. Denote by $\phi(\vec{r}, t)$ the quantity that describes the wave in space and time. In sound waves, for example, the "wave function" ϕ describes variations in air pressure. The two slits individually produce waves ϕ_1 and ϕ_2, and by the principle of superposition the two slits together produce a combined wave $\phi = \phi_1 + \phi_2$.

Two further points complete the picture. First we note that ϕ can take on either positive or negative values. By analogy to surface waves on water, the places where ϕ is greatest are called the wave "crests," while the places where ϕ is least (most negative) are called the wave "troughs." Second, the observed intensity of the wave at any place is related to the square of the magnitude of the wave function there: $I \propto |\phi|^2$.

At some points on the screen, the two partial waves ϕ_1 and ϕ_2 are "out of phase," so that a crest of ϕ_1 is coincident with a trough of ϕ_2 and vice versa. At these points, the waves cancel each other out, and $|\phi|^2$ is small. This phenomenon is called *destructive interference* and is responsible for the dark fringes.

At certain other points on the screen, the two partial waves ϕ_1 and ϕ_2 are "in phase," by which we mean that their crests and troughs arrive synchronously. When ϕ_1 is positive, so is ϕ_2, and so on. The partial waves reinforce each other, and $|\phi|^2$ is large. This phenomenon, *constructive interference*, is responsible for the bright fringes.

At intermediate points, ϕ_1 and ϕ_2 neither exactly reinforce one another nor exactly cancel, so the resulting intensity has an intermediate value.

Exercise 1.8 In the two slit experiment, in a particular region of the screen the light from a single slit has an intensity I, but when two slits are open, the intensity ranges over the interference fringes from 0 to $4I$. Explain this in terms of ϕ_1 and ϕ_2.

Young was able to use two-slit interference to determine the wavelength λ of light, which does turn out to be quite small. (For green light, λ is only 500 nm.) Later in the 19th Century, James Clerk Maxwell put the wave theory of light on a firm foundation by showing that light is a travelling disturbance of electric and magnetic fields – an *electromagnetic wave*.

But the wave theory of light was not the last word. In the first years of the 20th Century, Max Planck and Albert Einstein realized that the interactions of light with matter can only be explained by assuming that the energy of light is carried by vast numbers of discrete light *quanta* later called *photons*. These photons are like particles in that each has a specific discrete energy E and momentum p, related to the wave properties of frequency f and wavelength λ:

$$E = hf,$$
$$p = \frac{h}{\lambda},$$

(1.8)

where $h = 6.626 \times 10^{-34}$ J s, called *Planck's constant*. When matter absorbs or emits light, it does so by absorbing or creating a whole number of photons.

Einstein used this idea to explain the photoelectric effect. In this phenomenon, light falling on a metal in a vacuum can cause electrons to be ejected from the surface. If the light intensity is increased, the number of ejected electrons increases, but the kinetic energy of each photoelectron remains the same. In a simple wave theory, this is hard to understand. Why should a more intense light, with stronger electric and magnetic fields, not produce more energetic photoelectrons? Einstein reasoned that each ejected electron gets its energy from the absorption of one photon. A brighter light has more photons, but each photon still has the same energy as before.

Exercise 1.9 The "work function" W of a metal is the amount of energy that must be added to an electron to free it from the surface. Write down an expression for the kinetic energy K of a photoelectron in terms of W and the incident light frequency f. Also find an expression for the minimum frequency f_0 required for the photoelectric effect to take place. (This will depend on W, and so may be different for different metals.)

This "quantum theory" of light poses some perplexities. In view of Young's two-slit interference experiment, there can be no question of abandoning the wave theory entirely. Photons cannot be Newtonian corpuscles. Nevertheless, the fact that light propagates through space as a continuous wave (as seen in the two-slit experiment) does not prevent light from interacting with matter as a collection of discrete particles (as in the photo-electric effect). Furthermore, this bizarre situation is not limited to light. In 1924 Louis De Broglie discovered that the particles of matter – electrons and so forth – also have wave properties, with particle and wave quantities related by Eq. 1.8. It is possible to do a two-slit experiment with electrons and observe interference effects. The general principle that everything in nature has both wave and particle properties is sometimes called *wave – particle duality*.

The effort to put quantum ideas into a solid, consistent mathematical theory led to the development of *quantum mechanics* by Werner Heisenberg, Erwin Schrödinger, and Paul Dirac. Quantum mechanics has proved to be a superbly successful theory of phenomena ranging from elementary particles to solid state physics. It is also a very peculiar theory that challenges our intuitions on many levels. Quantum mechanics involves far-reaching alterations in our ideas about mechanics, probability theory, and even (as we shall see) the concept of information.

To illustrate this in a small way, let us re-examine Young's two-slit experiment with quantum eyes. First, we must understand that the intensity of light is a statistical phenomenon. When we say that light is more intense at one point than it is at another, we simply mean that more photons can be found there. But what can this mean when the number is very small? What can it mean if there is only one photon present?

In the single-photon case, the intensity of the wave at any point is proportional to the *probability* of finding the photon at that point. In general, quantum mechanics predicts only the probability of an event, not whether or not that event will definitely occur. So it is with photons. The behavior of any particular photon cannot be predicted exactly, but the

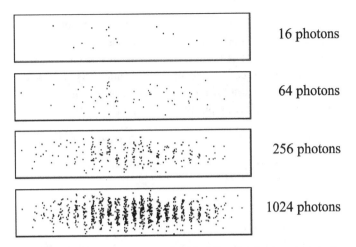

16 photons

64 photons

256 photons

1024 photons

Fig. 1.2   Photons fall randomly on a screen according to a probability distribution given by two-slit interference. Each image shows four times as many photons as the one before. After many photons, a smooth intensity pattern emerges statistically.

statistical behavior of a great many photons gives rise to a smooth intensity pattern. See Fig. 1.2 for an illustration of this.

In the single-photon case, therefore, the wave ϕ is actually a *probability amplitude*, a curious mathematical creature that is not itself a probability, but from which a probability may be calculated. Roughly speaking, the probability[3] P of finding a photon at a given point is just $P = |\phi|^2$. Probability is the square of the magnitude of a probability amplitude.

The probability amplitude wave ϕ obeys the principle of superposition. In the two-slit experiment, consider a particular point X on the screen. With only slit #1 open, the probability amplitude that the photon lands at X is ϕ_1, so that the probability of finding the photon there is $P_1 = |\phi_1|^2$. Opening only slit #2 yields an amplitude ϕ_2, which gives rise to a probability P_2 of finding the photon at X. But with both slits open, we have a combined probability amplitude $\phi = \phi_1 + \phi_2$, yielding a probability

$$P = |\phi|^2 = |\phi_1 + \phi_2|^2, \tag{1.9}$$

for the photon to wind up at X. The two probability amplitudes may reinforce one another or cancel each other out, enhancing or suppressing the probability that the photon lands at X.

If the photon can pass through only one slit, the probability of reaching X is P_1. If it can pass only through the other, it is P_2. In ordinary probability theory, if there are two possible mutually exclusive ways that an event can happen, then the combined probability is $P = P_1 + P_2$. For example, if we flip two coins, the probability that they land with the same face upward is

$$P(\text{same face}) = P(\text{both heads}) + P(\text{both tails}). \tag{1.10}$$

---

[3] In the two-slit experiment, where the photon can be found in a continuous range of positions, P is actually a probability *density* rather than a probability. This technical detail, and a great many others, will be worked out carefully in later chapters!

But quantum probabilities are not ordinary probabilities! In the two slit experiment, the combined likelihood may be either less than or greater than the sum $P_1 + P_2$, depending on the relative phase of the two amplitudes ϕ_1 and ϕ_2. In other words, quantum probabilities can exhibit destructive and constructive interference effects.

Suppose at a point X on the screen the probabilities P_1 and P_2 both equal p. This means that the probability amplitudes at this point satisfy

$$|\phi_1| = |\phi_2| = \sqrt{p}. \qquad (1.11)$$

If the two amplitudes constructively interfere at X, then the two amplitudes are "in phase" there: $\phi_1 = \phi_2$, and so

$$P = |\phi|^2 = |2\phi_1|^2 = 4p. \qquad (1.12)$$

If the two amplitudes destructively interfere at X, then $\phi_1 = -\phi_2$ (the amplitudes are "out of phase"). Then $\phi = 0$ and so $P = 0$. We can see that the probability P for finding a photon in this region of the screen will vary over the interference fringes between 0 and $4p$.

Exercise 1.10 Consider a point X on the screen at which $P_1 = p$ and $P_2 = 2p$. That is, with only slit #1 open, the photon has a probability p of reaching X, but with only slit #2 open this probability is twice as great. Now open both slits. What are the largest and smallest possible values for P at X due to interference effects?

When analyzing the behavior of a photon in the two-slit experiment, we find that $P = |\phi_1 + \phi_2|^2$. Yet the conventional probability law $P = P_1 + P_2$ does apply to the two-coin example. So we are faced with an apparent inconsistency. Sometimes we must add probabilities, and sometimes we must add probability amplitudes. How do we know which of these rules will apply in a given situation?

The difference cannot be mere size. Quantum interference effects have been observed in surprisingly large systems, including molecules more than a million times more massive than electrons (see Problem 1.4). Conversely, we can often apply ordinary probability rules to microscopic systems. The essential difference between the two situations must lie elsewhere.

Notice that, in the two-coin example, we can check to see which of the two contributing alternatives actually occurred. That is, we can examine the coins and tell whether they are both heads or both tails. But in the two-slit experiment, this is not possible. If the single photon arrives in one of the bright interference fringes, it could have passed through either of the slits. Even a very close examination of the apparatus afterward would not tell us which possible alternative occurred.

Suppose we were to modify the two-slit experiment so that we could tell which slit the photon passed through. We can for instance imagine a very sensitive photon detector placed beside one of the slits, which is able to register the passage of a photon without destroying it. This detector need not be a large device: a single atom would be enough in principle, if the state of that atom were sufficiently affected by the passing light quantum. With such a detector in place, we could perform the two-slit interference experiment and then afterwards determine which path the photon took, simply by checking whether or not a photon had been detected.

But, as Niels Bohr pointed out, this new experiment is *not* the same as the original two-slit experiment. If we analyze the proposed modification carefully, we will find that the presence of the detector modifies the behavior of the light. The consistent phase relationship between the partial waves from the two slits will be destroyed, and so no consistent interference effects will be observable. The pattern of light intensity (photon probability) on the screen will show no bright and dark interference fringes. In fact, the probability P of a photon arriving at a point X will be exactly the sum $P_1 + P_2$ for this experiment.

Exercise 1.11 Suppose that a particle detector is placed beside slit #2 in the two-slit experiment. As a simplified model, imagine that the effect of the detector on the quantum amplitude is to randomly multiply the partial wave ϕ_2 by $+1$ or -1. Show that, on average, the ordinary probability law holds – that is, that the average of $|\phi_1 + \phi_2|^2$ and $|\phi_1 - \phi_2|^2$ is exactly $P_1 + P_2$. (This is true whether the amplitudes are real or complex quantities.)

Bohr said that the interference experiment and the "which slit" experiment are *complementary* measurement procedures. We can do either of them, but choosing to perform one logically excludes performing the other on the same photon. We can *either* arrange the apparatus so that interference effects are present, *or* we can arrange it so that we find out which slit the photon passed; *but not both*.

The essential difference between the two-coin experiment (sum the probabilities) and the two-slit experiment (sum the amplitudes) is *information*. In each situation, two alternatives contribute to a final result. For the coins, there is no obstacle to obtaining information about which of the two possible alternatives (heads or tails) is realized. In that case, the total probability is given by $P = P_1 + P_2$. But for a photon in a two-slit interference experiment, such information is not available. Indeed, *it does not exist*, because any actual arrangement in which the photon's path is registered will show no interference effects at all, even if the information is never read by a human experimenter. The quantum rule for adding probability amplitudes applies when the system is *informationally isolated* and produces no physical record of any sort anywhere in the Universe about which possible intermediate alternative is realized.

Exercise 1.12 Explain the following slogan, which might be suitable for printing on a T-shirt: *Quantum mechanics is what happens when nobody is looking*.

The idea that a photon might pass through the slits and leave *no trace at all* of its precise route is slightly disturbing and does not accord with "classical" intuitions based on Newtonian mechanics. Imagine that a Newtonian particle can travel by one of two possible paths. This particle is continually interacting with all of the other particles in the Universe. The position of the planet Saturn, say, will be minutely affected by the gravitation of the particle, which will in turn depend upon the particle's position. Therefore, by an immensely precise determination of Saturn's motion, we should (in principle) be able to tell which path the particle followed. In classical mechanics, no system can really be informationally isolated.

In a slightly more realistic example, the path of the photon through the slits should produce a slight lateral recoil in the barrier, and a careful determination of this recoil should in principle allow us to figure out which slit was passed. Einstein proposed just such a thought-experiment to Bohr in the course of a years-long debate about the internal consistency of quantum theory. Bohr responded that quantum mechanics must apply to the barrier as well. The two possible final states of the barrier, which we wish to use to distinguish which slit the photon went through, do have slightly different quantum descriptions. Nevertheless, the two states are not reliably distinguishable by any possible measurement, and so cannot be counted as distinct physical situations.[4] So it remains true that no physical record exists of the photon's choice of slit, and the quantum probability law applies.

The concepts of information and distinguishability are at the heart of the theory of quantum mechanics. In the chapters that follow, we will develop that theory into a sophisticated mathematical structure and then apply it to many physical situations. Ideas about probability, measurement, and information will be our constant guides. Such guides will not make quantum mechanics seem less strange to our naive intuition, but they will help us begin to build a new quantum intuition, one that more nearly conforms to the strange and marvelous ways of nature.

Problems

Problem 1.1 We said that our definition of H applies when the possible messages are equally likely. Now consider a binary message in which 0 has probability 1/3 and 1 has probability 2/3. What value of H should we assign when the probabilities are not equal?

We determine this by "dividing" the message 1 into two messages, 1a and 1b, which are equally likely. Then the overall message has three equally likely possibilities (0,1a,1b). This message is composed of the original (0,1) message, followed (if the first message is 1) by the (1a,1b) message.

Next we *postulate* that

$$
\begin{bmatrix} \text{entropy} \\ \text{of } (0,1a,1b) \\ \text{message} \end{bmatrix} = \begin{bmatrix} \text{entropy} \\ \text{of } (0,1) \\ \text{message} \end{bmatrix}
$$

$$
+ \left(\begin{array}{c} \text{probability} \\ \text{of message 1} \end{array} \right) \times \begin{bmatrix} \text{entropy} \\ \text{of } (1a,1b) \\ \text{message} \end{bmatrix}.
$$

(Think about why this postulate might make sense.) This becomes

$$
\log 3 = H + \frac{2}{3} \log 2 \qquad \text{and thus} \qquad H = \log 3 - \frac{2}{3} \log 2 \approx 0.918.
$$

(a) Explain intuitively why H should be less than 1.0 in this situation.

---

[4] Bohr also considered the case where a barrier of very low mass is given a sufficient "kick" that the photon's slit can be determined. But in this case, the quantum indeterminacy in the barrier's own position is enough to "wipe out" any interference effects! (We analyze a related example in Section 10.4.) The Bohr–Einstein debate, with Einstein challenging and Bohr defending the principles of quantum theory and complementarity, played a vital role in clarifying the conceptual content of the quantum theory.

(b) Calculate H if message 0 has probability 1/6 and message 1 has probability 5/6.

(c) Generalize this idea to the following situation. Message 0 has a probability $p = k/n$ and message 1 has probability $q = k'/n$, where k, k', and n are positive integers with $k + k' = n$. Find an expression for H in this case that only involves p and q.

Problem 1.2 Five cards are dealt face-down from a 52-card deck.

(a) How many possible sets of five cards are there? How much information do we lack about the cards?

(b) The first three are turned over and revealed. Knowing these, how many possibilities remain?

(c) How much information was conveyed when the three cards were revealed? Is this 3/5 of the total? Why or why not?

(d) Repeat parts (a)–(c) if the five cards are dealt from five independent decks.

Problem 1.3 In his short story "The Library of Babel," Jorge Luis Borges imagines a seemingly infinite library containing books of random text. The language of the library has twenty-five characters, and

> ... each book is of four hundred and ten pages; each page, of forty lines, each line, of some eighty letters which are black in color.

Calculate the entropy of one of the books in Borges' library.

Problem 1.4 In 1999, a research group at the University of Vienna was able to observe quantum interference in a beam of C_{60} molecules. C_{60} is called *buckminsterfullerene*, and the soccerball-shaped C_{60} molecules are sometimes called *buckyballs*. A buckyball molecule has a mass of about 1.2×10^{-24} kg.

(a) The buckyball wavelength in the experiment was about 3 pm. How fast were the molecules moving?

(b) What would be the wavelength of an electron moving at the same speed?

Problem 1.5 The kinetic energy K of a particle is related to its momentum p by $K = p^2/2\mu$, where μ is the particle's mass. In a gas at absolute temperature T, the molecules have a typical kinetic energy of $3k_B T/2$. Derive an expression for the *thermal de Broglie wavelength*, a typical value for the de Broglie wavelength λ of a molecule in a gas. For helium atoms ($\mu = 6.7 \times 10^{-27}$ kg), calculate the thermal de Broglie wavelength at room temperature ($T = 300$ K) and at the boiling point of helium ($T = 4$ K).

Quantum effects become most significant in matter when the thermal de Broglie wavelength of the particles is greater than their separation. At atmospheric pressure, gas molecules are about 1–2 nm apart; in a condensed phase (liquid, solid) they are about ten times closer. How do these compare with the thermal de Broglie wavelengths you calculated for helium?

Problem 1.6 A single photon passes through a barrier with four slits and strikes a screen some distance away. Consider a point X on the screen. The probability amplitudes for reaching X via the four slits are ϕ_1, ϕ_2, ϕ_3, and ϕ_4.

(a) What is the net probability P that the photon is found at X if no measurement is made of which slit the photon passed through?

(b) A detector is placed by slit #4, which can register whether or not the photon passes that slit (but does not absorb the photon or deflect it). What is P in this case?

(c) The detector is now moved to a point between slits #3 and #4 and registers whether or not the photon passes through one of these slits. However, the detector does *not* record which of these two slits the photon passes. What is P in this case?

2 Qubits

2.1 The photon in the interferometer

This chapter introduces many of the ideas of quantum theory by exploring three specific "case studies" of quantum systems. Each is an example of a *qubit*, a generic name for the simplest type of quantum system. The concepts we develop will be incorporated into a rigorous mathematical framework in the next chapter. Our business here is to provide some intuition about why that mathematical framework is reasonable and appropriate for dealing with the quantum facts of life.

Interferometers

In Section 1.2 we discussed the two-slit interference experiment with a single photon. In that experiment, the partial waves of probability amplitude were spread throughout the entire region of space beyond the two slits. It is much easier to analyze the situation in an *interferometer*, an optical apparatus in which the light is restricted to a finite number of discrete *beams*. The beams may be guided from one point to another, split apart or recombined as needed, and when two beams are recombined into one, the result may show interference effects. At the end of the interferometer, one or more sensors can measure the intensity of various beams. (A beam is just a possible path for the light, so there is nothing paradoxical in talking about a beam of zero intensity.) Figure 2.1 shows the layout of a *Mach–Zehnder interferometer*, which is an example of this kind of apparatus.

What happens when we do an interferometer experiment with a single photon? We will consider this question for interferometers that contain only *linear* optical devices, which do not themselves create or absorb photons.[1] At the end of our interferometer, our light sensors are *photon detectors*, which can register the presence or absence of a single photon. Thus, in our calculations we will be interested in the probabilities that the various detectors will "click," recording the presence of the photon in the corresponding beam.

We learned in our discussion of the two-slit experiment in Section 1.2 that the probability of finding the photon at a particular location is the square of the magnitude of a probability

---

[1] These devices are also *unchanged* by the passage of a photon. For instance, we assume it is impossible to determine whether or not a photon has reflected from a given mirror, simply by examining the mirror afterward. The photon therefore remains *informationally isolated* during its passage through the interferometer. As we will see in Section 10.4, this is an entirely reasonable assumption for actual interferometer experiments.

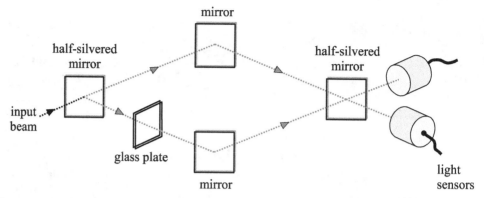

Fig. 2.1 Layout of a Mach–Zehnder interferometer. Light in the input beam is divided into two beams, which are later recombined. Light sensors measure the intensities of the two output beams.

amplitude. Each beam in our single-photon interferometer experiment will have an amplitude α, and the probability P that a detector would find the photon there (if we were to introduce such a detector) is just

$$P = |\alpha|^2. \tag{2.1}$$

Suppose at some stage of our interferometer we know for sure that the photon must be in one of two beams, which have amplitudes α and β respectively. Then it follows that $|\alpha|^2 + |\beta|^2 = 1$.

Complex amplitudes

One important kind of device that we can introduce into a beam is called a *phase shifter*. This could simply be a glass plate through which the beam travels. A phase shifter does not alter the probability that the photon is found in the beam, so the magnitude $|\alpha|$ is not changed. However, the *phase* of α can be altered. By introducing a particular thickness δ of glass, we can change the amplitude from α to $-\alpha$. (The exact value of δ depends on the index of refraction of the glass and the wavelength of the light.) This change in phase is highly significant, for it can turn constructive interference into destructive interference at a later stage of the interferometer.

If we have two such plates, or a single plate with thickness 2δ, the amplitude will become $-(-\alpha) = \alpha$, and the original amplitude is restored. But suppose we have a plate of thickness $\delta/2$? This plate would produce a change the amplitude α such that (1) the magnitude $|\alpha|$ is still the same, and (2) if the change were performed twice, the phase would be multiplied by -1.

Glass plates can be made in a continuous range of thicknesses, producing a continuous range of phase shifts. For this to be possible, *the beam phases α must be complex quantities, with both real and imaginary parts*. A plate with thickness $\delta/2$ may multiply the amplitude by a factor of $i = \sqrt{-1}$. This does not change the magnitude of the complex phase α, since $|\alpha| = |i\alpha|$. Two such plates (or a single plate of thickness δ) multiply the phase by $i^2 = -1$, as required.

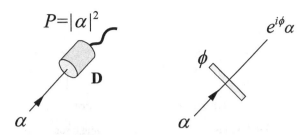

Fig. 2.2 Two important interferometer components. The photon detector D will register the presence of a photon in the beam with probability $P = |\alpha|^2$, where α is the probability amplitude. A phase shift of ϕ changes the amplitude from α to $e^{i\phi}\alpha$.

In general, a glass plate of some thickness will multiply the amplitude of the beam by $e^{i\phi}$, where ϕ (the *phase shift*) is proportional to the thickness of the glass. Changing α to $-\alpha$ could be accomplished by phase shifters with $\phi = \pi, 3\pi, 5\pi$, and so on. A phase shift of ϕ does not change the probability that the photon is found in the beam, since for any α, see Fig. 2.2,[2]

$$\left| e^{i\phi}\alpha \right|^2 = |\alpha|^2 . \tag{2.2}$$

The fact that quantum probability amplitudes are complex quantities is one of the oddest facts about quantum mechanics. Mathematicians introduced complex numbers in the 16th Century to help solve certain algebraic problems. Such numbers are often viewed as highly abstract entities, little connected to the physical world. The number i is, after all, said to be "imaginary." Complex numbers are sometimes used as an algebraic shortcut in Newtonian mechanics or electromagnetism. But in quantum mechanics, complex numbers are not just a convenient trick; they are inescapable and full of significance.

Exercise 2.1 Remind yourself of the rules of complex arithmetic. If α^* denotes the complex conjugate of α, show

(a) $|\alpha|^2 = \alpha^*\alpha$.
(b) $\alpha + \alpha^* = 2\Re(\alpha)$.
(c) For real ϕ, $\left(e^{i\phi} \right)^* = e^{-i\phi}$.

Exercise 2.2

(a) Suppose δ is the smallest thickness of glass that produces a phase shift of π – in other words, that multiplies the phase by -1. What is the phase shift if the glass plate has a thickness of $\delta/5$?

(b) Suppose δ is the *next-to-smallest* thickness of glass that produces the same change in phase (i.e. multiplying the phase by -1). What is the smallest thickness that would do so? What phase shift would be produced by a plate of thickness $\delta/5$?

The beam amplitudes in an interferometer obey the principle of superposition. We will illustrate this with a simple example. Suppose at some stage of the interferometer, there

---

[2] Anything that changes the optical path length of the beam, including a distance of empty space, will act as a phase shifter. In our simplified treatment here, we will ignore the effect of distance and think of all phase shifters as discrete objects that can be either put into or left out of the interferometer beam.

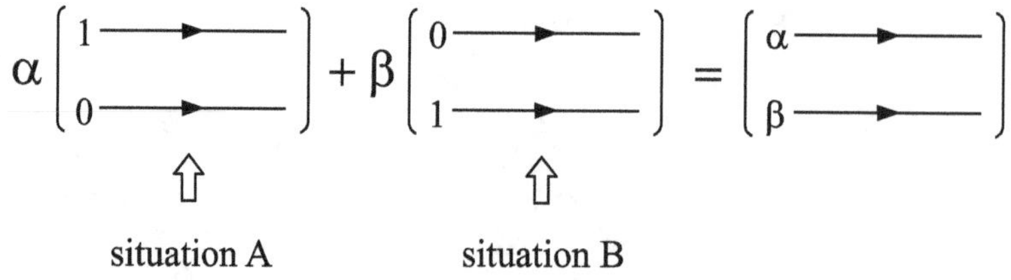

$$\alpha \begin{bmatrix} 1 \longrightarrow \\ 0 \longrightarrow \end{bmatrix} + \beta \begin{bmatrix} 0 \longrightarrow \\ 1 \longrightarrow \end{bmatrix} = \begin{bmatrix} \alpha \longrightarrow \\ \beta \longrightarrow \end{bmatrix}$$

$$\Uparrow \qquad\qquad\qquad \Uparrow$$

situation A   situation B

Fig. 2.3   A graphical representation of Eq. 2.4, showing a superposition of situation A and situation B.

are just two beams available for the photon, which we will call the "upper" beam and the "lower" beam. Consider two possible physical situations, denoted A and B. In situation A, the photon is certainly in the upper beam. The probability amplitude for this beam is 1 and the amplitude for the lower beam is 0. (The upper beam amplitude could be anything of the form $e^{i\phi}$, but we will consider the simplest case.) In situation B, the roles are reversed: the upper amplitude is 0 and the lower is 1, and so the photon is certainly in the lower beam.

The principle of superposition means that the existence of these two situations implies the existence of many other situations in which the beam amplitudes are linear combinations of the assignments for A and B. Given complex coefficients α and β, then there is a possible physical situation which we can formally write as

$$\alpha \,(\text{situation } A) + \beta \,(\text{situation } B). \tag{2.3}$$

In this combined situation, the amplitude for the upper beam is just $\alpha \cdot 1 + \beta \cdot 0 = \alpha$, while the lower beam amplitude is $\alpha \cdot 0 + \beta \cdot 1 = \beta$. Of course, to maintain a proper assignment of probabilities, we will have to require that $|\alpha|^2 + |\beta|^2 = 1$.

This is much easier to express if we describe each situation by a column vector whose entries are the beam amplitudes. Then the first situation could be written $\binom{1}{0}$ and the second one $\binom{0}{1}$. The principle of superposition tells us that

$$\begin{pmatrix} \alpha \\ \beta \end{pmatrix} = \alpha \begin{pmatrix} 1 \\ 0 \end{pmatrix} + \beta \begin{pmatrix} 0 \\ 1 \end{pmatrix}, \tag{2.4}$$

is also a possible physical situation, provided $|\alpha|^2 + |\beta|^2 = 1$, see Fig. 2.3 for an illustration. From this we note, first, that a physical situation for the photon in the interferometer can be summarized by a vector whose components are probability amplitudes. Second, the principle of superposition means that a complex linear combination of two such vectors also represents a possible physical situation, provided the amplitudes satisfy a *normalization* condition (meaning that all probabilities must add up to one).

Beamsplitters

Now we turn our attention to a key element of an interferometer, the *beamsplitter*. This is a device that takes an input beam and splits it into two beams of lower intensity. A typical beamsplitter is a half-silvered mirror. A beam incident on such a mirror will produce both a reflected beam and a transmitted beam, each having half the intensity of the original.

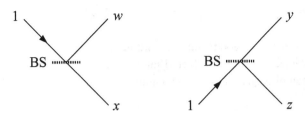

Fig. 2.4 At beamsplitter BS, input beams of unit amplitude produce output beams with amplitudes *w*, *x*, *y*, and *z*.

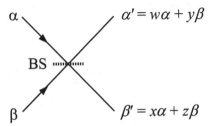

Fig. 2.5 The general situation for the beamsplitter BS. Input amplitudes α and β are transformed into output amplitudes α' and β', each of which is a linear combination of the input amplitudes.

What is the effect of a beamsplitter on the probability amplitudes when the incident beam has only a single photon? Figure 2.4 summarizes. There are two possible input beams for the beamsplitter. For an upper input beam with amplitude 1, we denote the resulting reflected and transmitted beam amplitudes by w and x respectively. A lower input beam with amplitude 1 yields output beam amplitudes y and z, as shown. If the beamsplitter is a half-silvered mirror, then the probability that the photon is reflected or transmitted at the mirror is one-half. That is,

$$|w|^2 = |x|^2 = |y|^2 = |z|^2 = \frac{1}{2}. \tag{2.5}$$

Now we can apply the principle of superposition to find how the beamsplitter works for situations in which the photon could be in either input beam. Suppose α and β are the amplitudes for the upper and lower input beam. The beamsplitter transforms these into amplitudes α' and β' for the corresponding output beams. By superposition, these are

$$\begin{aligned} \alpha' &= w\alpha + y\beta, \\ \beta' &= x\alpha + z\beta, \end{aligned} \tag{2.6}$$

as shown in Fig. 2.5. The relation between input and output amplitudes is easy to express in the amplitude-vector notation introduced above. It is

$$\begin{pmatrix} \alpha \\ \beta \end{pmatrix} \longrightarrow \begin{pmatrix} \alpha' \\ \beta' \end{pmatrix} = \begin{pmatrix} w & y \\ x & z \end{pmatrix} \begin{pmatrix} \alpha \\ \beta \end{pmatrix}. \tag{2.7}$$

This is pretty neat. We represent the photon amplitudes by column vectors $\begin{pmatrix} \alpha \\ \beta \end{pmatrix}$ and $\begin{pmatrix} \alpha' \\ \beta' \end{pmatrix}$. The beamsplitter is described by the 2×2 matrix $\begin{pmatrix} w & y \\ x & z \end{pmatrix}$. The action of the beamsplitter on the input amplitudes then corresponds to simple matrix multiplication.

Exercise 2.3 Verify that Eq. 2.7 is correct.

So far, so good. But what are the elements of the beamsplitter matrix for a particular device? For a half-silvered mirror, we know from Eq. 2.5 that the matrix elements are complex quantities with magnitude $\frac{1}{\sqrt{2}}$. The simplest possible choice would therefore be

$$w = x = y = z = \frac{1}{\sqrt{2}}.$$

What would be the properties of such a beamsplitter? Photons incident along one or the other of the two input beams yield

$$\begin{pmatrix} 1 \\ 0 \end{pmatrix} \longrightarrow \frac{1}{\sqrt{2}} \begin{pmatrix} 1 & 1 \\ 1 & 1 \end{pmatrix} \begin{pmatrix} 1 \\ 0 \end{pmatrix} = \begin{pmatrix} \frac{1}{\sqrt{2}} \\ \frac{1}{\sqrt{2}} \end{pmatrix}$$

$$\begin{pmatrix} 0 \\ 1 \end{pmatrix} \longrightarrow \frac{1}{\sqrt{2}} \begin{pmatrix} 1 & 1 \\ 1 & 1 \end{pmatrix} \begin{pmatrix} 0 \\ 1 \end{pmatrix} = \begin{pmatrix} \frac{1}{\sqrt{2}} \\ \frac{1}{\sqrt{2}} \end{pmatrix}.$$

These are perfectly reasonable amplitudes for the output beams. In either case, the photon has a probability $\left| \frac{1}{\sqrt{2}} \right|^2 = \frac{1}{2}$ of being found in each of the output beams. But suppose we consider an input that is a superposition of the two beams:

$$\begin{pmatrix} \frac{1}{\sqrt{2}} \\ \frac{1}{\sqrt{2}} \end{pmatrix} \longrightarrow \frac{1}{\sqrt{2}} \begin{pmatrix} 1 & 1 \\ 1 & 1 \end{pmatrix} \begin{pmatrix} \frac{1}{\sqrt{2}} \\ \frac{1}{\sqrt{2}} \end{pmatrix} = \begin{pmatrix} 1 \\ 1 \end{pmatrix}.$$

Now the photon has probability $|1|^2 = 1$ of being found in *each* output beam. This is certainly wrong! The "simplest possible" matrix elements for a beam splitter thus cannot correspond to any actual beamsplitter, because that matrix can lead to illegal probability assignments. It does not "conserve probability."

The output probabilities are too large because constructive interference of the amplitudes takes place in both output beams. This is not possible. If constructive interference happens in some places, destructive interference must happen elsewhere.

In other words, our "simplest possible" beamsplitter matrix fails because the phases of the matrix elements cannot be as proposed. On the other hand, this matrix works fine:

$$\begin{pmatrix} w & y \\ x & z \end{pmatrix} = \frac{1}{\sqrt{2}} \begin{pmatrix} 1 & 1 \\ 1 & -1 \end{pmatrix}. \tag{2.8}$$

Exercise 2.4 Show that, for any allowable input amplitudes $\begin{pmatrix} \alpha \\ \beta \end{pmatrix}$, a beamsplitter described by Eq. 2.8 yields output amplitudes such that $|\alpha'|^2 + |\beta'|^2 = 1$.

Equation 2.8 describes a device called a *balanced* beamsplitter. The negative sign in the lower-right (z) matrix element means that when the lower input beam is reflected,

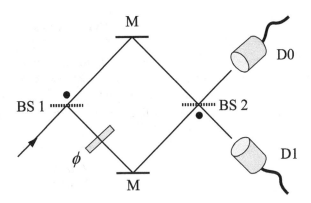

Fig. 2.6 A Mach–Zehnder interferometer. Compare Fig. 2.1.

it undergoes a phase shift of π, but other reflected and transmitted beams have zero net phase shift.

This accords with classical wave optics. A real half-silvered mirror is a slab of glass with a very thin metallic coating on one side. When light is reflected at an interface, the wave picks up a π phase shift whenever the incident beam is coming from a medium of lower refractive index to one of higher index – for instance, from air to glass. Thus, the beam that is reflected on the metal coating from outside the glass gets a negative sign, but not the one that reflects from the inside.[3]

When we include a balanced beamsplitter in our calculations, we will have to be careful to indicate on which side the reflected beam acquires the negative sign. In diagrams, we will do this by placing a dot (•) on one side of the beamsplitter. The reflected beam amplitude on the dotted side is multiplied by -1.

Consider Fig. 2.6, a diagram of the Mach–Zehnder interferometer sketched in Fig. 2.1 above. Two balanced beamsplitters BS1 and BS2 are present, as are a pair of mirrors (both labelled M) and a pair of photon detectors designated D0 and D1. A phase shifter is present on one of the beams, which introduces a phase shift of ϕ. We send photons into the interferometer along just one of the input beams, so that the amplitude of that beam can be taken to be 1.

Exercise 2.5 Consider the Mach–Zehnder interferometer set-up in Fig. 2.6, and suppose $\phi = 0$.

(a) Ignoring any effects of the mirrors M, show that the probabilities P_0 and P_1 of the photon being detected by D0 and D1, respectively, are just 1 and 0. In other words, there is constructive interference for D0 and destructive interference for D1.

(b) Is your answer in part (a) changed if you take into account that reflection from a mirror M introduces a phase shift of π into that beam?

See also Problem 2.1.

---

[3] For simplicity we are neglecting other phase shifts due to the thickness of the glass. However, if these are arranged to be integer multiples of 2π, or if the beamsplitter is built so that all beams undergo exactly the same phase shifts, these may be ignored.

Matrix methods

At any stage of a Mach–Zehnder interferometer, the photon may be in one of two possible beams. We have drawn our diagrams so that one beam is the "upper" beam and one is the "lower" beam. Devices such as phase shifters and beamsplitters alter the probability amplitudes of those beams in a linear way. This linearity is what permits us to describe the transformations by matrices.

The physical situation is described by a column vector of probability amplitudes:

$$\mathbf{v} = \begin{pmatrix} \alpha \\ \beta \end{pmatrix}. \tag{2.9}$$

The various elements of an interferometer apparatus are described by matrices acting on the amplitude vector $\mathbf{v}$. The balanced beamsplitter of Eq. 2.8 is described by:

$$\mathbf{B}_l = \frac{1}{\sqrt{2}} \begin{pmatrix} 1 & 1 \\ 1 & -1 \end{pmatrix}. \tag{2.10}$$

The subscript l indicates that the negative phase appears when the lower beam is reflected. This beamsplitter transforms the amplitude vector $\mathbf{v}$ to a new vector $\mathbf{v}'$ according to

$$\mathbf{v}' = \mathbf{B}_l \mathbf{v}. \tag{2.11}$$

A phase shifter can also be described by a matrix. Suppose the phase of the upper beam is shifted by ϕ. This can be represented by the matrix

$$\mathbf{P}_u(\phi) = \begin{pmatrix} e^{i\phi} & 0 \\ 0 & 1 \end{pmatrix}, \tag{2.12}$$

and the amplitude vector transforms by $\mathbf{v}' = \mathbf{P}_u(\phi)\mathbf{v}$. Once again, the subscript u indicates that the phase of the upper beam is shifted.

Exercise 2.6 Write down the matrices $\mathbf{B}_u$ and $\mathbf{P}_l(\phi)$ describing a beamsplitter with the opposite orientation (negative phase for upper beam reflection) and a phase shifter on the lower beam.

The full-silvered mirrors that guide the beam around the interferometer introduce phase shifts by π into the beam, so they can be represented by matrices

$$\mathbf{M}_{u,l} = \mathbf{P}_{u,l}(\pi). \tag{2.13}$$

We finish our inventory with two very simple cases. First, we can imagine an arrangement in which the beams are simply allowed to cross one another, without any beamsplitter intervening. This just exchanges the upper and lower amplitudes, and so can be represented by the matrix

$$\mathbf{X} = \begin{pmatrix} 0 & 1 \\ 1 & 0 \end{pmatrix}. \tag{2.14}$$

Simplest of all is a part of the interferometer in which the beams are not affected by any sort of optical element, and the amplitudes are unchanged. This is a sort of "device" as well! Its (trivial) action is represented by the identity matrix:

$$\mathbf{1} = \begin{pmatrix} 1 & 0 \\ 0 & 1 \end{pmatrix}. \tag{2.15}$$

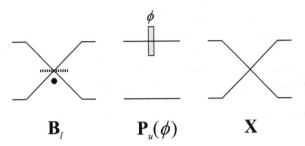

Fig. 2.7

Representations of various linear optical elements in an interferometer.

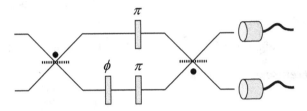

Fig. 2.8

The Mach–Zehnder interferometer. Compare Fig. 2.6.

Obviously, $\mathbf{1v} = \mathbf{v}$ for any amplitude vector $\mathbf{v}$.

We can represent each of these graphically using a modification of our previous diagrams. From now on we will draw the upper and lower beam paths as parallel lines, except where they are brought together at a beamsplitter or a beam crossing. The photon is assumed to go from left to right, see Fig. 2.7.[4]

What happens when the basic optical elements are assembled into a larger experiment? In a diagram, we simply string the pieces together in sequence, as in Fig. 2.8. How can we describe this sort of interferometer arrangement mathematically? Suppose a pair of beams with amplitude vector $\mathbf{v}$ pass through three optical elements. The first is described by a matrix $\mathbf{R}$, the second by $\mathbf{S}$, and the third by $\mathbf{T}$. To find the final amplitude vector $\mathbf{v}'$, we must first multiply $\mathbf{v}$ by $\mathbf{R}$, then by $\mathbf{S}$, then by $\mathbf{T}$:

$$\mathbf{v}' = \mathbf{TSRv}. \tag{2.16}$$

The effect of the entire complex apparatus is represented by a *single* 2×2 matrix, the product $\mathbf{TSR}$. This product is a sequence in time of successive transformations of the amplitude vector for the beams, with the *time order* from right to left: $\mathbf{R}$ occurs first and $\mathbf{T}$ occurs last. To put it another way, the order of the matrices in the product is the opposite of the order of the corresponding elements in our left-to-right diagrams.

Exercise 2.7 Write down a matrix product that represents the Mach–Zehnder interferometer shown in Fig. 2.8. (You may ignore the photon detectors at the end.)

---

[4] Do not be worried by the fact that our beams no longer go in straight lines in our diagrams. The diagrams are merely schematics of a real optical apparatus. But as a matter of fact, we can build interferometers in which the beams are guided in curved paths by optical fibers.

As you thought about Exercise 2.7, you may have noticed a difficulty. The two beams strike two different mirrors, each of which yields a phase shift of π. These reflections happen at about the same time, as suggested in Fig. 2.8. In which order should we write the corresponding matrices? Fortunately, it turns out that the order of these phase shifter matrices does not matter. We will cast the relevant fact as an exercise:

Exercise 2.8 Suppose $\mathbf{P}$ and $\mathbf{P}'$ are the matrices for two phase shifters. Show that $\mathbf{P}$ and $\mathbf{P}'$ *commute*:

$$\mathbf{PP}' = \mathbf{P}'\mathbf{P}$$

when (a) the two phase shifters are applied to the same beam, and (b) the two phase shifters are applied to different beams.

Some of the matrices commute with each other, but not all of them. For example:

Exercise 2.9 Show that

$$\mathbf{XP}_u(\pi) \neq \mathbf{P}_u(\pi)\mathbf{X}.$$

Explain in words why this makes sense.

The analysis of a two-beam interferometer system has now been boiled down to matrix calculations. The translation between the physical apparatus and the mathematical expression is straightforward. The following exercise should give you some easy practice at these calculations and translations. You will find more examples in the problems at the end of the chapter.

Exercise 2.10 Verify the following matrix facts, and explain each one in words and pictures as a fact about interferometer systems. (a) $\mathbf{XX} = \mathbf{1}$. (b) $\mathbf{B}_l\mathbf{B}_l = \mathbf{1}$. (c) $\mathbf{B}_l\mathbf{P}_l(\pi)\mathbf{P}_u(\pi)\mathbf{B}_l = -\mathbf{1}$. (d) $\mathbf{B}_l\mathbf{P}_l(\pi)\mathbf{B}_l = \mathbf{X}$. (e) $\mathbf{B}_l\mathbf{P}_l(\pi)\mathbf{B}_u = \mathbf{P}_u(\pi)$.

Because of the principle of superposition, any linear optical element will produce a linear transformation on the input amplitude vector $\mathbf{v}$, and can therefore be represented by a 2×2 matrix $\mathbf{R}$ acting on $\mathbf{v}$. But we saw in our analysis of beamsplitters that not all 2×2 matrices could possibly correspond to an actual optical device. The reason was that some matrices did not preserve the normalization of the probabilities. Which matrices $\mathbf{R}$ *do* preserve this normalization, and so might correspond to actual devices?

First, we need to express the normalization requirement in terms of matrices. The *Hermitian conjugate* operation is designated by the "dagger" symbol " $\dagger$ ". This indicates the complex conjugate of the transpose of the matrix. Thus,

$$\mathbf{v}^\dagger = \begin{pmatrix} \alpha^* & \beta^* \end{pmatrix}. \tag{2.17}$$

Our normalization requirement for the probability amplitudes can then be written as

$$\mathbf{v}^\dagger\mathbf{v} = 1. \tag{2.18}$$

(Note that we are equating the number 1 with the 1×1 matrix whose only entry is 1. This is a harmless abuse of mathematical notation.)

Exercise 2.11 Verify that this equation is the same as $|\alpha|^2 + |\beta|^2 = 1$.

The vector $\mathbf{v}' = \mathbf{R}\mathbf{v}$ contains the output amplitudes when the input is $\mathbf{v}$. We are thus requiring that $(\mathbf{v}')^\dagger \mathbf{v}' = 1$ for any input vector that has $\mathbf{v}^\dagger \mathbf{v} = 1$. In other words,

$$(\mathbf{v}')^\dagger \mathbf{v}' = \mathbf{v}^\dagger \mathbf{R}^\dagger \mathbf{R} \mathbf{v} = 1. \tag{2.19}$$

(We have used the fact that, for any complex matrices, $(\mathbf{UV})^\dagger = \mathbf{V}^\dagger \mathbf{U}^\dagger$. This, or at least the corresponding fact for the matrix transpose, should be familiar.)

We can view Eq. 2.19 as a property of the matrix $\mathbf{R}^\dagger \mathbf{R}$. Let

$$\mathbf{R}^\dagger \mathbf{R} = \begin{pmatrix} q & r \\ s & t \end{pmatrix}. \tag{2.20}$$

What can we say about these matrix elements? First, consider an input amplitude vector $\mathbf{v} = \begin{pmatrix} 1 \\ 0 \end{pmatrix}$. Then

$$\mathbf{v}^\dagger \mathbf{R}^\dagger \mathbf{R} \mathbf{v} = \begin{pmatrix} 1 & 0 \end{pmatrix} \begin{pmatrix} q & r \\ s & t \end{pmatrix} \begin{pmatrix} 1 \\ 0 \end{pmatrix} = q. \tag{2.21}$$

So Eq. 2.19 tells us that $q = 1$.

Exercise 2.12 Verify Eq. 2.21, and then repeat the calculation with $\mathbf{v} = \begin{pmatrix} 0 \\ 1 \end{pmatrix}$ to show that $t = 1$.

The two diagonal elements of $\mathbf{R}^\dagger \mathbf{R}$ must both equal 1. What about the other two elements? If we let $\mathbf{v} = \frac{1}{\sqrt{2}} \begin{pmatrix} 1 \\ 1 \end{pmatrix}$, we have

$$\mathbf{v}^\dagger \mathbf{R}^\dagger \mathbf{R} \mathbf{v} = \frac{1}{2} \begin{pmatrix} 1 & 1 \end{pmatrix} \begin{pmatrix} 1 & r \\ s & 1 \end{pmatrix} \begin{pmatrix} 1 \\ 1 \end{pmatrix} = 1 + \frac{1}{2}(r + s). \tag{2.22}$$

Since this must equal 1, we know that $s = -r$. Finally, we recall that the amplitudes are complex numbers, so that the input $\mathbf{v} = \frac{1}{\sqrt{2}} \begin{pmatrix} 1 \\ i \end{pmatrix}$ is possible. This yields

$$\mathbf{v}^\dagger \mathbf{R}^\dagger \mathbf{R} \mathbf{v} = \frac{1}{2} \begin{pmatrix} 1 & -i \end{pmatrix} \begin{pmatrix} 1 & r \\ -r & 1 \end{pmatrix} \begin{pmatrix} 1 \\ i \end{pmatrix} = 1 + ri. \tag{2.23}$$

From this, we conclude that $r = 0$.

Exercise 2.13 Verify Eq. 2.22 and Eq. 2.23.

Putting it all together, we have shown that, if the matrix $\mathbf{R}$ is to preserve the normalization of probabilities, it must have the property that

$$\mathbf{R}^\dagger \mathbf{R} = 1. \tag{2.24}$$

Matrices with this property are called *unitary* matrices. We have arrived at an important general fact: *Any physically possible linear optical element in a two-beam interferometer is represented by a 2×2 unitary matrix.*

Exercise 2.14 Here is what we have proved: If **R** is to preserve the normalization of probabilities for any input **v**, then it must be unitary.

Now you prove the (much easier) converse: If **R** is unitary, then it will preserve this normalization for any input **v**. (Be sure that you understand the distinction between these statements!)

We can further show that any unitary 2×2 matrix **R** may be physically realized as an interferometer set-up made out of beam splitters and phase-shifters, see Problem 2.3.

Testing bombs

The components of an interferometer do not register the passage of a photon, so that the photon remains informationally isolated. This is why the beams exhibit interference. Consider, for example, the simplified Mach–Zehnder arrangement in Fig. 2.9. The photon is introduced along the lower beam, so the input amplitude vector can be taken to be $\binom{0}{1}$. If nothing else is introduced into the apparatus, the matrix describing the interferometer's effect is just

$$\mathbf{B}_l \mathbf{B}_u = \begin{pmatrix} 0 & 1 \\ -1 & 0 \end{pmatrix}. \tag{2.25}$$

The output amplitude vector is thus

$$\mathbf{B}_l \mathbf{B}_u \mathbf{v} = \begin{pmatrix} 0 & 1 \\ -1 & 0 \end{pmatrix} \begin{pmatrix} 0 \\ 1 \end{pmatrix} = \begin{pmatrix} 1 \\ 0 \end{pmatrix}. \tag{2.26}$$

Exercise 2.15 Check this matrix arithmetic.

Therefore, the photon will always reach the upper detector D0. The probabilities are

outcome	P
photon reaches D0	1
photon reaches D1	0.

There is constructive interference in the beam that leads to D0, and destructive interference in the beam that leads to D1.

Now suppose that we change the interferometer slightly by sticking a hand into the lower beam at the point A. For simplicity, imagine that the photon is absorbed if it hits the hand.

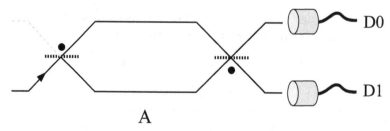

A

Simplified Mach–Zehnder interferometer.

This produces a physical change in the hand that could in principle be detected ("Ow!"). Thus, the hand is a photon detector that measures whether or not the photon travels along the lower beam at A.

This will, of course, destroy any interference effects. If we send a photon into the apparatus, it has a 50% probability of striking the hand. If it travels along the upper beam instead, when it reaches the second beamsplitter it will be equally likely to go toward D0 and D1. In short, we have

outcome	P
photon reaches D0	1/4
photon reaches D1	1/4
photon hits hand	1/2.

Notice that, by blocking one beam with a hand, we have actually *increased* the probability that the photon is detected by D1.

This paradoxical result is the basis for a remarkable thought-experiment proposed by Avshalom Elitzur and Lev Vaidman in 1993. Imagine a factory that produces a type of bomb triggered by light. So sensitive is the trigger that the passage of a *single photon* through its mechanism will explode a bomb.

Because of manufacturing defects, however, many bombs come off the assembly line without working triggers. Photons pass through these mechanisms without being registered at all, and the bombs are duds. The factory managers want to be able tell for sure that at least some bombs are in working order. How can they do this? Of course, if they send a photon through a given bomb, and it blows up, then they can be sure that the bomb was in working order – but they have also destroyed that bomb. What the managers want is a way to identify bombs that are explosive, but are not yet exploded. Since the bomb triggers are set off even by one photon, this appears impossible.

But in fact, the interferometer arrangement in Fig. 2.9 can do the job. A bomb is placed at the point A and then one photon is sent through. If the bomb is a dud, it will not register the passage of the photon, and there will be interference effects. If the bomb is working, it will function as a photon detector on the lower path. The results are

Bomb is a dud		Bomb is working	
outcome	P	outcome	P
photon reaches D0	1	photon reaches D0	1/4
photon reaches D1	0	photon reaches D1	1/4
bomb explodes	0	bomb explodes	1/2.

Suppose an unknown bomb is placed in the apparatus and one photon is sent through. If the bomb explodes, then it was in working order, but this bomb is now lost. If the photon is detected by D0, the test is inconclusive and may be repeated.[5] But if the photon ever arrives at D1, then the managers know that the unexploded bomb is in working order, *even though the bomb never detects the passage of the photon.*

---

[5] If the photon always arrives at D0 during many trials, the factory managers may confidently conclude that the bomb is a dud.

Exercise 2.16 If you do not find the previous paragraph strange and disturbing, re-read it.

Exercise 2.17 Suppose the interferometer test is performed on a large number of bombs from the factory. When the test is inconclusive on a particular bomb, it is repeated until the bomb's status is reasonably certain. What fraction of the working bombs are certified as working but not detonated?

The Elitzur–Vaidman thought-experiment is a good example of the sometimes perplexing behavior of quantum systems. It also illustrates why information is such a key idea in quantum theory. Whether or not a working bomb actually detects a photon in a given trial, its final state (intact or exploded) provides a record of which beam the photon has traversed. That means that the photon was not informationally isolated in the apparatus, and so there can be no interference between the beams.

2.2 Spin 1/2

Having analyzed in detail the problem of a single photon in a two-beam interferometer, we are in a position to identify a few key ideas:

- At any point, the photon can be in one of two distinct beams. Linear superpositions of the beams are also possible.
- The physical situation of the photon is described by a vector $\mathbf{v}$ of two complex probability amplitudes. If a given beam has an amplitude α, then $|\alpha|^2$ is the probability that a detector would find the photon in that beam. Normalization of probabilities means that $\mathbf{v}^\dagger \mathbf{v} = 1$.
- The effect of a linear optical device like a phase shifter or a beamsplitter is described by a matrix $\mathbf{R}$. The amplitude vector $\mathbf{v}$ is changed to a new vector $\mathbf{v}' = \mathbf{R}\mathbf{v}$. The matrix $\mathbf{R}$ must be unitary to guarantee that the final probabilities are normalized.
- Even a quantum system as simple as this can yield surprising results, as in the bomb-testing thought-experiment.

In this section, we will apply these same ideas to a quite different type of quantum system.

Particles with spin

A particle has angular momentum by virtue of its movement through space. It may also have an intrinsic angular momentum called *spin*. This term suggests an analogy to Newtonian physics, in which the angular momentum of an extended body like the Earth is due to both its translational and rotational motion. The quantum situation is a bit more subtle. Electrons, for instance, appear to be entirely point-like, without any spatial extent at all. We therefore cannot attribute the intrinsic spin of an electron to mere rotational motion.

Electrons, protons, and neutrons are all examples of *spin-1/2 particles*. Suppose we measure the z-component S_z of the spin angular momentum for one of these particles. The result of such a measurement is always either $+\hbar/2$ or $-\hbar/2$, where $\hbar$ is related to Planck's constant h by

$$\hbar = \frac{h}{2\pi}, \tag{2.27}$$

and has a value of 1.055×10^{-34} J s. In Newtonian mechanics the component S_z can take on a continuous range of values, but as an experimental fact only these two results are possible. There is, of course, nothing special about the z-axis. The same basic fact holds true for measurements of S_x, S_y, or any other component of the spin.

How can a component of a particle's intrinsic angular momentum be measured? This was the problem faced by Otto Stern and Walther Gerlach in the early days of quantum physics. They were testing Bohr's quantum theory of atomic structure, in which angular momentum has only discrete values. The angular momentum $\vec{S}$ is hard to probe directly, but the magnetic moment $\vec{\mu}$ of an atom is proportional to $\vec{S}$:

$$\vec{\mu} = \gamma \vec{S}, \tag{2.28}$$

where γ is the *gyromagnetic ratio*, a constant property of the particle. In an external magnetic field $\vec{B}$, the magnetic moment will contribute to the energy of an atom by

$$E = -\vec{\mu} \cdot \vec{B}. \tag{2.29}$$

Given a magnetic field in the z-direction, this becomes $E = -\mu_z B_z$. If E could be measured for a given B_z, then μ_z could be found and from this the angular momentum component S_z inferred.

Determining this energy directly was beyond the experimental capabilities available to Stern and Gerlach in the early 1920s. However, they realized that the trajectory of an atom in a *non-uniform* magnetic field depended on this energy. Consider an external magnetic field B_z that is increasing in the positive z-direction, so that $\frac{dB_z}{dz} > 0$. The energy of a particle with a given μ_z will depend on the z-coordinate of its position:

- If $\mu_z > 0$, then E decreases as z increases:
- If $\mu_z < 0$, then E increases as z increases.

This contributes to the effective potential energy for the atom, and so in either case there is a net force on it. The force is upward for $\mu_z > 0$ and downward for $\mu_z < 0$. (By "upward" and "downward" we mean in the positive and negative directions on the z-axis.)

Stern and Gerlach sent a stream of silver atoms through a region of space with a strong magnetic field gradient. The basic arrangement is shown in Fig. 2.10. If Newtonian physics were true (so that the S_z value is continuous), then the stream of atoms would spread out continuously in the z-direction. However, as Stern and Gerlach found, the atoms are actually separated into discrete streams, reflecting the discreteness of the possible values of S_z.

Although the original Stern–Gerlach experiment was performed on entire silver atoms, variations of it can be done with elementary particles. The magnetic fields involved can be oriented in any direction. In all cases, it is found that a measurement of any component of the spin of a particle can only produce certain discrete results.

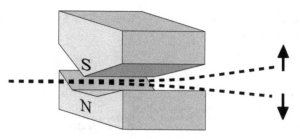

Fig. 2.10 The Stern–Gerlach experiment. A stream of atoms moving from the right passes between the asymmetric poles of a magnet. Particles with different values of μ_z are deflected in different directions. The final position of the atom determines its μ_z.

Amplitude vectors

In the two-beam interferometer, the photon can be found in one of two distinct beams. A spin-1/2 particle can be found to have one of two distinct values for S_z. The same quantum rules that apply to the photon also apply to the spin-1/2 particle. That is, in addition to the "spin up" and "spin down" situations, there are also situations which are complex superpositions of these two:

$$\alpha \text{ (spin up)} + \beta \text{ (spin down)} . \tag{2.30}$$

The coefficients α and β are probability amplitudes for finding the value of S_z to be $+\hbar/2$ or $-\hbar/2$, respectively. We can represent any of these superpositions by a column vector of the probability amplitudes. The amplitude vectors

$$\mathbf{z}_+ = \begin{pmatrix} 1 \\ 0 \end{pmatrix} \quad \text{and} \quad \mathbf{z}_- = \begin{pmatrix} 0 \\ 1 \end{pmatrix}, \tag{2.31}$$

represent situations where the spin component S_z definitely has either its positive or negative possible value. The superposition vector

$$\begin{pmatrix} \alpha \\ \beta \end{pmatrix} = \alpha \mathbf{z}_+ + \beta \mathbf{z}_-, \tag{2.32}$$

is also possible, but what does it mean?

It turns out[6] that the superposition vectors describe situations in which some spin component other than S_z has a definite value. For example, suppose we were to consider S_x. The amplitude vectors

$$\mathbf{x}_+ = \frac{1}{\sqrt{2}} \begin{pmatrix} 1 \\ 1 \end{pmatrix} \quad \text{and} \quad \mathbf{x}_- = \frac{1}{\sqrt{2}} \begin{pmatrix} 1 \\ -1 \end{pmatrix}, \tag{2.33}$$

---

[6] What do we mean by "It turns out"? When we use this phrase, we may be appealing to theoretical developments that we have not yet discussed, or to experimental results, or to both. Physics, unlike mathematics, cannot really be developed in a linear way from a set of explicit axioms. The justification for any theory lies in experiments, but experiments cannot be understood without a theory! The best we can hope for in empirical science is a consistent, testable, mutually reinforcing system of ideas and observations. When we say "It turns out," we are simply opening a door into that system.

describe situations in which $S_x = +\hbar/2$ or $S_x = -\hbar/2$ respectively. For the spin component S_y, the corresponding vectors are

$$\mathbf{y}_+ = \frac{1}{\sqrt{2}} \begin{pmatrix} 1 \\ i \end{pmatrix} \quad \text{and} \quad \mathbf{y}_- = \frac{1}{\sqrt{2}} \begin{pmatrix} 1 \\ -i \end{pmatrix}. \tag{2.34}$$

Exercise 2.18 (a) Suppose a particle is described by $\mathbf{x}_-$ and we measure S_z. What is the probability that we will obtain the result $-\hbar/2$? (b) Answer the same question if the amplitude vector is $\mathbf{y}_+$.

Indeed, suppose we choose a direction in the xz-plane that is inclined at an angle θ from the z-axis. Then the amplitude vectors

$$\theta_+ = \begin{pmatrix} \cos\frac{\theta}{2} \\ \sin\frac{\theta}{2} \end{pmatrix} \quad \text{and} \quad \theta_- = \begin{pmatrix} -\sin\frac{\theta}{2} \\ \cos\frac{\theta}{2} \end{pmatrix}, \tag{2.35}$$

describe situations where the spin component S_θ has the definite values $\pm\hbar/2$ respectively.

Exercise 2.19 If the amplitude matrix is θ_+, what is the probability that a measurement of S_z yields $+\hbar/2$?

Notice the pattern here. For any spin component S_u, we have two amplitude vectors describing situations in which that component has a definite value in a measurement. The amplitude vectors $\mathbf{u}_+$ and $\mathbf{u}_-$ satisfy

$$(\mathbf{u}_+)^\dagger \, \mathbf{u}_+ = (\mathbf{u}_-)^\dagger \, \mathbf{u}_- = 1$$
$$(\mathbf{u}_+)^\dagger \, \mathbf{u}_- = 0. \tag{2.36}$$

We recognize the first part of Eq. 2.36 as the normalization of probabilities for a measurement of S_z. The second relation is also quite significant, as we will now see.

Suppose our particle is described by the amplitude matrix $\mathbf{v} = \begin{pmatrix} \alpha \\ \beta \end{pmatrix}$, but the measurement we perform is S_x rather than S_z. The probabilities for the two possible outcomes will not simply be $|\alpha|^2$ and $|\beta|^2$. How can we find these probabilities? To do so, we must find amplitudes α' and β' so that

$$\mathbf{v} = \alpha'\mathbf{x}_+ + \beta'\mathbf{x}_-. \tag{2.37}$$

Then the probability of obtaining $+\hbar/2$ is $|\alpha'|^2$ and the probability of obtaining $-\hbar/2$ is $|\beta'|^2$. We can find α' and β' by solving a system of simultaneous equations. But there is an easier and more direct way:

$$(\mathbf{x}_+)^\dagger \, \mathbf{v} = (\mathbf{x}_+)^\dagger \, (\alpha'\mathbf{x}_+ + \beta'\mathbf{x}_-)$$
$$= \alpha' \, (\mathbf{x}_+)^\dagger \, \mathbf{x}_+ + \beta' \, (\mathbf{x}_+)^\dagger \, \mathbf{x}_-$$
$$= \alpha'. \tag{2.38}$$

(Notice how we have used both parts of Eq. 2.36.) Similarly, $(\mathbf{x}_-)^\dagger \, \mathbf{v} = \beta'$. We can find the probability amplitudes for any spin component by evaluating matrix products of the form $\mathbf{u}^\dagger\mathbf{v}$.

Exercise 2.20 (a) Find α' and β' in terms of α and β. (b) Find these and the probabilities for the outcomes of an S_x measurement in the case where $\mathbf{v} = \theta_+$.

The amplitude vectors here play a dual role. First, they describe the physical situation of the spin-1/2 particle as a superposition of $\mathbf{z}_+$ and $\mathbf{z}_-$. Second, the amplitude vectors $\mathbf{u}_+$ and $\mathbf{u}_-$ (describing situations where the spin component S_u has a definite value) also provide us with a way to compute probability amplitudes for the outcomes of a measurement of S_u.

Basis independence

So far, it appears that the z-axis has been granted special privileges. The amplitude vectors $\mathbf{z}_+ = \begin{pmatrix} 1 \\ 0 \end{pmatrix}$ and $\mathbf{z}_- = \begin{pmatrix} 0 \\ 1 \end{pmatrix}$ are especially simple; all other amplitude vectors are "superpositions" of these. But there is nothing special about the z-axis! We should be able to describe things equally well – and obtain the same predictions – with respect to any axis.

This is rather like the behavior of familiar three-dimensional spatial vectors. A vector $\vec{a}$ is often described by listing its components (a_x, a_y, a_z) in a Cartesian coordinate system. However, *a vector is not the same thing as the list of its components*. The vector has a definite geometrical or physical meaning, like the momentum $\vec{p}$ of a particular particle. The momentum $\vec{p}$ does not depend on how we choose to orient our coordinate axes to describe it.

We emphasize this by using a notation for vectors that does not refer to any specific coordinate system: $\vec{a}$ rather than (a_x, a_y, a_z). We do the same for vector operations such as $\vec{a} + \vec{B}$, $|\vec{a}|$ or $\vec{a} \cdot \vec{B}$. When we wish to compute these, we often resort to the components of the vectors involved. But we know that the *results* of our computations cannot depend on our choice of spatial coordinate system.

In quantum physics, we need a corresponding notation that is independent of our choice of "coordinate system." We use the term *state* to mean a physical situation for a given quantum system. The state is written as a *ket*, which consists of a straight line | and an angular parenthesis ⟩, between which we put a symbol or a short description of the state. For instance, a situation in which a spin-1/2 particle has $S_z = +\hbar/2$ might be written

$$|z_+\rangle \quad \text{or} \quad |\text{spin up}\rangle \quad \text{or} \quad |\uparrow\rangle . \tag{2.39}$$

Similarly, a state in which $S_x = -\hbar/2$ might be

$$|x_-\rangle \quad \text{or} \quad |\leftarrow\rangle \quad \text{or} \quad \left| S_x = -\frac{\hbar}{2} \right\rangle . \tag{2.40}$$

What sits inside the ket is merely a label for the state; the only requirement is clarity.

The state (written as a ket) is exactly what we have until now been representing as an amplitude vector. Therefore, the principle of superposition implies that kets can be written as complex linear combinations of other kets – that is, that the kets are elements of a complex vector space. For example, we can write

$$|x+\rangle = \frac{1}{\sqrt{2}} |z_+\rangle + \frac{1}{\sqrt{2}} |z_-\rangle . \tag{2.41}$$

Exercise 2.21 Write $|y_-\rangle$ as a linear combination of $|z_+\rangle$ and $|z_-\rangle$.

In order to calculate probability amplitudes, we found it handy to compute $\mathbf{u}^\dagger \mathbf{v}$, where $\mathbf{u}$ and $\mathbf{v}$ were amplitude matrices (see Eq. 2.38). In our new language, we denote the Hermitian conjugate matrix $\mathbf{u}^\dagger$ by a *bra* symbol, which is a reversed ket: $\langle u|$. The product is written

$$\mathbf{u}^\dagger \mathbf{v} = \langle u | v \rangle, \tag{2.42}$$

which is called a *bracket* (= *bra · ket*).[7]

Exercise 2.22 Show that $\langle u | v \rangle^* = \langle v | u \rangle$.

We have seen that a measurement of any component S_u of the spin of the particle is associated with two states $|u_+\rangle$ and $|u_-\rangle$. These are states in which S_u has definite values of $+\hbar/2$ and $-\hbar/2$. These states have the properties

$$\langle u_+ | u_+ \rangle = \langle u_- | u_- \rangle = 1,$$
$$\langle u_+ | u_- \rangle = 0. \tag{2.43}$$

The first line tells us that the probabilities are normalized; the second line is called the *orthogonality* of the states $|u_+\rangle$ and $|u_-\rangle$. We say that Eq. 2.43 expresses the *orthonormality* of the $|u_\pm\rangle$ states.

The states $|u_\pm\rangle$ form a *basis* for the set of all possible states. This means that any state $|v\rangle$ can be written as a superposition of them:

$$|v\rangle = \alpha_+ |u_+\rangle + \alpha_- |u_-\rangle. \tag{2.44}$$

The numbers $\alpha_\pm$ are the probability amplitudes for the outcomes of a measurement of S_u on a particle in the state $|v\rangle$. These amplitudes can be computed by

$$\alpha_\pm = \langle u_\pm | v \rangle. \tag{2.45}$$

In other words, we can represent the state $|v\rangle$ with respect to the $|u_\pm\rangle$ states as an amplitude vector

$$\mathbf{v} = \begin{pmatrix} \langle u_+ | v \rangle \\ \langle u_- | v \rangle \end{pmatrix}, \tag{2.46}$$

but this representation depends on the choice of the states $|u_\pm\rangle$.

Exercise 2.23 Show that the *only* kets that with certainty yield a value of $+\hbar/2$ in a measurement of S_u are of the form $e^{i\phi} |u_+\rangle$.

Measurements and filters

Suppose we have a Stern–Gerlach apparatus oriented along the z-axis. A stream of spin-1/2 particles passing through the apparatus will be split into two streams, according to the measured value of S_z. If we observe in which of the two streams an atom is found, we have measured S_z for that atom, see Fig. 2.11.

---

[7] This notation and this pun are both attributable to Paul Dirac.

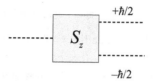

Fig. 2.11   A Stern–Gerlach apparatus.

A very simple sort of observation would be to block one of the two beams, say the one corresponding to $S_z = -\hbar/2$. The value of S_z is registered by whether or not the atom hits the barrier. This arrangement is not simply a measurement of S_z, but also an S_z filter. Atoms with $S_z = +\hbar/2$ are permitted to pass, but other atoms are stopped.

If we prepare a particle in the state $|z_-\rangle$ and send it through our apparatus, then it is blocked. If we prepare it in $|z_+\rangle$, then it will certainly pass through the apparatus. What will be its state afterwards? This will in general depend on the detailed physics of the apparatus, since magnetic fields and so forth might produce changes in the spin of the particle. For now we will consider the simplest case, in which the spin is unchanged: the particle will emerge with spin state $|z_+\rangle$.

Now suppose we introduce a particle in the state $|x_+\rangle$, given in Eq. 2.41. Such a particle will have a probability 1/2 of being blocked and probability 1/2 of passing through the apparatus. If the particle passes through, what will be its state afterwards?

We might be tempted to say that the spin will still be $|x_+\rangle$, since we have said that the spin is "unchanged" by the apparatus. But $|x_+\rangle$ is a superposition of $|z_+\rangle$ and $|z_-\rangle$ – in essence, an *interference* of these two states – and that interference cannot survive a measurement of S_z. We conclude instead that the final state of the spin, given that it passes through our S_z filter, is just $|z_+\rangle$.

This means that a second measurement of S_z would produce exactly the same result as the first measurement.[8] To put it a different way, consider two S_z filters in succession. The first one passes $S_z = +\hbar/2$ and the second one passes $S_z = -\hbar/2$, as shown in Fig. 2.12. Any particle that passes the first filter is then in a state $|z_+\rangle$, and so has probability zero for passing the second filter.

Naively, we might think that a filter merely removes particles which do not meet some specified criterion. If this were an adequate picture of how our filters work, then it would follow that adding additional filters to a series could never increase the likelihood that a particle would pass all the way through. This is indeed true if we add filters to the end of the series. But what if we insert one in the middle?

Let us modify the arrangement in Fig. 2.12 by inserting an S_x filter between the two S_z filters. This is shown in Fig. 2.13. A particle that passes through the first filter will then be in a state $|z_+\rangle$. In this state, it will pass the second filter with probability 1/2, and if it does, it will afterwards have a spin state $|x_+\rangle$. But a particle with this state will have some chance (again, probability 1/2) of passing the final filter and ending up in the state $|z_-\rangle$. By

---

[8] This observation, that successive measurements of the same observable quantity will yield identical results, is sometimes elevated to an axiom of quantum theory. However, as we will see in Section 4.3, this is only true in the most ideal cases, and is not a general fact about actual measurement procedures.

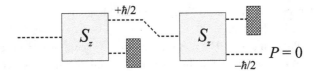

Fig. 2.12 Successive Stern–Gerlach filters for opposite values of S_z. The probability of passing through both filters is zero.

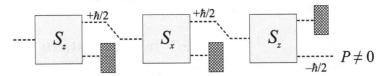

Fig. 2.13 If we insert an S_x filter between two opposite S_z filters, we can increase the probability of passing the whole series.

inserting an extra filter, we have increased the probability that the particle passes the whole series.[9]

Therefore, the filters, and the measurements they are based on, do more than just "read off" the value of some variable. They also have an effect on the state of the system that is being observed. A particle prepared with spin state $|z_+\rangle$ and subjected to a measurement of S_x, will afterwards be found in one of the states $|x_\pm\rangle$. The particle will retain no "memory" of its previous commitment to a definite value of S_z.

This is because S_z and S_x are *complementary* quantities. We must orient our Stern–Gerlach magnets one way or the other, choosing one spin measurement or the other. Measuring S_x precludes measuring S_z, and furthermore, any definite value of S_z the particle might have carried is destroyed by the measurement of S_x.

Exercise 2.24 Suppose the particle starts out with a spin state of $|x_-\rangle$. What is the probability that it will pass through all three filters in Fig. 2.13?

This has an interesting implication for the storage and retrieval of information using quantum systems. Imagine representing one bit of information by the state of a spin-1/2 particle. A simple code would be

Signal	Message
$|z_+\rangle$	0
$|z_-\rangle$	1.

When we wish to retrieve the information, all we need to do is measure S_z for the particle. However, suppose we measure S_x instead? The result of this measurement would tell us *nothing at all* about the bit encoded in the spin. Worse, once we have measured S_x,

---

[9] This is quite similar to a simple lecture demonstration involving polarizing filters. Polarization is a property of photons that is exactly analogous to particle spin. A polarizing filter blocks light of one polarization, but permits light of the other (perpendicular) polarization to pass. No light can get through a pair of "crossed" polarizing filters. But if a third tilted filter is introduced between the pair, some of the photons do pass through.

the particle will have "forgotten" all about its previous state. The information will be irretrievably lost.

Put it this way. The particle can carry a "secret message" encoded in its spin state. If we read the particle in the right way (measuring S_z), the message is revealed. But if we choose the wrong way (measuring S_x), then the message self-destructs! This suggests that the quantum physics of a spin-1/2 particle might be used to preserve the privacy of information. And so it can; we will return to this subject, called *quantum cryptography*, later on.

2.3 Two-level atoms

Energy levels and quantum states

Atoms of a given element can absorb or emit light only at certain discrete frequencies, the pattern of which is characteristic of that element. (This is the basis of *spectroscopy*.) That is, the atoms can absorb or emit photons only with certain discrete energies. As Niels Bohr realized, this must mean that the atoms themselves can only have certain discrete internal *energy levels*. As the atom "jumps" between levels, photons are absorbed or emitted. The atomic energy levels determine the pattern of light frequencies in the spectrum of the element.

Viewed abstractly, the atomic energy levels form a kind of irregular ladder, each rung of which is a possible value for the internal energy of the atom. Suppose an atom is occupying one rung or another on the ladder. When a photon is absorbed, the atom jumps to a higher rung. When it jumps to a lower rung, a photon is emitted. The lowest rung of all, the internal state with the smallest energy, is called the *ground state*, and all of the others are *excited states*. This is illustrated in Fig. 2.14.

Later in this book we will discuss at length how the complicated structure of energy levels emerges from the quantum physics of the interacting nucleus and electrons in an atom. For now, we will simply take that structure as a given and begin to explore what it entails. A particular type of atom generally has many different energy levels. In many experiments, though, only two energy levels – usually the ground state and one excited state – play any significant role. In this case, we can adopt a simplified model, the *two-level atom*, in which these are the only rungs present in the energy level ladder.

The two-level atom, like the two-beam interferometer and the spin-1/2 particle, is a simple quantum system. Let us apply some of the quantum ideas we have seen so far to this new example. Let E_0 and E_1 denote the energy of the ground state and the excited state of our two-level atom. Each of these, of course, corresponds to a state of the atom that is represented by a ket: $|E_0\rangle$ for the ground state, $|E_1\rangle$ for the excited state.

We could imagine measuring the energy E of the atom. Such a measurement would have only E_0 and E_1 as possible values, and the states $|E_0\rangle$ and $|E_1\rangle$ will have these values with certainty. They are thus analogous to the $|z_+\rangle$ and $|z_-\rangle$ states for a spin-1/2 particle, which are the states with definite values for S_z. We conclude that the energy level states must be orthonormal, as in Eq. 2.43:

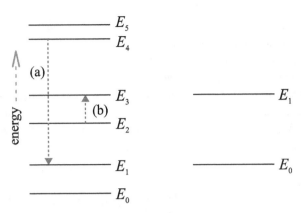

Fig. 2.14 On the left is the energy level "ladder" for an imaginary atom with six energy levels included. The jump (a) is accompanied by the emission of a photon with energy $E_4 - E_1$, while the transition shown in (b) absorbs a photon having much lower energy $E_3 - E_2$. On the right, the much simpler ladder of a two-level atom.

$$\langle E_0 | E_0 \rangle = \langle E_1 | E_1 \rangle = 1,$$
$$\langle E_0 | E_1 \rangle = 0. \tag{2.47}$$

The principle of superposition tells us that there are other states as well formed by complex linear combinations of these two. In general, then, the atom will be in a state

$$|\psi\rangle = \alpha_0 |E_0\rangle + \alpha_1 |E_1\rangle . \tag{2.48}$$

The amplitudes $\alpha_k = \langle E_k | \psi \rangle$. If the atom is in the state $|\psi\rangle$, then a measurement of its energy E will yield E_0 with probability $|\alpha_0|^2$ and E_1 with probability $|\alpha_1|^2$.

Exercise 2.25 Show that any state of the form $e^{i\phi} |E_k\rangle$ is a state with definite energy E_k.

The early quantum physicists thought that an atom must always be "in" one or another of its energy levels, and even today physicists, chemists, and others will often speak and write in this way. (To see an example, go back a few paragraphs and re-read our description of an atom "jumping" from one rung of the energy level ladder to another.) But it is not so! Superpositions such as Eq. 2.48 are perfectly possible quantum states, which means that we can have interference effects between different energy levels. We will have more to say on this point a little later. For now, we need to explain why this important fact can so often be ignored.

The superposition of energy levels in Eq. 2.48, with its potential for interference between the levels, only makes sense provided that the two-level atom remains informationally isolated. But if the atom emits a photon (with energy $E_1 - E_0$) then it has announced to the world that its energy was E_1 and has now become E_0. Since the surroundings contain a record (in the form of the photon) of the atom's energy E, the superposition can no longer apply. The same would be true if the atom absorbed a photon from its surroundings.

In a simple spectroscopic experiment, we probe an atom by studying the photons that it emits and absorbs. The atom is therefore not informationally isolated. No interference effects between the levels can be observed, so we can always imagine the atom to be in some particular energy level at any given time. Because spectroscopy is so important as a tool for probing quantum-level physics, it is easy to forget the deeper truth that energy levels are not the only possible quantum states.

When we discuss spectroscopic experiments, we too will happily adopt the simplified "one level or another" language. We, however, will *not* forget.

Time evolution

A measurement of energy on the two-level atom is associated with the two energy level states $|E_0\rangle$ and $|E_1\rangle$. Measurements of other variables will be associated with other orthonormal pairs of states. Suppose $|u\rangle$ is the state associated with some measurement outcome u. Then if the system is in the state $|\psi\rangle$, the probability P_u that this measurement results in u is

$$P_u = |\langle u|\psi\rangle|^2 . \tag{2.49}$$

Now (in principle) we can calculate the probabilities for measurement outcomes for measurable quantities other than energy. What are these other quantities exactly? Without a more detailed description of the atom, we cannot say. For the present, we will simply assume that we can measure them when required, and leave the details for later.

A two-level atom that is informationally isolated can exist in an arbitrary superposition state $|\psi\rangle$, as given in Eq. 2.48. But how does the state $|\psi\rangle$ change over time? In other words, what is $|\psi(t)\rangle$? The answer will involve a new principle of quantum physics. We will motivate this principle by recalling the Planck–De Broglie relations from Eq. 1.8. In particular, the energy E of a photon is related to the frequency f of the light by $E = hf$. It will be a little more convenient to write this in terms of the angular frequency $\omega = 2\pi f$:

$$E = hf = \frac{h}{2\pi} 2\pi f = \hbar\omega. \tag{2.50}$$

(**Note:** We use the quantity ω so much more often than f that from now on we will use the unadorned term "frequency" to refer to ω. Whenever we need to talk about f, we will call it the *circular frequency*.)

In Section 1.2 we said that Eq. 2.50 is a general relation between particle and wave properties for quantum systems. How does this apply to the two-level atom? If the atom is initially in an energy level state $|E_k\rangle$, its state should somehow oscillate in time with a frequency $\omega_k = E_k/\hbar$. But the energy of an isolated atom does not change. We conclude that it is the complex *phase* of the state that changes. If at $t = 0$ the state is $|\psi(0)\rangle = |E_k\rangle$, then at a later time

$$|\psi(t)\rangle = e^{-i\omega_k t} |E_k\rangle . \tag{2.51}$$

As Exercise 2.25 tells us, this is also a state with definite energy E_k, so the energy of the atom does not change.

In fact, for an atom in an energy level state, the time evolution in Eq. 2.51 makes no observable change at all. If u is an outcome of some measurement, the probability P_u at time t is

$$\begin{aligned}
P_u(t) &= |\langle u|\psi(t)\rangle|^2 \\
&= \left|e^{-i\omega_k t}\langle u|E_k\rangle\right|^2 \\
&= |\langle u|E_k\rangle|^2 = P_u(0).
\end{aligned} \tag{2.52}$$

If the atom is in an energy level state, the probability of every measurement outcome is constant over time. From the point of view of experiment, $|\psi(t)\rangle$ is indistinguishable from $|\psi(0)\rangle$. In fact, the states of definite energy are sometimes called *stationary states*.

Given this, why do we even bother with Eq. 2.51? The answer appears when we have more general states that are not states of definite energy. Suppose that our atom at $t = 0$ is in the superposition state $|\psi\rangle$ given in Eq. 2.48. We now make the assumption that the time evolution acts in a linear way, so that each term evolves independently according to Eq. 2.51. Then

$$|\psi(t)\rangle = \alpha_0 e^{-i\omega_0 t}|E_0\rangle + \alpha_1 e^{-i\omega_1 t}|E_1\rangle. \tag{2.53}$$

Not only does the overall phase of this state vary over time, but the relative phases of the two terms will also change, provided $\omega_0 \neq \omega_1$.

To see some of the implications of this, consider the state $|u\rangle$ of the two-level atom given by

$$|u\rangle = \frac{1}{\sqrt{2}}|E_0\rangle + \frac{1}{\sqrt{2}}|E_1\rangle. \tag{2.54}$$

We will suppose that $|u\rangle$ is the state associated with the outcome u of some measurement, and also that the initial state of the atom is $|\psi(0)\rangle = |u\rangle$. At time $t = 0$ the probability that our measurement would yield u is

$$P_u(0) = |\langle u|\psi(0)\rangle|^2 = 1. \tag{2.55}$$

Over time, however, the atom's state will evolve into something different:

$$|\psi(t)\rangle = \frac{1}{\sqrt{2}}e^{-i\omega_0 t}|E_0\rangle + \frac{1}{\sqrt{2}}e^{-i\omega_1 t}|E_1\rangle. \tag{2.56}$$

How does the probability $P_u(t)$ vary over time?

First, we compute the probability amplitude at time t:

$$\begin{aligned}
\langle u|\psi(t)\rangle &= \frac{1}{\sqrt{2}}\left(\langle E_0|\psi(t)\rangle + \langle E_1|\psi(t)\rangle\right) \\
&= \frac{1}{2}\left(e^{-i\omega_0 t} + e^{-i\omega_1 t}\right).
\end{aligned} \tag{2.57}$$

Notice that we have used the orthonormality of $|E_0\rangle$ and $|E_1\rangle$ to compute this amplitude. The probability is

$$\begin{aligned}
P_u(t) &= |\langle u|\psi(t)\rangle|^2 \\
&= \langle u|\psi(t)\rangle^* \langle u|\psi(t)\rangle \\
&= \frac{1}{2}\left(1 + \cos\left((\omega_1 - \omega_0)t\right)\right).
\end{aligned} \tag{2.58}$$

Exercise 2.26 Fill in the algebra to arrive at Eq. 2.58.

As time progresses, the probability $P_u(t)$ of the measurement outcome u changes from 1 to 0 and then back to 1 again with an angular frequency $\omega_2 - \omega_1$. At least some of the observable properties of the atom do vary over time.

Exercise 2.27 Show that, if $E_0 = E_1$, then every state of the two-level atom is a stationary state.

Exercise 2.28 Suppose that a two-level atom goes from its excited state energy E_1 to its ground state energy E_0 via the emission of a photon. What is the frequency ω of this light? Compare this to the frequency in Eq. 2.58.

Exercise 2.29 Let P_k be the probability that a measurement of the atom's energy will yield E_k. Show that, for any initial state of the two-level atom, P_0 and P_1 do not change over time. (This is what is meant by "conservation of energy" for an atom that might not have a definite energy.)

Operators

We have postulated that the time evolution of an informationally isolated two-level atom is linear. That is, if the initial states $|\phi(0)\rangle$ and $|\psi(0)\rangle$ evolve into later states $|\phi(t)\rangle$ and $|\psi(t)\rangle$, then a superposition of the two will evolve by:

$$\alpha\,|\phi(0)\rangle + \beta\,|\psi(0)\rangle \longrightarrow \alpha\,|\phi(t)\rangle + \beta\,|\psi(t)\rangle. \tag{2.59}$$

Another way of putting this is to say that the time evolution of the system is described by an operator on the initial state.

An *operator* is a linear mapping on a set of vectors, in this case the kets describing the states of the quantum system. Operators play exactly the same role as the 2×2 matrices from Section 2.1, which transformed the amplitude vectors for the photon in the two-beam interferometer. Operators, however, are objects that do not presume a particular basis for the space of states. If the operator A acts on the state $|\psi\rangle$, we write the result A $|\psi\rangle$.

The time evolution from 0 to t is described by the operator U(t), which satisfies:

- U$(t)\,|E_k\rangle = e^{-i\omega_k t}\,|E_k\rangle$ for an energy level state $|E_k\rangle$;
- U(t) acts on states in a linear way.

Since U(t) gives the time evolution of the energy level states, and since any state of the atom can be written as a superposition of these, it follows that

$$|\psi(t)\rangle = \mathsf{U}(t)\,|\psi(0)\rangle, \tag{2.60}$$

for any state of the informationally isolated atom.

Exercise 2.30 The product AB of two operators A and B is the operator that acts this way: $AB|\psi\rangle = A\left(B|\psi\rangle\right)$ for any $|\psi\rangle$. Show that, for the two-level atom,

$$U(t_2) = U(t_2 - t_1)\,U(t_1). \tag{2.61}$$

If we know the time evolution operator $U(t)$ for all t, then we know everything about how the two-level atom's state changes with time. To make an analogy with Newtonian physics, this is like knowing the future trajectory of a particle for any initial position $\vec{r}_0$ and velocity $\vec{v}_0$. In the Newtonian case, these trajectories are governed by the Newtonian equation of motion, usually written $\vec{F} = m\vec{a}$. What is the equation of motion for the quantum state of the two-level atom? That is, what equation describes how the state $|\psi(t)\rangle$ is changing at any given moment?

We recall Eq. 2.53, which gives the time evolution of an arbitrary superposition of energy levels:

$$|\psi(t)\rangle = \alpha_0 e^{-i\omega_0 t}|E_0\rangle + \alpha_1 e^{-i\omega_1 t}|E_1\rangle. \tag{Re 2.53}$$

The kets $|E_0\rangle$ and $|E_1\rangle$ are taken to be constant over time. The time derivative of $|\psi(t)\rangle$ is therefore

$$\frac{d}{dt}|\psi(t)\rangle = -i\omega_0\alpha_0 e^{-i\omega_0 t}|E_0\rangle - i\omega_1\alpha_1 e^{-i\omega_1 t}|E_1\rangle. \tag{2.62}$$

Multiplying both sides of this equation by $i\hbar$, and recalling the relation between frequency and energy, we have

$$i\hbar\frac{d}{dt}|\psi(t)\rangle = E_0\alpha_0 e^{-i\omega_0 t}|E_0\rangle + E_1\alpha_1 e^{-i\omega_1 t}|E_1\rangle. \tag{2.63}$$

The right-hand side of this equation is almost $|\psi(t)\rangle$, except that each term is multiplied by the energy. This suggests that we should define an *energy operator* (also known as the *Hamiltonian*) with the following properties:

- $H|E_k\rangle = E_k|E_k\rangle$ for an energy level state $|E_k\rangle$;
- H acts on states in a linear way.

Then we have shown that, for a two-level atom,

$$i\hbar\frac{d}{dt}|\psi(t)\rangle = H|\psi(t)\rangle. \tag{2.64}$$

Equation 2.64 is our first look at the *Schrödinger equation*, one of the most famous equations in all of physics. It tells us that the energy of a system (represented by the operator H) governs how that system evolves over time.

The time-evolving state $|\psi(t)\rangle = U(t)|\psi(0)\rangle$ is a *solution* to the quantum equation of motion, the Schrödinger equation. The time evolution operator $U(t)$ must therefore have some connection to the Hamiltonian energy operator H. Some such connection can be gleaned by comparing the definitions of $U(t)$ and H. A deeper look at this issue will have to wait for later.

There is another issue that merits discussion right away. The energy of the two-level atom is a measurable quantity, and a measurement of it can only yield the values E_0 and E_1. We have here found it convenient to represent the energy by an operator H. This operator multiplies energy level states by their energies; its action on other states is then determined by linearity.

We can do the same thing for *any* measurable quantity. For instance, recall the spin-1/2 particle of the previous section. A measurement of S_z, the z-component of the particle's spin, can only yield the values $\pm\hbar/2$. This motivates us to define an operator S_z on spin states of the particle thus:

- $S_z |z_+\rangle = +\dfrac{\hbar}{2} |z_+\rangle$ and $S_z |z_-\rangle = -\dfrac{\hbar}{2} |z_-\rangle$;
- S_z acts on states in a linear way.

Exercise 2.31 Evaluate $S_z |x_+\rangle$.

The association of observable quantities with operators acting on kets is an important piece of the mathematical machinery of quantum mechanics.

Spins as two-level atoms

We said that the magnetic moment $\vec{\mu}$ of a particle is related to its spin $\vec{S}$ by

$$\vec{\mu} = \gamma \vec{S}. \tag{Re 2.28}$$

The gyromagnetic ratio γ is positive for protons and most nuclei; it is negative for electrons, neutrons, and a few other nuclei. If the particle is in a uniform magnetic field, Eq. 2.29 tells us that it has an energy

$$E = -\vec{\mu} \cdot \vec{B}. \tag{Re 2.29}$$

If the magnetic field points in the positive z-direction, then the particle's energy is

$$E = -\gamma B S_z. \tag{2.65}$$

For a spin-1/2 particle, the only possible values of S_z are $\pm\hbar/2$. Therefore, a spin-1/2 particle in a magnetic field has two energy levels. The spin state $|z_+\rangle$ has energy $-\gamma B\hbar/2$ and $|z_-\rangle$ has energy $+\gamma B\hbar/2$.

Our analysis of the two-level atom now lets us write down how a general spin state evolves in time. We define the *Larmor frequency* $\Omega = \gamma B$. The two possible energies for the spin are therefore $\pm\hbar\Omega/2$. If we write the initial spin state as $|\psi(0)\rangle = \alpha_+ |z_+\rangle + \alpha_- |z_-\rangle$, the state at a time t will be

$$|\psi(t)\rangle = \alpha_+ e^{i\Omega t/2} |z_+\rangle + \alpha_- e^{-i\Omega t/2} |z_-\rangle, \tag{2.66}$$

(see Eq. 2.53). Suppose our particle initially starts out in a spin state $|\psi(0)\rangle = |x_+\rangle$. At a later time, the probability that a measurement of S_x would find the particle in $|x_+\rangle$ oscillates according to

$$P_{x+}(t) = \frac{1}{2}\left(1 + \cos \Omega t\right). \tag{2.67}$$

Exercise 2.32 Adapt the derivation of Eq. 2.58 to derive Eq. 2.67.

Since the two energy levels are separated by $\Delta E = \hbar\Omega$, the Larmor frequency Ω is the frequency of electromagnetic radiation that can be absorbed or emitted by the spin-1/2 particle. Consider a proton, which has a gyromagnetic ratio $\gamma_p = 2.675 \times 10^8$ s^{-1}T^{-1}. If we put this proton in a (very strong) 10.00 T magnetic field, the Larmor frequency will be

$$\Omega = \gamma_p B = 2.675 \times 10^9 \text{ s}^{-1}, \tag{2.68}$$

which corresponds to a circular frequency $f = 425.7$ MHz. Radio waves of this frequency will strongly interact with protons in a 10.00 T magnetic field. This is the basic idea behind *nuclear magnetic resonance*, which we will discuss in much more detail in Chapter 18.

Exercise 2.33 A ^{13}C nucleus is also a spin-1/2 particle. Its gyromagnetic ratio is almost exactly one-fourth that of the proton. What radio frequency will strongly interact with ^{13}C nuclei in a 10.00 T magnetic field? What if the magnetic field were only 5.00 T?

2.4 Qubits and isomorphism

In this chapter, we have discussed three very different types of quantum system, each described by the same mathematical machinery. Each system has two states – that is, states that can be distinguished from each other by some measurement – together with an infinite number of superpositions of these. In Section 1.1, we defined a "bit" to be a physical system with two possible distinguishable physical states. Each of our examples is a quantum generalization of this, having two distinguishable states and many more possible ones. Our generic term for this type of quantum system is a *qubit*.

Just as all bits are isomorphic, so too all qubits are isomorphic, in that they are described by the same mathematical structures. We have exploited this idea several times already. For any qubit system, we can choose two distinguishable states and denote them by $|0\rangle$ and $|1\rangle$. This pair of states is called the *standard basis*. The choice of the standard basis is arbitrary, and corresponds to the choice of a standard measurement on the qubit system. The states in the standard basis are orthonormal; that is, $\langle 0|0\rangle = \langle 1|1\rangle = 1$ and $\langle 0|1\rangle = 0$.

Any other state of the qubit is a superposition of the states in the standard basis:

$$|\psi\rangle = \alpha_0 |0\rangle + \alpha_1 |1\rangle. \tag{2.69}$$

The coefficients $\alpha_k = \langle k|\psi\rangle$, and so they are probability amplitudes for the two possible outcomes of the standard measurement. Probability amplitudes for other measurements are given by brackets of $|\psi\rangle$ with other basis states. If a qubit system is informationally isolated, the change of its state with time will be described by an evolution operator $\mathsf{U}(t)$.

This same template applies to each of our three examples, though we have used the examples to emphasize different aspects of it. For instance, in the example of the spin-1/2 particle, it is clear that the choice of basis states was arbitrary. Fixing our standard

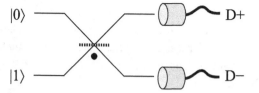

Fig. 2.15   Measurement apparatus for the $|\pm\rangle$ basis states in the two-beam interferometer.

measurement to be a measurement of S_z, we have

$$\begin{aligned}|0\rangle &= |z_+\rangle\,, \\ |1\rangle &= |z_-\rangle\,.\end{aligned} \qquad (2.70)$$

On the other hand, there are many other possible measurements on the system. For example, the basis states

$$\begin{aligned}|+\rangle &= \frac{1}{\sqrt{2}}\left(|0\rangle + |1\rangle\right), \\ |-\rangle &= \frac{1}{\sqrt{2}}\left(|0\rangle - |1\rangle\right),\end{aligned} \qquad (2.71)$$

are states of definite S_x. (Note that $|\pm\rangle = |x_\pm\rangle$.) The choice of S_z as the standard measurement, rather than S_x or some other spin component, is merely a matter of convention.

For the two-beam interferometer, our remarks about basis independence seem at first to be a bit strained. We let $|0\rangle$ represent the state in which the photon is in the upper beam and $|1\rangle$ the state in which the photon is in the lower beam. This choice of standard basis seems to be especially natural, since we can easily perform the corresponding measurement with a pair of photon detectors. In fact, we might be tempted to say that this is the only sort of measurement that makes sense. If the "which beam" measurement corresponds to S_z, what could possibly correspond to a measurement of S_x?

In fact, it is possible to measure the photon in any basis we choose. Consider the apparatus shown in Fig. 2.15. The two beams pass through a balanced beamsplitter and the resulting beams are directed into ordinary photon detectors D+ and D−. We should think of this combination of beamsplitter and detectors as a single measurement device, which can produce two possible results depending on which photon detector is triggered. What measurement is performed?

It is not the standard measurement on the input beams. If either $|0\rangle$ or $|1\rangle$ is the input state, the detectors D+ and D− will be triggered with equal probability. In fact, the apparatus is useless for determining which input beam contains the photon. On the other hand, suppose the superposition state $|+\rangle$ from Eq. 2.71 is introduced. It is not hard to show that the photon is found at D+ with probability one. Similarly, if the input state is $|-\rangle$, then D− must register the photon. The apparatus in Fig. 2.15 performs the measurement associated with the $|\pm\rangle$ basis, which is the analog of S_x. More complicated interferometer arrangements involving beamsplitters and phase shifters can perform other measurements.

Exercise 2.34 Devise an apparatus to measure the analog of S_y in the two-beam interferometer.

Thus we see that all sorts of measurements do after all "make sense" for the two-beam interferometer example. Because qubit systems are isomorphic to one another, commonplace observations about one can lead us to insights about the others.

Qubit systems can be used to perform tasks such as the storage of information. We can encode a one-bit message into a qubit in the obvious way, using $|0\rangle$ to represent 0 and $|1\rangle$ to represent 1. But qubits are not merely bits. There are many states available other than the standard basis, and many measurements possible other than the standard measurement. How do these new possibilities affect the qubit's capacity to perform information tasks? What new limitations are imposed – and new possibilities afforded – by quantum physics? These are important questions to which we will return many times as we develop the theory of quantum mechanics.

Problems

Problem 2.1 Find expressions for the probabilities P_0 and P_1 for the Mach–Zehnder interferometer of Fig. 2.6 for arbitrary values of ϕ. From your results, show that $P_0 + P_1 = 1$ for any ϕ, as expected.

Problem 2.2 Find all of the distinct interferometer arrangements – that is, arrangements represented by different matrices – that can be constructed out of only balanced beamsplitters $\mathbf{B}_l$ and $\mathbf{B}_u$. (Do not forget to include the arrangement that uses no beamsplitters at all!)

Problem 2.3 Constructing $U(2)$. The set of all 2×2 unitary matrices is called $U(2)$. We have shown that linear optical elements are always represented by members of $U(2)$. But do all of the members of $U(2)$ represent physically possible linear optical elements?

Show that this is so by proving that *any* matrix in $U(2)$ can be written as a product of the matrices for phase shifters and balanced beamsplitters. We can therefore use these basic devices as component parts to construct any imaginable linear optical element. Hint: Tackle a simpler problem first. Figure out how to construct any unitary matrix with only *real* entries of the form

$$\mathbf{U} = \left(\begin{array}{cc} \cos\alpha & \sin\alpha \\ -\sin\alpha & \cos\alpha \end{array} \right).$$

Problem 2.4 The magnetic moment μ of a small loop of electric current equals the current in the loop times its area. Suppose that a particle of mass M and charge Q is really a spinning charged ring with some tiny radius R. Find the gyromagnetic ratio γ for this ring in terms of Q and M. How well does this model work for the proton?

Problem 2.5 Consider a spin-1/2 particle passing through a series of spin filters, as in Fig. 2.13. Now, however, there are 13 filters (numbered 0 through 12). The nth filter selects for the state $|n_+\rangle$, the $+\hbar/2$ basis state for a measurement of spin along an axis in the *xz*-plane an angle $n\pi/12$ from z. From Eq. 2.35, we can write

$$|n_+\rangle = \cos\frac{n\pi}{24}|z_+\rangle + \sin\frac{n\pi}{24}|z_-\rangle.$$

What is the probability that a spin-1/2 particle, initially in the state $|z_+\rangle$, will pass all 13 filters?

Problem 2.6 Devise a spin-1/2 analog of the bomb-testing experiment in Section 2.1. Your bomb will have a trigger that is set off by the passage of a single particle with state $|z_-\rangle$.

Problem 2.7 If we represent the state $|\psi\rangle$ of a two-level atom as a complex column vector with respect to the $|E_0\rangle$ and $|E_1\rangle$ basis states, then the time evolution operator $U(t)$ will be a 2×2 complex matrix $\mathbf{U}(t)$. Find this matrix and show that it is unitary.

Problem 2.8 Two boxes each produce a stream of qubits. Box A produces the qubits all in the state $|+\rangle = \frac{1}{\sqrt{2}}(|0\rangle + |1\rangle)$. Box B randomly produces qubits in states $|0\rangle$ and $|1\rangle$, each with probability 1/2. We have one of the boxes, but it is unmarked and so we do not know which kind it is. Describe an experiment on the qubits that can tell the difference between box A and box B. Can you reliably tell the difference between the boxes by examining only one of the qubits?

3 States and observables

3.1 Hilbert space

The prototype qubit systems of the last chapter are very simple, but they can be generalized to more complicated versions. We can send a photon through an interferometer with three, four or more distinct beams. We can perform experiments on particles with higher intrinsic angular momentum than the spin-1/2 particles we have discussed. And we can analyze atomic systems in situations that involve more than two different energy levels. For these cases and others, we will need a more general version of quantum theory.

That theory will include two pieces. First, we will have a general mathematical structure that is applicable to many kinds of system. Here the qubit case will be our guide, since many of the basic concepts for other quantum systems are already present in the qubit case. Second, we will have to describe how to apply the quantum formalism to specific physical situations. Though the quantum systems we discuss will appear quite various, they share strong family resemblances that are expressed in the common mathematical framework. Keeping the framework in mind will help us understand specific examples; keeping the examples in mind will help us understand the framework.

The *states* of a quantum system are described by kets $|\psi\rangle$, which obey the principle of superposition. This means that the kets are elements of an abstract vector space $\mathcal{H}$ called a *Hilbert space*. The key property of a vector space is that any linear combination of vectors is also a vector. If kets $|\phi\rangle$ and $|\psi\rangle$ are both in $\mathcal{H}$, and a and b are complex coefficients, then

$$a|\phi\rangle + b|\psi\rangle \in \mathcal{H}$$

We remind ourselves that whatever we write within the ket symbol $|\cdots\rangle$ is merely a label. In particular, a ket $|0\rangle$ labelled by the symbol "0" is not the same thing as the zero vector in $\mathcal{H}$, which we will generally denote by 0 (with no ket). In other words,

$$|\psi\rangle - |\psi\rangle = 0 \neq |0\rangle. \tag{3.1}$$

A Hilbert space, however, is more than just a vector space with complex scalars. It also has an *inner product*, which is analogous to the familiar dot product for spatial vectors. Given $|\phi\rangle$ and $|\psi\rangle$ in $\mathcal{H}$, the combination $\langle\phi|\psi\rangle$ is a complex number. The Hilbert space inner product must have the following defining properties:[1]

---

[1] The technical definition of a Hilbert space also includes the property of *metric* or *Cauchy completeness*, which is related to the existence of limits for sequences. We will not have much to say about metric completeness. The property automatically holds for finite-dimensional vector spaces, and we will implicitly assume it otherwise.

Symmetry. For any vectors $|\phi\rangle$ and $|\psi\rangle$, $\langle\phi|\psi\rangle = \langle\psi|\phi\rangle^*$. (This means that $\langle\phi|\phi\rangle$ is real for any $|\phi\rangle$.)

Linearity. For any vectors $|\phi\rangle$, $|\psi_1\rangle$, and $|\phi_2\rangle$, and any scalars a_1 and a_2,

$$\langle\phi|\left(a_1|\psi_1\rangle + a_2|\psi_2\rangle\right) = a_1\langle\phi|\psi_1\rangle + a_2\langle\phi|\psi_2\rangle. \tag{3.2}$$

Positive-definiteness. For any non-zero vector $|\phi\rangle$, $\langle\phi|\phi\rangle > 0$. (Inner products involving the zero vector are always zero.) We call $\sqrt{\langle\phi|\phi\rangle}$ the *norm* of the vector $|\phi\rangle$.

The kets representing physical states of quantum systems are normalized to be "unit" vectors: $\langle\psi|\psi\rangle = 1$. Two kets are said to be *orthogonal* if their inner product is zero: $\langle\phi|\psi\rangle = 0$.

We usually describe a particular element $|\psi\rangle$ in $\mathcal{H}$ by means of a set of *basis* vectors. If $\{|\phi_n\rangle\}$ is a basis, then $|\psi\rangle$ can be written as

$$|\psi\rangle = \sum_n c_n |\phi_n\rangle. \tag{3.3}$$

Furthermore, given $|\psi\rangle$ and the basis, the coefficients c_n are unique. These numbers are called the *components* of $|\psi\rangle$ with respect to the basis set $\{|\phi_n\rangle\}$.

If we agree upon a basis set, then any vector is exactly specified by its components. This is neat, because the various vector operations – vector addition, scalar multiplication, and the inner product – can be worked out using complex arithmetic on the vector components.

Any two bases for the same Hilbert space will have exactly the same number of elements. This number is called the *dimension* of the space. Therefore, the sum over n in Eq. 3.3 ranges from 1 to $d = \dim\mathcal{H}$. (As we have done here, we usually omit the limits of a summation when it is clear from context that the summed index ranges over all of its possible values.) All of the qubit systems that we discussed in the previous chapter had $\dim\mathcal{H} = 2$. In fact, we will sometimes use the special symbol $\mathcal{Q}$ to denote the qubit Hilbert space with dimension 2. More complicated systems may be associated with spaces of higher dimension. For the present we will suppose that the Hilbert space dimension is finite; later on we will discuss how to deal with infinite-dimensional spaces.

It can happen that only part of the overall Hilbert space $\mathcal{H}$ applies to a particular problem. We saw this in our discussion of "two-level atoms" in Section 2.3. Real atoms have many energy levels, so the Hilbert space $\mathcal{H}$ for the atomic system has a high dimension. In the case we considered, however, only two energy levels (and their superpositions) had any physical relevance, so the atom effectively behaved as a qubit.

The Hilbert space $\mathcal{Q}$ for the simplified two-level atom is a *subspace* of a larger Hilbert space $\mathcal{H}$ for the atom. A subspace is a subset of a vector space that is also a vector space in itself – i.e. is closed under linear combinations. The subspaces of $\mathcal{H}$ include two trivial extreme cases: the null space (which contains only the zero vector) and $\mathcal{H}$ itself. In between the extremes, however, there are some interesting cases. For spatial vectors in 3-D, subspaces include all the straight lines and flat planes that pass through the origin.

The dimension of a subspace is no larger than that of the parent space: If $\mathcal{T}$ is a subspace of $\mathcal{H}$, then $\dim\mathcal{T} \leq \dim\mathcal{H}$. This means that a basis for $\mathcal{T}$ has no more vectors than a basis for $\mathcal{H}$. In fact, it is always possible to extend a basis for $\mathcal{T}$ (by including more vectors) into a basis for $\mathcal{H}$. Two different subspaces $\mathcal{T}_1$ and $\mathcal{T}_2$ are said to be *orthogonal subspaces* if every vector in $\mathcal{T}_1$ is orthogonal to every vector in $\mathcal{T}_2$.

Exercise 3.1 Describe in words a pair of orthogonal subspaces for spatial vectors in 3-D.

Exercise 3.2 Show that, if $\mathcal{T}_1$ and $\mathcal{T}_2$ are two orthogonal subspaces, then the only vector that lies in both subspaces is the zero vector. Is the converse true? That is, if two subspaces contain only the zero vector in common, are they necessarily orthogonal?

The restriction of a state to a subspace of $\mathcal{H}$ may have some physical basis, such as a limitation on the total energy of an atom; or it might be assumed simply as a mathematical convenience. In either case, the subspace acts as an effective Hilbert space for the system.

Orthonormal bases

In a Hilbert space, the most convenient sort of basis set is one that is *orthonormal*. This means that the basis vectors are orthogonal to each other and normalized. We can summarize this condition by

$$\langle \phi_m | \phi_n \rangle = \delta_{mn}, \tag{3.4}$$

where δ_{mn} is the *Kronecker delta symbol*, defined as

$$\delta_{mn} = \begin{cases} 1 & m = n \\ 0 & m \neq n. \end{cases} \tag{3.5}$$

From now on, the term "basis" for a Hilbert space is assumed to mean an orthonormal basis, unless otherwise noted.

We can easily compute the inner product $\langle \alpha | \beta \rangle$ whenever we know the components of the two kets with respect to an orthonormal basis. Suppose that $|\alpha\rangle = \sum_n a_n |\phi_n\rangle$ and $|\beta\rangle = \sum_n b_n |\phi_n\rangle$. Then

$$\langle \alpha | \beta \rangle = \sum_n a_n^* b_n. \tag{3.6}$$

The proof of Eq. 3.6 is worth going over in detail.

We begin with $\langle \alpha | \beta \rangle$, expanding $|\beta\rangle$ using the $\{|\phi_n\rangle\}$ basis and remembering that the inner product is linear:

$$\langle \alpha | \beta \rangle = \sum_n b_n \langle \alpha | \phi_n \rangle. \tag{3.7}$$

In each term of this sum, we want to expand $|\alpha\rangle$ using our basis. To do this, we have to bear in mind two things. First, we must choose a different index for the new sum, because the index n is already in use. Second, we recall that the inner product is conjugate-linear in $|\alpha\rangle$. This yields

$$\langle \alpha | \beta \rangle = \sum_n b_n \left(\sum_m a_m^* \langle \phi_m | \phi_n \rangle \right). \tag{3.8}$$

Each b_n is a constant with respect to the m-sum, so we can "bring it into" the m-sum (distribute the b_n factor over the terms of this sum). We then wind up with the double sum

$$\langle \alpha | \beta \rangle = \sum_n \sum_m b_n a_m^* \langle \phi_m | \phi_n \rangle. \qquad (3.9)$$

We can clean this up in several ways. First, we will often write multiple sums in a compact way, as in $\sum_{nm}$. Second, because the b_ns and the a_m^*s are complex numbers, their multiplication is commutative and we can write them in any order. Finally, orthonormality of the basis makes everything simpler:

$$\langle \alpha | \beta \rangle = \sum_{nm} a_m^* b_n \delta_{mn}. \qquad (3.10)$$

The δ_{mn} allows us to do the m-sum immediately. Each term in that sum will be zero except for the term in which $m = n$. And so we arrive at our destination, namely

$$\langle \alpha | \beta \rangle = \sum_n a_n^* b_n. \qquad (\text{Re } 3.6)$$

This derivation illustrates a number of the common "tricks of the trade" for dealing with sums. Some of these may seem obscure at first, but as you think them over, they should seem plausible, then rigorous, then obvious. (Try to get all the way to "obvious.")

Exercise 3.3 Show that, if $|\alpha\rangle = \sum_n a_n |\phi_n\rangle$ for an orthonormal basis $\{|\phi_n\rangle\}$, then

$$a_m = \langle \phi_m | \alpha \rangle. \qquad (3.11)$$

Orthonormal bases are associated with the simplest type of measurement procedure, which we will call a *basic measurement*. Each result of a basic measurement is associated with a basis vector. Qubit examples included measurements of the spin components S_x, S_z, and so on for a spin-1/2 particle, as well as a measurement of the energy of a two-level atom.

For now, we do not necessarily have to associate numerical values with the various possible outcomes of an experiment. The outcomes might be "red," "orange," "yellow," etc. We only require that distinct outcomes are given distinct labels, and that each outcome is associated with exactly one member of an orthonormal basis of the Hilbert space $\mathcal{H}$ for the system.

Suppose the basis vector $|\phi_k\rangle$ is associated with the kth possible outcome for the measurement, and the quantum state of the system is given by $|\psi\rangle$. Then the probability of the kth outcome is

$$p(k) = |\langle \phi_k | \psi \rangle|^2. \qquad (3.12)$$

The inner product $\langle \phi_k | \psi \rangle$, which is the kth component of $|\psi\rangle$ with respect to the basis $\{|\phi_k\rangle\}$, is also the *probability amplitude* for the kth possible outcome of the measurement associated with that basis.

Exercise 3.4 Show that, if the state $|\psi\rangle = |\phi_m\rangle$ for some basis vector $|\phi_m\rangle$, the measurement will yield the mth result with certainty.

Exercise 3.5 Show that, for any quantum state $|\psi\rangle$ and any basic measurement, the probabilities satisfy the usual laws:

$$p(k) \geq 0,$$

$$\sum_k p(k) = 1. \tag{3.13}$$

Note that the second of these facts relies on the normalization of the physical state vector: $\langle \psi | \psi \rangle = 1$.

Suppose two state vectors differ only by an overall phase factor: $|\psi\rangle$ and $|\psi'\rangle = e^{i\phi} |\psi\rangle$. Equation 3.12 tells us that $|\psi\rangle$ and $|\psi'\rangle$ lead to exactly the same probabilities for the outcomes of any basic measurement. In fact, two vectors that differ only by an overall phase factor are equivalent, representing the same physical situation. This does not mean that phases are irrelevant. In general, $a|\alpha\rangle + b|\beta\rangle$ and $a|\alpha\rangle + be^{i\phi}|\beta\rangle$ are quite different states that lead to different predictions. For example, consider the states of the spin-1/2 particle

$$|x_+\rangle = \frac{1}{\sqrt{2}} \left(|z_+\rangle + |z_-\rangle \right) \quad \text{and} \quad |x_-\rangle = \frac{1}{\sqrt{2}} \left(|z_+\rangle - |z_-\rangle \right). \tag{3.14}$$

These two states, which give opposite results in an S_x measurement, differ only in the sign of the superposition terms – a phase factor of $e^{i\pi}$. We usually express this by saying that the *global* phase of $|\psi\rangle$ does not have any physical significance, but the *relative* phases among the terms in a superposition do.

Equation 3.12 is one of the fundamental postulates of quantum mechanics, because it tells us how the quantum state $|\psi\rangle$ is connected to measurement outcomes. We will illustrate how it works by considering an extended example.

Spin one

We previously studied the behavior of spin-1/2 particles. Examples included electrons, protons, and neutrons. Now we will introduce *spin-1 particles*. These include several types of atomic nuclei, as well as the elementary particles associated with the strong and weak nuclear forces.[2] If we measure a spin component of a spin-1 particle, perhaps by a Stern–Gerlach experiment, we get three possible results: $+\hbar$, 0, and $-\hbar$.

Just as the spin of a spin-1/2 particle is described by a Hilbert space Q of dimension 2, the spin of a spin-1 particle is described by a Hilbert space with dimension 3. One possible basis for this space is associated with a measurement of the z-component of spin S_z. We denote these three orthonormal vectors $|z_+\rangle$, $|z_0\rangle$, and $|z_-\rangle$. Any state is a superposition of these three:

$$|\psi\rangle = a_+ |z_+\rangle + a_0 |z_0\rangle + a_- |z_-\rangle. \tag{3.15}$$

Given this state, the probability that a measurement of S_z will yield $+\hbar$, for instance, is exactly $|a_+|^2$.

---

[2] The photon is spin-1, but it is also massless and so does not have a "rest frame." This complicates matters, and our analysis here does not apply to photons.

The z-axis is not sacred. We can equally well consider the trio of basis vectors associated with a measurement of some other component of spin, say S_x. These basis vectors are related to the S_z basis by

$$|x_+\rangle = \frac{1}{2}|z_+\rangle + \frac{1}{\sqrt{2}}|z_0\rangle + \frac{1}{2}|z_-\rangle,$$

$$|x_0\rangle = \frac{1}{\sqrt{2}}|z_+\rangle - \frac{1}{\sqrt{2}}|z_-\rangle,$$

$$|x_-\rangle = \frac{1}{2}|z_+\rangle - \frac{1}{\sqrt{2}}|z_0\rangle + \frac{1}{2}|z_-\rangle. \tag{3.16}$$

Here are two exercises about spin-1 particles.

Exercise 3.6 Verify that the S_x basis vectors are orthonormal. (How many inner products do you have to evaluate to show this?)

Exercise 3.7 The state of a spin-1 particle is $|\psi\rangle = |z_0\rangle$. What is the probability that a measurement of S_x yields zero? Explain this in terms of a stream of spin-1 particles passing through various Stern–Gerlach devices.

See also Problem 3.1.

Of course, we have not yet said how we come to identify a particular orthonormal basis with a particular measurement on a system. At this stage, Eq. 3.16 must be taken on trust! But once we have made that identification, Eq. 3.12 allows us to predict the results of experiments on the quantum system.

Matrices and dual vectors

A good way to organize these calculations is to use matrices. Suppose we fix an orthonormal basis in a d-dimensional Hilbert space $\mathcal{H}$, and the ket $|\alpha\rangle$ has components a_n with respect to this basis. Then we can represent $|\alpha\rangle$ by the $d \times 1$ matrix

$$\mathbf{a} = \begin{pmatrix} a_1 \\ \vdots \\ a_d \end{pmatrix}. \tag{3.17}$$

As we emphasized in Section 2.2, the column matrix $\mathbf{a}$ is not the same thing as the ket $|\alpha\rangle$. The latter might have an intrinsic physical meaning (e.g. the state of a system at a given moment), whereas the former depends on our choice of orthonormal basis. Nevertheless, we can use the matrix representations of vectors to perform operations in the Hilbert space.

Exercise 3.8 Write matrix representations with respect to the S_z basis for the $|x_+\rangle$, $|x_0\rangle$, and $|x_-\rangle$ states of a spin-1 particle. See Eq. 3.16.

For instance, let **a** and **b** be column matrices representing the kets $|\alpha\rangle$ and $|\beta\rangle$ with respect to some orthonormal basis. Then the inner product is given by

$$\langle \alpha | \beta \rangle = \mathbf{a}^{\dagger} \mathbf{b}, \tag{3.18}$$

where $\dagger$ denotes the Hermitian conjugate of the matrix, as introduced in Section 2.1. Examining Eq. 3.18, we are tempted to say that the row matrix $\mathbf{a}^{\dagger}$ represents a "bra" $\langle \alpha |$, which appears in the inner product. This is just what we said in the context of qubit quantum theory in Chapter 2. But what kind of mathematical object is the bra $\langle \alpha |$?

A *linear functional* is a function that maps vectors in $\mathcal{H}$ to complex scalars, acting in a linear way. In other words, if f is a linear functional, then

$$f\left(a_1 |\psi_1\rangle + a_2 |\psi_2\rangle\right) = a_1 f\left(|\psi_1\rangle\right) + a_2 f\left(|\psi_2\rangle\right). \tag{3.19}$$

Many familiar aspects of vector algebra can be expressed using linear functionals.

Exercise 3.9 Suppose we fix a basis set $\{|\phi_k\rangle\}$ and define a function f_n that maps each vector to its nth component. Show that f_n is a linear functional.

Linear functionals are sometimes called *dual vectors*, since they themselves form a vector space.

By the bra $\langle \alpha |$, we simply mean the dual vector (linear functional) that maps the vector $|\beta\rangle$ to the scalar $\langle \alpha | \beta \rangle$. The inner product on $\mathcal{H}$ thus allows us to turn regular vectors into dual vectors. As it turns out, every dual vector on $\mathcal{H}$ is the "bra" of some "ket," as the following exercise illustrates.

Exercise 3.10 Write the linear functional f_n from the previous exercise as a bra (dual vector).

Note that the relation between ket and bra is conjugate-linear. That is, if $|\phi\rangle = a_1 |\phi_1\rangle + a_2 |\phi_2\rangle$, then

$$\langle \phi | = a_1^* \langle \phi_1 | + a_2^* \langle \phi_2 |. \tag{3.20}$$

Exercise 3.11 Use the properties of the inner product to show Eq. 3.20.

From this it follows that, if $|\alpha\rangle = \sum_n a_n |\phi_n\rangle$, then

$$\langle \alpha | = \sum_n a_n^* \langle \phi_n |. \tag{3.21}$$

The coefficients a_n^* are just the entries of the row matrix $\mathbf{a}^{\dagger}$. Thus, we can take $\mathbf{a}^{\dagger}$ to be the matrix representation of $\langle \alpha |$, as expected.

Is this talk of dual vectors more than just a new language for talking about inner products? Not really. We have introduced no additional properties for the Hilbert space $\mathcal{H}$. On the other hand, the dual vector language will make it easier to describe the properties of linear operators on $\mathcal{H}$ – a topic to which we now must turn.

3.2 Operators

In our development of qubit quantum theory in Chapter 2, we found that we needed to use operators to describe both the time evolution of a quantum system and a system's measurable quantities. This will also be true for more general systems in quantum theory. Since we are keenly interested in how quantum systems change over time and what quantitative predictions can be made about them, we clearly need to learn a lot about operators and their properties!

Elementary concepts

Formally, an *operator* G on the Hilbert space $\mathcal{H}$ is a function that maps vectors to vectors in a linear way:

$$G\left(a_1 \left|\phi_1\right\rangle + a_2 \left|\phi_2\right\rangle\right) = a_1 G \left|\phi_1\right\rangle + a_2 G \left|\phi_2\right\rangle. \tag{3.22}$$

An operator is completely defined by how it acts upon "input" vectors to produce "output" vectors. Indeed, because an operator is a linear map, it is enough to specify how it acts on a basis of input vectors. Let $\{|n\rangle\}$ be a basis for $\mathcal{H}$, and consider an arbitrary input vector $|\psi\rangle$:

$$G \left|\psi\right\rangle = G \left(\sum_n c_n \left|n\right\rangle\right) = \sum_n c_n G \left|n\right\rangle. \tag{3.23}$$

If we know the output vectors $G |n\rangle$ for basis vector inputs, we can calculate the effect of G on any input $|\psi\rangle$.

For any operator G there is a set of vectors that map to the zero vector – a set that always contains the zero vector itself. This set is called the *kernel* of G, denoted ker G.

Exercise 3.12 For any operator G on $\mathcal{H}$, show that the kernel ker G is actually a subspace of $\mathcal{H}$.

If ker G is the zero subspace of $\mathcal{H}$ – if it contains the zero vector and no other – then we say that ker G is "trivial." The kernel is a useful tool for analyzing the action of an operator. Here is a good example.

Exercise 3.13 Suppose the operator G has the property that $G |x\rangle = G |y\rangle$ only if $|x\rangle = |y\rangle$. Then G is said to be *one-to-one*. Prove that G is one-to-one if and only if ker G is trivial.

The simplest operator is the *identity* operator **1**, which maps every vector to itself:

$$\mathbf{1} \left|\psi\right\rangle = \left|\psi\right\rangle. \tag{3.24}$$

This is obviously one-to-one, so ker **1** is trivial. Scalar multiplication can also be regarded as a kind of operator: the map that takes $|\psi\rangle$ to $a |\psi\rangle$ for some scalar a is just the operator $a\mathbf{1}$.

We can multiply operators by scalars and add operators together to make new operators.[3] We can also multiply operators together. Given A and B, the operator AB is defined by

$$(AB) \, |\psi\rangle = A \left(B \, |\psi\rangle \right). \tag{3.25}$$

The operator product AB is thus read from right to left: first the operator B acts on the input, and then A acts on the result. Note that this is *not* necessarily the same process as acting first with A and then B. In other words, operator multiplication is not commutative, and in general $AB \neq BA$. The *commutator* [A, B] is just the difference between AB and BA:

$$[A, B] = AB - BA. \tag{3.26}$$

This is an operator in its own right. The commutator of any operator with itself is zero: $[A, A] = 0$.

Another simple type of operator is the *outer product* of two vectors in $\mathcal{H}$. The outer product is written $|\alpha\rangle\langle\beta|$. This looks a bit mysterious at first, but its meaning is easy to understand. When we apply this operator to an input vector $|\psi\rangle$, we get

$$|\alpha\rangle\langle\beta| \left(|\psi\rangle \right) = |\alpha\rangle \, \langle\beta \,|\, \psi\rangle. \tag{3.27}$$

That is, we compute the scalar $\langle\beta \,|\, \psi\rangle$ and then multiply the vector $|\alpha\rangle$ by it.[4]

Exercise 3.14 Consider the qubit Hilbert space $\mathcal{Q}$ with standard basis vectors $|0\rangle$ and $|1\rangle$. Let $A = |0\rangle\langle1|$ and $B = |1\rangle\langle0|$. Show that $AB \neq BA$, illustrating the fact that operator multiplication is not commutative. (Hint: Show that the two operator products act differently on the input vector $|0\rangle$.)

If we have a normalized vector $|\alpha\rangle$, then the outer product $\Pi_\alpha = |\alpha\rangle\langle\alpha|$ is called the *projection* on $|\alpha\rangle$. Let us consider a specific qubit example, the projection $\Pi_0 = |0\rangle\langle0|$ on $\mathcal{Q}$. Clearly, $\Pi_0 \, |0\rangle = |0\rangle$ and $\Pi_0 \, |1\rangle = 0$. For a general superposition of $|0\rangle$ and $|1\rangle$,

$$\Pi_0 \, |\psi\rangle = |0\rangle\langle0| \left(c_0 \, |0\rangle + c_1 \, |1\rangle \right) = c_0 \, |0\rangle. \tag{3.28}$$

Intuitively, the projection Π_0 on $|0\rangle$ just picks out the part of $|\psi\rangle$ that is parallel to $|0\rangle$. The projection Π_1 works in the same way for the part of $|\psi\rangle$ that is parallel to $|1\rangle$. Thus, if we add these two projections together, we get

---

[3] The set of operators on $\mathcal{H}$ is sometimes denoted $\mathcal{B}(\mathcal{H})$. Since the linear combination of two operators is also an operator, $\mathcal{B}(\mathcal{H})$ is a vector space in its own right.

[4] The only oddity here is a less-than-customary order for the scalar multiplication. The expression is of course equal to $\langle\beta \,|\, \psi\rangle \, |\alpha\rangle$.

$$(\Pi_0 + \Pi_1) \, |\psi\rangle = c_0 \, |0\rangle + c_1 \, |1\rangle = |\psi\rangle. \tag{3.29}$$

In other words, for qubits we have the operator relation

$$|0\rangle\langle 0| + |1\rangle\langle 1| = \mathbf{1}. \tag{3.30}$$

This fact can be generalized from qubits to any type of quantum system. Suppose $\{\,|n\rangle\,\}$ is an orthonormal basis for $\mathcal{H}$, which may have any dimension. Then

$$\sum_n |n\rangle\langle n| = \mathbf{1}. \tag{3.31}$$

This is called the *completeness relation* for an orthonormal basis, and it will be an extremely important tool in our analysis of operators.

Exercise 3.15 Prove the completeness relation.

For the qubit Hilbert space $\mathcal{Q}$, the identity operator $\mathbf{1} = |0\rangle\langle 0| + |1\rangle\langle 1|$. We now introduce three other useful operators on $\mathcal{Q}$, called the *Pauli operators*. These are:

$$X = |0\rangle\langle 1| + |1\rangle\langle 0|,$$
$$Y = -i \, |0\rangle\langle 1| + i \, |1\rangle\langle 0|,$$
$$Z = |0\rangle\langle 0| - |1\rangle\langle 1|. \tag{3.32}$$

(Note that the Pauli operators depend on the standard basis. If we choose other standard basis vectors $|0'\rangle$ and $|1'\rangle$, then we will obtain a different set of Pauli operators.) The Pauli operators have many useful and interesting properties, a few of which are suggested by the following exercises.

Exercise 3.16 Calculate $X\,|0\rangle$, $X\,|1\rangle$, $Y\,|0\rangle$, $Y\,|1\rangle$, $Z\,|0\rangle$, and $Z\,|1\rangle$.

Exercise 3.17 Show that $X^2 = Y^2 = Z^2 = \mathbf{1}$.

Exercise 3.18 Show that $XY = iZ$. Also show that we can cyclically permute X, Y, and Z in this relation.

Operators and matrices

Let G be an operator and $\{\,|n\rangle\,\}$ be an orthonormal basis for the d-dimensional Hilbert space $\mathcal{H}$. As we saw in Eq. 3.23, if we specify the d vectors $G\,|n\rangle$, then we have completely described the operator G. We can specify each of these vectors by giving d components. Thus, we can describe G by giving d^2 complex quantities.

Given two kets $|\alpha\rangle$ and $|\beta\rangle$, we define the corresponding *matrix element* of G to be the inner product $\langle\alpha|\,G\,|\beta\rangle$. The operator G can be described by giving the d^2 matrix elements for basis vectors:

$$G_{mn} = \langle m|\,G\,|n\rangle = \text{ the } m\text{th component of } G\,|n\rangle. \tag{3.33}$$

These can be conveniently arranged as the entries in a $d \times d$ square matrix:

$$(G_{mn}) = \begin{pmatrix} G_{11} & G_{12} & \cdots & G_{1d} \\ G_{21} & G_{22} & & G_{2d} \\ \vdots & & \ddots & \vdots \\ G_{d1} & G_{d2} & \cdots & G_{dd} \end{pmatrix}. \tag{3.34}$$

This is called the *matrix representation* of G with respect to the basis $\{\,|n\rangle\}$.

For instance, in $\mathcal{Q}$ the Pauli operators have matrix representations (with respect to the standard basis):

$$(X) = \begin{pmatrix} 0 & 1 \\ 1 & 0 \end{pmatrix}, \qquad (Y) = \begin{pmatrix} 0 & -i \\ i & 0 \end{pmatrix}, \qquad (Z) = \begin{pmatrix} 1 & 0 \\ 0 & -1 \end{pmatrix}. \tag{3.35}$$

These are called the *Pauli matrices*.

Exercise 3.19 Start with Eq. 3.32 and confirm Eq. 3.35.

Exercise 3.20 Suppose $|\alpha\rangle$ and $|\beta\rangle$ have components a_n and b_n with respect to the $\{\,|n\rangle\}$ basis. Show that the matrix representation of the outer product $|\alpha\rangle\langle\beta|$ has components

$$\left(|\alpha\rangle\langle\beta|\right)_{mn} = a_m b_n^*. \tag{3.36}$$

We can compute the action of the operator G on the vector $|\alpha\rangle$ using the matrix representation of G. Suppose $|\beta\rangle = G\,|\alpha\rangle$, and that the input and output vectors $|\alpha\rangle$ and $|\beta\rangle$ have components a_n and b_n, respectively. Then

$$\begin{aligned} b_m &= \langle m\,|\beta\rangle \\ &= \langle m|\, G\,|\alpha\rangle \\ &= \langle m|\, G\left(\sum_n |n\rangle\langle n|\right)|\alpha\rangle \\ &= \sum_n \langle m|\, G\,|n\rangle\,\langle n\,|\alpha\rangle \\ b_m &= \sum_n G_{mn} a_n. \end{aligned} \tag{3.37}$$

(Notice how we introduced the identity operator **1** in the form given by the completeness relation for the $\{\,|n\rangle\}$ basis. A handy trick!) We recognize this as simple matrix multiplication:

$$\begin{pmatrix} b_1 \\ \vdots \\ b_d \end{pmatrix} = \begin{pmatrix} G_{11} & \cdots & G_{1d} \\ \vdots & \ddots & \vdots \\ G_{d1} & \cdots & G_{dd} \end{pmatrix} \begin{pmatrix} a_1 \\ \vdots \\ a_d \end{pmatrix}. \tag{3.38}$$

Exercise 3.21 Show that Eq. 3.38 is the same as Eq. 3.37.

Similarly, consider the matrix representation for the product operator AB.

$$(AB)_{mn} = \langle m| \text{ AB } |n\rangle$$

$$= \sum_k \langle m| \text{ A } |k\rangle \langle k| \text{ B } |n\rangle$$

$$(AB)_{mn} = \sum_k A_{mk}B_{kn}, \tag{3.39}$$

which is just the matrix product

$$\begin{pmatrix} (AB)_{11} & \cdots & (AB)_{1d} \\ \vdots & \ddots & \vdots \\ (AB)_{d1} & \cdots & (AB)_{dd} \end{pmatrix} = \begin{pmatrix} A_{11} & \cdots & A_{1d} \\ \vdots & \ddots & \vdots \\ A_{d1} & \cdots & A_{dd} \end{pmatrix} \begin{pmatrix} B_{11} & \cdots & B_{1d} \\ \vdots & \ddots & \vdots \\ B_{d1} & \cdots & B_{dd} \end{pmatrix}. \tag{3.40}$$

Exercise 3.22 Fill in the details of the derivation of Eq. 3.39 by writing AB = A1B for a suitable form of the identity **1**.

Exercise 3.23 Verify that the matrix product in Eq. 3.40 expresses the same fact as Eq. 3.39.

In short, we can do computations with operators and vectors by doing matrix computations with the matrix representations of those operators and vectors.

The matrix elements G_{mn} for the operator G are actually the "components" of the operator with respect to a "basis set" of operators formed by the outer products of the vector basis $\{|n\rangle\}$. That is,

$$\text{G} = \sum_{mn} |m\rangle \langle m| \text{ G } |n\rangle \langle n| = \sum_{mn} G_{mn} |m\rangle \langle n|. \tag{3.41}$$

Although not every operator is an outer product of two vectors, we see here that every operator can be written as a sum of outer products.

Matrices and indices. The connection between component expressions (such as Eq. 3.37 and 3.39) and the corresponding matrix expressions (Eq. 3.38 and 3.40) deserves some comment. By convention, G_{mn} represents the matrix entry in the mth row and nth column, as in Eq. 3.34. When we write a matrix multiplication in terms of the matrix entries, the *column* index of the left matrix factor is equal to the *row* index of the right matrix factor, and this index is summed over. This can be seen in Eq. 3.39:

$$(AB)_{mn} = \sum_k A_{mk}B_{kn}. \tag{Re 3.39}$$

The index that is summed over, called a "bound" or "dummy" index, can be freely renamed. Also, the quantities that appear in a component expression are simply complex numbers and can be multiplied in any order. Thus,

$$\sum_k A_{mk}B_{kn} = \sum_k B_{kn}A_{mk} = \sum_j A_{mj}B_{jn}. \tag{3.42}$$

When introducing a sum or renaming a bound index, it is important to avoid using an index that is already used in the expression. In general,

$$\sum_k A_{mk} B_{kn} \neq \sum_n A_{mn} B_{nn}.$$ (3.43)

Exercise 3.24 Construct an explicit example using 2×2 matrices that shows why these two expressions do not mean the same thing.

Exercise 3.25 Suppose

$$D_{ij} = \sum_{kl} A_{ik} B_{lj} C_{kl}.$$

Rewrite this as a matrix expression – that is, the matrix **D** is a product of **A**, **B**, and **C**. The question is, in what order do these matrix factors appear?

The trace

Another operator idea having close connections to matrices is the *trace* of an operator, denoted Tr A. The trace is a linear functional on operators. It can be defined by the following property:

$$\text{Tr} \, |\alpha\rangle\langle\beta| = \langle\beta \, |\alpha\rangle.$$ (3.44)

The trace turns an outer product into an inner product, keeping the bra and ket parts the same. Since any operator can be written as a sum of outer products, this property is enough to completely define the trace. Using a matrix representation,

$$\text{Tr} \, A = \text{Tr} \left(\sum_{mn} A_{mn} |m\rangle\langle n| \right)$$

$$= \sum_{mn} A_{mn} \left(\text{Tr} \, |m\rangle\langle n| \right)$$

$$= \sum_{mn} A_{mn} \delta_{nm}$$

$$\text{Tr} \, A = \sum_n A_{nn}.$$ (3.45)

The trace of an operator is equal to the sum of the diagonal components of a matrix representation – which, not coincidentally, is also called the "trace" of that matrix. From this it is easy to see that the traces of the Pauli operators, which have matrix representations given in Eq. 3.35, are all zero.

Another way to write Eq. 3.45 is

$$\text{Tr} \, A = \sum_n \langle n| \, A \, |n\rangle.$$ (3.46)

We should emphasize that, although we can calculate the trace via Eq. 3.45 or 3.46 in terms of a particular basis $\{ |n\rangle \}$, the trace itself is independent of the choice of basis. Suppose that $\{\,|\alpha_n\rangle\,\}$ and $\{\,|\beta_n\rangle\,\}$ are two orthonormal bases. Then

$$
\begin{aligned}
\sum_n \langle\alpha_n|\,A\,|\alpha_n\rangle &= \sum_n \langle\alpha_n|\,A\left(\sum_m |\beta_m\rangle\langle\beta_m|\right)|\alpha_n\rangle \\
&= \sum_{nm} \langle\alpha_n|\,A\,|\beta_m\rangle\,\langle\beta_m\,|\alpha_n\rangle \\
&= \sum_{nm} \langle\beta_m\,|\alpha_n\rangle\,\langle\alpha_n|\,A\,|\beta_m\rangle \\
&= \sum_m \langle\beta_m|\left(\sum_n |\alpha_n\rangle\langle\alpha_n|\right)A\,|\beta_m\rangle \\
&= \sum_m \langle\beta_m|\,A\,|\beta_m\rangle .
\end{aligned}
\tag{3.47}
$$

Exercise 3.26 Write a sentence explaining each step in this derivation.

A handy fact about the trace is that, for operators A and B,

$$
\mathrm{Tr}\,AB = \mathrm{Tr}\,BA.
\tag{3.48}
$$

We can see this most easily by using matrix representations:

$$
\sum_n (AB)_{nn} = \sum_{nm} A_{nm}B_{mn} = \sum_{nm} B_{mn}A_{nm} = \sum_m (BA)_{mm}.
\tag{3.49}
$$

This is called the *cyclic property* of the trace – cyclic because we can easily extend it to a product of three or more operators,

$$
\mathrm{Tr}\,A\cdots BC = \mathrm{Tr}\,CA\cdots B.
\tag{3.50}
$$

Be careful! We are *not* asserting that the order of the operators in a product is irrelevant to the trace, but only that we can "cyclically permute" the factors. Other sorts of rearrangements might indeed change the overall trace.

Exercise 3.27 In a qubit Hilbert space, let $A = |0\rangle\langle0|$, $B = |0\rangle\langle1|$, and $C = |1\rangle\langle0|$. Show that

$$
\mathrm{Tr}\,ABC \neq \mathrm{Tr}\,ACB.
\tag{3.51}
$$

Exercise 3.28 Show that the trace of any commutator is zero.

3.3 Observables

An *observable* A is a basic measurement in which each outcome is associated with a numerical value. Suppose A_n is the numerical value associated with the nth outcome, which has a basis element $|n\rangle$. The operator A associated with the observable A is defined by

$$A \,|n\rangle = A_n \,|n\rangle, \tag{3.52}$$

for all $|n\rangle$. (As we have already pointed out, knowing how the linear operator A acts on basis vectors is sufficient for its definition.)

In the measurement basis, which is orthonormal, the matrix representation of the operator A has components

$$A_{mn} = \langle m|\, A\, |n\rangle = A_n \delta_{mn}, \tag{3.53}$$

which yield a diagonal matrix

$$(A_{mn}) = \begin{pmatrix} A_1 & & \\ & \ddots & \\ & & A_d \end{pmatrix}. \tag{3.54}$$

Equation 3.41 therefore gives us the following expression for the operator A:

$$A = \sum_n A_n \,|n\rangle\langle n|. \tag{3.55}$$

We can illustrate these relations by returning to a familiar qubit example.

Spin components

For a spin-1/2 particle, the S_x, S_y, and S_z components of spin are all observables. Each one is a basic measurement that associates a numerical value $\pm\hbar/2$ with the measurement outcomes. What are the operators associated with these observables, and what are their matrix representations?

As our standard basis for matrix representation, we will choose the $\{\,|z_\pm\rangle\,\}$ basis of states with definite S_z. In this basis, the observable S_z is very simple. Equation 3.55 tells us that

$$S_z = \frac{\hbar}{2}\,|z_+\rangle\langle z_+| - \frac{\hbar}{2}\,|z_-\rangle\langle z_-|. \tag{3.56}$$

This has a matrix representation

$$(S_z) = \frac{\hbar}{2} \begin{pmatrix} 1 & 0 \\ 0 & -1 \end{pmatrix}. \tag{3.57}$$

In other words, with respect to the S_z measurement basis, the operator

$$S_z = \frac{\hbar}{2}\, Z, \tag{3.58}$$

where Z is the Pauli operator from Eq. 3.32.

How about S_x? Once again, we can write

$$S_x = \frac{\hbar}{2}\,|x_+\rangle\langle x_+| - \frac{\hbar}{2}\,|x_-\rangle\langle x_-|. \tag{3.59}$$

We would like to express everything in terms of our standard $\{|z_\pm\rangle\}$ basis. The S_x measurement basis states are

$$|x_\pm\rangle = \frac{1}{\sqrt{2}}\left(|z_+\rangle \pm |z_-\rangle\right), \tag{3.60}$$

(see Eq. 2.33 and 2.41). The projections on these states are

$$|x_+\rangle\langle x_+| = \left[\frac{1}{\sqrt{2}}\left(|z_+\rangle + |z_-\rangle\right)\right]\left[\frac{1}{\sqrt{2}}\left(\langle z_+| + \langle z_-|\right)\right]$$

$$= \frac{1}{2}\left(|z_+\rangle\langle z_+| + |z_+\rangle\langle z_-| + |z_-\rangle\langle z_+| + |z_-\rangle\langle z_-|\right),$$

$$|x_-\rangle\langle x_-| = \frac{1}{2}\left(|z_+\rangle\langle z_+| - |z_+\rangle\langle z_-| - |z_-\rangle\langle z_+| + |z_-\rangle\langle z_-|\right). \tag{3.61}$$

Thus, the operator S_x is

$$S_x = \frac{\hbar}{2}\left(|z_+\rangle\langle z_-| + |z_-\rangle\langle z_+|\right). \tag{3.62}$$

This has a matrix representation

$$(S_x) = \frac{\hbar}{2}\begin{pmatrix} 0 & 1 \\ 1 & 0 \end{pmatrix}, \tag{3.63}$$

and therefore

$$S_x = \frac{\hbar}{2}\,\mathsf{X}, \tag{3.64}$$

for the Pauli operator X in the $\{|z_\pm\rangle\}$ basis.

Exercise 3.29 Check the algebra in Eq. 3.62. Also show that $|x_+\rangle\langle x_+| + |x_-\rangle\langle x_-| = |z_+\rangle\langle z_+| + |z_-\rangle\langle z_-|$. (What should each of these be equal to?)

Exercise 3.30 Do all this for the observable S_y.

Exercise 3.31 Refer to the spin-1 example in Section 3.1. Adopt the S_z basis states as the standard basis. Find the operators S_z and S_x and their matrix representations with respect to this basis. (You will need Eq. 3.16.)

Expectation values

Suppose our quantum system is in a state $|\psi\rangle$ and we measure the observable A. The nth outcome occurs with probability

$$p(n) = |\langle n|\psi\rangle|^2 = \langle\psi|n\rangle\langle n|\psi\rangle. \tag{3.65}$$

The *mean* or *expectation value* $\langle A\rangle$ is the average of the numerical values A_n, weighted by their probabilities. That is,

$$\langle A\rangle = \sum_n A_n\,p(n). \tag{3.66}$$

As discussed in Appendix A, the term "expectation value" for $\langle A \rangle$, though entirely standard, is a little misleading. The value $\langle A \rangle$ is not necessarily a value that we *expect* to obtain in any particular experiment! Instead, we imagine an *ensemble*, a huge number of similar quantum systems all prepared in the state $|\psi\rangle$. The observable A is measured on the members of this ensemble. Statistically, we then do expect the average of our measurement results to be about $\langle A \rangle$.

The expectation value $\langle A \rangle$ is easily expressed in terms of the operator A:

$$\langle A \rangle = \sum_n A_n \langle \psi | n \rangle \langle n | \psi \rangle$$

$$= \langle \psi | \left(\sum_n A_n |n\rangle \langle n| \right) |\psi\rangle$$

$$\langle A \rangle = \langle \psi | A | \psi \rangle . \tag{3.67}$$

If the state $|\psi\rangle$ has a column matrix $\boldsymbol{\psi}$ of components and operator A has matrix representation $\mathbf{A}$, then

$$\langle A \rangle = \boldsymbol{\psi}^\dagger \mathbf{A} \boldsymbol{\psi} . \tag{3.68}$$

Equation 3.67 tells us that we can make a statistical prediction (the mean of the possible measurement outcomes) by computing the single matrix element $\langle \psi | A | \psi \rangle$. We can use the matrix expression in Eq. 3.68 to do this computation with respect to some basis.[5]

For example, suppose our spin-1/2 particle is in the quantum state $|\psi\rangle = 0.6 |z_+\rangle - 0.8 |z_-\rangle$. (You should verify that this is properly normalized.) Then the expectation value $\langle S_x \rangle$ for a measurement of the x-component of the spin is

$$\langle S_x \rangle = \begin{pmatrix} 0.6 & -0.8 \end{pmatrix} \frac{\hbar}{2} \begin{pmatrix} 0 & 1 \\ 1 & 0 \end{pmatrix} \begin{pmatrix} 0.6 \\ -0.8 \end{pmatrix} = -0.48\hbar. \tag{3.69}$$

(Note that this is close to $-\hbar/2$, which makes sense because $|\psi\rangle$ is close to $|x_-\rangle$ in the Hilbert space.)

Exercise 3.32 A two-level atom has possible energies E_0 and E_1. Write down the Hamiltonian (energy operator) H as a sum of outer products. Now suppose that the atom is in the state $|\psi\rangle = \frac{1}{2} |E_0\rangle + \frac{\sqrt{3}}{2} |E_1\rangle$. Use your operator H to find the expectation value of the atom's energy.

Given the observable A, what is the square A^2 of the operator A?

$$A^2 = \left(\sum_m A_m |m\rangle \langle m| \right) \left(\sum_n A_n |n\rangle \langle n| \right)$$

$$= \sum_{mn} A_m A_n |m\rangle \, \delta_{mn} \, \langle n| = \sum_n A_n^2 |n\rangle \langle n| . \tag{3.70}$$

---

[5] Where convenient, we will sometimes write $\langle A \rangle$ – using the operator A – to represent the expression in Eq. 3.67. Despite this apparently suggestive notation, we understand that $\langle A \rangle$ is a scalar, not an operator.

This is the operator associated with an observable having the same measurement basis as A, but whose numerical values are the squares of the A-values – in other words, the observable A^2.

The same idea works for the third and higher powers as well. In general, the observable A^k is associated with the operator A^k. Given a quantum state $|\psi\rangle$, its expectation value is

$$\left\langle A^k \right\rangle = \langle \psi | \mathsf{A}^k | \psi \rangle . \tag{3.71}$$

Note that this is *not* generally the same as $\langle A \rangle^k$!

Exercise 3.33 A spin-1/2 particle is in the state $|x_+\rangle$. Show that $\langle S_z \rangle^2 = 0$, but $\left\langle S_z^2 \right\rangle \neq 0$.

Finally, we note that expectation values can be conveniently written in terms of the operator trace. Given a state $|\psi\rangle$, we define the *density operator* ρ to be the projection $|\psi\rangle\langle\psi|$. Then

$$\langle A \rangle = \mathrm{Tr}\, \rho\mathsf{A}. \tag{3.72}$$

For now, this is simply a mathematical curiosity. Later on in Chapter 8, we will use this idea to extend quantum theory to situations in which we cannot assign a definite quantum state vector to a system.

3.4 Adjoints

Given any operator A on $\mathcal{H}$, there is another operator called the *adjoint* (or *Hermitian conjugate*), denoted $\mathsf{A}^\dagger$. The adjoint is defined so that, for any vectors $|\alpha\rangle$ and $|\beta\rangle$ in $\mathcal{H}$,

$$\langle \alpha | \mathsf{A}^\dagger | \beta \rangle = \left(\langle \beta | \mathsf{A} | \alpha \rangle \right)^* . \tag{3.73}$$

The adjoint of an operator is related to the Hermitian conjugate of a matrix. The matrix representation of the operator $\mathsf{A}^\dagger$ has entries

$$\left(A^\dagger \right)_{mn} = \langle m | \mathsf{A}^\dagger | n \rangle = \left(\langle n | \mathsf{A} | m \rangle \right)^* = A_{nm}^*. \tag{3.74}$$

The matrix representation of $\mathsf{A}^\dagger$ is the complex conjugate of the transpose of the matrix for A.

The adjoint has two very useful properties. First, and pretty obviously, the adjoint of the adjoint is the original operator: $\left(\mathsf{A}^\dagger \right)^\dagger = \mathsf{A}$. The second property gives the adjoint of a product of operators:

$$(\mathsf{AB})^\dagger = \mathsf{B}^\dagger \mathsf{A}^\dagger. \tag{3.75}$$

Exercise 3.34 Prove Eq. 3.75 by considering matrix representations of A, B, and AB.

In this and many other cases, the easiest way to prove a fact about abstract operators is to work with their matrix representations. This is analogous to proving a relation among 3-D spatial vectors by a calculation involving their x, y, and z components.

We can extend the adjoint to any sort of expression involving scalars, vectors, and operators. Here are the basic rules:

- The adjoint of a scalar is its complex conjugate. The adjoint of a ket is a bra and vice versa.
- The adjoint distributes over addition:

$$(\text{this} + \text{that})^\dagger = (\text{this})^\dagger + (\text{that})^\dagger. \tag{3.76}$$

- The adjoint reverses the order of a multiplication:

$$(\text{this that})^\dagger = (\text{that})^\dagger \, (\text{this})^\dagger. \tag{3.77}$$

The abstract definition in Eq. 3.73 is a straightforward application of these rules. To give another illustration, suppose we have an operator A with a matrix representation A_{mn}. Then

$$
\begin{aligned}
A^\dagger &= \left(\sum_{mn} A_{mn} \, |m\rangle\langle n| \right)^\dagger \\
&= \sum_{mn} \left(A_{mn} \, |m\rangle\langle n| \right)^\dagger \\
&= \sum_{mn} \left(|m\rangle\langle n| \right)^\dagger A_{mn}^* \\
&= \sum_{mn} A_{mn}^* \, |n\rangle\langle m|.
\end{aligned}
$$

Since m and n are bound indices, we can rename m as n and n as m. Then we find that

$$A^\dagger = \sum_{mn} A_{nm}^* \, |m\rangle\langle n|, \tag{3.78}$$

from which we see that $(A^\dagger)_{mn} = A_{nm}^*$, as already shown in Eq. 3.74.

Exercise 3.35 If $|\beta\rangle = B\,|\alpha\rangle$, show that $\langle\beta| = \langle\alpha|\,B^\dagger$. Comment on the meaning of this expression.

Exercise 3.36 Suppose that A has the form

$$A = \sum_n A_n \, |n\rangle\langle n|, \tag{Re 3.55}$$

for a basis $\{\,|n\rangle\,\}$. Show that

$$A^\dagger = \sum_n A_n^* \, |n\rangle\langle n|. \tag{3.79}$$

The operators used in quantum mechanics usually have some special properties with respect to the adjoint operation †. Below we describe types of operator with one or another of these properties.

Hermitian operators

Exercise 3.36 tells us something important about the operators associated with observables. If the observables are *real-valued*, so that $A_n^* = A_n$ for all n, then the associated operator satisfies

$$A^\dagger = A. \tag{3.80}$$

Operators with this property are called *Hermitian*. From now on, we will assume that our observables are real-valued, so that the operators associated with them are always Hermitian operators.

Of course, the expectation value $\langle A \rangle$ for a real observable will always be real, as the following exercise demonstrates:

Exercise 3.37 If A is Hermitian, show that $\langle \psi | A | \psi \rangle$ is real for any $| \psi \rangle$.

Positive operators

Another important type of operator is a *positive* operator: P is said to be positive if it is Hermitian (so that $P^\dagger = P$) and, for all $| \psi \rangle$,

$$\langle \psi | P | \psi \rangle \geq 0. \tag{3.81}$$

(This inequality only makes sense because $\langle \psi | P | \psi \rangle$ must be real.) The projection $\Pi_\alpha = | \alpha \rangle \langle \alpha |$ on any state $| \alpha \rangle$ is positive, since

$$\langle \psi | \Pi_\alpha | \psi \rangle = \langle \psi | \alpha \rangle \langle \alpha | \psi \rangle = | \langle \alpha | \psi \rangle |^2 \geq 0. \tag{3.82}$$

This, together with Eq. 3.12, means that the probabilities for the outcomes of a basic measurement can be written in terms of the projections on the orthonormal basis states $| n \rangle$. For a state $| \psi \rangle$,

$$p(n) = \langle \psi | \Pi_n | \psi \rangle. \tag{3.83}$$

The probability $p(n)$ is the expectation value for an observable that assigns 1 to the nth outcome and zero to all the others, an observable given by Π_n. Since Π_n is a positive operator, $p(n)$ can never be negative.

Exercise 3.38 What property of the projections Π_n for a basic measurement guarantees that the probabilities always sum to 1?

Exercise 3.39 Given any operator B, show that $B^\dagger B$ is positive.

Anti-Hermitian operators

An operator B is called *anti-Hermitian* if $B^\dagger = -B$. It follows that an anti-Hermitian operator B can always be rewritten iA, where A is Hermitian. Any operator at all can be written as the sum of a Hermitian and an anti-Hermitian operator:

Exercise 3.40 Let G be an operator on $\mathcal{H}$. Show that (a) $H = \frac{1}{2}\left(G + G^{\dagger}\right)$ is Hermitian, (b) $K = -\frac{1}{2}\left(G - G^{\dagger}\right)$ is Hermitian, and (c) $G = H + iK$.

The Hermitian and anti-Hermitian operators are analogous to the purely real and purely imaginary numbers in the complex plane. The adjoint is then analogous to complex conjugation of numbers.

Exercise 3.41 Recall the definition of the commutator of two operators: $[A, B] = AB - BA$. Show that the commutator of two Hermitian operators is an anti-Hermitian operator.

Exercise 3.42 In the qubit Hilbert space $\mathcal{Q}$, write the operator $|0\rangle\langle 1|$ as the sum of a Hermitian and an anti-Hermitian operator. Express these in terms of the Pauli operators given in Eq. 3.32.

Unitary operators

An operator U is *unitary* if its adjoint equals its inverse:

$$U^{\dagger}U = UU^{\dagger} = 1. \tag{3.84}$$

In a finite-dimensional Hilbert space, it turns out that the condition $U^{\dagger}U = 1$ is enough to guarantee unitarity, since the rest of Eq. 3.84 follows, (see Problem 3.6).

Exercise 3.43 As we said in Section 2.1, a matrix **R** is said to be unitary when

$$\mathbf{R}^{\dagger}\mathbf{R} = 1. \tag{Re 2.24}$$

Show that a matrix representation for a unitary operator must be unitary.

The action of a unitary operator preserves the inner product between vectors. Suppose $|\alpha\rangle$ and $|\beta\rangle$ are in $\mathcal{H}$, and let $|\alpha'\rangle = U |\alpha\rangle$ and $|\beta'\rangle = U |\beta\rangle$. If U is unitary,

$$\langle \alpha' |\beta'\rangle = \langle\alpha| U^{\dagger}U |\beta\rangle = \langle\alpha| 1 |\beta\rangle = \langle\alpha |\beta\rangle. \tag{3.85}$$

This means that a unitary operator U takes any orthonormal basis set to another orthonormal basis – that is, given a basis $\{ |\alpha_n\rangle\}$, the vectors $|\beta_n\rangle = U |\alpha_n\rangle$ are also orthonormal (and thus form a basis). It follows that

$$U = U \left(\sum_n |\alpha_n\rangle\langle\alpha_n| \right) = \sum_n |\beta_n\rangle\langle\alpha_n|. \tag{3.86}$$

The next exercise establishes the converse fact.

Exercise 3.44 If $\{ |\alpha_n\rangle\}$ and $\{ |\beta_n\rangle\}$ are orthonormal bases, show that

$$V = \sum_n |\beta_n\rangle\langle\alpha_n|$$

is a unitary operator.

As we will see in Section 5.1, unitary operators play a central role in describing the time evolution of isolated quantum systems.

Normal operators

"Normal" may be the most overused word in mathematics. In the present context, an operator N is said to be *normal* if it commutes with its adjoint:

$$N^\dagger N = NN^\dagger. \tag{3.87}$$

The first thing to note is that not every operator is normal.

Exercise 3.45 In a qubit Hilbert space, show that $|0\rangle\langle1|$ is not normal.

On the other hand, the special types of operator that we have mentioned so far – Hermitian (including positive) operators, anti-Hermitian operators, and unitary operators – are all normal.

As we have pointed out, we can write $N = H + iK$, where H and K are Hermitian. The adjoint $N^\dagger = H - iK$. Then

$$N^\dagger N = (H + iK)(H - iK)$$
$$= H^2 + K^2 - i[H, K], \tag{3.88}$$
$$NN^\dagger = (H - iK)(H + iK)$$
$$= H^2 + K^2 + i[H, K]. \tag{3.89}$$

The operator N is normal if and only if these two expressions are equal – that is, provided $[H, K] = 0$. Thus, N is normal if and only if its Hermitian and anti-Hermitian parts commute with each other. See Problem 3.3 for more about normal operators.

3.5 Eigenvalues and eigenvectors

An observable A is associated with an orthonormal measurement basis $\{|n\rangle\}$ and a set of numerical values A_n. This has motivated us to introduce the operator A defined by

$$A|n\rangle = A_n|n\rangle. \tag{Re 3.52}$$

Equation 3.52 has the form of an *eigenvalue equation*:

$$(\text{operator})|\text{vector}\rangle = \text{scalar}|\text{vector}\rangle. \tag{3.90}$$

In Eq. 3.52, the scalar A_n is called an *eigenvalue* of the operator A, and $|n\rangle$ is the corresponding *eigenvector* of A. (Eigenvectors are required to be non-zero; otherwise, any scalar at all would be an "eigenvalue" of any operator. An eigenvalue, however, may happen to be zero.)

The German word "eigen" suggests that the eigenvalue of an operator A is proper to or characteristic of A. It is common practice to attach "eigen" as a prefix to various words associated with an eigenvalue problem (or *eigenproblem*). A normalized eigenvector in $\mathcal{H}$ may be called an *eigenstate* of the operator. A basis composed of eigenvectors of A is often called an *eigenbasis*. And so on!

Exercise 3.46 Fix a scalar α and an operator A. Show that the set of vectors for which $A |\psi\rangle = \alpha |\psi\rangle$ forms a subspace of $\mathcal{H}$. This is called the *eigenspace* for α. It includes the null vector 0 and all of the eigenvectors of A having eigenvalue α.

Equation 3.52 tells us that the operator A associated with the real observable A has a complete basis of eigenvectors $|n\rangle$, each one associated with a real eigenvalue A_n. As we have seen, this means that the operator A is Hermitian. Our goal in this section is to strengthen this connection between observables and Hermitian operators. We will show that every Hermitian operator has only real eigenvalues and a complete orthonormal basis of eigenvectors associated with them.

Existence of eigenvalues

The first thing we must note is that every operator on a finite-dimensional $\mathcal{H}$ has at least one eigenvalue and corresponding eigenvector. That is, for any operator A, we can find a complex scalar α and a non-zero ket $|\psi\rangle$ such that $A |\psi\rangle = \alpha |\psi\rangle$. We can rewrite this as

$$(A - \alpha \mathbf{1}) |\psi\rangle = 0. \tag{3.91}$$

In the language of Exercise 3.13, this is equivalent to saying the operator $(A - \alpha \mathbf{1}) |\psi\rangle$ has a non-trivial kernel.

If $d = \dim \mathcal{H}$, consider the set of vectors $\{ |\psi\rangle, A |\psi\rangle, A^2 |\psi\rangle, \ldots, A^d |\psi\rangle \}$. Since there are $d + 1$ vectors, this must be a linearly dependent set. Thus, for any vector $|\phi\rangle$, the equation

$$\alpha_0 |\phi\rangle + \alpha_1 A |\phi\rangle + \ldots + \alpha_n A^n |\phi\rangle$$
$$= (\alpha_0 + \alpha_1 A + \ldots + \alpha_n A^n) |\phi\rangle = 0, \tag{3.92}$$

has a solution with at least one nonzero α_i. Let α_n be the last (largest n) of these nonzero coefficients. We can divide the equation by α_n and obtain

$$(\beta_0 + \beta_1 A + \ldots + \beta_{n-1} A^{n-1} + A^n) |\phi\rangle = 0. \tag{3.93}$$

We can use the same coefficients from Eq. 3.93 to write down the polynomial equation

$$\beta_0 + \beta_1 z + \ldots + z^n = 0. \tag{3.94}$$

The Fundamental Theorem of Algebra tells us that Eq. 3.94 can be completely factored:

$$\beta_0 + \beta_1 z + \ldots + z^n = (z - \lambda_n) \ldots (z - \lambda_2)(z - \lambda_1). \tag{3.95}$$

The algebraic relation between the β_i coefficients in Eq. 3.94 and the λ_is in Eq. 3.95 will still hold if we replace the variable z with the operator A. It follows that Eq. 3.92 also factors:

$$(\beta_0 + \beta_1 A + \ldots + A^n) \, |\phi\rangle = (A - \lambda_n 1)(A - \lambda_{n-1} 1) \ldots (A - \lambda_1 1) \, |\phi\rangle = 0. \quad (3.96)$$

Now we define a sequence of vectors $|e_0\rangle = |\phi\rangle \neq 0$, $|e_1\rangle = (A - \lambda_1 1) \, |e_0\rangle$, and in general $|e_{i+1}\rangle = (A - \lambda_i 1) \, |e_i\rangle$. According to Eq. 3.96, $|e_n\rangle = 0$, so there must be a smallest k (with $0 < k \leq n$) for which $|e_{k-1}\rangle \neq 0$ and $|e_k\rangle = (A - \lambda_k 1) \, |e_{k-1}\rangle = 0$. Therefore the operator $A - \lambda_k 1$ has a non-trivial kernel. The value λ_k is an eigenvalue of A with associated eigenvector $|e_{k-1}\rangle$.

Exercise 3.47 This argument only works in a *complex* Hilbert space. Show that the operation represented by the matrix

$$\mathbf{A} = \begin{pmatrix} \cos(\pi/4) & \sin(\pi/4) \\ -\sin(\pi/4) & \cos(\pi/4) \end{pmatrix}, \quad (3.97)$$

has no real eigenvalue.

Our argument demonstrates the existence of an eigenvalue, but it does not provide a very efficient method for calculating eigenvalues and their associated eigenvectors. One method, especially useful when $d = \dim \mathcal{H}$ is small, is based on the idea of a determinant.

Suppose the operator A has a $d \times d$ matrix representation $\mathbf{A}$. An eigenvalue λ of the operator A is also an eigenvalue of the matrix $\mathbf{A}$, so there is a non-zero column vector $\boldsymbol{\psi}$ with

$$(\mathbf{A} - \lambda 1) \, \boldsymbol{\psi} = 0. \quad (3.98)$$

Recall that the *determinant* of a $d \times d$ matrix $\mathbf{M}$ has the property that $\det \mathbf{M} \neq 0$ if and only if $\mathbf{M}$ has an inverse matrix $\mathbf{M}^{-1}$. The inverse exists if and only if the mapping described by $\mathbf{M}$ is one-to-one. Exercise 3.13 tells us that this mapping is one-to-one if and only if the kernel of $\mathbf{M}$ is trivial. Therefore, the kernel of $\mathbf{A} - \lambda 1$ is *non-trivial* (as in Eq. 3.98) provided that its determinant is zero:

$$\det(\mathbf{A} - \lambda 1)) = \det \begin{pmatrix} A_{11} - \lambda & \cdots & A_{1d} \\ \vdots & \ddots & \vdots \\ A_{d1} & \cdots & A_{dd} - \lambda \end{pmatrix} = 0. \quad (3.99)$$

This is called the *characteristic equation* for the matrix $\mathbf{A}$. The determinant is a polynomial in λ of degree d, and the eigenvalues of A are exactly the roots of this polynomial. Since every polynomial has at least one root, we can conclude that every matrix has at least one eigenvalue.

Determinants are especially simple in the 2×2 case:

$$\det \begin{pmatrix} a & b \\ c & d \end{pmatrix} = ad - bc. \quad (3.100)$$

The characteristic equation (Eq. 3.99) is thus a quadratic

$$(A_{11} - \lambda)(A_{22} - \lambda) - A_{12}A_{21} = 0, \quad (3.101)$$

whose solution is straightforward.

Exercise 3.48 Find all the eigenvalues of the Pauli matrices of Eq. 3.35, and of the operator $|0\rangle\langle 1|$.

Hermitian operators

It is easy to see that the eigenvalues of any Hermitian operator must be real. If A is Hermitian and $A|\psi\rangle = \lambda|\psi\rangle$ for a non-zero $|\psi\rangle$, then

$$\lambda = \frac{\langle\psi|A|\psi\rangle}{\langle\psi|\psi\rangle}. \tag{3.102}$$

Exercise 3.49 Verify Eq. 3.102.

Since both inner products on the right-hand side are real, λ must be real as well. The set of eigenvalues for a given operator is called the *spectrum* of the operator. The spectrum of a Hermitian operator contains only real quantities.

Now suppose that λ_1 and λ_2 are two eigenvalues for Hermitian A, with respective eigenvectors $|\psi_1\rangle$ and $|\psi_2\rangle$. Then

$$\langle\psi_2|A|\psi_1\rangle = \lambda_1\langle\psi_2|\psi_1\rangle, \tag{3.103}$$

$$\langle\psi_1|A|\psi_2\rangle = \lambda_2\langle\psi_1|\psi_2\rangle. \tag{3.104}$$

Since A is Hermitian, these two inner products must be complex conjugates of one another. Since the eigenvalues of A are real,

$$\lambda_1\langle\psi_2|\psi_1\rangle = \lambda_2\langle\psi_2|\psi_1\rangle. \tag{3.105}$$

If $\lambda_1 = \lambda_2$, this is clearly true. But suppose the two eigenvalues are not equal. Then this equation can only be satisfied if $\langle\psi_2|\psi_1\rangle = 0$. Therefore, given a Hermitian operator A, the eigenvectors of A corresponding to distinct eigenvalues must be orthogonal.

We will now turn to the main business of this section: to establish that, for any Hermitian A on a finite-dimensional Hilbert space $\mathcal{H}$, there exists a complete orthonormal basis of eigenvectors of A.

We proceed by mathematical induction on the dimension d of the Hilbert space $\mathcal{H}$. If $d = 1$, then any normalized vector $|1\rangle$ forms an orthonormal basis for $\mathcal{H}$. The vector $|1\rangle$ also must be an eigenvector for any operator A on $\mathcal{H}$, because the vector $A|1\rangle$ has to be some multiple of $|1\rangle$. Therefore, for any Hermitian operator A on the trivial Hilbert space with dim $\mathcal{H} = 1$, we can find an orthonormal basis of eigenvectors of A.

Now suppose we have established that a Hermitian operator on a space of dimension d gives rise to an orthonormal basis of eigenvectors. Consider an operator A on a space $\mathcal{H}$ of dimension $d + 1$. This operator must have at least one eigenvalue; call it A_{d+1} and the corresponding normalized eigenvector $|d + 1\rangle$.

Let $\mathcal{T}_d$ be the set of all vectors in $\mathcal{H}$ that are orthogonal to $|d+1\rangle$. This is a d-dimensional subspace of $\mathcal{H}$. How does the operator A act on this subspace? That is, if $|\psi\rangle$ is in $\mathcal{T}_d$, what can we say about $A|\psi\rangle$? This vector is also orthogonal to $|d+1\rangle$, since

$$\langle d+1|\,A\,|\psi\rangle = \left(\langle\psi|\,A\,|d+1\rangle\right)^* = A_{d+1}\,\langle d+1|\psi\rangle = 0. \tag{3.106}$$

(Note how we have used the fact that A is Hermitian.) The operator A therefore maps vectors in $\mathcal{T}_d$ to vectors in $\mathcal{T}_d$. Within this subspace, A effectively acts as a Hermitian operator on a d-dimensional space. Therefore, by assumption there exists an orthonormal basis $\{|1\rangle,\ldots,|d\rangle\}$ for $\mathcal{T}_d$ composed of eigenvectors of A. Adding $|d+1\rangle$ to this set gives an orthonormal basis of A eigenvectors spanning all of $\mathcal{H}$.

We have proven our claim for the special case $d=1$, and we have shown that if the claim holds for spaces of dimension d then it must also hold for spaces of dimension $d+1$. It follows that there is an orthonormal eigenbasis for any Hermitian operator on any Hilbert space of finite dimension.

Real observables are associated with Hermitian operators. But our results also establish that every Hermitian operator can be associated with a possible observable. For Hermitian operator A, we can find an orthonormal eigenbasis $\{|n\rangle\}$ together with eigenvalues A_n so that

$$A|n\rangle = A_n|n\rangle. \tag{Re 3.52}$$

This is exactly what we would expect from an observable whose measurement basis is $\{|n\rangle\}$ and whose possible values are the elements A_n of the spectrum of A.

It follows from our argument that we can write any Hermitian operator in the form

$$A = \sum_n A_n\,|n\rangle\langle n|. \tag{Re 3.55}$$

This is called the *spectral decomposition* of A. With respect to the eigenbasis $\{|n\rangle\}$, A has the diagonal matrix representation

$$(A_{mn}) = \begin{pmatrix} A_1 & & \\ & \ddots & \\ & & A_d \end{pmatrix}. \tag{Re 3.54}$$

Note that the diagonal entries in this matrix are the eigenvalues A_n.

In a diagonal representation, it is easy to see that the trace $\mathrm{Tr}\,A$ of a Hermitian operator A is exactly the sum of its eigenvalues. This is sometimes useful. We can compute the trace using any matrix representation, diagonal or not. The trace then gives us a significant piece of information about the spectrum of eigenvalues of A – namely, the sum of that spectrum – without the work of finding the eigenvalues.

To sum up: we previously showed that every real observable – every basic measurement whose outcomes are real numerical quantities – is associated with a Hermitian operator. Now we have shown that every Hermitian operator can be associated with an observable. Given the operator A, we can determine both the measurement basis and the spectrum of possible measurement results. So close is this correspondence between (real) observable and (Hermitian) operator that we often say that the operator "is" the observable.

Spin-1/2 revisited

Let us see how this works out for the case of a spin-1/2 particle, which we discussed in Section 2.2. In that discussion, several facts about the states and observable properties of spins were offered as mere assertions. We are now in a position to give them a better justification.

We will still take as given the operator forms for the S_x, S_y, and S_z observables. In the standard (S_z) basis:

$$S_x = \frac{\hbar}{2} X, \qquad S_y = \frac{\hbar}{2} Y, \qquad S_z = \frac{\hbar}{2} Z . \qquad (3.107)$$

These operators have eigenvalues and eigenvectors that we already know. But what about the spin component S_θ, where $\hat{\theta}$ is a unit vector in the xz-plane, tilted at an angle of θ from the z-axis?

We can write $\hat{\theta} = \sin\theta \hat{x} + \cos\theta \hat{z}$. Given a spin vector $\vec{S}$, the spin component

$$S_\theta = \hat{\theta} \cdot \vec{S} = \sin\theta \left(\hat{x} \cdot \vec{S} \right) + \cos\theta \left(\hat{z} \cdot \vec{S} \right)$$
$$= \sin\theta \, S_x + \cos\theta \, S_z. \qquad (3.108)$$

It is reasonable to assume that this relation also holds for the operators of the spin component observables:

$$S_\theta = \sin\theta \, S_x + \cos\theta \, S_z = \frac{\hbar}{2} \left(\sin\theta \, X + \cos\theta \, Z \right). \qquad (3.109)$$

This has a matrix representation

$$\mathbf{S}_\theta = \frac{\hbar}{2} \begin{pmatrix} \cos\theta & \sin\theta \\ \sin\theta & -\cos\theta \end{pmatrix}. \qquad (3.110)$$

What are the possible measured values for the observable S_θ? They can be worked out most easily using the characteristic equation.

Exercise 3.50 Use Eq. 3.99 to show that the eigenvalues of $\mathbf{S}_\theta$ are $\pm\hbar/2$.

The spectrum of eigenvalues of S_θ, and thus the set of possible measurement outcomes, is exactly as we expect for a spin component. In fact, this is true for any component of spin, not just those in the xz-plane.

Exercise 3.51 Suppose we have a general unit vector $\hat{u}$ that does not lie in the xz-plane:

$$\hat{u} = \sin\theta \cos\phi \, \hat{x} + \sin\theta \sin\phi \, \hat{y} + \cos\theta \, \hat{z}. \qquad (3.111)$$

Show again that the possible outcomes for a measurement of S_u are just $\pm\hbar/2$.

Now let us return to the spin component S_θ in the xz-plane. We can find an orthonormal basis of S_θ eigenvectors corresponding to the eigenvalues of that operator.

Exercise 3.52 Find a pair of states $|\theta_+\rangle$ and $|\theta_-\rangle$ that constitute an orthonormal basis of eigenstates of the spin component operator S_θ. You can find these by solving the eigenvalue equations directly (not too hard); or you could consult Eq. 2.35 and verify that the measurement basis states presented there are solutions to the eigenvalue equations.

Degeneracy

The eigenvalues A_n of a Hermitian operator A need not be distinct. Two or more orthogonal eigenvectors may have equal eigenvalues. This is called *degeneracy*. If A has degeneracy, then A will not uniquely determine an eigenbasis.

For example, consider a three-level atom with an energy observable (the Hamiltonian operator) H. The energy spectrum consists of E_1, E_2, and E_3. But it might happen that two of these energy values are degenerate, e.g. $E_2 = E_3$.

If an atom has only two possible energies E_1 and E_2, in what sense can we still call it a "three-level atom"? First, there may be other measurements (besides energy) which make it clear that dim $\mathcal{H} = 3$ for the atom. Another possibility is that the atom's energy depends on an adjustable external parameter, such as the magnetic field $\vec{B}$ experienced by the atom. For most values of $\vec{B}$, the atom would have three distinct possible energies, but for some special value of the field ($\vec{B} = 0$, say) two of the levels might be degenerate.

An eigenbasis for H consists of three states $|1\rangle$, $|2a\rangle$, and $|2b\rangle$, having eigenvalues E_1, E_2, and E_2, respectively. But this is not the only possible eigenbasis that can be formed from the operator H. We could replace the degenerate states $|2a\rangle$ and $|2b\rangle$ with, say,

$$|2a'\rangle = \frac{1}{\sqrt{2}} \Big(|2a\rangle + |2b\rangle \Big),$$

$$|2b'\rangle = \frac{1}{\sqrt{2}} \Big(|2a\rangle - |2b\rangle \Big). \tag{3.112}$$

The new set $\{ |1\rangle, |2a'\rangle, |2b'\rangle \}$ is also an orthonormal basis of eigenvectors of H.

There is no real surprise here. Exercise 3.46 told us that the set of eigenvectors of H with a given eigenvalue form a subspace. The E_2 eigenspace in this example has dimension 2. There are infinitely many possible basis sets for this eigenspace, including $\{ |2a\rangle, |2b\rangle \}$ and $\{ |2a'\rangle, |2b'\rangle \}$.

We conclude that there are many different basic measurements which could be an "energy measurement" of the atom. All of these different ways of measuring energy, however, correspond to the same Hermitian operator H:

$$H = E_1 |1\rangle\langle 1| + E_2 \Big(|2a\rangle\langle 2a| + |2b\rangle\langle 2b| \Big)$$

$$= E_1 |1\rangle\langle 1| + E_2 \Big(|2a'\rangle\langle 2a'| + |2b'\rangle\langle 2b'| \Big). \tag{3.113}$$

Since these two spectral decompositions represent the same operator, it must be true that

$$\Pi_2 = |2a\rangle\langle 2a| + |2b\rangle\langle 2b| = |2a'\rangle\langle 2a'| + |2b'\rangle\langle 2b'|. \tag{3.114}$$

We have called this operator Π_2 by analogy to the projection $\Pi_1 = |1\rangle\langle 1|$ that also appears in the spectral decomposition. In fact, Π_2 is a projection operator under a more general definition than we have yet given. An operator Π is called a projection if it is Hermitian and if $\Pi^2 = \Pi$.

Exercise 3.53 Prove that Π_1 and Π_2 are both projections under the general definition.

Exercise 3.54 Show that the only possible eigenvalues for a projection operator are 0 and 1.

Problem 3.2 provides a geometric understanding of the projection operator Π. The projection "collapses" the Hilbert space $\mathcal{H}$ to a subspace $\mathcal{G}$, which is itself unchanged by the operator Π. Intuitively, the action of Π gives the "shadow" of any vector on the subspace $\mathcal{G}$ – which is exactly why operators like Π are known as "projections."[6] For the projection Π_2, the corresponding subspace is the eigenspace associated with the eigenvalue E_2.

The spectral decomposition of the Hamiltonian in Equation 3.113 is thus

$$\mathsf{H} = E_1 \Pi_1 + E_2 \Pi_2. \tag{3.115}$$

Completeness tells us that $\Pi_1 + \Pi_2 = \mathbf{1}$. Both the spectral decomposition and the completeness relation are independent of which basic measurement for energy we may choose.

Suppose the atom is in the state $|\psi\rangle$ and we make a basic measurement of its energy. The probability that we obtain the result E_2 is

$$
\begin{aligned}
p\,(E_2) &= p(2a) + p(2b) \\
&= \langle \psi\,|2a\rangle\,\langle 2a\,|\psi\rangle + \langle \psi\,|2a\rangle\,\langle 2a\,|\psi\rangle \\
&= \langle \psi|\,\Big(\,|2a\rangle\langle 2a|\, + \,|2b\rangle\langle 2b|\,\Big)\,|\psi\rangle = \langle \psi|\,\Pi_2\,|\psi\rangle. \tag{3.116}
\end{aligned}
$$

In other words, we can calculate the probability of finding E_2 directly from the operator Π_2, without specifying the particular basic measurement. (Indeed, as we will discuss later, we can in principle carry out a *non-basic* measurement procedure in which only the projection Π_2 matters.)

What we have found for an energy measurement on a three-level atom holds for any sort of observable A. If the operator A has degeneracy, more than one basic measurement could be used to measure A. By using projection operators, though, we can do calculations without specifying which basic measurement (if any) we might be using.

Suppose $\{A_\alpha\}$ is the set of distinct eigenvalues of A (so that no value appears twice). Then each eigenvalue A_α is associated with an eigenspace (which need not be one-dimensional) and a projection Π_α. The spectral decomposition for A is

$$\mathsf{A} = \sum_\alpha A_\alpha \Pi_\alpha. \tag{3.117}$$

---

[6] In mathematics books, you may find that an operator is called a projection if $\Pi^2 = \Pi$, whether or not it is Hermitian. Our definition actually gives an *orthogonal* projection. Since we will not need this more generalized view, however, we have adopted the term "projection" to mean only the orthogonal (Hermitian) ones.

In a measurement of A on a system in state $|\psi\rangle$, the probability of obtaining the result A_α is

$$p(A_\alpha) = \langle\psi|\,\Pi_\alpha\,|\psi\rangle. \tag{3.118}$$

Finally, the completeness relation for the projection operators is

$$\sum_\alpha \Pi_\alpha = \mathbf{1}. \tag{3.119}$$

The measurement of a degenerate observable is called an *incomplete* measurement, because the result does not always completely specify a particular basis element.

Such measurements can arise in two ways. It could be that the real physical process is a basic measurement, but the numerical read-out is "coarse-grained."[7] In this case, the mathematical ambiguity of the eigenbasis due to the degeneracy of the observable is only apparent. There is a single "real" basis for the measurement. It could also happen that the physical measurement process is not a basic measurement at all, but something weaker. This possibility is described in the next chapter. Yet whatever the exact physical nature of the measurement process, we can calculate probabilities and expectation values using the operator description of the observable.

Compatible observables

Two observables A and B are said to be *compatible* if there is a basic measurement that can determine the values of both A and B together. That is, they are compatible if they have a common eigenbasis. In this basis,

$$A = \sum_n A_n\,|n\rangle\langle n| \quad \text{and} \quad B = \sum_n B_n\,|n\rangle\langle n|. \tag{3.120}$$

Compatible observables are always commuting operators.

Exercise 3.55 Show that $[A, B] = 0$ for the operators above.

This means that, if two observables are *not* commuting, then they are *not* compatible. Consider, for instance, the observables S_x and S_y for the spin components of a spin-1/2 particle. These operators do not commute.

Exercise 3.56 Show that $\left[S_x, S_y\right] = i\hbar S_z$.

Therefore, there can be no single basic measurement that provides us with full information about both S_x and S_y for the particle. To use Bohr's term, S_x and S_y are *complementary* observables. An experimental procedure that determines the value of S_x is logically incompatible with a procedure that determines S_y.

---

[7] *Coarse-graining* means that several distinct possibilities are grouped together. This happens, for instance, when a continuous quantity is specified with finite precision. If we know the x-coordinate of a particle's position to the nearest meter, we have grouped "14.66 m" and "15.021 m" and "15.30421448 m" all under the general heading of "15 m."

We have shown that compatible observables commute – or, equivalently, that non-commuting observables are not compatible. We will now show the converse: any two commuting observables are compatible.

Suppose $[A, B] = 0$. Every distinct eigenvalue A_α of A is associated with an eigenspace, a subspace that contains the eigenvectors with that eigenvalue. Distinct eigenspaces are orthogonal to each other. If $|\psi\rangle$ is in the A_α eigenspace for A, what about the vector $B|\psi\rangle$? This is also in the same eigenspace, since

$$A\left(B|\psi\rangle\right) = BA|\psi\rangle = A_\alpha B|\psi\rangle. \tag{3.121}$$

Thus, the operator B acts as a Hermitian operator on each A_α eigenspace individually. For each of these subspaces, therefore, we can find a basis of B eigenstates. If we collect these B basis sets for each of the A_α eigenspaces, we obtain a basis for the entire space $\mathcal{H}$, each of whose elements is an eigenstate both of A and B. Therefore, A and B are compatible.

This is a remarkable connection between the algebraic properties of operators – specifically, whether or not they commute – and the fundamental physical compatibility of measurement processes. In the next chapter, we will deepen this connection by deriving a general limit on our simultaneous knowledge of two quantum observables. There, as here, the commutator $[A, B]$ will play a central role.

Problems

Problem 3.1 Recall the spin-1 S_x basis vectors defined in Eq. 3.16.

(a) Find a quantum state $|\psi\rangle$ such that neither a measurement of S_z nor one of S_x could possibly yield the result zero. (This state happens to be $|y_0\rangle$. In fact, the three states $|x_0\rangle$, $|y_0\rangle$, and $|z_0\rangle$ form yet another orthonormal basis, as you can easily check.)
(b) Write the S_z basis vectors as superpositions of the S_x basis vectors.

Problem 3.2 Suppose Π is a projection operator on $\mathcal{H}$. Let $\mathcal{G}$ be the set of all vectors in $\mathcal{H}$ that are images of the operator Π – in other words, the set of all $|\psi\rangle = \Pi|\phi\rangle$ for some $|\phi\rangle$ in $\mathcal{H}$.

(a) Show that $\mathcal{G}$ is a subspace of $\mathcal{H}$.
(b) Show that, if $|\psi\rangle$ is in $\mathcal{G}$, then $|\psi\rangle = \Pi|\psi\rangle$.
(c) We denote by $\mathcal{G}^\perp$ the set of all vectors in $\mathcal{H}$ that are orthogonal to every element of $\mathcal{G}$. (This is called "$\mathcal{G}$-perp.") Show that $\mathcal{G}^\perp$ is a subspace.
(d) Show that, if $|\phi\rangle$ is in $\mathcal{G}^\perp$, then $\Pi|\phi\rangle = 0$.

Problem 3.3 In a finite-dimensional Hilbert space $\mathcal{H}$, show that an operator N is normal if and only if $\mathcal{H}$ has a complete orthonormal basis of eigenvectors of N. (Hint: "If" is easy. For "only if," use the fact that the Hermitian and anti-Hermitian parts of N must commute.)

Problem 3.4 Show that any Hermitian operator is the difference of two positive operators.

Problem 3.5

(a) For any Hermitian operator G and unitary operator U, show that $UGU^\dagger$ is Hermitian and has the same eigenvalues as G.

(b) Show that if two Hermitian operators A and B have exactly the same eigenvalues with the same degeneracies, then there must exist a unitary operator U such that $B = UAU^\dagger$.

(c) For any operator K, show that $K^\dagger K$ and $KK^\dagger$ are positive operators with the same eigenvalues. (Provide an example to show that they are not necessarily the *same* operator.)

Problem 3.6

(a) Suppose $\dim \mathcal{H} = d$, and also suppose that U satisfies $U^\dagger U = 1$. Pick an orthonormal basis $\{|k\rangle\}$, and show that the set $\{U|k\rangle\}$ is also an orthonormal basis. Use this fact to prove that $UU^\dagger = 1$ also, so that U must be unitary.

(b) Now imagine an infinite dimensional Hilbert space, with an infinite orthonormal basis set $\{|1\rangle, |2\rangle, |3\rangle, \ldots\}$. Define the operator U by

$$U = \sum_{n=1}^{\infty} |n+1\rangle\langle n|.$$

Show that $U^\dagger U = 1$, but $UU^\dagger \neq 1$. Thus U is not unitary.

Problem 3.7

(a) Suppose G is Hermitian, and $\langle \phi| G |\phi\rangle = 0$ for all $|\phi\rangle$. Show that $G = 0$. (Hint: Consider $|\phi\rangle$ to be an eigenvector of G.)

(b) Prove part (a) for an arbitrary operator G by writing G as the sum of a Hermitian and an anti-Hermitian operator.

(c) Prove that, given operators A and B, if $\langle \phi| A |\phi\rangle = \langle \phi| B |\phi\rangle$ for all $|\phi\rangle$, then $A = B$. (Note: This is quite a useful mathematical fact!)

(d) Prove that, if $\langle \phi| G |\phi\rangle = 1$ for all normalized $|\phi\rangle$, then $G = 1$.

Distinguishability and information

4.1 Quantum communication

We now return our focus to the idea of information. A quantum system may be used as a medium for the storage, transmission, and retrieval of information. The rules of quantum physics have implications for this process. In a larger sense, too, the state vector $|\psi\rangle$ of a quantum system provides information about the outcomes of possible future measurements. In this chapter, we will use the quantum theory developed so far to explore these issues.

We introduced the notion of information in Section 1.1 by discussing the process of communication. A sender encodes various possible messages by preparing various states ("signals") of a physical system. We said that distinct messages should be represented by distinct physical states. In this way, the receiver could reconstruct the original message from the signal in a unique way.

This discussion presumed that *distinct* states of the communication medium were also *distinguishable* by the receiver. But this is not the case if the communication medium is a quantum system. Suppose Alice wishes to send a message to Bob via a single spin-1/2 particle. She uses the following code (see Fig. 4.1):

Signal	Message
$|z_+\rangle$	The British are coming by land.
$|z_-\rangle$	The British are coming by sea.
$|x_+\rangle$	The British are not coming.

This appears to be a good code, since each message is represented by a distinct physical state of the qubit. Yet something seems to be amiss. How can Bob recover the message that Alice has encoded? If, for example, he makes a measurement of S_z, he could draw the following conclusions:

Result	Inference
$+\hbar/2$	Either the British are coming by land, or not at all.
$-\hbar/2$	Either the British are coming by sea, or not at all.

Exercise 4.1 Explain how Bob could arrive at these inferences from the result of an S_z measurement on the qubit. What could he infer from the result of an S_x measurement?

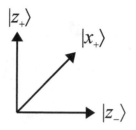

Fig. 4.1 Alice's three signal states in the Hilbert space. Note that any diagram of vectors in $\mathcal{H}$ must necessarily be schematic, since $\mathcal{H}$ is a *complex* space, while the Euclidean space of the page is not.

Bob gains some information – he may, for instance, be able to exclude the possibility that the British are coming by sea – but he does not learn everything that Alice is trying to tell him.

The problem is that *no* measurement exists that can distinguish between the states in Alice's code. It is not possible, *even in principle*, to read all of Alice's message. This is not because quantum systems are somehow noisy and unreliable as carriers of information. If Alice had only wished to encode a single bit of information into the particle (instead of $\log 3 \approx 1.58$ bits), then she could have used the $|z_+\rangle$ and $|z_-\rangle$ basis states in her code. By measuring S_z, Bob would be able to read the message exactly. Alternatively, Alice could have used a quantum system with a larger Hilbert space, like a spin-1 particle. Then her three possible messages could have been encoded in three orthogonal states. With a suitable choice of decoding measurement, Bob would be able to distinguish these perfectly.

Information encoded in orthogonal quantum states can be perfectly retrieved by some possible measurement. If non-orthogonal quantum signals are used, then the information will not be perfectly retrieved. In this section and the next, we will give more mathematical substance to these intuitive points by proving a pair of theorems. The theorems provide limits to how reliably a set of quantum states may be identified by a measurement. They will give us a quantitative look at the difference between distinguishability and distinctness for quantum states.

We want to frame our theorem in as general a context as possible. Alice is sending a message to Bob using a quantum system with Hilbert space $\mathcal{H}$ of dimension $\dim \mathcal{H} = d$. She prepares the quantum state in one of N possible states $|\alpha\rangle$, corresponding to the N possible messages α that she might want to transmit. We assume that each message has an equal likelihood $1/N$.

Bob will decode the signal by making a basic measurement using a basis $\{ |k\rangle \}$. Now, it might be that the Hilbert space $\mathcal{H}$ containing Alice's signals is only a subspace of a larger Hilbert space. For instance, Alice might use a two-level atom, a qubit, as her communication medium. Then $d = 2$. But the two-level atom is just a simplified picture of the real atom, which has a much larger Hilbert space. We want to give Bob every possible advantage in decoding the signal, so we will allow him to make any basic measurement on this larger space. The number of elements in the basis $\{ |k\rangle \}$ could be much larger than d.

Since $\mathcal{H}$ may be a subspace of a larger Hilbert space, we introduce the projection operator Π that projects onto $\mathcal{H}$. For every $|\alpha\rangle$ in $\mathcal{H}$, $\Pi |\alpha\rangle = |\alpha\rangle$. We can write Π as

$$\Pi = \sum_r |\phi_r\rangle\langle\phi_r|, \tag{4.1}$$

where $\{|\phi_r\rangle\}$ is an orthonormal basis for the subspace $\mathcal{H}$. This subspace basis has d elements, so that $\mathrm{Tr}\,\Pi = d$.

Once Bob obtains a measurement result k, he uses it to infer what Alice's message was. That is, the set of possible measurement results is divided into subsets C_α, one for each possible message α. Each measurement result k must be in exactly one of the subsets. If the measurement result k is in C_α, then Bob will infer the message α from the result k.

A measurement confined to the Hilbert space $\mathcal{H}$ would have only d possible results, and thus no more than d of the subsets C_α could be non-empty. An empty C_α would mean that Bob could never correctly identify the message α, whatever his measurement result. This would be a problem if $N > d$. But we are allowing Bob to make his measurements on a larger space containing $\mathcal{H}$, so it is possible for all of the C_αs to be non-empty even if $N > d$. In principle, then, any of the messages α *might* be correctly identified by Bob. But how likely is he to come to the correct conclusion?

To proceed, we will need a few facts about positive operators. Recall that an operator P is positive if $\langle\psi|\,\mathsf{P}\,|\psi\rangle \geq 0$ for all $|\psi\rangle$. Then:

- $|\beta\rangle\langle\beta|$ is positive for any $|\beta\rangle$;
- If P is positive and π is a projection operator, then $\pi\mathsf{P}\pi$ is also positive;
- For a positive operator P and a normalized vector $|\psi\rangle$,

$$\langle\psi|\,\mathsf{P}\,|\psi\rangle \leq \mathrm{Tr}\,\mathsf{P}. \tag{4.2}$$

Exercise 4.2 Prove the properties of positive operators just listed.

We are interested in the probability P_S that Bob's measurement result allows him to successfully identify the message α. This will be

$$P_S = \sum_\alpha p(\text{message } \alpha) \times \left(\sum_{k\in C_\alpha} p(\text{result } k \mid \text{message } \alpha)\right). \tag{4.3}$$

(Recall that $p(A|B)$ is the conditional probability that A occurs under the condition that B occurs.) Noting that $p(\text{message } \alpha) = 1/N$ for all α,

$$P_S = \sum_\alpha \frac{1}{N}\left(\sum_{k\in C_\alpha} |\langle k|\alpha\rangle|^2\right). \tag{4.4}$$

Then, noting that $\Pi |\alpha\rangle = |\alpha\rangle$,

$$P_S = \frac{1}{N}\sum_\alpha \sum_{k\in C_\alpha} \langle\alpha|k\rangle\langle k|\alpha\rangle$$

$$= \frac{1}{N}\sum_\alpha \sum_{k\in C_\alpha} \langle\alpha|\,\Pi\,|k\rangle\langle k|\,\Pi\,|\alpha\rangle. \tag{4.5}$$

By Eq. 4.2, $\langle \alpha | \, \Pi \, | k \rangle \langle k | \, \Pi \, | \alpha \rangle \leq \operatorname{Tr} \Pi \, | k \rangle \langle k | \, \Pi$, and so

$$P_S \leq \frac{1}{N} \sum_\alpha \sum_{k \in C_\alpha} \operatorname{Tr} \left(\Pi \, | k \rangle \langle k | \, \Pi \right)$$

$$= \frac{1}{N} \operatorname{Tr} \Pi \left(\sum_\alpha \sum_{k \in C_\alpha} | k \rangle \langle k | \right) \Pi. \tag{4.6}$$

The double sum is actually equivalent to the single sum $\sum_k$, since every k is in exactly one of the C_αs. The expression in parentheses is therefore 1, and

$$P_S \leq \frac{1}{N} \operatorname{Tr} \Pi^2 = \frac{1}{N} \operatorname{Tr} \Pi = \frac{d}{N}. \tag{4.7}$$

The probability of error $P_E = 1 - P_S$. We have now proven the following theorem:

Basic decoding theorem. If Alice encodes N equally likely messages as states in a quantum system with $\dim \mathcal{H} = d$, and if Bob decodes this by performing a basic measurement (perhaps on a larger Hilbert space) and inferring the message from the result, Bob's probability of error is bounded by

$$P_E \geq 1 - \frac{d}{N}. \tag{4.8}$$

If the number N of possible signals exceeds the dimension d of the Hilbert space containing the signals, Bob will not be able to reliably ($P_S = 1$) distinguish between the signals by any basic measurement. Since $H = \log N$ tells us the amount of information in the message, we conclude that a quantum system described by a Hilbert space of dimension d has a communication or information *capacity* of $\log d$. This is the maximum number of bits that can be stored in the system so that the data can be reliably read by a measurement.

Exercise 4.3 A quantum system has an information capacity of $\log d$. Suppose that Alice tries to send one additional bit of information above the capacity. Use the basic decoding theorem to find a lower bound for Bob's probability of error P_E.

Clearly, if $N \leq d$, we can choose signal states that are all orthogonal to one another, which can be perfectly distinguished by a measurement. In this way we can achieve $P_E = 0$.

The capacity of a quantum system for conveying information is not limited by the number of different states available. Nor is it limited by the number of possible outcomes of some measurement procedure. Instead, it is limited by the number of states that can be reliably distinguished, which is given by the dimension d of the system's Hilbert space.

The lesson of the basic decoding theorem is that quantum systems have a limited capacity to carry information. Our mathematical machinery so far has allowed us to prove this for the case of equally likely messages (for which we have a definition of H) and basic measurements (the only kind that we have analyzed). We will generalize both conditions later in the book. The lesson holds in these more general situations as well.

4.2 Distinguishability

Let us put ourselves in Bob's situation. In the previous section, we were faced with the task of distinguishing reliably between an entire set of possible signals that Alice might use. Now suppose that we are trying to distinguish between only two quantum states. Alice has prepared the system in either $|\alpha_0\rangle$ (representing 0) or $|\alpha_1\rangle$ (representing 1), assumed to be equally likely. We shall make a basic measurement to try to determine which state is present. The probability that this measurement succeeds in distinguishing the states is P_S.

The two states lie in a subspace $\mathcal{T}$ of dimension 2, so the basic decoding theorem only tells us that $P_S \leq 1$. If the two states happen to be orthogonal, then we can achieve $P_S = 1$ by choosing a measurement basis with $|0\rangle = |\alpha_0\rangle$ and $|1\rangle = |\alpha_1\rangle$. Now imagine that the two states are not orthogonal, so that $\langle\alpha_0|\alpha_1\rangle \neq 0$. We will not be able to distinguish these states perfectly. But if perfection is out of reach, just how well can we do?

As before, we make a basic measurement that may extend to a larger Hilbert space containing the two-dimensional subspace $\mathcal{T}$. The measurement results are partitioned into two sets C_0 and C_1, based on the inference that we will draw. The probability P_S that we infer the right message is

$$P_S = \frac{1}{2} \sum_{k \in C_0} \langle\alpha_0|k\rangle \langle k|\alpha_0\rangle + \frac{1}{2} \sum_{k \in C_1} \langle\alpha_1|k\rangle \langle k|\alpha_1\rangle, \qquad (4.9)$$

(see Eq. 4.4). Since the sets C_0 and C_1 include all of the outcomes k,

$$\sum_{k \in C_1} |k\rangle\langle k| = \mathbf{1} - \sum_{k \in C_0} |k\rangle\langle k|. \qquad (4.10)$$

This lets us rewrite our expression for P_S as

$$P_S = \frac{1}{2} + \frac{1}{2} \sum_{k \in C_0} \langle k| \left(|\alpha_0\rangle\langle\alpha_0| - |\alpha_1\rangle\langle\alpha_1| \right) |k\rangle. \qquad (4.11)$$

Look at the operator $\mathsf{D} = |\alpha_0\rangle\langle\alpha_0| - |\alpha_1\rangle\langle\alpha_1|$ that appears in this expression: D is Hermitian, so it will have a complete basis of eigenstates. Furthermore, for any $|\psi\rangle$ that is orthogonal to both $|\alpha_0\rangle$ and $|\alpha_1\rangle$, $\mathsf{D}|\psi\rangle = 0$. This tells us that D has at most two non-zero eigenvalues, whose eigenvectors must lie in $\mathcal{T}$.

The sum of the eigenvalues of D is $\operatorname{Tr}\mathsf{D} = 0$. We can therefore write the eigenvalues as $\pm\delta$, and

$$\mathsf{D} = \delta|+\rangle\langle+| - \delta|-\rangle\langle-|, \qquad (4.12)$$

for an orthonormal pair of D eigenstates $|\pm\rangle$. Now we write

$$\begin{aligned}
P_S &= \frac{1}{2} + \frac{1}{2} \sum_{k \in C_0} \langle k|\mathsf{D}|k\rangle \\
&= \frac{1}{2} + \frac{1}{2}\delta \sum_{k \in C_0} \langle k|+\rangle\langle+|k\rangle - \frac{1}{2}\delta \sum_{k \in C_0} \langle k|-\rangle\langle-|k\rangle \\
&\leq \frac{1}{2} + \frac{1}{2}\delta \sum_{k \in C_0} \langle k|+\rangle\langle+|k\rangle.
\end{aligned} \qquad (4.13)$$

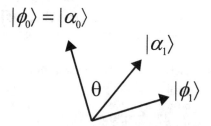

$$|\phi_0\rangle = |\alpha_0\rangle$$

Fig. 4.2 The states $|\alpha_0\rangle$ and $|\alpha_1\rangle$, together with the orthogonal basis used for the matrix representation of D.

Since $|+\rangle$ is normalized, $\sum_{k \in C_0} |\langle k | + \rangle|^2 \leq 1$. Thus

$$P_S \leq \frac{1}{2}(1 + \delta).$$ (4.14)

Exercise 4.4 Show that we can choose a measurement so that equality is achieved in Eq. 4.14.

The problem boils down to finding the positive eigenvalue δ for the operator D. We do this by solving the characteristic equation for a matrix representation of D. For that we need a basis for the subspace $\mathcal{T}$. Choose the first basis vector $|\phi_0\rangle = |\alpha_0\rangle$ and let the $|\phi_1\rangle$ be a vector orthogonal to this (see Fig. 4.2). Then

$$|\alpha_1\rangle = c_0 |\phi_0\rangle + c_1 |\phi_1\rangle.$$ (4.15)

Of course, $c_0 = \langle \alpha_0 | \alpha_1 \rangle$. A convenient way of writing the two components is suggested by the following exercise:

Exercise 4.5 Suppose $|c_0|^2 + |c_1|^2 = 1$ for complex numbers c_0 and c_1. Show that we can find an angle θ with $0 \leq \theta \leq \pi/2$ together with real phases β_0 and β_1 so that

$$c_0 = e^{i\beta_0} \cos \theta \quad \text{and} \quad c_1 = e^{i\beta_1} \sin \theta.$$ (4.16)

We can interpret the angle θ as the angle between $|\alpha_0\rangle$ and $|\alpha_1\rangle$. It satisfies

$$|\langle \alpha_0 | \alpha_1 \rangle| = \cos \theta.$$ (4.17)

With respect to this basis, the matrix representation for D is

$$\mathbf{D} = \begin{pmatrix} 1 - \cos^2 \theta & -e^{i(\beta_0 - \beta_1)} \cos \theta \sin \theta \\ -e^{i(\beta_1 - \beta_0)} \cos \theta \sin \theta & -\sin^2 \theta \end{pmatrix}.$$ (4.18)

This gives us, after a little simplification, the characteristic equation

$$0 = \det(\mathbf{D} - \delta \mathbf{1}) = \delta^2 - \sin^2 \theta.$$ (4.19)

Exercise 4.6 Confirm the matrix representation of D in Eq. 4.18, and obtain from it the characteristic equation in Eq. 4.19.

The positive eigenvalue is therefore $\sin \theta$. We have arrived at the following theorem:

Basic distinguishability theorem. Given two equally likely states $|\alpha_0\rangle$ and $|\alpha_1\rangle$ such that $|\langle\alpha_0|\alpha_1\rangle| = \cos\theta$, the probability P_S of correctly identifying the state by a basic measurement is bounded by

$$P_S \leq \frac{1}{2}(1 + \sin\theta). \tag{4.20}$$

Equality is achievable by a particular choice of observable.

We could of course manage $P_S = 1/2$ simply by guessing, without any measurement at all. If two quantum states are close together in the Hilbert space, then $\sin\theta$ is small and we cannot do much better than guessing. Such nearby states are *distinct*, in that they are different vectors in the Hilbert space and represent different physical preparations of the system. But they are not very well *distinguishable*. Only orthogonal quantum states of a system are completely distinguishable by measurement.

4.3 The projection rule and its limitations

In a measurement process, we extract information from a quantum system. But what sort of information is it? Exactly what does a measurement tell us?

A naive answer is, "When we measure an observable A, we find out the value of A." Yet this is at best only a partial answer. If the quantum system did originally have a definite value of A – that is, if it were prepared in an A-eigenstate – then the measurement would reveal that A-value. On the other hand, the quantum system might not have a definite value of A before the measurement. A spin-1/2 particle prepared in the state $|x_+\rangle$ does not have a definite value for S_z. In this case, it is more difficult to say what the measurement tells us.

Another answer often proposed is, "When we measure an observable A, we find out about the results of future measurements." The idea is that, immediately after an A-measurement, the system is in an eigenstate of A corresponding to the result obtained. A second A-measurement performed immediately after the first must yield the same value.

This assertion is sometimes called the *projection rule*. It does give a straightforward answer to our question: A measurement provides a definite prediction of the result of a subsequent measurement of the same type. The only problem with the projection rule is that, in this simple form, it is usually false.

To see why, recall our example of the two-beam interferometer. When a single photon is in the interferometer, the system behaves like a qubit, with basis states $|u\rangle$ (photon in upper beam) and $|l\rangle$ (photon in lower beam). Arbitrary superpositions of these are also possible quantum states.

There are of course many other possible states of the light in the interferometer. Like the two-level atom, the one-photon interferometer is a simplification of a more complicated real system with more possible states. One of these other states, orthogonal to those already mentioned, is $|0\rangle$, the "no photon" state. If our system is a two-beam interferometer with exactly one photon, it is described by a two-dimensional Hilbert space spanned by $|u\rangle$

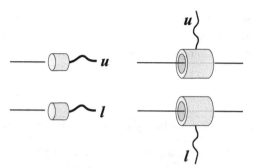

Fig. 4.3   **Absorbing (left) and non-destructive (right) photon detectors in the interferometer beams.**

and $|l\rangle$. This was our qubit example. Now, however, we will consider the system of a two-beam interferometer that has *at most* one photon. Our Hilbert space is three-dimensional and includes $|0\rangle$.

Exercise 4.7 The *photon number observable N* counts the total number of photons present in our interferometer system. For our new situation, show that N has two eigenvalues, one of which is degenerate. Describe the two eigenspaces for these eigenvalues.

We can make a measurement in the $\{|u\rangle, |l\rangle, |0\rangle\}$ basis by putting a pair of photon detectors in the beams. The states $|u\rangle$ and $|l\rangle$ correspond to photon detections by the upper or lower detectors, respectively. The state $|0\rangle$ corresponds to an outcome in which neither detector "clicks." (Since we have at most one photon in the interferometer, these are the only possible outcomes.)

But an ordinary photon detector *absorbs the photon* upon detection. Whatever the measurement result, therefore, the interferometer system is always in the state $|0\rangle$ immediately afterward. A second measurement using a second set of detectors would always yield the "no photon" result, whatever the result of the first measurement. The projection rule in this case would fail.

For the projection rule to hold, we must imagine photon detectors that register the passage of a photon without destroying it or scattering it out of the beam. Such "non-destructive" detectors are possible in principle, though they are hard to build (see Fig. 4.3). (We imagined photon detectors of this type for the two-slit experiment in Section 1.2.) Then if the upper or lower detectors register a photon, the system is afterward known to be in the $|u\rangle$ or $|l\rangle$ state, as the projection rule requires. If no photon is detected, the system is in $|0\rangle$.[1]

We have now described two possible measurement processes for the same observable, one of which violates the projection rule and one of which obeys it. The validity of the projection rule depends, not on *which* measurement we make, but rather on *how* we make it – the details of the physical measurement process. The projection rule is a special assumption about that process. Roughly speaking, the projection rule works when this

---

[1] We are assuming that even these "non-disturbing" photon detectors have 100% efficiency, so that they will certainly detect a photon if it is present. Real photon detectors are not so perfect and require a more sophisticated analysis. See Chapter 9, especially Problem 9.11.

process disturbs the system as little as possible, e.g. by simply registering a passing photon instead of absorbing it completely.

Why is this assumption called the "projection rule"? The outcomes of the measurement are represented by the projections

$$\Pi_u = |u\rangle\langle u|, \qquad \Pi_l = |l\rangle\langle l|, \qquad \Pi_0 = |0\rangle\langle 0|. \tag{4.21}$$

If the interferometer starts out in the state $|\psi\rangle$, the probabilities of the various outcomes are given by $P(U) = \langle\psi|\Pi_u|\psi\rangle$ and so on. After the measurement, if the result is x then the state has changed according to

$$|\psi\rangle \to K\,\Pi_x|\psi\rangle, \tag{4.22}$$

where K is a normalization factor.

Exercise 4.8 Since the overall phase of a state vector is physically irrelevant, we can choose K to be real and positive. Show that the choice $K = P(x)^{-1/2}$ works.

The projection rule says that the action of the measurement process on the state of the system is described by the projection operator associated with the outcome.

We could make a measurement in the interferometer using a photon detector in only one of the beams. Suppose for instance we place a non-destructive photon detector in the upper beam. This detector would "click" if the system were prepared in state $|u\rangle$, but not if it were in either $|l\rangle$ or $|0\rangle$ (or a superposition of these). This detector arrangement measures a "click number" observable C with eigenvalues 1 and 0. The eigenvalue 0 is degenerate, and the corresponding eigenspace is spanned by $|l\rangle$ and $|0\rangle$.

If the initial state is a multiple of $|u\rangle$ or any state of the form $a|l\rangle + b|0\rangle$, it will be unaffected by this measurement procedure. A general input state will produce the following results:

- With probability $P(1) = \langle\psi|\Pi_1|\psi\rangle$, the result 1 is obtained, and the final state will be of the form $K\,\Pi_1|\psi\rangle$ (which is a multiple of $|u\rangle$);
- With probability $P(0) = \langle\psi|\Pi_0|\psi\rangle$, the result 0 is obtained, and the final state will be of the form $K'\,\Pi_0|\psi\rangle$ (which is a superposition of $|l\rangle$ and $|0\rangle$).

This measurement process satisfies the projection rule, as it applies to an incomplete measurement. This means that a second such measurement on the same system would always obtain the same result.

Exercise 4.9 Consider the overall photon number operator N and the upper-beam click number operator C. What is [N, C]?

Exercise 4.10 Suppose the interferometer is initially in the state

$$|\psi\rangle = \frac{1}{\sqrt{3}}\left(|u\rangle - |l\rangle - |0\rangle\right). \tag{4.23}$$

We introduce an absorbing photodetector into the lower beam and count the number of times it "clicks" (either zero or one). Find the probabilities of the two outcomes and the state of the interferometer after each one. Does this measurement satisfy the projection rule?

4.4 Quantum cryptography

Suppose Alice wishes to send Bob a secret message, one that can be read by Bob but not by an unauthorized third party, the eavesdropper Eve. If there is a private mode of communication to which Eve has no access, then Alice can employ it. But what if there is no such private channel? After all, Eve might bug Alice's telephone calls, open the letters she sends, or intercept her outgoing e-mail. What if the only available mode of communication from Alice to Bob is effectively a *public* channel? In order to send secret information over a public communication channel, Alice must *encrypt* the information that she sends to Bob.

We will describe one particular procedure for encryption. Let us say that Alice's message consists of a string of n binary digits, called the *plaintext*. (As we discussed in Section 1.1, any sort of message can be represented in this format.) We also suppose that Alice and Bob already share a secret string of n bits, called the *key*. This key string contains no message in itself, but at least it is completely unknown to Eve. Alice encrypts her information by adding together the plaintext and the key, creating a new string of n bits called the *ciphertext*:

$$\begin{array}{rll} & \text{plaintext} & \texttt{00110011000011011101100010011} \\ + & \text{key} & \texttt{10011010010011101100100000110} \\ \hline = & \text{ciphertext} & \texttt{10101001010000110001000010101} \end{array}$$

(The addition is modulo 2, so that $1 + 1 = 0$.) Symbolically, we write

$$P + K = C, \tag{4.24}$$

where P represents the plaintext, K the key, and C the ciphertext. It is the ciphertext C that Alice sends over the public channel to Bob, and to which Eve potentially has access.

To decrypt the message, Bob (who also possesses K) adds the ciphertext and the key string:

$$\begin{array}{rll} & \text{ciphertext} & \texttt{10101001010000110001000010101} \\ + & \text{key} & \texttt{10011010010011101100100000110} \\ \hline = & \text{plaintext} & \texttt{00110011000011011101100010011} \end{array}$$

This works because, in modulo 2 arithmetic, $X + X = 0$ for any string X, where "0" means a string of all zeroes. Therefore,

$$C + K = P + K + K = P + 0 = P. \tag{4.25}$$

Bob can thus use the key to recover Alice's message from the ciphertext string.

What about Eve? As far as she is concerned, the key string K might be any n-bit string. Even if she possesses the ciphertext C, the plaintext $P = C + K$ could still be anything at all. In other words, at the outset Eve must consider that there are 2^n possible plaintext messages P. With the ciphertext in hand, there are still 2^n possible Ps. From the ciphertext alone, Eve has gained no information at all (see Eq. 1.7).

Exercise 4.11 Eve, in desperation, simply guesses at the key K, and then decides that the plaintext P must be 00000101100101101000111000011. What key did she guess? What are the odds of successfully decrypting the message by this method?

This method allows Alice to send Bob a secret message over a public channel. But you may have noticed a difficulty. In order for Alice to send the secret message P to Bob, they must already share a secret key K of the same length. To send a secret, they must already have shared a secret. How did they do that?

This is the problem of *key distribution*, and it is a formidable challenge in practical cryptography. The usual approach is to use some especially secure form of communication for the key K – a trusted courier with a locked briefcase, perhaps. The problem is that, despite all precautions, Alice and Bob cannot be completely sure that the secrecy of K has not been compromised.

If a particular key K were known to be insecure – if Eve's fingerprints were discovered inside the locked briefcase, for example – then Alice and Bob would simply agree not to use the "blown" key. Since K contains no message and thus has no value in itself, Eve has gained nothing for her pains. Eve only poses a danger if she can *indetectably* figure out the key K.

In 1984, Charles Bennett and Gilles Brassard invented a protocol for secure key distribution that relies on the special properties of quantum systems. Their scheme, sometimes called "BB84," is the simplest protocol for *quantum key distribution*, and was the starting point for the subject of *quantum cryptography*. We will now describe the BB84 protocol[2] and discuss some reasons why it works.

In the BB84 protocol, Alice begins by generating two random strings of bits, the "basis string" and the "parent string." While we can represent the parent string by 0s and 1s, it is more convenient to represent the basis string by Zs and Xs. From each corresponding pair of bits from the two strings, Alice chooses a quantum state for a qubit system, which she then passes on to Bob. The states are

basis	parent	qubit state
Z	0	$\lvert 0 \rangle$
Z	1	$\lvert 1 \rangle$
X	0	$\lvert + \rangle = \frac{1}{\sqrt{2}}\left(\lvert 0 \rangle + \lvert 1 \rangle \right)$
X	1	$\lvert - \rangle = \frac{1}{\sqrt{2}}\left(\lvert 0 \rangle - \lvert 1 \rangle \right)$

In other words, Alice generates a sequence of qubits that are in one of the four states $\lvert 0 \rangle$, $\lvert 1 \rangle$, $\lvert + \rangle$, or $\lvert - \rangle$, with equal probability. Be sure to note that these states are eigenstates of the Pauli observables Z or X for the qubit, and that the basis string tells us which. Figure 4.4 lays out an example; so far, we have discussed the first three lines.

---

[2] We will not here give a full proof of the security of BB84 against every sort of "attack" that Eve might mount. Such a general proof does exist, but it requires ideas and techniques of quantum information theory beyond our present discussion.

Alice's basis string	X	Z	Z	X	Z	X	X	Z	$\cdots$
Alice's parent string	1	0	1	1	0	1	0	1	$\cdots$
Qubit states	$|-\rangle$	$|0\rangle$	$|1\rangle$	$|-\rangle$	$|0\rangle$	$|-\rangle$	$|+\rangle$	$|1\rangle$	$\cdots$
Bob's basis string	X	X	Z	Z	X	Z	X	Z	$\cdots$
Bob's results	$|-\rangle$	$|+\rangle$	$|1\rangle$	$|1\rangle$	$|-\rangle$	$|0\rangle$	$|+\rangle$	$|1\rangle$	$\cdots$
Bob's parent string	1	0	1	1	1	0	0	1	$\cdots$
Right basis?	yes	no	yes	no	no	no	yes	yes	$\cdots$
Key string K	1		1				0	1	$\cdots$

Fig. 4.4 An example of the BB84 protocol in action.

The qubits are now delivered to Bob. If Bob somehow knew Alice's basis string, then it would be possible for him to determine the parent string as well. He would simply measure Z or X on the qubit, as appropriate. But Bob does not know the basis string. So instead, he just makes up his own random basis string of Zs and Xs, then measures accordingly. He then records his version of the parent string bits as determined by these measurements.

Note that when Bob guesses the right basis (which happens about half the time), he correctly learns the corresponding parent string bit. But when he guesses wrong, he is equally likely to get the parent right or wrong. Thus, Bob's version of the parent string is only about 75% right. (Examine the next group of rows in Fig. 4.4.)

Next, Bob contacts Alice over a public communication channel and tells her his own basis string. That is, he announces which measurements he made, but not the results of those measurements. Alice replies by telling Bob which of his measurements were "correct." Alice and Bob discard those qubits on which Bob measured the wrong basis – about half of the whole sequence. On the remaining qubits, Alice prepared and Bob measured in the same way. The parent sequences of Alice and Bob therefore agree on these, and so they can use this subsequence as their key K. (See the last rows of Fig. 4.4.)

Exercise 4.12 Bob's version of the parent string includes about 25% errors. Yet in the protocol, Alice and Bob throw away 50% of the bits. Why do they dispose of so many bits that are, in fact, correct?

What about the eavesdropper Eve? Alice and Bob's public communication only reveals their basis strings, not the parent strings from which the key K is extracted. To have any hope of learning this key string, Eve will have to intercept and examine the qubits using some measurement. But she cannot reliably determine which of the four BB84 states Alice is sending for each qubit. The basic decoding theorem in Eq. 4.8 tells us that her probability of error must be at least

$$P_E \geq 1 - \frac{2}{4} = \frac{1}{2}. \tag{4.26}$$

She will get it wrong half the time.

Since Eve must sometimes misidentify the state, then the qubit that she sends on to Bob will sometimes be prepared incorrectly. This will show up as additional errors in the BB84 protocol. For example, suppose Alice has basis bit Z and parent bit 0. She sends the state $|0\rangle$ to Bob. Eve, intercepting the qubit, misidentifies this as the state $|+\rangle$, and sends that state on to Bob. Bob happens to measure in the Z-basis, but by chance obtains the result $|1\rangle$ instead of $|0\rangle$. Even though Alice and Bob have measured in the same basis, the key bits they extract do not agree in this case.

Alice and Bob can detect the presence of Eve by checking for errors. They choose some of their key bits at random and compare them over the public channel. These bits are now no longer useful as key bits, since they are no longer secret; but they are useful for finding out whether extra errors are occurring in the protocol. If, after sacrificing a few hundred bits in this way, they find no cases of unexplained errors, they can be very confident that Eve is not intercepting and measuring the qubits. The remainder of the key is therefore secure.

Exercise 4.13 Suppose Eve simply measures X on every qubit, inferring that the state is either $|+\rangle$ or $|-\rangle$. What is the probability that a given bit in the final key K will actually be an error? If 500 bits of K are sacrificed to check for errors, about how many errors should be found?

The disturbance of the qubit system due to Eve's measurement process acts as Eve's "fingerprints," alerting Alice and Bob to her activities and warning them that their proposed key is not secure. Of course, in the real world, there will always be some noise in the system, and some errors will always be present. To be on the safe side, Alice and Bob have to assume that any noise might indicate the effect of an eavesdropper. But if the error rate is small enough, it turns out that Alice and Bob are still able to arrive at a shared secret key that they are confident is completely unknown to Eve.

Distinct quantum states are not necessarily perfectly distinguishable, and this fact has led to limits on our ability to use quantum systems to perform information tasks. We saw in Section 4.1, for instance, that a quantum system with Hilbert space dimension d can only convey up to $\log d$ bits reliably as a communication medium. In quantum cryptography, the lack of distinguishability among the BB84 states is not a limit, but an opportunity. We can use it to perform a task (establishing a secret key) that may be difficult or impossible otherwise.

A loophole?

Poor Eve. She does not know what observable to measure in order to determine the state of the intercepted qubits. What makes this particularly maddening is that, when Alice and Bob later communicate over the public channel, Eve will actually learn what measurements to use! By then, of course, it is too late, since the qubits have already been delivered to Bob.

However, suppose Eve has a device that could *copy* or *clone* qubits exactly. That is, when the device is presented with a single qubit in any quantum state $|\psi\rangle$, it produces two qubits with the same state $|\psi\rangle$. Such a machine is not a measurement device, since Eve does not extract any information from the system.

If a quantum cloning machine is available, then Eve can beat the BB84 protocol. She makes perfect copies of the qubits and sends the originals on to Bob. Then she *waits* until Alice and Bob discuss their measurement bases before making her own measurements. In those cases when Alice prepared and Bob measure in the same basis, Eve can do the same and be sure that her results agree with Bob's. Eve would end up with the same key that Alice and Bob share, without introducing any disturbance on the qubits.

Therefore, the security of quantum key distribution depends on the unavailability of quantum cloning machines. Such machines are in fact impossible, as we will prove in Section 7.2.

4.5 The uncertainty relation

Knowing the quantum state of a system lets us assign probabilities to the possible results of various future measurements. In some cases we may be certain of the outcome, but for other variables the probabilities are spread out over several possibilities. If a spin-1/2 particle is prepared in the state $|x_+\rangle$, then a measurement of S_x would surely yield $+\hbar/2$, but the result of an S_z measurement is not certain. On the other hand, if the state is $|z_-\rangle$, we know that a measurement of S_z would find $-\hbar/2$, but do not know what an S_x measurement would find. Certainty about one observable comes at the price of uncertainty about the other. There is no quantum state of a spin-1/2 particle that predicts a certain outcome for both S_x and S_z.

Our aim here is to give a mathematical formulation of this heuristic point about quantum physics. This will lead us to a general theorem of quantum physics with far-reaching implications. We begin by proving a handy inequality about vectors in a Hilbert space.

Suppose $|a\rangle$ and $|b\rangle$ are two vectors in a Hilbert space $\mathcal{H}$. We can pick a vector $|b'\rangle = e^{i\alpha} |b\rangle$, where the phase α is chosen so that $\langle a | b' \rangle$ is real and non-negative. That means that $\langle a | b' \rangle = |\langle a | b \rangle|$. Now define

$$|\lambda\rangle = |a\rangle + \lambda |b'\rangle, \tag{4.27}$$

where λ is an adjustable real parameter. For any value of λ, it must be true that $\langle \lambda | \lambda \rangle \geq 0$. Let us define

$$\begin{aligned} f(\lambda) &= \langle \lambda | \lambda \rangle \\ &= \langle a | a \rangle + \lambda^2 \langle b' | b' \rangle + \lambda \left(\langle a | b' \rangle + \langle b' | a \rangle \right) \\ &= \langle a | a \rangle + \lambda^2 \langle b | b \rangle + 2\lambda |\langle a | b \rangle|. \end{aligned} \tag{4.28}$$

The function f is quadratic in λ. We know that $f(\lambda) \geq 0$ for any value of λ; in particular, we know that this is true for the value $\lambda_{\min}$ that minimizes f. We can find this minimal value in the usual way by setting $f'(\lambda_{\min}) = 0$:

$$0 = 2\lambda_{\min} \langle b | b \rangle + 2 |\langle a | b \rangle|$$

$$\lambda_{\min} = -\frac{|\langle a | b \rangle|}{\langle b | b \rangle}. \tag{4.29}$$

Substituting this into f, we find that

$$f(\lambda_{\min}) = \langle a \,|\, a \rangle + \left(\frac{|\langle a \,|\, b \rangle|}{\langle b \,|\, b \rangle} \right)^2 \langle b \,|\, b \rangle - 2 \frac{|\langle a \,|\, b \rangle|}{\langle b \,|\, b \rangle} |\langle a \,|\, b \rangle| \geq 0, \qquad (4.30)$$

and therefore

$$\langle a \,|\, a \rangle \langle b \,|\, b \rangle \geq |\langle a \,|\, b \rangle|^2 . \qquad (4.31)$$

Equation 4.31 is called the *Schwartz inequality*, and it is a remarkably useful result about vectors in $\mathcal{H}$.

Exercise 4.14 Here are a couple of quick questions about this derivation. (a) How do we know that $\lambda_{\min}$ in Eq. 4.29 gives us a *minimum* of f? Does it matter? (b) We have tacitly assumed that $|b\rangle \neq 0$. Where? Does the Schwartz inequality hold true in the case $|b\rangle = 0$?

With the Schwartz inequality in hand, we next turn to analyzing the uncertainties of quantum observables. First, though, we have to give a precise mathematical definition of what we mean by "uncertainty." The mean or expectation value of a random variable X is

$$\langle X \rangle = \sum_x x \, p(x). \qquad (\text{Re A.12})$$

The expectation value $\langle X \rangle$ is not necessarily the value (or even *a* value) that we expect to really occur. Then how far from $\langle X \rangle$ is an actual outcome x likely to be?

As Appendix A explains, the variance is defined as the mean of the square of the deviation from the mean. That is,

$$\left\langle \Delta X^2 \right\rangle = \left\langle (X - \langle X \rangle)^2 \right\rangle = \left\langle X^2 \right\rangle - \langle X \rangle^2 , \qquad (4.32)$$

(see Eq. A.14).

Exercise 4.15 A random variable Q takes on the values ± 1 with equal probability. Find the mean and the variance of Q. How would these change if $P(+1) = 3/4$ and $P(-1) = 1/4$?

We define the *uncertainty* of X to be the standard deviation $\Delta X = \sqrt{\langle \Delta X^2 \rangle}$. This is a good measure of how "spread out" the probability distribution is around the value $\langle X \rangle$. The actual value of the variable X in an experiment is unlikely to be many times further than ΔX away from the mean $\langle X \rangle$. (Appendix A develops this idea in more technical detail.)

Exercise 4.16 Show that $\Delta Q > 0$ if and only if more than one distinct value q has $p(q) > 0$. That is, $\Delta Q > 0$ whenever there is actually some uncertainty about the value of Q.

Exercise 4.17 Suppose all of the possible values of Q lie in an interval on the real line whose length is L. Show that $\Delta Q \leq L$.

Applying this idea of uncertainty to a quantum observable A is at once trivial and subtle. It is trivial, because any state $|\psi\rangle$ leads to a probability distribution over the spectrum of A eigenvalues, from which we can calculate the mean and variance. The subtleties arise when we try to interpret the average in $\langle \Delta A^2 \rangle$ as an expectation value given by some operator.

Suppose we define the deviation operator ΔA by

$$\Delta \mathsf{A} = \mathsf{A} - \langle A \rangle \, \mathbf{1}. \tag{4.33}$$

This would be a strange definition for an operator, because the expectation value $\langle A \rangle$ already presumes a particular quantum state $|\psi\rangle$ for which $\langle A \rangle = \langle \psi | \mathsf{A} | \psi \rangle$. If $|\psi\rangle$ does happen to be the state of the system, then

$$\langle \psi | \, \Delta \mathsf{A}^2 \, | \psi \rangle = \langle \psi | \mathsf{A}^2 | \psi \rangle - 2 \langle A \rangle \langle \psi | \mathsf{A} | \psi \rangle + \langle A \rangle^2 \langle \psi | \psi \rangle$$
$$= \left\langle A^2 \right\rangle - \langle A \rangle^2 = \left\langle \Delta A^2 \right\rangle, \tag{4.34}$$

as we wish. But if the system is in another state $|\phi\rangle$ with a different expectation value for A, then the variance of A will not be $\langle \phi | \, \Delta \mathsf{A}^2 \, | \phi \rangle$.

We have to be more careful. For a real value α, define

$$\Delta_\alpha \mathsf{A} = \mathsf{A} - \alpha \mathbf{1}. \tag{4.35}$$

This is the deviation of the observable A from a fixed value α that is independent of any choice of state. If it happens that $\alpha = \langle A \rangle$ for a particular $|\psi\rangle$, then

$$\langle \psi | \, (\Delta_\alpha \mathsf{A})^2 \, | \psi \rangle = \left\langle \Delta A^2 \right\rangle, \tag{4.36}$$

for the distribution of A-values given by the state $|\psi\rangle$. But if $\alpha \neq \langle A \rangle$, then this equality will not hold. (See Problem 4.2 for a more general relation.)

Now suppose we have two observables A and B for a quantum system. We pick out values α and β and define $\Delta_\alpha \mathsf{A}$ and $\Delta_\beta \mathsf{B}$ as above. If the system is in the quantum state $|\psi\rangle$, we can consider the vectors

$$|a\rangle = \Delta_\alpha \mathsf{A} | \psi \rangle \quad \text{and} \quad |b\rangle = \Delta_\beta \mathsf{B} | \psi \rangle. \tag{4.37}$$

Applying the Schwartz inequality (Eq. 4.31) to $|a\rangle$ and $|b\rangle$, we find

$$\langle \psi | \, (\Delta_\alpha \mathsf{A})^2 \, | \psi \rangle \langle \psi | \, (\Delta_\beta \mathsf{B})^2 \, | \psi \rangle \geq \left| \langle \psi | \, (\Delta_\alpha \mathsf{A})(\Delta_\beta \mathsf{B}) \, | \psi \rangle \right|^2. \tag{4.38}$$

The right-hand side is of the form $\langle \psi | \mathsf{FG} | \psi \rangle$ for Hermitian operators F and G. What is this matrix element? The product FG can be written

$$\mathsf{FG} = \frac{1}{2}[\mathsf{F}, \mathsf{G}] + \frac{1}{2}\{\mathsf{F}, \mathsf{G}\}, \tag{4.39}$$

where $\{\mathsf{F}, \mathsf{G}\} = \mathsf{FG} + \mathsf{GF}$ is the *anticommutator* of F and G.

Exercise 4.18 Given Hermitian F and G, show that (a) the commutator $[\mathsf{F}, \mathsf{G}]$ is anti-Hermitian, and (b) the anticommutator $\{\mathsf{F}, \mathsf{G}\}$ is Hermitian. (Part (a) was also Exercise 3.41.)

It follows that

$$\langle \psi | \mathsf{FG} | \psi \rangle = \frac{1}{2} \langle \psi | [\mathsf{F}, \mathsf{G}] | \psi \rangle + \frac{1}{2} \langle \psi | \{\mathsf{F}, \mathsf{G}\} | \psi \rangle. \tag{4.40}$$

The commutator term is purely imaginary and the anticommutator term is purely real. Therefore,

$$\left| \langle \psi | \, \mathsf{FG} \, | \psi \rangle \right|^2 \geq \frac{1}{4} \left| \langle \psi | \, [\mathsf{F}, \mathsf{G}] \, | \psi \rangle \right|^2. \tag{4.41}$$

Exercise 4.19 Carefully explain the line of reasoning from Eq. 4.40 to Eq. 4.41.

To apply this general operator fact, we note that

$$[\Delta_\alpha \mathsf{A}, \Delta_\beta \mathsf{B}] = [\mathsf{A}, \mathsf{B}]. \tag{4.42}$$

Therefore, Eq. 4.38 implies that

$$\langle \psi | \, (\Delta_\alpha \mathsf{A})^2 \, | \psi \rangle \, \langle \psi | \, (\Delta_\beta \mathsf{B})^2 \, | \psi \rangle \geq \frac{1}{4} \left| \langle \psi | \, [\mathsf{A}, \mathsf{B}] \, | \psi \rangle \right|^2. \tag{4.43}$$

The right-hand side does not depend on α and β. If we choose $\alpha = \langle A \rangle$ and $\beta = \langle B \rangle$, then the left-hand side is the product of the variances $\langle \Delta A^2 \rangle$ and $\langle \Delta B^2 \rangle$. Therefore,

$$\langle \Delta A^2 \rangle \langle \Delta B^2 \rangle \geq \frac{1}{4} \left| \langle \psi | \, [\mathsf{A}, \mathsf{B}] \, | \psi \rangle \right|^2. \tag{4.44}$$

It is convenient to write everything in terms of uncertainties.

General uncertainty relation. Suppose A and B are two observables on a quantum system and let K be the observable such that

$$[\mathsf{A}, \mathsf{B}] = i\mathsf{K}. \tag{4.45}$$

Then, for any quantum state $| \psi \rangle$,

$$\Delta A \; \Delta B \geq \frac{1}{2} \left| \langle K \rangle \right|. \tag{4.46}$$

Equation 4.46 tells us that there is a trade-off between the uncertainties ΔA and ΔB, and that this trade-off is governed by the commutator $[\mathsf{A}, \mathsf{B}]$ of the two operators. If $|\langle K \rangle| > 0$, then it is not possible for both ΔA and ΔB to be close to zero.

If two operators commute (K = 0), then we know that there are simultaneous eigenstates for A and B. Equation 4.46 therefore gives a lower bound of zero for $\Delta A \; \Delta B$.

We shall illustrate the meaning of the general uncertainty relation by considering our old friend, the spin-1/2 particle. Recall from Exercise 3.56 that the spin component operators S_x, S_y, and S_z satisfy

$$[\mathsf{S}_x, \mathsf{S}_y] = i\hbar \mathsf{S}_z. \tag{4.47}$$

(If you have not worked out Exercise 3.56 for yourself, you should go back and do it.) This means that, for any state of the particle,

$$\Delta S_x \; \Delta S_y \geq \frac{\hbar}{2} \left| \langle S_z \rangle \right|. \tag{4.48}$$

We can extract a lot of meaning from this uncertainty relation. If we have an eigenstate of S_x, then $\Delta S_x = 0$. This can only happen provided $\langle S_z \rangle = 0$. The same must be true of S_y eigenstates. If the expectation $\langle S_z \rangle \neq 0$, then neither S_x nor S_y can have a definite value.

We have called Eq. 4.46 an "uncertainty" relation, but a better name might be an "indeterminacy" relation. The word "uncertainty" seems to refer to our own state of mind, to our lack of sureness about the value of A. This leaves open the possibility that A actually has a definite value. We may be somewhat uncertain about the exact area of Brazil, but we do not doubt that the territory of Brazil comprises some definite number of square kilometers. We merely happen to be ignorant of that number.

The situation for quantum systems appears to be quite different. The measured value of an observable A is uncertain, not because of some ignorance on our part, but because the particular quantum state does not *have* a definite value of A. The quantities ΔA and ΔB in Eq. 4.46 describe the indeterminacy of the observables A and B, not merely our uncertainty about them. Nevertheless, the word "uncertainty" has become so standard in connection with Eq. 4.46 that we will just have to live with it.

Problems

Problem 4.1 In Section 4.2, suppose that the two states $|\alpha_0\rangle$ and $|\alpha_1\rangle$ have a real, positive inner product $\langle \alpha_0 | \alpha_1 \rangle = \cos\theta$. On a diagram like Fig. 4.2, locate the eigenstates $|\pm\rangle$ of the operator $D = |\alpha_0\rangle\langle\alpha_0| - |\alpha_1\rangle\langle\alpha_1|$.

Problem 4.2 Consider a random variable Z with a given probability distribution. For a parameter q, define the *mean square deviation from q* to be $D(q) = \langle (Z - q)^2 \rangle$. Show that

$$\langle \Delta Z^2 \rangle \leq D(q),$$

for any choice of q, with equality if $q = \langle Z \rangle$. (Hint: Write out the algebraic form of $D(q)$ and minimize the function.)

Problem 4.3 Suppose $|\psi\rangle$ is an eigenstate of both A and B. (This may be the only such eigenstate of both operators; we are *not* assuming that A and B commute.) Show that $|\psi\rangle$ is also an eigenstate of [A, B] with eigenvalue 0. How does the $\Delta A \, \Delta B$ uncertainty relation work out for this particular state?

Problem 4.4 For a spin-1/2 particle, consider the generic state

$$|\psi\rangle = e^{i\phi} \cos\theta \, |z_+\rangle + e^{-i\phi} \sin\theta \, |z_-\rangle. \tag{4.49}$$

Find ΔS_x, ΔS_y, and $\langle S_z \rangle$, and show that the spin component uncertainty relation in Eq. 4.48 holds.

Problem 4.5 Derive an uncertainty relation that involves the anticommutator of the operators A and B instead of the commutator. Your derivation will be the same as ours until about Eq. 4.40, after which it will take a slightly different turn.

Problem 4.6 Find and prove the necessary and sufficient condition for equality in the Schwartz inequality (Eq. 4.31).

Problem 4.7 In 1992, Bennett proposed an alternative scheme for quantum key distribution that uses only two non-orthogonal states. In the "B92" protocol, Alice sends to Bob the states $|0\rangle$ and $|+\rangle$ with equal likelihood, and Bob randomly measures them using the X or Z bases. For some combinations (which ones?) of state and measurement, Bob can deduce Alice's state. Rather than announcing his measurements, Bob announces which qubits had their states determined with certainty.

Complete the description of the protocol, including an analysis of the difficulties Eve faces in eavesdropping. What percentage of the transmitted qubits are discarded? On which of our basic information results does B92 depend?

Problem 4.8 Stephen Wiesner has proposed the following idea for *quantum money*, which cannot be reliably counterfeited. Each quantum banknote contains a string of N qubits, which maintain their quantum states unchanged indefinitely. It also has a printed random N-bit serial number. The bank keeps on file a secret N-bit basis string for each banknote. When the money is manufactured, the qubit states are set using the BB84 scheme, based on the basis string and serial number.

The bank tests any banknote presented to it by using the secret basis string to carry out a series of measurements on the qubits. A genuine note will produce results that exactly match the serial number. A counterfeiter would like to create fake banknotes that can pass this test, without knowing the correct basis string. (He must do this by making copies of an existing note, since the bank has records of which of the notes are in circulation.)

(a) Is a genuine banknote still valid after the bank's inspection? Explain what assumptions you make to answer this question.
(b) Consider the kth qubit on the banknote. Given the kth serial number bit printed on the note, what is the maximum probability that a counterfeiter can determine the bank's kth basis bit by a basic measurement on the qubit?
(c) The counterfeiter tries to identify the basis bit as in part (b), then uses this to set the kth qubit state on a fake banknote with the same serial number. If the fake note is inspected by the bank, with what probability will the kth qubit pass inspection?
(d) The whole banknote has $N = 100$ qubits. What is the likelihood that the counterfeiter's fake banknote will pass the bank's test? (Remember, the bank will reject the banknote if a single error is found.)

5 Quantum dynamics

5.1 Unitary evolution

Isolated systems

During a measurement, information is extracted from a quantum system, and in this process the state changes, either via the projection rule or not. But the state of a quantum system may also change over time even when the system is not being measured, when it is informationally isolated. Understanding this time evolution means understanding the dynamical laws governing quantum systems. To this indispensible task, we now turn.

We have already discussed time evolution in the context of qubit systems in Chapter 2. Section 2.1 analyzed the effect of successive optical elements on the state of a one-photon interferometer system. Section 2.3 examined in detail how the state of a two-level atom evolves, as well as the behavior of a spin-1/2 system in an external magnetic field. Much of what we will develop here has already been foreshadowed there. (If you have by now forgotten all about Chapter 2, this would be an excellent time to turn back and browse through it for a few minutes.)

Our basic principle is this: If a quantum system is informationally isolated, then its evolution respects the principle of superposition. If the states $|\psi\rangle$ and $|\phi\rangle$ would evolve over a given interval of time according to

$$|\psi\rangle \longrightarrow |\psi'\rangle,$$
$$|\phi\rangle \longrightarrow |\phi'\rangle, \tag{5.1}$$

then a superposition of the two states would evolve over the same interval by

$$a|\psi\rangle + b|\phi\rangle \longrightarrow a|\psi'\rangle + b|\phi'\rangle. \tag{5.2}$$

The time evolution of the state is a linear map from vectors in $\mathcal{H}$ to vectors in $\mathcal{H}$ – in other words, an operator.

Let t_1 and t_2 be two times, and suppose the operator $\mathsf{U}(t_2, t_1)$ describes the time evolution of an informationally isolated system over the interval from t_1 to t_2. If $|\psi(t_1)\rangle$ and $|\psi(t_2)\rangle$ are the quantum states of the system at the two times, then

$$|\psi(t_2)\rangle = \mathsf{U}(t_2, t_1)|\psi(t_1)\rangle. \tag{5.3}$$

From the operator $U(t_2, t_1)$ and the initial state $|\psi(t_1)\rangle$, we can calculate the final state $|\psi(t_2)\rangle$ of the system.[1]

What are the properties of the time evolution operator $U(t_2, t_1)$? Simplifying our notation slightly, let us suppose the old state $|\psi\rangle$ evolves to the new state $U|\psi\rangle$. The first thing that we note is that the new state must be normalized, just like the old state. So if $\langle\psi|\psi\rangle = 1$, then

$$\langle\psi| U^\dagger U |\psi\rangle = 1. \tag{5.4}$$

The operator $U^\dagger U$ must be Hermitian, so it has an orthonormal eigenbasis. If the old state $|\psi\rangle$ is chosen to be any eigenstate of $U^\dagger U$, then its eigenvalue must be 1. It follows that

$$U^\dagger U = \mathbf{1}. \tag{5.5}$$

In a Hilbert space of finite dimension, this means that the time evolution operator U must be *unitary*, and so $UU^\dagger = \mathbf{1}$ also. We will postulate that U is unitary in the infinite-dimensional case as well (see Problem 3.6).

Unitary time evolution preserves, not just the normalization of the quantum states, but also the inner products between states. Suppose $|\psi'\rangle = U|\psi\rangle$ and $|\phi'\rangle = U|\phi\rangle$. Then

$$\langle\psi'|\phi'\rangle = \langle\psi| U^\dagger U |\phi\rangle = \langle\psi|\phi\rangle. \tag{5.6}$$

This has an important consequence. The basic distinguishability theorem of Section 4.2 (Eq. 4.20) tells us that the inner product $\langle\psi|\phi\rangle$ governs how well the two states may be distinguished by a measurement. Their distinguishability is not changed by unitary evolution. In particular, two states for which $|\langle\psi|\phi\rangle| > 0$ – which are therefore not fully distinguishable – cannot become more distinguishable as they evolve in time.

Exercise 5.1 Show that, if a set of signal states of a system lies within a d-dimensional subspace initially, after a period of unitary time evolution the states will still lie within a subspace having the same dimension d. Comment on the following claim: *The communication capacity of an informationally isolated system is not increased by its time evolution.*

Consider a system that evolves in a unitary way through two successive stages. The first stage runs from t_1 to t_2, and the second from t_2 to t_3. The overall evolution of the system from t_1 to t_3 is described by a single unitary operator:

$$|\psi(t_3)\rangle = U(t_3, t_1) |\psi(t_1)\rangle. \tag{5.7}$$

But we can also treat this as evolution over the first stage followed by evolution over the second stage:

$$|\psi(t_3)\rangle = U(t_3, t_2) |\psi(t_2)\rangle = U(t_3, t_2) U(t_2, t_1) |\psi(t_1)\rangle. \tag{5.8}$$

Since these two expressions must give the same $|\psi(t_3)\rangle$ for any initial state $|\psi(t_1)\rangle$,

$$U(t_3, t_1) = U(t_3, t_2) U(t_2, t_1). \tag{5.9}$$

---

[1] The future state of an isolated quantum system is determined by its past state. In this sense, the quantum dynamical laws are *completely deterministic*. Indeterminacy and probabilities only arise in the context of measurement, when the system ceases to be isolated.

This is the rule for composing the evolution operators for two successive intervals of time. It has the form

$$U(\text{whole period}) = U(\text{second stage}) \, U(\text{first stage}). \qquad (5.10)$$

So the composition is right-to-left, with the rightmost operator representing the earliest stage of the overall evolution. This obviously generalizes to time evolution consisting of three, four or more successive stages.

This result might sound familiar from the discussion of the two-beam interferometer in Section 2.1. With a single photon present, the system is a qubit whose Hilbert space has basis states $|u\rangle$ and $|l\rangle$, representing the photon's presence in the upper or lower beam. We imagine that the light moves through the interferometer and successively encounters the various optical elements, the phase shifters and beamsplitters and so forth, each of which acts to alter the quantum state in some way.

Each optical element is described by an operator. The matrix representations for these operators were worked out in Section 2.1. For instance, a phase shifter by ϕ in the upper beam is described by $P_u(\phi)$, which has the properties

$$P_u(\phi) \, |u\rangle = e^{i\phi} \, |u\rangle \, ,$$

$$P_u(\phi) \, |l\rangle = |l\rangle \, . \qquad (5.11)$$

Thus, the operator can be written $P_u(\phi) = e^{i\phi} |u\rangle\langle u| + |l\rangle\langle l|$. (Compare the matrix representation in Eq. 2.12.)

Exercise 5.2 Review the various optical elements described in Section 2.1. Write the beamsplitter operator B_l and the beam exchange operator X in terms of outer products of the $|u\rangle$ and $|l\rangle$ states.

Exercise 5.3 Now consider an interferometer with *at most* one photon, as we did in Section 4.3. How do you suppose a beamsplitter would affect the $|0\rangle$ (no photon) state? Write down an expression for the B_l operator in this larger Hilbert space, and then show that it is unitary.

An interferometer arrangement with several optical elements in succession can be described by a single operator, which is the product of the operators for each element. These elements appear in the product in time-order from right to left, as in Equation 5.10. (Refer to Eq. 2.16 and the surrounding discussion.)

For a unitary operator, $U^\dagger U = U U^\dagger$. This means that all unitary operators are also *normal*. In Problem 3.3, we showed that normal operators always have an orthonormal basis of eigenstates, just like Hermitian operators. If $|k\rangle$ is an eigenstate of U with eigenvalue U_k, then

$$1 = \langle k \, | k \rangle = \langle k | \, U^\dagger U \, | k \rangle = U_k^* U_k \, \langle k \, | k \rangle = |U_k|^2 \, . \qquad (5.12)$$

Every eigenvalue of U is a complex number of magnitude 1, which can be conveniently written $U_k = e^{i\alpha_k}$. Thus we can write

$$U = \sum_k e^{i\alpha_k} \, |k\rangle\langle k| \, . \qquad (5.13)$$

Uniform dynamics

Physically, the unitary time evolution of a system is determined by the internal properties of the system and various external parameters. For instance, an atom's behavior depends on the internal electrostatic interaction among the electrons and the nucleus, and also on any externally applied electric and magnetic fields. If the external fields were time-varying, then the dynamics of the atom would be different at different moments, depending on the field values. Yet it often happens that the basic dynamical laws governing a quantum system do not change over time. When this is true, we say that the system experiences *uniform dynamics*.

In the case of uniform dynamics, the time evolution operator $U(t_2, t_1)$ only depends on the *difference* of the initial and final times:

$$U(t_2, t_1) = U(t_2 - t_1). \tag{5.14}$$

The unitary operator describing the evolution of the system over a period of ten seconds is the same regardless of *when* that ten-second interval begins. Even though the state of the system changes over time, the rules describing that change are the same from one moment to the next.

The time evolution of a system having uniform dynamics can be written $U(t)$. This represents a whole family of unitary operators, one for each possible duration t. (This is often called a "one-parameter family" of operators.) Obviously, $U(0) = 1$. We will also assume that the evolution of the system is smooth and continuous. This implies that, for a very short time interval ϵ, the operator $U(\epsilon)$ is close to 1.

Any two operators $U(t)$ and $U(s)$ in the one-parameter family must commute with one another, since

$$U(t)U(s) = U(s)U(t) = U(t + s). \tag{5.15}$$

Exercise 5.4 If we take Equation 5.15 to hold for all t, both positive and negative, how is $U(-t)$ related to $U(t)$?

Now consider the very short time interval ϵ. Then

$$U(2\epsilon) = U(\epsilon)^2, \quad U(3\epsilon) = U(\epsilon)^3, \quad U(4\epsilon) = U(\epsilon)^4, \quad \text{etc.} \tag{5.16}$$

These obviously all commute and share a common eigenbasis, the eigenbasis for $U(\epsilon)$. If ϵ is chosen small enough, then any time t is very close to a multiple of ϵ.

We conclude that we can find a common eigenbasis for every member of the one-parameter family of operators $U(t)$. If $|k\rangle$ is one of these eigenstates, then for any time t,

$$U(t)|k\rangle = e^{i\alpha_k(t)}|k\rangle. \tag{5.17}$$

The eigenstates of $U(t)$ change over time only by an overall phase factor, which has no effect on any physical property of the system. Such states are called *stationary states* of the system. Every informationally isolated quantum system with uniform dynamics has a basis of stationary states.

There is some ambiguity in the phase functions $\alpha_k(t)$, since we may add any multiple of 2π and leave $e^{i\alpha_k(t)}$ unchanged. However, this ambiguity can be resolved by insisting that $\alpha_k(0) = 0$ and that the functions are all continuous.[2] With these requirements, it must be that

$$\alpha_k(t + s) = \alpha_k(t) + \alpha_k(s). \tag{5.18}$$

In other words, the phase functions are *linear* functions of time. We can write $\alpha_k(t) = -\omega_k t$ for some ω_k. (The negative sign is purely conventional, since ω_k itself may be either positive or negative.) Therefore, for the stationary state $|k\rangle$,

$$\mathsf{U}(t) |k\rangle = e^{-i\omega_k t} |k\rangle. \tag{5.19}$$

Knowing how $\mathsf{U}(t)$ acts on a basis tells us everything about the operator. We see that

$$\mathsf{U}(t) = \sum_k e^{-i\omega_k t} |k\rangle\langle k|. \tag{5.20}$$

If the system follows uniform dynamics, then we only need to know two things: a basis $\{|k\rangle\}$ of stationary states, and a corresponding set of (angular) frequencies ω_k. With this, we can find the evolution operator $\mathsf{U}(t)$ for any time interval, and from this calculate the dynamical evolution for any quantum state.

Exercise 5.5 In Section 2.3 we studied the evolution of a spin-1/2 particle in an external magnetic field directed along the z-axis. We defined the Larmor frequency to be $\Omega = \gamma B$, where B is the field strength and γ is the gyromagnetic ratio for the particle. The $|z_\pm\rangle$ eigenstates were stationary states with frequencies $\pm\Omega/2$. Write down the general evolution operator $\mathsf{U}(t)$ for this system, and use it to derive the evolution of an arbitrary spin state given in Eq. 2.66.

As this exercise reminds us, the superposition of two or more stationary states is not generally a stationary state. The phase changes in Eq. 5.19 produce relative phase changes in the superposition states, which do have physical consequences.

It seems paradoxical that we can understand the time evolution of a quantum system through a knowledge of its stationary states, which apparently do not change at all. It is also remarkable that these "fixed point" states are numerous enough to form a complete orthonormal basis. Compare the situation of a classical harmonic oscillator, for which only one situation – the system at rest at the equilibrium point – is time-independent. The stationary states are the key to analyzing a quantum system with uniform dynamics.

5.2 The Schrödinger equation

The Hamiltonian operator

Suppose a quantum system evolves in a unitary way, so that at any time t the state of the system is $|\psi(t)\rangle = \mathsf{U}(t, t_0) |\psi(t_0)\rangle$.

---

[2] We have already assumed that the operator function $\mathsf{U}(t)$ is continuous; here we are just making sure that the phase functions have no sudden jumps by multiples of 2π.

How is $|\psi(t)\rangle$ changing at a given moment of time? The derivative $\frac{d}{dt}|\psi(t)\rangle$ is itself a vector; and furthermore, it is a vector that depends linearly on $|\psi(t)\rangle$ itself. There is therefore an operator G, possibly time-dependent, such that

$$\frac{d}{dt}|\psi(t)\rangle = G|\psi(t)\rangle. \tag{5.21}$$

What sort of operator is G? The overall normalization of the state does not change, so

$$\begin{aligned}
0 = \frac{d}{dt}\langle\psi(t)|\psi(t)\rangle &= \left(\frac{d}{dt}\langle\psi(t)|\right)|\psi(t)\rangle + \langle\psi(t)|\left(\frac{d}{dt}|\psi(t)\rangle\right) \\
&= \left(\langle\psi(t)|G^\dagger\right)|\psi(t)\rangle + \langle\psi(t)|\left(G|\psi(t)\rangle\right) \\
&= \langle\psi(t)|\left(G^\dagger + G\right)|\psi(t)\rangle.
\end{aligned} \tag{5.22}$$

This is zero for any choice of $|\psi(t)\rangle$, so we conclude that $G^\dagger + G = 0$. The operator G must be anti-Hermitian. It is more common to consider the *Hamiltonian* operator $H = i\hbar G$, which is Hermitian. Then

$$H|\psi(t)\rangle = i\hbar\frac{d}{dt}|\psi(t)\rangle. \tag{5.23}$$

Equation 5.23 (which we have seen before as Eq. 2.64 in the context of qubit quantum theory) is called the *Schrödinger equation* for an informationally isolated quantum system. The Schrödinger equation is one of the most significant equations in all of physics. On the other hand, as we have introduced it here, it appears to be nothing but a *definition* of the operator H. We need to add something more, and that something is an understanding of the physical meaning of the Hamiltonian operator.

Exercise 5.6 First, dimensional analysis. What are the units of the constant $\hbar$? What are the units of the operator G? What are the units of the operator H?

Suppose now that the system has uniform dynamics. This means that the Hamiltonian operator H cannot vary with time. If we apply the Schrödinger equation to a stationary state $|k\rangle$, we find that

$$H\left(e^{-i\omega_k t}|k\rangle\right) = i\hbar\frac{d}{dt}\left(e^{-i\omega_k t}|k\rangle\right) = (i\hbar)(-i\omega_k)e^{-i\omega_k t}|k\rangle,$$

and so

$$H|k\rangle = \hbar\omega_k|k\rangle. \tag{5.24}$$

The stationary state $|k\rangle$ is an eigenstate of H with eigenvalue $\hbar\omega_k$. We interpret this by recalling the Planck relation between angular frequency and energy: $E = \hbar\omega$ (Eq. 2.50). For a system with uniform dynamics, we come to three highly significant conclusions:

- The Hamiltonian operator in the Schrödinger equation (Eq. 5.23) is the energy operator of the system;
- The stationary states of the system are its energy eigenstates;
- The frequency ω_k of the stationary state is related to the energy eigenvalue E_k by $E_k = \hbar\omega_k$.

The energy eigenvalue equation is sometimes known as the *time-independent Schrödinger equation*:

$$H \ket{\psi} = E \ket{\psi}. \tag{5.25}$$

The energy eigenvalue problem has a unique importance, because the stationary states and their frequencies completely determine the dynamical evolution of the system. As we shall see, a great deal of mathematical effort and ingenuity has been invested in discovering exact or approximate solutions to Eq. 5.25 for a wide variety of physical systems.

Even when the dynamics of the system is not uniform, we can still interpret the Hamiltonian H as the energy operator for the system. The system's energy may depend on external fields and so forth that vary over time, so that H is time-dependent. External forces, in other words, may do work on the system, but the dynamics of the system may still be unitary. In such a time-dependent situation, the momentary energy eigenstates of H are not "stationary states" of the unitary evolution. In non-uniform dynamics there may not even be any states that remain physically unchanged for all times.

Exercise 5.7 Write down the adjoint of the Schrödinger equation. Use the Schrödinger equation and its adjoint to verify that the inner product $\braket{\psi | \phi}$ remains constant as both $\ket{\psi}$ and $\ket{\phi}$ evolve over time.

A remark is in order about terminology and notation. Why do we call the energy operator the "Hamiltonian" and denote it by H? This seems unnecessarily confusing (particularly in light of our use of the letter "H" for other important things, like a Hilbert space $\mathcal{H}$ and the entropy H). The reason is historical. The Hamiltonian of classical mechanics is, roughly speaking, the energy of a system written in terms of position and momentum variables. For systems with continuous degrees of freedom (such as those described in Chapter 10), the classical Hamiltonian function tells us the Hamiltonian operator for the corresponding quantum system. This is no small matter, and such an important connection is enshrined in our nomenclature.

Observables in time

The Schrödinger equation tells us how a quantum state changes over time. This in turn will lead to changes in the observable properties of the system.

Suppose A is an observable, and let us assume for now that the operator A does not itself depend on time. The expectation value $\langle A \rangle = \braket{\psi | A | \psi}$ will change as the quantum state $\ket{\psi}$ evolves.

$$\begin{aligned}
\frac{d}{dt} \langle A \rangle &= \frac{d}{dt} \braket{\psi | A | \psi} \\
&= \left(\frac{d}{dt} \bra{\psi} \right) A \ket{\psi} + \bra{\psi} A \left(\frac{d}{dt} \ket{\psi} \right) \\
&= \frac{1}{-i\hbar} \braket{\psi | HA | \psi} + \frac{1}{i\hbar} \braket{\psi | AH | \psi}.
\end{aligned} \tag{5.26}$$

That is,

$$\frac{d}{dt} \langle A \rangle = \frac{1}{i\hbar} \langle \psi | [A, H] | \psi \rangle,$$
(5.27)

for an observable A that is not an explicit function of time.

What if A itself does depend on time? We could, for example, imagine a Stern–Gerlach apparatus that slowly rotates about the axis along which the particle beam travels. At different times, such a device would measure different components of the spin of a particle that passed through it. The resulting observable would require a modification of Eq. 5.27.

Exercise 5.8 Suppose the observable A is itself time-dependent. Show that Eq. 5.27 becomes

$$\frac{d}{dt} \langle A \rangle = \frac{1}{i\hbar} \langle \psi | [A, H] | \psi \rangle + \langle \psi | \left(\frac{dA}{dt} \right) | \psi \rangle.$$
(5.28)

We will assume, unless otherwise stated, that the observables we are considering are not time-dependent. The simpler form in Eq. 5.27 will be the result we use.

According to Eq. 5.27, the rate of change of an observable A (or rather, of its expectation value) is governed by the commutator [A, H]. One immediate consequence of this concerns operators that commute with the Hamiltonian: [A, H] = 0. If this is true, then the expectation $\langle A \rangle$ has zero time derivative; and if the Hamiltonian is time-independent, the expectation $\langle A \rangle$ will always remain constant over time. In a system with uniform dynamics, observable quantities that commute with the Hamiltonian are *conserved*.

The most obvious example is the Hamiltonian itself, in the case of uniform dynamics. Since [H, H] = 0, the energy of the system is conserved.

Exercise 5.9 Since the Hamiltonian operator commutes with itself whether or not the dynamics is uniform, why do we add this "uniform dynamics" proviso to our statement about energy conservation?

If [A, H] = 0, we know that A and H are compatible observables. This means that there is a basis of eigenvectors of both A and H. The states of definite A-value (A-eigenstates) can also be taken to be stationary states (energy eigenstates). This is another way of looking at the conservation of A.

The commutator [A, B] is turning out to be of considerable significance. It is a key quantity in describing the complementarity of two observables – whether or not they are compatible, and how their indeterminacies are related. The commutator with the Hamiltonian operator H governs how an observable property of a system varies over time. In the next section, we will join these two ideas together and arrive at an important physical principle.

5.3 Quantum clock-making

It is a curious thing that *time* is not itself an observable quantity in quantum theory. Nevertheless, we do make measurements of time. We do this by building *clocks*, which are systems whose dynamics causes their state to change in a predictable way. Making a measurement on such a system is "reading the dial" on the clock, and from the result we can infer an approximate present value of the parameter t.

Let us consider the situation in general. Denote by C the *clock variable* from which we must determine the time t. This might, for example, be the position of a moving hand on the face of an analog clock. Now imagine that the clock variable C is subject to an uncertainty ΔC in its actual value.[3] Such uncertainty would lead to an uncertainty Δt in the inferred time.

Exercise 5.10 Suppose our clock variable is the angular position θ of a hand on an analog clock, and suppose this is uncertain by $\Delta\theta = 6°$ (1/60 of a full circle). What would be the uncertainty Δt in the measured time if the hand is the second hand of the clock? If it is the minute hand? How about the hour hand?

From this exercise, we can see that the relation of ΔC and Δt depends on the rate of change of C over time. In fact,

$$\Delta C = \left| \frac{dC}{dt} \right| \Delta t. \tag{5.29}$$

Although this resembles a relation between *changes* in C and t, we should remind ourselves that it is actually a relation between their uncertainties. This accounts for the absolute value function, since it does not matter whether C is increasing or decreasing. Strictly speaking, Eq. 5.29 holds only if the uncertainty ΔC is sufficiently small, but that will be good enough for our purposes.

Now consider the quantum mechanics of the situation. The clock is a quantum system whose evolution is given by a Hamiltonian H, which we assume is time-independent. The clock variable C is an observable on the quantum system represented by an operator C. We are interested in how the quantum uncertainty ΔC in the clock observable affects the precision of our time measurement. The quantum version of Eq. 5.29 is

$$\Delta C = \left| \frac{d\langle C \rangle}{dt} \right| \Delta t, \tag{5.30}$$

where $\langle C \rangle$ is the expectation value of the clock observable.

Both sides of Eq. 5.30 are mathematically related to the commutator of C with the Hamiltonian H. From the general uncertainty relation in Eq. 4.46, remembering that H is the energy observable, we find

$$\Delta C \, \Delta E \geq \frac{1}{2} \left| \langle \psi | \, [C, H] \, | \psi \rangle \right|. \tag{5.31}$$

Similarly, the rate of change of $\langle C \rangle$ is given by Eq. 5.27

$$\left| \frac{d\langle C \rangle}{dt} \right| = \frac{1}{\hbar} \left| \langle \psi | \, [C, H] \, | \psi \rangle \right|. \tag{5.32}$$

Combining these equations with Eq. 5.30, we discover that

$$\Delta E \, \Delta t \geq \frac{\hbar}{2}. \tag{5.33}$$

---

[3] For the moment, we do not inquire about the reason for this uncertainty – poor eyesight, perhaps!

This is called the *time–energy uncertainty relation*. Notice that the clock observable C is nowhere to be seen. Equation 5.33 is independent of the observable used for time measurement.

A clock in an energy eigenstate would be useless, since its observable properties would not change over time. A clock in a stationary state is a *stopped* clock. Thus, a working clock must have a "spread" in its energy values. Equation 5.33 gives us a quantitative trade-off between the system's energy indeterminacy and its possible resolution as a time indicator.

We have derived Eq. 5.33 for a "clock," but any physical system with a non-stationary state can function as a clock in our sense. The uncertainty Δt measures how long it takes for the system to evolve to a new state that is sufficiently distinguishable to indicate a new "clock time." (Problem 5.4 works out a simple example; Problem 5.5, a more complicated one.) The time–energy uncertainty relation is therefore a general, fundamental fact about quantum systems.

To get some idea of the scope of Eq. 5.33, consider an unstable particle such as a radioactive nucleus. The particle, which we can take to be at rest, decays into two or more other particles. Though the exact moment of decay cannot be predicted, the particle has a mean lifetime of τ. The decay process may be used as a sort of primitive clock, but because of the randomness of the decay, this clock has a time resolution of about $\Delta t \approx \tau$. This means that the energy of the system must satisfy $\Delta E \gtrsim \hbar/2\tau$. But the initial energy of the system is just $E = mc^2$, where m is the particle's rest mass and c is the speed of light. This means that the mass of an unstable particle is not a perfectly determinate quantity, but has an inescapable quantum uncertainty Δm related to its lifetime τ.

Exercise 5.11 The Δ^+ is a very short-lived elementary particle with a mean lifetime of 6×10^{-24} seconds. Find the uncertainty of the Δ^+ mass in MeV/c^2, and compare it to the Δ^+ mass of 1230 MeV/c^2. ("MeV" is a unit of energy equal to 1.60×10^{-13} J. The Δ^+ is closely related to the proton, whose mass is about 938 MeV/c^2.)

5.4 Operators and symmetries

Operators in the exponent

We now have two different ways of looking at the dynamical evolution of an informationally isolated system. Over an interval of time from t_1 to t_2 we can describe the evolution by a unitary operator:

$$|\psi(t_2)\rangle = \mathsf{U}(t_2, t_1)\,|\psi(t_1)\rangle. \tag{Re 5.3}$$

At any given moment t, we can describe the instantaneous change in the state by the Schrödinger equation:

$$\mathsf{H}\,|\psi(t)\rangle = i\hbar\,\frac{d}{dt}\,|\psi(t)\rangle. \tag{Re 5.23}$$

Let us take a closer look at the connection between these two descriptions. For convenience, we will assume uniform dynamics. The evolution operators can be written $U(t)$ and the Hamiltonian H is time-independent.

The first thing to note is that the Schrödinger equation actually applies to the operator function $U(t)$. The Schrödinger equation as we have seen it can be written

$$H\, U(t)\, |\psi(0)\rangle = i\hbar\, \frac{d}{dt} U(t)\, |\psi(0)\rangle . \tag{5.34}$$

Since the vector $|\psi(0)\rangle$ can be anything (but includes no time-dependence), we find an operator version of the Schrödinger equation:

$$H\, U(t) = i\hbar\, \frac{d}{dt} U(t). \tag{5.35}$$

This is of the general form

$$\frac{d}{dt}(\text{something}) = \text{constant} \times (\text{something}). \tag{5.36}$$

If the quantities involved were scalars, then we would know that the solution would be an exponential:

$$\text{something}(t) = \text{something}(0)\, \exp(\text{constant} \times t). \tag{5.37}$$

Recalling that $U(0) = \mathbf{1}$, we are therefore tempted to write down the solution to Eq. 5.35 as follows:

$$U(t) = \exp\left(-\frac{i}{\hbar}Ht\right). \tag{5.38}$$

But this looks rather weird. How can we *exponentiate* an operator?

There are some functions of operators that are easy to define. We can multiply operators by scalars and add them together. We can find any integer power A^n of an operator A. Using only this set of "natural" operations, we define the exponential function e^A by the *power series*

$$\exp(A) = e^A = \mathbf{1} + A + \frac{A^2}{2!} + \frac{A^3}{3!} + \cdots \tag{5.39}$$

In fact, for any function $f(x)$ given by a power series with coefficients a_n, we can define the operator function by a similar power series:

$$f(A) = a_0\mathbf{1} + a_1 A + a_2 A^2 + \cdots \tag{5.40}$$

There are issues of convergence for these series, just as there are for power series in the context of numerical variables. It is easiest to address these by computing the power series using a matrix representation of the operator. Assuming that the operator is normal, it must have an eigenbasis. In this basis, the representation of A is diagonal:

$$(A) = \begin{pmatrix} A_1 & & \\ & \ddots & \\ & & A_d \end{pmatrix}. \tag{5.41}$$

When we work out the power series for $f(A)$ in this representation, we find that

$$(f(A)) = \begin{pmatrix} f(A_1) & & \\ & \ddots & \\ & & f(A_d) \end{pmatrix}.$$ (5.42)

Exercise 5.12 Verify Eq. 5.42.

Exercise 5.13 If f is given by a power series, show that $[A, f(A)] = 0$. Must you assume that A is normal?

We can see that the power series definition of $f(A)$ will converge provided that the entire spectrum of A-eigenvalues is within the region of convergence of the power series for f acting on a numerical variable. (The power series for the exponential function converges everywhere.)

So it does after all make sense to write

$$U(t) = \exp\left(-\frac{i}{\hbar}Ht\right).$$ (Re 5.38)

Not only does this make sense, it is even the right answer! Let $\{|k\rangle\}$ be a basis of energy eigenstates. Then

$$H = \sum_k E_k |k\rangle\langle k|,$$ (5.43)

which as a matrix is diagonal with diagonal entries E_k. The exponential solution in Eq. 5.38 tells us that

$$U(t) = \sum_k e^{-iE_k t/\hbar} |k\rangle\langle k|.$$ (5.44)

This is a correct expression for $U(t)$, as a comparison with Eq. 5.20 quickly shows. We can therefore go from the Hamiltonian operator H to an expression for the time evolution operator $U(t)$ for the system.

The relation between unitary and Hermitian operators is quite general.

Exercise 5.14 If the operator B is Hermitian, show that $U = e^{iB}$ is unitary.

Exercise 5.15 If U is unitary, show that there is a Hermitian operator B so that $U = e^{iB}$.

(Equation 5.20 is helpful for both of these exercises.)

Before we go any further, we had better issue a warning. The operator exponential e^A does not have all of the familiar properties of the complex exponential function. For example, if two operators A and B do not commute with each other, then $e^A e^B \neq e^{A+B}$. On the other hand:

Exercise 5.16 If $[A, B] = 0$, show that $e^A e^B = e^{A+B}$.

We should be careful not to apply too quickly facts from *commutative* arithmetic to the analysis of non-commuting operators.

Let us make one final point about Eq. 5.38. If we consider a small time interval ϵ, the power series for $U(t)$ is

$$U(\epsilon) = 1 - \frac{i}{\hbar}H\epsilon + \mathcal{O}\left(\epsilon^2\right). \tag{5.45}$$

When ϵ is small enough, we may be able to ignore the higher-order terms represented by $\mathcal{O}\left(\epsilon^2\right)$. This is yet another way of looking at the connection between $U(t)$ and the Hamiltonian H.

Rotating a spin

The time evolution of our system is described by a smooth and continuous family of unitary operators $U(t)$, all of which are "generated" by the Hamiltonian H. But time evolution is only one kind of continuous transformation of the system. We will now apply the ideas we have developed to another example.

The example we have in mind is the *rotation* of a system, specifically a spin-1/2 particle. For every state $|\psi\rangle$ of the spin system, there is another state $|\psi'\rangle$ describing a situation in which the system has been rotated about the z-axis. The principle of superposition is respected by this rotation, as is the normalization of the states. Therefore, the rotation is described by an operator, and this operator must be unitary.

Exercise 5.17 Review the reasoning leading up to Eq. 5.5, and fill in the argument that the rotation operator is unitary.

Let $R_z(\theta)$ be the operator describing the rotation of the spin-1/2 system through an angle θ about the z-axis. (We consider the positive direction of θ to be a counter-clockwise rotation, when viewed from a point along the positive z-axis.) The rotation operators $R_z(\theta)$ form a smooth and continuous one-parameter family of unitary operators. Indeed, because we can apply smaller rotations in succession to obtain rotations through larger angles, the family should satisfy a "uniformity" requirement:

$$R_z(\alpha)R_z(\beta) = R_z(\alpha + \beta). \tag{5.46}$$

A close mathematical parallel is emerging between rotation and time evolution. But what plays the role of the Hamiltonian operator for these transformations? What is the "generator" of rotations?

If we rotate the system about the z-axis, then the z-component of the spin should not be affected. An eigenstate of S_z should not undergo any physical change. The basis states $|z_+\rangle$ and $|z_-\rangle$ are therefore the "stationary" states of the $R_z(\theta)$ family. They change only by overall phase factors, which by Eq. 5.46 must be linear functions of θ. (Compare the derivation of Eq. 5.19 for details.) We write

$$R_z(\theta) |z_+\rangle = e^{i\Lambda_+ \theta} |z_+\rangle,$$
$$R_z(\theta) |z_-\rangle = e^{i\Lambda_- \theta} |z_-\rangle. \tag{5.47}$$

We do not yet know the constant scalars $\Lambda_{\pm}$. We can begin to determine them, however, by noting that a rotation by $\pi/2$ will transform the $|x_+\rangle$ state to a multiple of $|y_+\rangle$:

$$R_z(\theta)\,|x_+\rangle = \frac{1}{\sqrt{2}}\left(e^{i\Lambda_+ \pi/2}\,|z_+\rangle + e^{i\Lambda_- \pi/2}\,|z_-\rangle\right). \tag{5.48}$$

For this to be a multiple of $|y_+\rangle = \dfrac{1}{\sqrt{2}}(\,|z_+\rangle + i\,|z_-\rangle)$, the constants $\Lambda_{\pm}$ must satisfy

$$e^{i\Lambda_- \pi/2} = e^{i(\Lambda_+ + 1)\pi/2}. \tag{5.49}$$

We have some freedom of choice here, corresponding to an overall choice of phase for the operator $R_z(\theta)$. The simplest choice is the "symmetric" one, in which $\Lambda_+ = -\Lambda_-$. Then $\Lambda_{\pm} = \mp 1/2$, and so

$$R_z(\theta) = e^{-i\theta/2}\,|z_+\rangle\langle z_+| + e^{i\theta/2}\,|z_-\rangle\langle z_-|. \tag{5.50}$$

Exercise 5.18 Show that $R_z(\pi) = -iZ$.

Exercise 5.19 Show that, when a spin is rotated by one full turn around the z-axis ($\theta = 2\pi$), then

$$R_z(2\pi) = e^{-i2\pi S_z/\hbar} = -1. \tag{5.51}$$

In other words, a spin state $|\psi\rangle$, rotated by 2π, will yield the spin state $-|\psi\rangle$.

This may seem like an artifact of our choice of phase for the rotation operators $R_z(\theta)$, but Problem 5.7 tells us that it is more or less inescapable. The overall phase of a state vector has no observational meaning in itself, so the negative sign in Eq. 5.51 seems to be of only mathematical interest. Nevertheless, it turns out to have some remarkable physical implications, which are discussed in Section 6.3.

What we have done so far corresponds to our "transformation over an interval of time" picture of time evolution. Now let us see if we can develop something analogous to the Schrödinger equation, the "instantaneous rate of change" picture. Let $|\psi(\theta)\rangle$ be the family of rotated versions of the state $|\psi(0)\rangle$:

$$|\psi(\theta)\rangle = R_z(\theta)\,|\psi(0)\rangle. \tag{5.52}$$

We find that

$$\frac{d}{d\theta}\,|\psi(\theta)\rangle = \frac{d}{d\theta}R_z(\theta)\,|\psi(0)\rangle$$
$$= \left(-\frac{i}{2}\,e^{-i\theta/2}\,|z_+\rangle\langle z_+| + \frac{i}{2}\,e^{i\theta/2}\,|z_-\rangle\langle z_-|\right)|\psi(0)\rangle. \tag{5.53}$$

If we multiply both sides by $i\hbar$ to make it look more like the Schrödinger equation, we obtain

$$i\hbar\frac{d}{d\theta}\,|\psi(\theta)\rangle = \left(\frac{\hbar}{2}\,|z_+\rangle\langle z_+| - \frac{\hbar}{2}\,|z_-\rangle\langle z_-|\right)$$
$$\times \left(e^{-i\theta/2}\,|z_+\rangle\langle z_+| + e^{i\theta/2}\,|z_-\rangle\langle z_-|\right)|\psi(0)\rangle. \tag{5.54}$$

We recognize both of the operators in parentheses. Our "Schrödinger equation for rotation" becomes

$$i\hbar\frac{d}{d\theta}\,|\psi(\theta)\rangle = \mathsf{S}_z\,|\psi(\theta)\rangle. \tag{5.55}$$

The spin component operator S_z generates rotations about the z-axis exactly as the Hamiltonian H generates time evolution. We can also write

$$\mathsf{R}_z(\theta) = e^{-i\mathsf{S}_z\theta/\hbar}. \tag{5.56}$$

Note that, if we rotate the system through a tiny angle ϵ, the rotation operator is

$$\mathsf{R}_z(\epsilon) = \mathbf{1} - \frac{i}{\hbar}\mathsf{S}_z\epsilon + \mathcal{O}\left(\epsilon^2\right). \tag{5.57}$$

For very small angles, we can ignore the $\mathcal{O}\left(\epsilon^2\right)$ terms.

There is nothing special about the z-axis. This link between the spin component and rotation operators holds for any spatial axis. Furthermore, the link holds for systems of spin-1 and higher spins as well. The Schrödinger equation tells us that energy (the Hamiltonian operator) is the generator of time evolution. Now we also learn that angular momentum is the generator of rotations. (We will return to this idea in Section 12.3.)

Symmetries

So far, operators have played a double role. Hermitian operators describe observable quantities assigned to a system; unitary operators describe transformations of the state of the system, including the transformation due to time evolution. As we saw in Exercises 5.14 and 5.15, these two types of operator are related by the exponential function:

$$\text{unitary} = e^{i\,\text{Hermitian}}. \tag{5.58}$$

We have already seen this in the case of time evolution and rotation, where the unitary state transformations were given by the Hamiltonian H and the spin component operator S_z.

A *symmetry* (or *dynamical symmetry*) is a unitary transformation on a system that does not affect the system's dynamics. That is, a unitary V is a symmetry provided it commutes with the time evolution of the system. For all t,

$$[\mathsf{V}, \mathsf{U}(t)] = 0. \tag{5.59}$$

Let us explore what this means. We consider two possible initial states, $|\psi(0)\rangle$ and its transformed cousin $\mathsf{V}\,|\psi(0)\rangle$. The symmetry condition tells us that

$$\mathsf{U}(t)\Big(\mathsf{V}\,|\psi(0)\rangle\Big) = \mathsf{V}\Big(\mathsf{U}(t)\,|\psi(0)\rangle\Big). \tag{5.60}$$

In words, "the future evolution of the transformed state is the same as the transformed future state." In other words, the dynamical evolution of the system is "transparent" to the symmetry operator V, evolving $|\psi(0)\rangle$ and $\mathsf{V}\,|\psi(0)\rangle$ in exactly the same way.

A dynamical symmetry V must also commute with the Hamiltonian:

$$[\mathsf{V}, \mathsf{H}] = 0. \tag{5.61}$$

In fact, Eq. 5.59 and 5.61 are equivalent.

Exercise 5.20 (a) Obtain Eq. 5.61 from Eq. 5.59 by taking a time derivative. (b) Obtain Eq. 5.59 from Eq. 5.61 by examining the power series for the operator exponential.

If V is a dynamical symmetry, then we can always find a basis of stationary states (energy eigenstates) that are also eigenstates of V. These are "symmetric" states, states that remain physically unchanged by the transformation V.

Symmetry operators V are sometimes also observables, but the conditions under which this can happen are very special.

Exercise 5.21 If V is both unitary and Hermitian, show that the only possible eigenvalues for V are ± 1 and that $V^2 = 1$. (For extra fun, show that any unitary V for which $V^2 = 1$ must be Hermitian.)

Most of the time, the connection between symmetries and observables is more indirect.

Suppose that A is a Hermitian operator for which $[A, H] = 0$. Then both of the following are true:

- A is a conserved quantity;
- $V(s) = e^{isA}$ is a symmetry for any real value of s.

Therefore, a conserved quantity generates a family of dynamical symmetries of the system.

For an example of this, we need look no further than the rotation of a spin system. The z-component of spin angular momentum will be conserved provided $[S_z, H] = 0$. In this case, the rotation operator

$$R_z(\theta) = e^{-iS_z\theta/\hbar}, \tag{Re 5.56}$$

is a symmetry of the system. The system is *rotationally invariant* about the z-axis.

The converse is also true. Suppose we have a smooth and continuous one-parameter family $V(s)$ of unitary operators for which

$$V(r)\,V(s) = V(r + s). \tag{5.62}$$

Then we can always write $V(s) = e^{isA}$ for some Hermitian A. For a very small value ϵ of the parameter, we have that

$$V(\epsilon) = 1 + i\epsilon A + \mathcal{O}\left(\epsilon^2\right). \tag{5.63}$$

Now suppose that all of the transformations $V(s)$ in the family are symmetries, so that $[V(s), H] = 0$. For the very small parameter ϵ, ignoring terms of order $\mathcal{O}\left(\epsilon^2\right)$,

$$0 = [V(\epsilon), H] = [1 + i\epsilon A, H] = i\epsilon\,[A, H]. \tag{5.64}$$

It follows that $[A, H] = 0$. Therefore, the observable represented by A is conserved in the system.

The connection between symmetries and conserved quantities is now apparent. Every conserved quantity generates a family of symmetry transformations on the system. Every smooth and continuous family of symmetries satisfying Eq. 5.62 is generated by a conserved quantity. Whenever an observable satisfies $[A, H] = 0$, there is a basis of energy eigenstates that are also A-eigenstates. These states are also *symmetric* under $V(s) = e^{isA}$, in that the transformation yields no observable change in the state. All of these ideas will be immensely useful in the chapters ahead.

Exercise 5.22 It is certainly *not* the case that if a unitary operator V satisfies $[V, H] = 0$, and if we can write $V = e^{iB}$ for some Hermitian B, then B also commutes with H. To see why, show that

$$e^{i 2\pi X} = 1,   \tag{5.65}$$

for the Pauli operator X on $\mathcal{Q}$. Although **1** is a unitary operator that certainly commutes with H, $2\pi X$ does not necessarily do so.

Problems

Problem 5.1 In Equation 5.22 we make an implicit assumption that the "product rule" for derivatives applies to the inner product of two Hilbert space vectors. Prove it. Let $|\psi(t)\rangle$ and $|\phi(t)\rangle$ be a pair of vectors that depend smoothly on the parameter t. (Do *not* assume that this is unitary time evolution!) By writing each vector in terms of a basis $\{|k\rangle\}$ that does not depend on t, show that the product rule applies to $\dfrac{d}{dt}\langle\psi|\phi\rangle$.

Problem 5.2 The commutation relations for the spin components of a spin-1/2 particle are

$$[S_x, S_y] = i\hbar S_z,   \qquad [S_y, S_z] = i\hbar S_x,   \qquad [S_z, S_x] = i\hbar S_y.$$

The Hamiltonian for the spin in an external magnetic field B in the z-direction is $H = -\gamma BS_z$. Find a set of three coupled differential equations that relate $\langle S_x\rangle$, $\langle S_y\rangle$, and $\langle S_z\rangle$ as functions of t.

Problem 5.3 We have described time evolution in the *Schrödinger picture*, in which a state $|\psi(t)\rangle$ evolves according to the unitary operator $U(t)$ but observables A are typically time-independent. An equally meaningful view is the *Heisenberg picture*, which redefines states and operators by

$$\left|\hat{\psi}\right\rangle = U(t)^\dagger |\psi(t)\rangle \qquad \text{and} \qquad \hat{A}(t) = U(t)^\dagger A U(t).   \tag{5.66}$$

(Assume uniform dynamics. What is $\hat{H}$?)

(a) Show that $\left|\hat{\psi}\right\rangle$ is independent of time.
(b) The change from one picture to the other preserves the mathematical relations among operators. Sums and scalar multiples are clearly unaffected. You prove that $AB = C$ if and only if $\hat{A}\hat{B} = \hat{C}$.
(c) Show that observable properties – i.e. the expectations of all observables at all times – are the same in the two pictures.

(d) Since all of the dynamics now resides in the observables, derive a "Schrödinger equation" governing $\hat{A}(t)$.

(e) What does it mean for $\hat{A}(t_1)$ and $\hat{A}(t_2)$ not to commute with each other? Explain this in physical (rather than mathematical) terms.

(f) Show that

$$[\hat{A}(t+\delta), \hat{A}(t)] = \frac{\delta}{i\hbar}[\hat{A}(t)^2, \hat{H}] + \mathcal{O}\left(\delta^2\right). \tag{5.67}$$

Problem 5.4 Consider a qubit used as a clock. The Hamiltonian for the system is H $=$ $\hbar\Omega\,|1\rangle\langle1|$, where Ω is some constant frequency.

(a) What are the energy eigenstates and their energies?

(b) Write down the time evolution operator U(t) for this system.

(c) We prepare the clock in an initial state $|\psi(0)\rangle = \frac{1}{\sqrt{2}}(|0\rangle + |1\rangle)$. What is the state $|\psi(t)\rangle$ at a later time?

(d) What is ΔE for this system?

(e) Let Δt denote the time it takes for the system to evolve from $|\psi(0)\rangle$ to some other distinguishable state. Calculate $\Delta E\,\Delta t$ and compare your result to the time–energy uncertainty relation in Eq. 5.33.

Problem 5.5 An atom has basis states $|0\rangle$, $|1\rangle$, ..., $|d-1\rangle$ and a Hamiltonian operator such that

$$H\,|n\rangle = n\varepsilon\,|n\rangle,$$

where ε is a constant with units of energy. Let the initial state of the atom be

$$|\psi(0)\rangle = \frac{1}{\sqrt{d}}\sum_n |n\rangle.$$

(a) Calculate $\langle E\rangle$ and ΔE for this state.

(b) Find an explicit expression for $|\psi(t)\rangle$.

(c) Let Δt be the smallest length of time for which $\langle\psi(\Delta t)|\psi(0)\rangle = 0$. That is, Δt is the time it takes for the initial state to evolve into something distinguishable from the initial state. Find Δt.

(d) Under this definition of Δt, calculate $\Delta E\,\Delta t$ and compare it to the time–energy uncertainty relation in Eq. 5.33.

Problem 5.6 The operator describing a rotation about the x-axis will be $R_x(\theta) =$ $\exp(-iS_x\theta/\hbar)$. Starting with

$$S_x = \frac{\hbar}{2}(\,|z_+\rangle\langle z_-| + |z_-\rangle\langle z_+|),$$

calculate S_x^2, S_x^3, and so on, until you figure out the general pattern for S_x^n. Then use this to evaluate the power series and find $R_x(\theta)$. Note: You may need to recall that

$$\cos\theta = 1 - \frac{\theta^2}{2!} + \frac{\theta^4}{4!} - \cdots \qquad \sin\theta = \theta - \frac{\theta^3}{3!} + \frac{\theta^5}{5!} - \cdots$$

Problem 5.7 In Exercise 5.19, an unexpected negative sign arose in a rotation by 2π about the z-axis. But perhaps this was merely due to some unwise choice on our part. The purpose of this problem is to suggest that, on the contrary, negative signs inevitably arise in rotation of spin-1/2 systems.

(a) Suppose the operator R describes a rotation by $\pi/2$ about the y-axis. We might hope that such an operator would change the spin states like so:

$$R\,|z_+\rangle = |x_+\rangle, \quad R\,|x_+\rangle = |z_-\rangle,$$
$$R\,|z_-\rangle = |x_-\rangle, \quad R\,|x_-\rangle = |z_+\rangle, \tag{5.68}$$

with no negative signs. Given the relationship between $|x_\pm\rangle$ and $|z_\pm\rangle$ (Eq. 3.14), show that no linear operator R could act in this way.

(b) Suppose instead that the operator R acts by

$$R\,|z_+\rangle = |x_+\rangle, \quad R\,|x_+\rangle = \alpha\,|z_-\rangle,$$
$$R\,|z_-\rangle = \beta\,|x_-\rangle, \quad R\,|x_-\rangle = \gamma\,|z_+\rangle, \tag{5.68}$$

with various phase factors α, β, and γ. (We have chosen the overall phase of R to eliminate any phase factor in the first rotation.) Find all of the phase factors and show that $R^4\,|z_+\rangle = -\,|z_+\rangle$.

Problem 5.8 For operators A and B, we saw that $e^A e^B \neq e^{A+B}$ unless $C = [A, B] = 0$. Now suppose merely that the commutator C commutes with both A and B, though it may not be zero. Prove that

$$e^A e^B = e^{A+B}\,e^{C/2}. \tag{5.68}$$

You will need to examine the power series expansions for each side. Equation 5.68 is sometimes called the *Baker–Campbell–Hausdorff identity* (although this name is also applied to a more general expression of which this is a special case).

Entanglement

6.1 Composite systems

In this chapter we will discuss the quantum theory of *composite* systems, systems composed of two or more distinct subsystems. When our theory is extended to these, we encounter the remarkable phenomenon of quantum entanglement. As we shall see, the statistical correlations between entangled quantum systems may be very different from the correlations that are possible between separated classical systems.

Suppose the composite quantum system AB is composed of two distinct subsystems A and B. There are certainly situations in which the subsystems A and B can individually be assigned definite quantum state vectors $|\psi^{(A)}\rangle$ and $|\phi^{(B)}\rangle$. This would occur, for instance, if A and B were prepared separately and independently. In this situation, we describe the joint state of the composite system simply by enumerating the states of the subsystems, like so: $|\psi^{(A)}, \phi^{(B)}\rangle$.

There are many different measurement procedures that we can perform on the composite system AB. Among these are separate basic measurements on the individual subsystems. We know how to calculate probabilities for the outcomes of such measurements. If $|a^{(A)}\rangle$ and $|b^{(B)}\rangle$ are basis vectors associated with the outcomes a and b of measurements on A and B respectively, then the probabilities are

$$P^{(A)}(a) = \left| \langle a^{(A)} | \psi^{(A)} \rangle \right|^2,$$
$$P^{(B)}(b) = \left| \langle b^{(B)} | \phi^{(B)} \rangle \right|^2, \tag{6.1}$$

(see Eq. 3.12). If the systems have been prepared completely independently, we expect that there would be no correlation between measurement results on them. Therefore, the probability that both outcome a and outcome b will occur is

$$P^{(AB)}(a, b) = P^{(A)}(a) P^{(B)}(b). \tag{6.2}$$

How can we express this as a quantum mechanical rule? The state vectors $|a^{(A)}\rangle$ and $|b^{(B)}\rangle$ for the subsystems yield the joint state $|a^{(A)}, b^{(B)}\rangle$. Our quantum probability rule (Eq. 3.12) says that

$$P^{(AB)}(a, b) = \left| \langle a^{(A)}, b^{(B)} | \psi^{(A)}, \phi^{(B)} \rangle \right|^2. \tag{6.3}$$

We can reconcile Eq. 6.2 and 6.3 by requiring that the inner product of the joint states is simply

$$\langle a^{(A)}, b^{(B)} | \psi^{(A)}, \phi^{(B)} \rangle = \langle a^{(A)} | \psi^{(A)} \rangle \langle b^{(B)} | \phi^{(B)} \rangle. \tag{6.4}$$

Exercise 6.1 Show that Eq. 6.2 follows from Eq. 6.3 and 6.4.

The principle of superposition tells us that a linear combination of state vectors is also an allowable state vector. Thus, there must be composite system states of the form

$$\left|\Psi^{(AB)}\right\rangle = \alpha_1 \left|\psi_1^{(A)}, \phi_1^{(B)}\right\rangle + \alpha_2 \left|\psi_2^{(A)}, \phi_2^{(B)}\right\rangle, \tag{6.5}$$

and so on. As we will see, such state vectors may not be of the form $|\psi^{(A)}, \phi^{(B)}\rangle$ for any subsystem states $|\psi^{(A)}\rangle$ and $|\phi^{(B)}\rangle$. They do nevertheless represent possible quantum states – actual physical situations – for the composite system AB.

With this discussion as background, we can now give a more formal account of the states of composite quantum systems. The subsystems A and B are described by Hilbert spaces $\mathcal{H}^{(A)}$ and $\mathcal{H}^{(B)}$. The Hilbert space for the composite system is $\mathcal{H}^{(AB)}$, which is constructed from $\mathcal{H}^{(A)}$ and $\mathcal{H}^{(B)}$. We write

$$\mathcal{H}^{(AB)} = \mathcal{H}^{(A)} \otimes \mathcal{H}^{(B)}. \tag{6.6}$$

This is the *tensor product* of the subsystem Hilbert spaces, defined by the following properties:

- For any $|a\rangle$ in $\mathcal{H}^{(A)}$ and $|b\rangle$ in $\mathcal{H}^{(B)}$, the space $\mathcal{H}^{(AB)}$ contains the *product vector* $|a, b\rangle = |a\rangle \otimes |b\rangle$.[1]
- Linear combinations distribute over the product operation $\otimes$. For example,

$$|a\rangle \otimes \Big(c_1 |b_1\rangle + c_2 |b_2\rangle\Big) = c_1 \Big(|a\rangle \otimes |b_1\rangle\Big) + c_2 \Big(|a\rangle \otimes |b_2\rangle\Big). \tag{6.7}$$

(A similar property holds for linear combinations in the first vector.)
- The inner product between two product vectors is the product of the inner products of the constituent vectors. This sounds complicated in words, but is simple in symbols:

$$\langle a_1, b_1 | a_2, b_2 \rangle = \langle a_1 | a_2 \rangle \langle b_1 | b_2 \rangle. \tag{6.8}$$

The inner products on the right-hand side, of course, are those in $\mathcal{H}^{(A)}$ and $\mathcal{H}^{(B)}$. We will often write $\langle a, b| = \langle a| \otimes \langle b|$ to denote the adjoint of a product vector.
- Since $\mathcal{H}^{(AB)}$ must be a vector space, it contains arbitrary superpositions of product vectors. Our final defining property for the tensor product space is that $\mathcal{H}^{(AB)}$ contains *only* these vectors, i.e. that any $|\Psi^{(AB)}\rangle$ in $\mathcal{H}^{(AB)}$ is either a product vector or a linear combination of product vectors.

To get some facility with the idea of a tensor product space, you should work out the following exercises:

Exercise 6.2 Is $(\eta |a\rangle) \otimes |b\rangle$ the same vector as $|a\rangle \otimes (\eta |b\rangle)$, for a scalar η? Explain.

---

[1] Note that we use the same symbol $\otimes$ for the combination of two vectors that we use for the tensor product of two Hilbert spaces. We will use it again for the tensor product of two operators. Luckily, this never leads to confusion, just as we are never bothered by using the same symbol $+$ to represent the addition of scalars, vectors, operators, etc. Context always tells us what sort of operation we mean.

Notice also that we sometimes omit the part of the label that indicates to which system a state, operator, or probability distribution refers. We will simplify our notation in this way when our meaning is unmistakable.

Exercise 6.3 If $|a_1\rangle$ is orthogonal to $|a_2\rangle$ in $\mathcal{H}^{(A)}$, show that $|a_1, b_1\rangle$ is orthogonal to $|a_2, b_2\rangle$ in $\mathcal{H}^{(AB)}$. Does this still hold if $|b_1\rangle = |b_2\rangle$?

Exercise 6.4 Suppose $\{|n\rangle\}$ is a basis for $\mathcal{H}^{(A)}$ and $\{|k\rangle\}$ is a basis for $\mathcal{H}^{(B)}$. Consider the set $\{|n, k\rangle\}$ that includes all products $|n, k\rangle = |n\rangle \otimes |k\rangle$ of basis vectors. Show that this set of vectors is a basis for $\mathcal{H}^{(AB)}$. To do this, you must show (1) that the set is orthonormal, and (2) that any vector in $\mathcal{H}^{(AB)}$ can be written as a linear combination of members of the set. The basis set formed in this way is called a *product basis* for $\mathcal{H}^{(AB)}$.

Exercise 6.5 Suppose $\dim \mathcal{H}^{(A)} = d^{(A)}$ and $\dim \mathcal{H}^{(B)} = d^{(B)}$. Show that $\dim \mathcal{H}^{(AB)} = d^{(A)} d^{(B)}$.

Exercise 6.6 You might be tempted to think that the Hilbert space $\mathcal{H}^{(A)}$ for the *subsystem* A is a *subspace* of the composite system Hilbert space $\mathcal{H}^{(AB)}$. Explain, as clearly as possible, why you must resist this temptation!

What are the operators on $\mathcal{H}^{(AB)}$? Given operators F on $\mathcal{H}^{(A)}$ and G on $\mathcal{H}^{(B)}$, we can naturally define the *product operator* $F \otimes G$ on $\mathcal{H}^{(AB)}$ by

$$(F \otimes G) |\psi, \phi\rangle = \left(F |\psi\rangle\right) \otimes \left(G |\phi\rangle\right), \tag{6.9}$$

for any product state[2] $|\psi, \phi\rangle = |\psi^{(A)}\rangle \otimes |\phi^{(B)}\rangle$. Linear combinations of operators distribute over the $\otimes$-product for operators.

Exercise 6.7 Given $|a\rangle$, $|c\rangle$ in $\mathcal{H}^{(A)}$ and $|b\rangle$, $|d\rangle$ in $\mathcal{H}^{(B)}$, show that

$$|a, b\rangle\langle c, d| = |a\rangle\langle c| \otimes |b\rangle\langle d|. \tag{6.10}$$

Exercise 6.8 Show that any operator D on $\mathcal{H}^{(AB)}$ can be written as a sum of product operators. To do this, first choose a product basis $\{|n, k\rangle\}$ for $\mathcal{H}^{(AB)}$ and write D as a sum of outer products, as in Eq. 3.41.

When particular clarity is needed, we may denote the operator F on $\mathcal{H}^{(A)}$ by $F^{(A)}$. There is a natural extension of F to an operator on $\mathcal{H}^{(AB)}$ that acts according to

$$|\psi, \phi\rangle \longrightarrow \left(F |\psi\rangle\right) \otimes |\phi\rangle. \tag{6.11}$$

This is just the operator $F \otimes \mathbf{1}$. For brevity, and when our meaning is unmistakeable, we will sometimes denote this AB operator by the same name $F^{(A)}$.

Exercise 6.9 Show that $[F^{(A)}, G^{(B)}] = 0$ for any operators for distinct subsystems A and B. (To make sense of this equation, of course, you will have to read $F^{(A)}$ and $G^{(B)}$ as operators on the tensor product space $\mathcal{H}^{(AB)}$.)

The states of the composite system AB are normalized vectors in $\mathcal{H}^{(AB)}$. Some of these are $\otimes$-products of normalized state vectors for A and B individually. We call these *product states*. However, some quantum states of AB are not product states, as we now show.

---

[2] Since every vector in $\mathcal{H}^{(AB)}$ is a superposition of product vectors, it is sufficient to define how $F \otimes G$ acts on product vectors – everything else follows by linearity.

We can write any joint state $|\Psi^{(AB)}\rangle$ using a product basis, like so:

$$|\Psi^{(AB)}\rangle = \sum_{n,k} \alpha_{nk} |n\rangle \otimes |k\rangle = \sum_n |n\rangle \otimes \left(\sum_k \alpha_{nk} |k\rangle \right). \qquad (6.12)$$

Thus, $|\Psi^{(AB)}\rangle$ can always be written

$$|\Psi^{(AB)}\rangle = \sum_n |n\rangle \otimes |\phi_n\rangle, \qquad (6.13)$$

where the $|\phi_n\rangle$ are vectors in $\mathcal{H}^{(B)}$ that need not be normalized. For a product state, these vectors have a special form. Let $|\Psi^{(AB)}\rangle = |\psi\rangle \otimes |\phi\rangle$, and expand the A-state using the $\{|n\rangle\}$ basis:

$$|\Psi^{(AB)}\rangle = \sum_n |n\rangle \otimes \left(\alpha_n |\phi\rangle \right). \qquad (6.14)$$

This has the same form as Eq. 6.13 with $|\phi_n\rangle = \alpha_n |\phi\rangle$. Each of the B-vectors in this expansion is a multiple of the same B-state $|\phi\rangle$.

Exercise 6.10 Show the converse to this. That is, show that, if $|\phi_n\rangle = \alpha_n |\phi\rangle$ for all n in Eq. 6.13, then $|\Phi^{(AB)}\rangle$ must be a product state.

This gives us a simple criterion for identifying which joint states are product states and which are not. We simply choose a basis for $\mathcal{H}^{(A)}$ and write the state according to Eq. 6.13. We have a product state if and only if the various $|\phi_n\rangle$ vectors are all multiples of a single vector.

Exercise 6.11 For what value α is $K\left(2|0,0\rangle + 3|0,1\rangle - |1,0\rangle + \alpha|1,1\rangle\right)$ a product state of two qubits? (K is a normalization constant.)

Not all states of the composite system AB are product states. For instance, consider the state of two qubits

$$|\Phi_+^{(AB)}\rangle = \frac{1}{\sqrt{2}} \left(|0,0\rangle + |1,1\rangle \right). \qquad (6.15)$$

This is already of the form in Eq. 6.13, but the B-vectors $|0\rangle$ and $|1\rangle$ are linearly independent. Therefore, $|\Phi_+^{(AB)}\rangle$ cannot be a product state.

The physical states for the composite system AB include some states that are not product states. These non-product states are called *entangled states* of AB. The existence of entangled states is a remarkable consequence of quantum theory. Remember, whenever state vectors can be assigned to the subsystems A and B, the composite system is in a product state. If AB is in an entangled state, *it is not possible to assign state vectors to the individual subsystems*.

This is very different from classical physics. When we specify the joint state of a pair of classical particles, we must always specify the classical state (position and momentum) of each individual particle. The quantum state vector, however, belongs only to an entire system, and it is not always true that the subsystems have definite state vectors of their own.

Quantum entanglement is one of the strangest and most far-reaching aspects of quantum theory. Later in this chapter, we will explore some of its implications.

6.2 Interaction and entanglement

The principle of superposition, applied to the states of a composite system, implies the possibility of entanglement between systems. But how do entangled states arise in real physical situations? The short answer is that entanglement is generally the result of dynamical *interaction*.

What do we mean by interaction? It is easier to define the opposite. A pair of quantum systems A and B is said to be *non-interacting* if the Hamiltonian of the composite system AB is just the sum of subsystem Hamiltonians:

$$H^{(AB)} = H^{(A)} + H^{(B)}. \tag{6.16}$$

We can draw several conclusions about non-interacting systems, including these facts:

- We can find an energy eigenbasis for the system AB that is composed of product states $|\alpha, \beta\rangle$, where $|\alpha\rangle$ is an eigenstate of $H^{(A)}$ in $\mathcal{H}^{(A)}$ and $|\beta\rangle$ is an eigenstate of $H^{(B)}$ in $\mathcal{H}^{(B)}$.
- The energy of a product eigenstate is additive:

$$E_{\alpha\beta} = E_\alpha + E_\beta, \tag{6.17}$$

where E_α is the energy of $|\alpha\rangle$ for A and E_β is the energy of $|\beta\rangle$ for B.

Exercise 6.12 Verify these results.

We sometimes say that non-interacting systems are *dynamically isolated* or *uncoupled* from each other. Two systems are *interacting* if they are not non-interacting.

If A and B are non-interacting, then they cannot become entangled simply by their own autonomous time evolution. We can show this through a sequence of simple exercises:

Exercise 6.13 Show that $e^{A \otimes 1} = \left(e^A\right) \otimes 1$.

Exercise 6.14 If A and B are non-interacting, show that

$$\exp\left(-iH^{(AB)}t/\hbar\right) = \exp\left(-iH^{(A)}t/\hbar\right) \otimes \exp\left(-iH^{(B)}t/\hbar\right). \tag{6.18}$$

(Hint: Use the result from Exercise 5.16.)

Exercise 6.15 Suppose the time evolution of a composite system AB is described by

$$U^{(AB)} = U^{(A)} \otimes U^{(B)}. \tag{6.19}$$

(Exercise 6.14 tells us that this must be the case whenever A and B are dynamically isolated from each other.) If AB is initially in a product state, show that it remains in a product state after the time evolution.

In other words, if the Hamiltonian for AB is additive, and if the two systems are prepared independently, then they will remain in a product state. For entanglement to arise from a product state, the systems must interact.

We illustrate this by a definite example. Suppose two spin-1/2 systems interact with each other so that their Hamiltonian has the form

$$H^{(AB)} = \varepsilon \, Z \otimes Z, \tag{6.20}$$

where ε is a constant that determines the strength of the interaction, and Z is the Pauli spin operator. Even though these spins are interacting, the energy eigenstates turn out to be product states.

Exercise 6.16 Show that $|z_+, z_+\rangle$ and $|z_-, z_-\rangle$ are both eigenstates of $H^{(AB)}$ with eigenvalue ε. What is the eigenvalue associated with the eigenstates $|z_+, z_-\rangle$ and $|z_-, z_+\rangle$?

These are all stationary states, so they cannot evolve into entangled states over time. However, suppose we begin with the product state

$$
\begin{aligned}
\left|\Psi^{(AB)}(0)\right\rangle &= |x_+, x_+\rangle \\
&= \frac{1}{2}\left(|z_+\rangle + |z_-\rangle\right) \otimes \left(|z_+\rangle + |z_-\rangle\right) \\
&= \frac{1}{2}\left(|z_+, z_+\rangle + |z_+, z_-\rangle \right. \\
&\qquad \left. + |z_-, z_+\rangle + |z_-, z_-\rangle\right).
\end{aligned}
\tag{6.21}
$$

If we allow the Hamiltonian $H^{(AB)}$ to act for a time $t = \pi\hbar/4\varepsilon$, then the evolution operator is

$$
U^{(AB)}(t) = \exp\left(-i\frac{\pi}{4} Z \otimes Z\right).
\tag{6.22}
$$

The later state will be

$$
\begin{aligned}
\left|\Psi^{(AB)}(t)\right\rangle &= U^{(AB)}(t)\left|\Psi^{(AB)}(0)\right\rangle \\
&= \frac{1}{2}\left(e^{-i\pi/4}|z_+, z_+\rangle + e^{i\pi/4}|z_+, z_-\rangle \right. \\
&\qquad \left. + e^{i\pi/4}|z_-, z_+\rangle + e^{-i\pi/4}|z_-, z_-\rangle\right).
\end{aligned}
\tag{6.23}
$$

With a little algebra, this becomes

$$
\left|\Psi^{(AB)}(t)\right\rangle = \frac{e^{-i\pi/4}}{2}\left(|z_+\rangle \otimes \left(|z_+\rangle + i|z_-\rangle\right) + i|z_-\rangle \otimes \left(|z_+\rangle - i|z_-\rangle\right)\right).
\tag{6.24}
$$

Exercise 6.17 Verify this algebra. Simplify further to show that

$$
\left|\Psi^{(AB)}(t)\right\rangle = \frac{1-i}{2}\left(|z_+, y_+\rangle + i|z_-, y_-\rangle\right).
\tag{6.25}
$$

The final state $|\Psi^{(AB)}(t)\rangle$ is clearly an entangled state of AB.

Exercise 6.18 Suppose we allowed the time evolution to run for a total time $2t = \pi\hbar/2\varepsilon$, just twice as long as before. Is the state $|\Psi^{(AB)}(2t)\rangle$ entangled?

One nice feature of this example is that the Hamiltonian $\varepsilon Z \otimes Z$ actually arises as the interaction between nearby nuclear spins for a molecule in an NMR experiment. As we will see in Chapter 18, this allows us to create entangled states of the spins and also to perform many types of quantum information processing.

Two systems need not interact directly to become entangled; it is enough if they both interact with a third system. To give an example of this, and to show how ubiquitous entanglement is, we will return to our archetypical example, the two-beam interferometer.

In our highly simplified model, the interferometer has basis states $|0\rangle$, $|u\rangle$, and $|l\rangle$, corresponding to physical situations with no photon or with one photon in the upper or lower beams. We will now introduce a pair of two-level atoms, one in each beam. Each atom has a ground state $|g\rangle$ and an excited state $|e\rangle$. We suppose that the overall unitary dynamics of the system produce the following state changes:

$$|0\rangle \otimes |g,g\rangle \rightarrow |0\rangle \otimes |g,g\rangle ,$$

$$|u\rangle \otimes |g,g\rangle \rightarrow |0\rangle \otimes |e,g\rangle ,$$

$$|l\rangle \otimes |g,g\rangle \rightarrow |0\rangle \otimes |g,e\rangle . \tag{6.26}$$

In words, if there are no photons, no atoms become excited. A photon in the upper beam is absorbed by the first atom, which becomes excited. A photon in the lower beam does the same thing to the second atom. The atoms do not interact with each other, but both of them are coupled to the light in the interferometer.

Now suppose that we prepare the interferometer in the superposition state $\alpha |u\rangle + \beta |l\rangle$, while the atoms are in their ground states. Then

$$\left(\alpha |u\rangle + \beta |l\rangle\right) \otimes |g,g\rangle \longrightarrow |0\rangle \otimes \left(\alpha |e,g\rangle + \beta |g,e\rangle\right). \tag{6.27}$$

The two atoms are now in the entangled state $\alpha |e,g\rangle + \beta |g,e\rangle$, even though they have not directly interacted.

We have assumed here that the time evolution is unitary. But is this reasonable? Do not the atoms "measure" the location of the photon in the interferometer? Not necessarily. As long as the composite system (interferometer + atoms) remains informationally isolated, the dynamics will remain linear. If the atoms later re-emit the light and no physical trace remains to record *which* atom was excited, the light may still exhibit interference at some later stage of the experiment. On the other hand, the atoms could be part of a measurement device, and the absorption of light could be the first part of a measurement process from which a physical record results. In this case, no interference of the beams would occur.

6.3 A 4π world

In the examples we have discussed so far, composite systems comprised distinct physical "things" – a pair of spins, or a couple of separated two-level atoms, etc. However, the quantum mechanical idea is more general than this. Consider, for instance, a particle like an electron that both moves through space and has an intrinsic spin. The Hilbert space for the electron is of the general form

$$\mathcal{H} = \mathcal{H}_{\text{space}} \otimes \mathcal{Q}_{\text{spin}}. \tag{6.28}$$

We cannot have an electron without both of these characteristics. However, it is possible in principle to measure or interact with either one without affecting the other, so in an abstract sense the electron's spatial location and internal spin are separate "subsystems" of the particle. Our next example is of this kind.

In Exercise 5.19 it was shown that the rotation of a spin-1/2 system by 2π about the z-axis introduced an unexpected negative sign in the spin state vector. The rotation operator was

$$R_z(2\pi) = e^{-i2\pi S_z/\hbar} = -1. \tag{Re 5.51}$$

To obtain the original spin state, the system must be rotated through an angle of 4π, i.e. through two complete turns about the axis. (Problem 5.7 showed that such negative signs were more or less inevitable.)

On the one hand this looks paradoxical, since it seems that a rotation by 2π around an axis should have the same effect as no rotation at all. This is certainly true for rotations of an ordinary geometrical shape. How could a spin-1/2 system see a "4π world?" On the other hand, the apparent paradox does not appear to be very serious. The overall phase of the state vector has no observational meaning, so that the kets $|\psi\rangle$ and $-|\psi\rangle$ are physically equivalent.

Nevertheless, the negative sign in Eq. 5.51 does lead to physical effects, and these have been observed in real experiments with neutrons.

Neutrons are spin-1/2 particles, and like protons they have a magnetic moment that can couple to an external magnetic field. A magnetic field in the z-direction can therefore be used to rotate the spin of the neutron about that axis. By controlling the strength of the field and the time over which it is applied, we can apply $R_z(\theta)$ for any angle θ we wish.

Exercise 6.19 The neutron magnetic moment is $\mu = -1.91\mu_N$. Here μ_N is the nuclear magneton:

$$\mu_N = \frac{e\hbar}{2m_p}, \tag{6.29}$$

where m_p is the proton mass, 1.67×10^{-27} kg. The negative sign indicates that the magnetic moment vector is opposite to the spin vector of the neutron. Suppose a neutron experiences a magnetic field with magnitude 0.01 T. How much time is required for the neutron spin to rotate by an angle 2π?

Neutrons also move through space. Like other particles, they have a de Broglie wavelength and exhibit constructive and destructive interference. It is possible to exhibit these phenomena in a *neutron interferometer* apparatus. Neutron interferometers are constructed from large, carefully machined single crystals of silicon. For present purposes, we can adopt a simplified picture of a two-beam neutron interferometer analogous to the optical interferometers of Chapter 2.

The neutron in the interferometer is essentially a pair of qubits, one describing the spin state of the neutron and one describing its location among the beams of the interferometer. This Hilbert space will be spanned by the product basis states

$$|u, z_+\rangle \qquad |u, z_-\rangle \qquad |l, z_+\rangle \qquad |l, z_-\rangle \; . \tag{6.30}$$

The elements of the interferometer only affect the spatial part of the state vector and are thus described by unitary operators of the form $U_{space} \otimes 1_{spin}$. For instance, a balanced neutron beamsplitter would act according to

$$B_l \otimes \mathbf{1} \, |u, s\rangle = \frac{1}{\sqrt{2}} \left(|u\rangle + |l\rangle \right) \otimes |s\rangle ,$$

$$B_l \otimes \mathbf{1} \, |l, s\rangle = \frac{1}{\sqrt{2}} \left(|u\rangle - |l\rangle \right) \otimes |s\rangle , \qquad (6.31)$$

for any spin state $|s\rangle$. Similarly, if the interferometer is placed in a uniform magnetic field, the spin will be rotated. The spatial part of the state, however, will be unaffected by the field, since neutrons have no net electric charge.

Things get more interesting if we arrange for *one* of the interferometer beams to pass through a magnetic field. This can be accomplished by sending the lower beam between the poles of an electromagnet. The spin state of the neutron would be rotated only if it passed along this beam. If the field is aligned along the z-axis and its strength is adjusted to give a spin rotation by θ, then the operator describing this device is

$$U = |u\rangle\langle u| \otimes \mathbf{1} + |l\rangle\langle l| \otimes R_z(\theta). \qquad (6.32)$$

With this *conditional rotation*, the spatial and spin "subsystems" of the neutron are no longer dynamically isolated.

Exercise 6.20 Suppose the magnetic field produces $\theta = \pi$. How does the resulting conditional rotation operator affect the four product basis states in Eq. 6.30? Describe a situation in which the neutron evolves from an initial product state to an entangled state of the space and spin subsystems. (Recall Exercise 5.18.)

Consider now the Mach–Zehnder neutron interferometer shown in Fig. 6.1. The neutron is introduced through the lower beam, so that its initial state is $|l, s\rangle$ for some spin state $|s\rangle$. With no magnetic field, $\theta = 0$ and spins in both beams are unrotated. It is easy to see that the constructive and destructive interference at the second beamsplitter will lead to detection probabilities $p(u) = 1$ and $p(l) = 0$ – in other words, the neutron is certain to be detected in the upper beam.

Exercise 6.21 Fill in the glib phrase "It is easy to see that" with a real derivation. If you need to, go back and review the discussion of two-beam interferometers in Section 2.1.

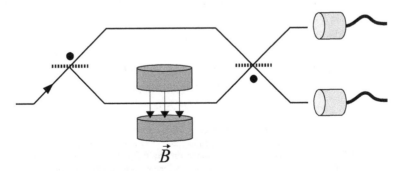

Fig. 6.1   A neutron Mach–Zehnder interferometer.

Now we set the magnetic field so that a neutron in the lower beam is rotated through an angle of 2π. The neutron state between the two beamsplitters evolves in the field by

$$\frac{1}{\sqrt{2}}\Big(|u, s\rangle + |l, s\rangle\Big) \longrightarrow \frac{1}{\sqrt{2}}\Big(|u, s\rangle - |l, s\rangle\Big). \qquad (6.33)$$

The negative sign from the 2π rotation of the neutron spin induces a relative phase in the spatial superposition. In other words, there has been no observable net effect on the neutron spin, but there has been an observable effect on the *location* of the neutron. At the end of the interferometer, the detection probabilities are now $p(u) = 0$ and $p(l) = 1$, exactly the opposite of the probabilities with no 2π rotation.

A rotation of the neutron spin by an angle of 2π is therefore *not* physically equivalent to a rotation by zero. We must rotate the spin by 4π to restore the original interference pattern. Actual neutron interferometry confirms this prediction.[3] We must conclude that spin-1/2 particles like neutrons do in some sense "see a 4π world."

Exercise 6.22 Calculate the probability $p(u)$ as a function of the spin rotation angle θ in the neutron Mach–Zehnder experiment.

The negative sign produced by a 2π rotation would be completely invisible if only the spin itself were involved in the experiment. It is the composite nature of the neutron – both space *and* spin – that leads to an observable effect. Or, to put it another way, it is the *relative* 2π rotation of one beam with respect to the other that differs from zero rotation.[4]

6.4 Conditional states

Composite systems follow the usual quantum rules for calculating the expectations of observables and the probabilities of measurement outcomes. The trick is to identify how the mathematical objects (operators, orthonormal bases, etc.) correspond to physical measurement procedures. Here are a few guidelines.

An observable on the system A is of course also an observable on the composite system AB. If Q is the observable's Hermitian operator on $\mathcal{H}^{(A)}$, then the corresponding operator on $\mathcal{H}^{(AB)}$ is $Q^{(A)} = Q \otimes \mathbf{1}$, where $\mathbf{1}$ here is the identity operator on $\mathcal{H}^{(B)}$. Eigenstates of $Q^{(A)}$ are of the form $|q_i\rangle \otimes |\psi\rangle$, where $|q_i\rangle$ is an eigenstate of Q in $\mathcal{H}^{(A)}$ and $|\psi\rangle$ can be any B-state.

Exercise 6.23 Show that there can be an entangled eigenstate of $Q \otimes \mathbf{1}$ if and only if the operator Q has one or more degenerate eigenvalues.

---

[3] This experiment was first suggested by H. J. Bernstein in 1967, and was actually done by S. A. Werner *et al.*, in 1975 – a particularly elegant demonstration of a subtle and surprising quantum effect.
[4] For a beautiful geometrical discussion of why this is true, and of its further implications for the physics of spin-1/2 particles, see Richard Feynman's lecture in *Elementary Particles and the Laws of Physics: The 1986 Dirac Memorial Lectures* by Richard Feynman and Steven Weinberg.

It is also useful to consider algebraic combinations of subsystem observables. Fix an A-observable Q and a B-observable R. The eigenvalues of Q are $\{q\}$, corresponding to eigenstates $|q\rangle$, while the eigenvalues of R are $\{r\}$ with eigenstates $|r\rangle$. Then:

- The sum

$$Q^{(A)} + R^{(B)} = Q \otimes \mathbf{1} + \mathbf{1} \otimes R, \qquad (6.34)$$

  has eigenvalues $\{q + r\}$ (including all combinations) corresponding to product eigenstates $|q, r\rangle = |q\rangle \otimes |r\rangle$.
- The product

$$Q^{(A)}R^{(B)} = Q \otimes R, \qquad (6.35)$$

  has eigenvalues $\{qr\}$ and the same product eigenstates.

We saw these combinations in Section 6.2, when discussing how systems may interact. Two systems are non-interacting if the composite Hamiltonian is simply the sum of Hamiltonians for the individual systems. A product term in the Hamiltonian, on the other hand, describes an interaction between the systems.

Each of these combined observables could be measured by performing two separate experimental procedures, one for each subsystem, after which the numerical results are added or multiplied. The two subsystem measurements may be performed in either order, or simultaneously.[5]

Suppose now that AB is initially in the joint state $|\Psi^{(AB)}\rangle$ and that basic measurements are performed on the individual subsystems. We measure A using basis $\{|a\rangle\}$ and B using basis $\{|b\rangle\}$. The joint probability $p(a, b)$ is

$$p(a, b) = \left|\langle a, b | \Psi^{(AB)}\rangle\right|^2. \qquad (6.36)$$

From the joint distribution over a and b values, we can calculate the distributions over a and b individually:

$$p(a) = \sum_b p(a, b), \qquad p(b) = \sum_a p(a, b). \qquad (6.37)$$

Exercise 6.24 Consider the joint state $\left|\Phi_+^{(AB)}\right\rangle$ for qubits A and B, defined in Eq. 6.15, and suppose the two qubits are each measured in the standard $\{|0\rangle, |1\rangle\}$ basis. Find the probability distribution for the joint measurement results and also for the individual subsystem results.

According to Eq. 6.13, we can write the joint state as

$$\left|\Psi^{(AB)}\right\rangle = \sum_b |\phi_b\rangle \otimes |b\rangle, \qquad (6.38)$$

where the $|\phi_b\rangle$ are non-normalized vectors in $\mathcal{H}^{(A)}$. These vectors are related to the joint probabilities as follows:

---

[5] Indeed, if the systems A and B are widely separated in space, special relativity tells us that the time order of the two measurement events may not be the same in different reference frames.

Exercise 6.25 Show that

$$p(a, b) = |\langle a | \phi_b \rangle|^2 . \tag{6.39}$$

The total probability $p(b)$ for the outcome b is therefore

$$p(b) = \sum_a \langle \phi_b | a \rangle \langle a | \phi_b \rangle = \langle \phi_b | \phi_b \rangle . \tag{6.40}$$

Notice that this expression for $p(b)$ depends only on the decomposition of $|\Psi^{(AB)}\rangle$ in Eq. 6.38, and not on the particular A-basis $\{ |a\rangle \}$. The probability $p(b)$ is independent of the choice of basic measurement performed on A. Furthermore, suppose that $|\Psi^{(AB)}\rangle$ is initially changed by applying a unitary operator for the A system. We have

$$\left| \Psi^{(AB)'} \right\rangle = (U \otimes 1) \left| \Psi^{(AB)} \right\rangle = \sum_b U | \phi_b \rangle \otimes | b \rangle . \tag{6.41}$$

For this state, the probability $p'(b)$ is

$$p'(b) = \langle \phi_b | U^\dagger U | \phi_b \rangle = \langle \phi_b | \phi_b \rangle = p(b). \tag{6.42}$$

The probability $p(b)$ is unaffected by a unitary operator affecting only system A.

Think about this as a communication problem. Suppose Alice wishes to send some information to Bob. To do this, she must choose one of a set of possible physical operations (encoding her possible messages), and then Bob must make some physical observation. For information to be transferred, Alice's choice must have at least a statistical effect on Bob's measurement results. Unless Alice's operations can affect Bob's probabilities, no information is conveyed. We have shown the following:

> **No-communication theorem.** Suppose Alice and Bob have access to systems A and B respectively, and that the joint state of AB is $|\Psi^{(AB)}\rangle$, which may be entangled. Then Alice cannot convey information to Bob by either
> - choosing from a set of possible basic measurements on A, or
> - choosing from a set of possible unitary evolution operators for A.

Exercise 6.26 Explain why this theorem remains true even if Alice involves a third system C in her operations.

This theorem expresses an important general principle. If it were not true, then Alice could send information to Bob by some purely local choice of measurement or unitary operation.[6] Since Alice and Bob could be very far apart, this would mean that information could be transmitted faster than the speed of light – instantaneously in fact. Quantum mechanics does not permit this.

Alice's choice of measurement cannot affect the probability $p(b)$ of Bob's measurement result. On the other hand, the value a that Alice obtains *can* affect Bob's probabilities. (This cannot be used for communication, since Alice does not control the outcome of her

---

[6] We will use the term *local* to mean that a given operation or statement only pertains to one part of a composite system – in this case, system A.

measurement and so cannot use it to encode a message.) The *conditional probability* of b given a is

$$p(b|a) = \frac{p(a,b)}{p(a)}. \tag{6.43}$$

Conditional probabilities are discussed in Appendix A. (You may also recall the idea from Eq. 4.3.) For a fixed value of a, $p(b|a)$ is a probability distribution over the various possible values of b.

Exercise 6.27 Show that, for any a, the conditional probabilities $p(b|a)$ are normalized.

Exercise 6.28 Show that $p(b|a) = p(b)$ for all a and b if and only if a and b are independent random variables (see Eq. 6.2).

The joint state $|\Psi^{(AB)}\rangle$ can be written

$$\left|\Psi^{(AB)}\right\rangle = \sum_a |a\rangle \otimes |\psi_a\rangle, \tag{6.44}$$

where the non-normalized B-vectors $|\psi_a\rangle$ give the joint distribution via $p(a,b) = |\langle b|\psi_a\rangle|^2$. The conditional probabilities are

$$p(b|a) = \frac{1}{p(a)} |\langle b|\psi_a\rangle|^2, \tag{6.45}$$

where $p(a) = \langle\psi_a|\psi_a\rangle$. Notice that these are exactly the probabilities predicted by the B-state

$$\left|\hat{\psi}_a\right\rangle = \frac{1}{\sqrt{p(a)}} |\psi_a\rangle. \tag{6.46}$$

Therefore, this is the quantum state that we would assign to B when Alice's measurement results in a. We call $\left|\hat{\psi}_a\right\rangle$ the *conditional state* of B given the A-measurement outcome a.

Like the projection rule discussed in Section 4.3, the conditional state rule is a way of assigning a new quantum state to a system based on the result of a measurement. But the projection rule applies only in the special case when the measurement process is "minimally disturbing." Real measurement processes are seldom of this type. The conditional state rule, on the other hand, is more generally applicable. It assumes only that the measurement process involves system A alone, i.e. that the apparatus does not interact with subsystem B. If, for instance, A and B are far apart in space, this condition may be easy to achieve.

We can express the idea of conditional states in a particularly elegant way by introducing the *partial inner product*. Suppose $|\alpha^{(A)}\rangle$ is a vector in $\mathcal{H}^{(A)}$ and $|\Phi^{(AB)}\rangle$ is a vector in $\mathcal{H}^{(AB)} = \mathcal{H}^{(A)} \otimes \mathcal{H}^{(B)}$. Then $\left|\phi_\alpha^{(B)}\right\rangle = \left\langle\alpha^{(A)}\middle|\Phi^{(AB)}\right\rangle$ is defined to be a vector in $\mathcal{H}^{(B)}$ such that, for any other $|\beta^{(B)}\rangle$ in $\mathcal{H}^{(B)}$,

$$\left\langle\beta^{(B)}\middle|\phi_\alpha^{(B)}\right\rangle = \left\langle\alpha^{(A)}, \beta^{(B)}\middle|\Phi^{(AB)}\right\rangle. \tag{6.47}$$

This is the partial inner product.

Exercise 6.29 Given non-null vectors $|\alpha^{(A)}\rangle$ and $|\Phi^{(AB)}\rangle$, show that there is a vector $\left|\phi_\alpha^{(B)}\right\rangle$ in $\mathcal{H}^{(B)}$ that satisfies Eq. 6.47. Then show that there is only one such vector $\left|\phi_\alpha^{(B)}\right\rangle$ in $\mathcal{H}^{(B)}$. Therefore, Eq. 6.47 defines a unique vector.

Another exercise makes the relevance of the partial inner product a bit more clear:

Exercise 6.30 Given $|\Psi^{(AB)}\rangle$ and an orthonormal A-basis $\{|a^{(A)}\rangle\}$, show that

$$|\Psi^{(AB)}\rangle = \sum_a |a^{(A)}\rangle \otimes \langle a^{(A)}|\Psi^{(AB)}\rangle. \tag{6.48}$$

(Does this remind you of the completeness relation?)

If we measure A using the basis $\{|a^{(A)}\rangle\}$, we let $|\psi_a^{(B)}\rangle = \langle a^{(A)}|\Psi^{(AB)}\rangle$. The probability of the result a is $p(a) = \langle \psi_a^{(B)}|\psi_a^{(B)}\rangle$, and the conditional state of B given the result a is

$$|\hat{\psi}_a\rangle = \frac{1}{\sqrt{p(a)}} \langle a^{(A)}|\Psi^{(AB)}\rangle. \tag{6.49}$$

The conditional state of B yields all of the correct quantum conditional probabilities for future B-measurement results, given that the A-result is a.

Let us consider an interesting example. A pair of qubits A and B is in the following entangled quantum state:

$$|\Psi_-^{(AB)}\rangle = \frac{1}{\sqrt{2}}\left(|0,1\rangle - |1,0\rangle\right). \tag{6.50}$$

This is sometimes called the "singlet" state of the two qubits, for reasons that will be explained in Chapter 12. If Alice makes a measurement of Z on qubit A, then the conditional states of Bob's system B will be

$$\langle 0^{(A)}|\Psi_-^{(AB)}\rangle = \frac{1}{\sqrt{2}}|1\rangle \quad \text{and} \quad \langle 1^{(A)}|\Psi_-^{(AB)}\rangle = -\frac{1}{\sqrt{2}}|0\rangle. \tag{6.51}$$

Therefore, the two measurement outcomes have equal probability (1/2), and after the measurement the qubit B will be in a Z eigenstate opposite to the A outcome.

Something similar happens if Alice instead measures X on qubit A. Here the eigenstates of X are $|\pm\rangle = \frac{1}{\sqrt{2}}\left(|0\rangle \pm |1\rangle\right)$ and

$$\langle +^{(A)}|\Psi_-^{(AB)}\rangle = -\frac{1}{\sqrt{2}}|-\rangle \quad \text{and} \quad \langle -^{(A)}|\Psi_-^{(AB)}\rangle = \frac{1}{\sqrt{2}}|+\rangle. \tag{6.52}$$

Again, the outcomes are equally likely, and after the measurement the state of B will be an X eigenstate opposite to the A outcome. In either case, a measurement performed on A allows one to predict with certainty the outcome of an identical measurement performed on B.

Suppose Alice could choose to measure either Z or X on A. After Alice's measurement (but before he learns of its result), Bob is presented with one of two situations:

Situation Z. If Alice measures Z, then the state of B is either $|0\rangle$ or $|1\rangle$, each of which occurs with probability 1/2.

Situation X. If Alice measures X, then the state of B is either $|+\rangle$ or $|-\rangle$, each of which occurs with probability 1/2.

The no-communication theorem tells us that Alice's choice of A-measurement cannot have any effect on the probabilities for any B-measurement Bob may perform. Thus, Situation Z and Situation X must be *experimentally indistinguishable* to Bob. Otherwise, Alice and Bob could use the singlet state $\left| \Psi_-^{(AB)} \right\rangle$ for instantaneous communication.

The situations outlined for B – after one of the possible A-measurements but before any particular result is disclosed – are cases in which there is incomplete knowledge of the quantum state of B. We know only that the state vector is one of a set of possibilities. As we will see in Chapter 8, this kind of thing is most naturally described by a *density operator*. It will turn out that Situations X and Z, though they are two quite different collections of B-states, nevertheless have exactly the same density operator description.

Exercise 6.31 Consider the entangled state $\left| \Phi_+^{(AB)} \right\rangle$ in Eq. 6.15. For Z and X measurements on A, find the probabilities and the conditional B-states for the two possible results. How do the details differ from the singlet state?

6.5 EPR

In a famous paper in 1935, Albert Einstein, Boris Podolsky, and Nathan Rosen (collectively denoted EPR) called attention to the strange properties of entangled states of quantum systems. They made the startling claim that the statistical correlations between entangled systems prove that quantum theory is an *incomplete* description of nature. Their argument is worth exploring in detail.

Consider a single qubit system. No quantum state of a qubit can be an eigenstate of both Z and X observables. If the qubit has a definite value of Z, its value of X must be indeterminate, and *vice versa*. If we take quantum theory to be a complete description of nature, then we must say that it is impossible for both Z and X to have definite values for the same qubit at the same time.

But we might have a different view. We might consider quantum theory to be an *incomplete* description, in which case we could imagine that Z and X both do have simultaneous definite values, although we only have knowledge of one of them at a time. This is sometimes called the hypothesis of "hidden variables." In this view, the indeterminacy found in quantum theory is merely due to our ignorance of underlying variables that are present in nature but not accounted for in the theory.

In order to decide whether quantum theory is a complete description of physical reality, EPR proposed a definite criterion of what constitutes "physical reality":

> If, without in any way disturbing a system, we can predict with certainty (i.e. with probability equal to unity) the value of a physical quantity, then there exists an element of physical reality corresponding to this physical quantity.

In other words, if we can exactly predict the result of a measurement without "touching" the system in any way, that measurement result must in some sense already exist in physical reality before the measurement.

Recall the singlet state of a pair of qubits:

$$\left| \Psi_-^{(AB)} \right\rangle = \frac{1}{\sqrt{2}} \left(|0, 1\rangle - |1, 0\rangle \right). \tag{Re 6.50}$$

Our previous analysis of this state tells us that, if qubit A is subjected to a Z-measurement, then the conditional state of B is always the opposite Z-eigenstate. The same thing is true of an X-measurement: after an X-measurement on A, qubit B will have a conditional state that is the X-eigenstate opposite to the measurement result.

Now apply EPR's criterion of reality. If the pair AB is in the singlet state $\left| \Psi_-^{(AB)} \right\rangle$, we can come to predict either Z or X on qubit B by performing one measurement or the other on qubit A. If we assume that our *choice* of A-measurement does not constitute a *disturbance* of B – a reasonable intuition, given that A and B could be as far apart as we like – then we are forced to conclude that the values of Z and X for B must simultaneously exist as elements of reality before any measurement of B. Since the quantum description of B cannot include simultaneous definite values for Z and X, the quantum description of the situation must be incomplete. This is the EPR argument.

We may summarize this argument in a schematic way as follows:

$$\begin{pmatrix} \textbf{Quantum} \\ \textbf{mechanics} \\ \text{Correlations} \\ \text{in } \left| \Psi_-^{(AB)} \right\rangle \end{pmatrix} + \begin{pmatrix} \textbf{Locality} \\ \text{A-measurement} \\ \text{choice does not} \\ \text{disturb B} \end{pmatrix} \Rightarrow \begin{pmatrix} \textbf{Hidden variables} \\ \text{Both } Z \text{ and } X \\ \text{can have} \\ \text{definite values} \end{pmatrix}.$$

According to EPR, the properties of the entangled quantum state $\left| \Psi_-^{(AB)} \right\rangle$, together with our assumption that the choice of A-measurement creates no disturbance at B, imply that there are hidden variables in nature that the theory does not account for.

Niels Bohr responded to the EPR paper with characteristic subtlety. He said that the concept of "disturbance" in the EPR criterion of reality is ambiguous when applied to quantum systems. True, there is no possibility that the measurement applied to qubit A exerts any sort of force on or causes any mechanical interaction with qubit B. Nevertheless, the two proposed procedures – measuring either Z or X on A – are still complementary, because they cannot both be done in the same experiment on the same qubit pair. They are *logically* exclusive, like the "which slit" or "interference" options in the two-slit experiment (see Section 1.2). It is therefore not legitimate to conclude that both the Z and X values of B exist simultaneously in the same experiment.

Although Bohr's critique showed that the EPR argument was not logically compelling, the EPR point of view retained considerable intuitive force. It stood for three decades as the most significant theoretical challenge to the completeness of quantum theory. Then, in 1964, John Bell turned the EPR argument upside-down. He showed that the statistical correlations between entangled quantum systems are far stranger and more mysterious than EPR – or anyone – had previously supposed.

6.6 Bell's theorem

We will give an elegant version of Bell's argument that was introduced in 1969 by John Clauser, Michael Horne, Abner Shimony, and Richard Holt (CHSH). Imagine a source that produces pairs of physical systems, which are then routed to separated observers Alice and Bob. Alice can measure one of two possible observables on her system, which we will denote A_1 and A_2. Similarly, Bob can measure either B_1 or B_2 on his system. There are thus four possible joint measurements on the composite system: (A_1, B_1), (A_1, B_2), (A_2, B_1) and (A_2, B_2). In any given run of the experiment, we can only measure one of these, but by repeating the experiment many times we can sample the statistics of all of the possible joint measurements. To make things as simple as possible, we will assume that all of the observables have only two possible measured values, $+1$ and -1.

We now make two hypotheses about the behavior of the composite system:

Hidden variables. We assume that the results of any measurement on any individual system are predetermined. Any probabilities we may use to describe the system merely reflect our ignorance of these hidden definite values, which may vary from one experimental run to another.

Locality. We assume that Alice's choice of measurement does not affect the outcomes of Bob's measurements, and *vice versa*.

As we have seen, these two hypotheses more or less capture the point of view advocated by EPR.

The role of the locality hypothesis deserves some comment. A statement like "The value of B_1 is $+1$," means, "If Bob were to measure B_1, then the result $+1$ would be obtained." We want to assume that, prior to the measurements, such a statement is definitely true or false. Without the locality hypothesis, though, the statement as it stands is ambiguous, since the value of B_1 could depend on whether A_1 or A_2 will be measured by Alice. The locality hypothesis allows us to speak of the predetermined definite values of A_1, A_2, B_1, and B_2 without further qualification.

Given that all four observables have definite values, we can form the combination

$$Q = A_1 (B_1 - B_2) + A_2 (B_1 + B_2). \tag{6.53}$$

Since each value can only be $+1$ or -1, the quantity $Q = \pm 2$. Therefore, any statistical average of Q must satisfy $-2 \le \langle Q \rangle \le +2$. Therefore,

$$-2 \le \langle A_1 B_1 \rangle - \langle A_1 B_2 \rangle + \langle A_2 B_1 \rangle + \langle A_2 B_2 \rangle \le +2. \tag{6.54}$$

Equation 6.54 is called the *CHSH inequality*. It is a special case of a *Bell inequality*, a generic name for a statistical inequality implied by the local hidden variable hypotheses.

Alice and Bob cannot perform a measurement to determine the value of Q in any particular experiment, but they can measure any of the products $A_1 B_1$, etc. By repeating the experiment many times, each of the statistical averages in the CHSH inequality can be estimated. Thus, Eq. 6.54 is an experimentally testable statement that must hold if the statistical behavior of the systems can be explained by any theory of local hidden variables.

What about quantum systems? Let us consider a quantum experiment that matches the general arrangement we are considering. Our source produces two qubit systems in the entangled singlet state $\left|\Psi_-^{(AB)}\right\rangle$, after which the A qubit is delivered to Alice and the B qubit to Bob. It is easy to confirm the following:

Exercise 6.32 Let $\left|\Psi_-^{(AB)}\right\rangle$ and $\left|\Phi_+^{(AB)}\right\rangle$ be defined according to Eq. 6.50 and 6.15, respectively. Show that

$$Z^{(A)}Z^{(B)}\left|\Psi_-^{(AB)}\right\rangle = -\left|\Psi_-^{(AB)}\right\rangle,$$

$$X^{(A)}X^{(B)}\left|\Psi_-^{(AB)}\right\rangle = -\left|\Psi_-^{(AB)}\right\rangle,$$

$$X^{(A)}Z^{(B)}\left|\Psi_-^{(AB)}\right\rangle = -\left|\Phi_+^{(AB)}\right\rangle,$$

$$Z^{(A)}X^{(B)}\left|\Psi_-^{(AB)}\right\rangle = +\left|\Phi_+^{(AB)}\right\rangle. \tag{6.55}$$

Alice and Bob will measure observables of the form W_θ, where

$$W_\theta = \sin\theta\, X + \cos\theta\, Z. \tag{6.56}$$

If the qubits are spin-1/2 systems, we recognize W_θ as the component of the spin along an axis in the xz-plane at the angle θ from the z-axis, measured in units of $\hbar/2$ (compare Eq. 3.109).

Exercise 6.33 Verify that the eigenvalues of W_θ are $+1$ and -1, so that the possible results of a W_θ measurement are like those assumed in the derivation of the CHSH inequality.

Suppose Alice measures W_θ and Bob measures $W_{\theta'}$. Then for the quantum state $\left|\Psi_-^{(AB)}\right\rangle$, the expectation $\langle W_\theta W_{\theta'}\rangle$ will be

$$\langle W_\theta W_{\theta'}\rangle = \sin\theta\sin\theta'\left\langle X^{(A)}X^{(B)}\right\rangle + \sin\theta\cos\theta'\left\langle X^{(A)}Z^{(B)}\right\rangle$$

$$+ \cos\theta\sin\theta'\left\langle Z^{(A)}X^{(B)}\right\rangle + \cos\theta\cos\theta'\left\langle Z^{(A)}Z^{(B)}\right\rangle$$

$$= -\sin\theta\sin\theta' - \cos\theta\cos\theta'. \tag{6.57}$$

Therefore,

$$\langle W_\theta W_{\theta'}\rangle = -\cos\left(\theta - \theta'\right). \tag{6.58}$$

If $\theta = \theta'$, we see that $\langle W_\theta W_{\theta'}\rangle = -1$. This occurs because the two parallel measurements always produce exactly opposite results, as we have already noted for Z and X.

Alice and Bob choose the following observables for their experiment:

$$A_1 = W_0, \quad B_1 = W_{\pi/4},$$

$$A_2 = W_{\pi/2}, \; B_2 = W_{3\pi/4}, \tag{6.59}$$

(see Fig. 6.2). With these choices,

$$\langle A_1 B_1\rangle = -\frac{1}{\sqrt{2}}, \; \langle A_1 B_2\rangle = \frac{1}{\sqrt{2}},$$

$$\langle A_2 B_1\rangle = -\frac{1}{\sqrt{2}}, \; \langle A_2 B_2\rangle = -\frac{1}{\sqrt{2}}, \tag{6.60}$$

and so

$$\langle A_1 B_1\rangle - \langle A_1 B_2\rangle + \langle A_2 B_1\rangle + \langle A_2 B_2\rangle = -2\sqrt{2}. \tag{6.61}$$

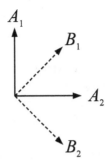

Fig. 6.2   Spin components measured by Alice and Bob.

This obviously violates the CHSH inequality (Eq. 6.54)! Therefore, the expectations predicted by quantum theory are *inconsistent* with the assumptions that led to that inequality.

Exercise 6.34 Find other observables A_1, B_1, etc., so that

$$\langle A_1 B_1 \rangle - \langle A_1 B_2 \rangle + \langle A_2 B_1 \rangle + \langle A_2 B_2 \rangle > 2. \tag{6.62}$$

The general lesson here can be summarized as follows:

Bell's theorem. The statistical properties of entangled quantum states cannot be accounted for by any theory of local hidden variables.

This is an amazing result, and a stunning reversal of the reasoning of EPR. We can summarize Bell's theorem this way:

$$
\begin{pmatrix} \textbf{Hidden variables} \\ \text{Observables have} \\ \text{definite values} \end{pmatrix} + \begin{pmatrix} \textbf{Locality} \\ \text{A-measurement} \\ \text{choice does not} \\ \text{disturb B} \end{pmatrix} \Rightarrow \begin{pmatrix} \textit{Not}\ \textbf{quantum} \\ \textbf{mechanics} \\ \text{CHSH inequality} \\ \text{always holds} \end{pmatrix}.
$$

EPR saw entanglement as evidence of the incompleteness of quantum theory. We can now come to almost the opposite conclusion: the phenomena of quantum systems *cannot* be explained by hidden variables that function locally.

What about actual laboratory experiments? Entangled quantum systems can be produced in a variety of ways, and several careful tests of Bell's theorem have been done. The observed data agree with the quantum mechanical predictions to a very high degree of statistical confidence. For example, a 1998 experiment by G. Weihs *et al.* found that the CHSH inequality was violated by 30 standard deviations, with "Alice" and "Bob" located on opposite sides of a university campus, 400 meters apart.

From a philosophical point of view, Bell's theorem and the experimental confirmation of quantum theory constitute one of the most remarkable results in all of physics. The dual hypotheses of hidden variables and locality have a pretty good claim to be the "common sense" view of the world. Nevertheless, *at least one of them must be wrong.* We must either give up the idea that hidden determinate variables underlie the indeterminacy of quantum systems, or we must give up the idea that widely separated parts of the universe act independently.

6.7 GHZ

In Bell's theorem, we find that the *statistical* properties of entangled systems are inconsistent with any local hidden variable theory. A very striking argument due to Daniel Greenberger, Michael Horne, and Anton Zeilinger (GHZ) goes further, and shows that the individual *deterministic* outcomes of experiments on entangled systems – outcomes that occur with probability 0 or 1 – are likewise inconsistent with local hidden variables.

We imagine three systems distributed among Alice, Bob, and Charles. Alice can measure either $X^{(A)}$ or $Y^{(A)}$ on her system, Bob can measure either $X^{(B)}$ or $Y^{(B)}$, and Charles can measure either $X^{(C)}$ or $Y^{(C)}$. Each of these measurements can produce only the result $+1$ or -1.

It happens that the systems are correlated in such a way that, whenever one X-measurement and two Y-measurements are made, the product of the results is always $+1$. That is, either all three results are $+1$ or exactly two of them are -1. Call this fact "Condition XYY." The question is exactly how the correlation described by Condition XYY comes about. If we assume that the measurement results are due to local hidden variables, then there are simultaneous definite values for $X^{(A)}$, $Y^{(A)}$, $X^{(B)}$, etc. It follows that

$$\left(X^{(A)}Y^{(B)}Y^{(C)}\right)\left(Y^{(A)}X^{(B)}Y^{(C)}\right)\left(Y^{(A)}Y^{(B)}X^{(C)}\right) = X^{(A)}X^{(B)}X^{(C)}, \qquad (6.63)$$

since $(Y^{(A)})^2 = (Y^{(B)})^2 = (Y^{(C)})^2 = +1$, no matter what the values are. Since all three factors on the left equal $+1$, it follows that $X^{(A)}X^{(B)}X^{(C)} = +1$. If we make an X-measurement on all three systems, the product must always be $+1$. We can call this "Condition XXX." We have shown that

$$\left(\begin{array}{c}\textbf{Local hidden}\\\textbf{variables}\end{array}\right) + \left(\begin{array}{c}\textbf{Condition}\\\textbf{XYY}\end{array}\right) \Rightarrow \left(\begin{array}{c}\textbf{Condition}\\\textbf{XXX}\end{array}\right).$$

Now consider a quantum-mechanical situation in which Alice, Bob, and Charles each possess a qubit. The joint state is

$$|\text{GHZ}\rangle = \frac{1}{\sqrt{2}}\left(|0,0,0\rangle - |1,1,1\rangle\right). \qquad (6.64)$$

The observables X and Y can be measured on each qubit. Next, we note the following fact:

Exercise 6.35 Show that

$$\text{X} \otimes \text{Y} \otimes \text{Y} |\text{GHZ}\rangle = |\text{GHZ}\rangle. \qquad (6.65)$$

Thus, the state $|\text{GHZ}\rangle$ is an eigenstate of $\text{X} \otimes \text{Y} \otimes \text{Y}$ with eigenvalue $+1$. Show also that $|\text{GHZ}\rangle$ is an eigenstate of $\text{Y} \otimes \text{X} \otimes \text{Y}$ and $\text{Y} \otimes \text{Y} \otimes \text{X}$ with the same eigenvalue.

It follows that the three-qubit system in the state $|\text{GHZ}\rangle$ satisfies Condition XYY above. But what if we measure $\text{X} \otimes \text{X} \otimes \text{X}$? We find that

$$\text{X} \otimes \text{X} \otimes \text{X} |\text{GHZ}\rangle = -|\text{GHZ}\rangle, \qquad (6.66)$$

so a measurement of $\text{X} \otimes \text{X} \otimes \text{X}$ must always yield the result -1. In short, the qubit triple in the $|\text{GHZ}\rangle$ state exactly contradicts Condition XXX every time. Therefore the correlations exhibited by this system cannot be ascribed to the effect of hidden variables that act locally.

Exercise 6.36 Show that the qubit operators satisfy

$$(X \otimes Y \otimes Y)(Y \otimes X \otimes Y)(Y \otimes Y \otimes X) = -X \otimes X \otimes X. \qquad (6.67)$$

Compare this with Eq. 6.63, which refers to classical variables with values ± 1.

Problems

Problem 6.1 Suppose the state $|\Psi^{(AB)}\rangle$ has the form given by Eq. 6.14 for some particular A-basis $\{|n\rangle\}$. Show that it must also have the same general form for any other choice of A-basis. Thus, we can recognize a product state by expanding it in any basis we choose.

Problem 6.2 This problem provides an alternate proof for the existence of entangled (non-product) states. Consider the two-qubit state $\left|\Phi_+^{(AB)}\right\rangle$ defined in Eq. 6.15. If this were a product state, then there would be normalized states $|\psi\rangle$ of A and $|\phi\rangle$ of B such that

$$\left|\left\langle \Phi_+^{(AB)} \middle| \psi, \phi \right\rangle\right| = 1.$$

Expand $|\psi\rangle$ and $|\phi\rangle$ in terms of the standard basis and use the Schwartz inequality to show that this equality cannot hold.

Problem 6.3

(a) Suppose X and Y are two observables, and that $\langle X \rangle = \langle Y \rangle$ for every state $|\psi\rangle$. Show that the operators satisfy $X = Y$.

(b) Let $H^{(AB)}$ be the Hamiltonian for a composite system AB, and suppose there are subsystem observables $G^{(A)}$ and $G^{(B)}$ such that

$$\left\langle H^{(AB)} \right\rangle = \left\langle G^{(A)} \right\rangle + \left\langle G^{(B)} \right\rangle,$$

for all joint AB states. Show that A and B are dynamically independent.

Problem 6.4 Consider random variables A, B, C, etc., all of which have two possible values ± 1. Define the "distance" between any two of these to be the likelihood that they disagree:

$$d(A, B) = \Pr(A \neq B). \qquad (6.68)$$

(a) First, show that $d(\cdot, \cdot)$ has the usual properties of a distance function: $d(A, A) = 0$, $d(A, B) = d(B, A)$, and $d(A, B) + d(B, C) \geq d(A, C)$. (The last property is called the "triangle inequality.")

(b) Find a relation between $d(A, B)$ and $\langle AB \rangle$.

(c) Show that one-half of the CHSH inequality (Eq. 6.54) can be written

$$d(A_1, B_1) + d(B_1, A_2) + d(A_2, B_2) \geq d(A_2, B_2). \qquad (6.69)$$

Show that this "quadrilateral inequality" is a consequence of the basic properties of $d(\cdot, \cdot)$. Draw a simple picture to illustrate this relation of distances.

(d) Find a set of spin component measurements on the state $|\Psi_-\rangle$ that violate this inequality.

Ponder where the concept of "local hidden variables" enters into this version of the Bell argument.

Problem 6.5 Artur Ekert suggested that entangled pairs of qubits could be used for quantum key distribution. (Review Section 4.4 if necessary.) Suppose Alice and Bob possess many pairs of qubits in the state $|\Psi_-\rangle$. Since these qubits are in a pure state, they have no correlations with any other systems, including those available to the eavesdropper Eve. By making appropriate measurements, they can arrive at a shared secret string of key bits.

Explain in detail how this should be done. Also explain how Alice and Bob can make sure that their qubits are really in the $|\Psi_-\rangle$ state, and not, for example, in a GHZ state with a qubit of Eve. What percentage of the qubits do they have to discard (not counting the ones sacrificed to check for Eve's activity)?

Problem 6.6 Consider the GHZ state of Eq. 6.64:

(a) Suppose that Alice makes a Z measurement on her qubit. Show that the qubits of Bob and Charlie are in a product state, regardless of the measurement result.

(b) Suppose Alice makes an X measurement on her qubit. Show that Bob and Charlie's qubits end up in an entangled state, regardless of the measurement result.

Problem 6.7 Synchronized quantum clocks. A qubit subject to a Hamiltonian $\hbar\omega Y$ evolves according to

$$|\psi(t)\rangle = \cos\omega t\,|0\rangle + \sin\omega t\,|1\rangle. \tag{6.70}$$

(a) Verify that this state is a solution to the Schrödinger equation.

(b) We use this qubit as a simple clock. Suppose the value of the time t is equally likely to be anything between $t = 0$ and $t = T$, so that $\Pr(t < T/2) = \Pr(t > T/2) = 1/2$. At this unknown time, we measure the qubit clock using the $\{|\pm\rangle\}$ basis. Show that $p(+) = p(-) = 1/2$, and calculate the conditional probabilities $\Pr(t < T/2\,|\,+)$, etc. Show that the result of this measurement does provide information about the value of t.

(c) Consider two identical, independent clocks that are both in the same time-dependent state:

$$\left|\Psi^{(12)}(t)\right\rangle = \left|\psi^{(1)}(t)\right\rangle \otimes \left|\psi^{(2)}(t)\right\rangle. \tag{6.71}$$

If we make simultaneous $|\pm\rangle$ measurements on both clocks, show that the clocks can sometimes disagree. The clocks are not quite "synchronized."

(d) Now imagine that the clocks are prepared in a "superposition of times," which we can write

$$\left|\Phi^{(12)}\right\rangle = A \int_0^T \left|\Psi^{(12)}(t)\right\rangle dt. \tag{6.72}$$

Evaluate this integral and find the normalization constant A. Show that the result is an entangled state of the two clocks, and show that $|\pm\rangle$ measurements on the clocks will always exactly agree.

(e) Show on the other hand that $|\Phi^{(12)}\rangle$ is a stationary state of the two-qubit Hamiltonian $\hbar\omega Y^{(1)} + \hbar\omega Y^{(2)}$. Although the clocks are now perfectly synchronized with each other, they are no longer of much use for time-telling!

Problem 6.8 Suppose Alice and Bob each possess a subsystem of a larger, entangled system. Alice makes a basic measurement on her subsystem, a process which obeys the projection rule of Section 4.3. Afterwards, both Alice and Bob make further measurements on their subsystems. Show that, conditioned on the result of Alice's initial measurement, further measurement results from the two subsystems are independent. (Thus, no correlations "survive" Alice's basic measurement.)

Problem 6.9 In Section 6.6 we showed that, if two qubits are in the singlet state $|\Psi_-\rangle$, then measurements of parallel spin components (in the XZ plane) always yield opposite results.

(a) Show that this is also true for Y measurements on $|\Psi_-\rangle$.
(b) Does there exist an "anti-singlet" state of two qubits, for which measurements of parallel spin components always yield *identical* results? If so, write down the state vector. If not, give a proof that no state, product or entangled, can do this.

Problem 6.10
(a) Starting with the definition of the tensor product of operators (Eq. 6.9), show that $(A \otimes B)^\dagger = A^\dagger \otimes B^\dagger$.
(b) If A and B are both Hermitian, prove that $A \otimes B$ is also Hermitian.
(c) Do part (b) again, replacing the word "Hermitian" with "positive," "unitary," and "normal."
(d) If $A \otimes B$ is Hermitian, does it follow that both A and B must be Hermitian also?

Information and ebits

7.1 Decoding and distinguishability

Communication

The quantum description of composite systems raises (and resolves!) a few issues about distinguishability and information transfer. We shall deal with these in this section.

The basic decoding theorem from Section 4.1 tells us that, if we try to encode more than d different messages in a quantum system with dim $\mathcal{H} = d$, then we will not be able to decode the message reliably by any basic measurement. This allows us to say that a quantum system with Hilbert space dimension d has an information capacity of $\log d$ bits.

But $\log d$ may not be an integer. For a spin-1 particle with $d = 3$, the capacity would be $\log d = 1.58$ bits. As we saw in Section 1.1, this only makes sense when we consider coding our messages "wholesale," using many quantum systems to represent many messages. That is, we need to consider coding messages by the states of a composite quantum system.

Suppose we have a quantum system composed of n identical subsystems, each one described by a Hilbert space $\mathcal{H}$ of dimension d. The Hilbert space describing the whole thing is

$$\mathcal{H}^{\otimes n} = \underbrace{\mathcal{H} \otimes \cdots \otimes \mathcal{H}}_{n \text{ times}}. \tag{7.1}$$

The dimension of this space is d^n. Thus, its information capacity is $n \log d$, exactly n times as large as the capacity for an individual subsystem.

This fact is exactly what we need. Suppose we wish to represent H bits of information per subsystem. This means that we will represent nH bits in a composite system of n subsystems. This information corresponds to 2^{nH} distinct messages. The basic decoding theorem tells us that the probability of error is at least

$$P_E \geq 1 - \frac{d^n}{2^{nH}} = 1 - \left(2^{(\log d - H)}\right)^n. \tag{7.2}$$

For reliable communication ($P_E = 0$), it follows that $H \leq \log d$. In fact, though, we have an even stronger result. When $H > \log d$, then $P_E \to 1$ as $n \to \infty$. If we try to represent "too much" information in each quantum system, the probability of error is non-zero, and in fact approaches unity as n becomes large. In the long run, an error is essentially guaranteed.

Exercise 7.1 Suppose we attempt to encode 1.01 bits of information per qubit system. How many qubits can be used before P_E is guaranteed to be at least 50%? At least 99%?

Multiple copies

Now consider just two states, $|\alpha_0\rangle$ and $|\alpha_1\rangle$, which we imagine to be equally likely. The distinguishability theorem of Section 4.2 tells us that, if $|\langle\alpha_0|\alpha_1\rangle| = \cos\theta$, then the probability of correctly distinguishing the two states by a measurement is no larger than

$$P_S \leq \frac{1}{2}(1 + \sin\theta). \qquad \text{(Re 4.20)}$$

By the best choice of distinguishing measurement, we can achieve equality.

But suppose we have not just a single quantum system, but n copies of the quantum system in the same state. Then our problem is to distinguish two different states of the composite system, namely $|\alpha_0\rangle^{\otimes n} = |\alpha_0\rangle \otimes \cdots \otimes |\alpha_0\rangle$ and $|\alpha_1\rangle^{\otimes n} = |\alpha_1\rangle \otimes \cdots \otimes |\alpha_1\rangle$. The inner product of these joint states is

$$\Big(\langle\alpha_0| \otimes \cdots \otimes \langle\alpha_0|\Big)\Big(|\alpha_1\rangle \otimes \cdots \otimes |\alpha_1\rangle\Big) = \langle\alpha_0|\alpha_1\rangle^n. \qquad (7.3)$$

Let $|\langle\alpha_0|\alpha_1\rangle|^n = \cos^n\theta = \cos\theta_n$. Then we can distinguish the two situations (n copies of $|\alpha_0\rangle$ versus n copies of $|\alpha_1\rangle$) with probability $P_S = (1 + \sin\theta_n)/2$. There are now two possibilities:

- If $|\langle\alpha_0|\alpha_1\rangle| = 1$, then the two states are physically equivalent, and $\theta_n = 0$ for all n. Then $P_S = 1/2$ is the best we can do – no better than guessing.
- If $|\langle\alpha_0|\alpha_1\rangle| < 1$, then $\theta_n \to \pi/2$ as $n \to \infty$, and we can arrange $P_S \to 1$.

Therefore, if $|\alpha_0\rangle$ and $|\alpha_1\rangle$ represent physically distinct states, we can pretty well distinguish n copies of $|\alpha_0\rangle$ from n copies of $|\alpha_1\rangle$, provided that n is large enough.

What does this mean? First, the mathematical meaning. Even if the angle θ between $|\alpha_0\rangle$ and $|\alpha_1\rangle$ in $\mathcal{H}$ is small, the angle θ_n between $|\alpha_0\rangle^{\otimes n}$ and $|\alpha_1\rangle^{\otimes n}$ in $\mathcal{H}^{\otimes n}$ will be nearly $\pi/2$ for sufficiently large n. The exception, of course, is when the angle is actually 0. In every other case, sufficiently many copies of the two states can be distinguished pretty well by some measurement.

Now a physical explanation. Imagine a machine that emits spin-1/2 particles, one after the other. The machine has two settings: one prepares particles in the state $|z_+\rangle$ and the other prepares them in $|x_+\rangle$. The control that determines this setting is hidden from view. It might with equal likelihood be set either way, but the setting does not change during the experiment. Our job is to determine the unknown setting by examining the output of the machine.

If we examine only one particle from the machine, we will not with certainty be able to determine the machine's setting, since the states $|z_+\rangle$ and $|x_+\rangle$ are not perfectly distinguishable.

Exercise 7.2 In this situation, show that $P_S \leq 0.854$ for any measurement.

On the other hand, if we have $n > 1$ particles from the machine, we can do much better. For instance, imagine that we measure S_z for each particle. If every one yields the result $+\hbar/2$, then we conclude that the state is $|z_+\rangle$, but otherwise we know that it is $|x_+\rangle$. The only possible error is if the setting is for $|x_+\rangle$, but by chance every S_z measurement happens to yield $+\hbar/2$. This happens with probability

$$P_E = \frac{1}{2}\left(\frac{1}{2}\right)^n = \frac{1}{2^{n+1}}. \tag{7.4}$$

As $n \to \infty$, $P_S = 1 - P_E \to 1$. (This is not the optimal measurement for distinguishing n copies of $|z_+\rangle$ from n copies of $|x_+\rangle$, but it works pretty well.)

Exercise 7.3 How many particles do we need to distinguish the two machine settings with a probability better than 99.9%, using the S_z measurement? What if we use an optimal measurement?

7.2 The no-cloning theorem

Cloning machines

At the end of our discussion of quantum key distribution in Section 4.4, we pointed out a potential problem based on a quantum cloning machine, a hypothetical device that exactly duplicates the state of a quantum system. Such a machine would allow an eavesdropper to break the security of the BB84 key distribution protocol. But we can now see that a quantum cloning machine would have an even more fundamental consequence.

Essentially, a cloning machine would allow us to produce the state $|\alpha_0\rangle^{\otimes n}$ from $|\alpha_0\rangle$ and $|\alpha_1\rangle^{\otimes n}$ from $|\alpha_1\rangle$. Since the multiple-copy states can be distinguished more reliably than the single-copy states, the machine would be able to evade the limitation of the basic distinguishability theorem. Any two distinct quantum states would be distinguishable with any desired reliability. Quantum cloning machines, if they existed, would manufacture distinguishability.

This would have quite disturbing consequences. Suppose Alice and Bob share a qubit pair in the singlet state $\left|\Psi_-^{(AB)}\right\rangle$ described above in Section 6.4. Alice's choice of Z or X measurement on her qubit leads to two different situations for Bob's qubit. Either Bob has $|0\rangle$ and $|1\rangle$ with equal probability (Situation Z), or Bob has $|+\rangle$ and $|-\rangle$ with equal probability (Situation X). These two situations cannot be distinguished by Bob.

But if Bob possesses a cloning machine, the situations can be distinguished. If Alice measures Z and Bob uses a cloning machine on his qubit to create n copies, they will either be in the state $|0\rangle^{\otimes n}$ or $|1\rangle^{\otimes n}$. A repeated Z-measurement on these will yield n identical results. If Alice measures X and Bob uses the cloning machine, he has either the state $|+\rangle^{\otimes n}$ or $|-\rangle^{\otimes n}$. A repeated Z-measurement on these will not necessarily produce n identical results – indeed, if $n \gg 1$, this would be very unlikely. Thus, with a cloning machine Bob can pretty reliably distinguish Situation Z and Situation X – and therefore, as we noted, Alice and Bob can use the entangled state $\left|\Psi_-^{(AB)}\right\rangle$ to send information instantaneously. A quantum cloning machine would permit faster-than-light communication!

However, it turns out that *quantum cloning machines cannot exist*. This is the well-known "no-cloning theorem" of quantum theory, first proved by William Wootters and Wojciech Zurek, and independently by Dennis Dieks, in 1982. The theorem is remarkable both for its simplicity and its fundamental importance. Even more amazing is the fact that the theorem was not formulated and proven until quantum mechanics was over a half-century old.

To state and prove the theorem, we need a more precise notion of what we mean by a "cloning" process. Think first of an ordinary photocopying machine. Its function involves three distinct physical components: a piece of paper to be copied; a second piece of paper, initially blank, on which the copy is to be printed; and the machine itself, which starts out in a "ready" condition.

By analogy, we posit that the cloning process involves three subsystems: an input A, an output B, and the cloning machine M. Note that A and B are systems described by identical Hilbert spaces, while M may be very much more complicated than either one. The initial state of the input system A is variable, but the output system B begins in a fixed "blank paper" state $|0\rangle$, and the cloning machine M starts out in a standard "ready" state $|\mathcal{M}_0\rangle$. We assume that the overall system is informationally isolated. After all, we could in principle isolate the entire cloning system, including all necessary equipment, power supplies, etc., from the rest of the world, and this should not affect its operation.[1] Then the action of the cloning machine will be represented by a unitary time evolution operator U_c acting on the joint state of the composite system ABM.

The important thing to notice is that the initial quantum state $|0\rangle \otimes |\mathcal{M}_0\rangle$ of BM and the overall time evolution U_c are both independent of the initial state of the input A. Consider two initial states of the composite system

$$|\Psi\rangle = |\psi\rangle \otimes |0\rangle \otimes |\mathcal{M}_0\rangle,$$
$$|\Phi\rangle = |\phi\rangle \otimes |0\rangle \otimes |\mathcal{M}_0\rangle, \tag{7.5}$$

where $|\psi\rangle$ and $|\phi\rangle$ are two possible initial A-states. After the cloning process, the state of the input is unchanged, but the output system B must now be in an exact copy of that state. Thus, the two final states must be

$$|\Psi'\rangle = U_c |\Psi\rangle = |\psi\rangle \otimes |\psi\rangle \otimes |\mathcal{M}_\psi\rangle,$$
$$|\Phi'\rangle = U_c |\Phi\rangle = |\phi\rangle \otimes |\phi\rangle \otimes |\mathcal{M}_\phi\rangle. \tag{7.6}$$

We allow for the possibility that the final states of the cloning machine depend on the exact input state that is cloned. A system that functions as described is called a "unitary cloning machine." Our main result can be stated thus:

No-cloning theorem. No unitary cloning machine exists that works on arbitrary initial states of the input A.

Our next task is to prove this theorem. In fact, we will prove it in two different ways.

---

[1] We are arguing here that informational isolation is a relatively harmless assumption. Nevertheless, this question involves some important issues about open quantum systems, which we will discuss at greater length in Chapter 9 and Appendix D.

First proof

Our first proof was devised by Wootters and Zurek. Suppose that our unitary cloning machine works as advertised for two orthogonal input states $|\psi\rangle$ and $|\phi\rangle$, and consider its action on the superposition input state

$$|\sigma\rangle = \frac{1}{\sqrt{2}}\left(|\psi\rangle + |\phi\rangle\right). \tag{7.7}$$

The initial state of the composite system is

$$|\Sigma\rangle = |\sigma\rangle \otimes |0\rangle \otimes |\mathcal{M}_0\rangle = \frac{1}{\sqrt{2}}\left(|\Psi\rangle + |\Phi\rangle\right). \tag{7.8}$$

Because the time evolution of an informationally isolated system is linear, a superposition of initial states evolves to a superposition of output states. That is, $|\Sigma\rangle$ evolves to the final joint state

$$|\Sigma'\rangle = \frac{1}{\sqrt{2}}\left(|\Psi'\rangle + |\Phi'\rangle\right)$$

$$= \frac{1}{\sqrt{2}}\left(|\psi\rangle \otimes |\psi\rangle \otimes |\mathcal{M}_\psi\rangle + |\phi\rangle \otimes |\phi\rangle \otimes |\mathcal{M}_\phi\rangle\right). \tag{7.9}$$

But this is certainly *not* the correct cloned state $|\sigma\rangle \otimes |\sigma\rangle \otimes |\mathcal{M}_\sigma\rangle$. How can we be so sure? For one thing, $|\Sigma'\rangle$ is not even a product state between A and BM. It has the general form of Eq. 6.13, since $|\psi\rangle$ and $|\phi\rangle$ are orthogonal, but it clearly fails the product state criterion, since $|\psi\rangle \otimes |\mathcal{M}_\psi\rangle$ and $|\phi\rangle \otimes |\mathcal{M}_\phi\rangle$ are not multiples of each other (and in fact are orthogonal).

At the end of the process, we see that our input and output systems are entangled – with each other, and possibly with the cloning machine as well. They do not even *have* definite quantum states of their own, much less exact copies of the initial input state $|\sigma\rangle$. Therefore, our unitary quantum cloning machine fails for at least some input states of A.

Exercise 7.4 Show that the unitary quantum cloning machine fails for any superposition of $|\psi\rangle$ and $|\phi\rangle$ with non-zero coefficients.

Second proof

The second proof was discovered independently of the first, and is due to Dieks. For this proof, we suppose that $|\psi\rangle$ and $|\phi\rangle$ are distinct but non-orthogonal states of A, so that

$$0 < |\langle \psi | \phi \rangle| < 1. \tag{7.10}$$

Since the initial state of BM does not depend on the initial A-state, the overall initial states have the same inner product: $\langle \Psi | \Phi \rangle = \langle \psi | \phi \rangle$.

Because the time evolution operator U_c is unitary, it preserves the inner products of states. In terms of magnitudes,

$$\left|\left\langle \Psi' \middle| \Phi' \right\rangle\right| = \left|\left\langle \Psi \middle| \Phi \right\rangle\right| = \left|\left\langle \psi \middle| \phi \right\rangle\right|. \tag{7.11}$$

On the other hand, consider the inner product of the cloned final states $\left|\Psi'\right\rangle$ and $\left|\Phi'\right\rangle$ of Eq. 7.6:

$$\left|\left\langle \Psi' \middle| \Phi' \right\rangle\right| = \left|\left\langle \psi \middle| \phi \right\rangle\right|^2 \left|\left\langle \mathcal{M}_\psi \middle| \mathcal{M}_\phi \right\rangle\right| < \left|\left\langle \psi \middle| \phi \right\rangle\right|. \tag{7.12}$$

Therefore $\left|\left\langle \Psi' \middle| \Phi' \right\rangle\right| \neq \left|\left\langle \psi \middle| \phi \right\rangle\right|$, which is a contradiction. It follows that no unitary cloning machine can work on two distinct, non-orthogonal input states.

Implications

Each proof emphasizes a different aspect of quantum theory. In the first proof, it is the principle of superposition itself that defeats the cloning machine. Any machine that functioned properly for two orthogonal input states would inevitably produce entangled outputs from a superposition input state. Since the principle of superposition is arguably the most basic idea in quantum theory, the no-cloning theorem is quite a basic result of that theory.

The second proof emphasizes the idea of distinguishability. The quantity $\left|\left\langle \psi \middle| \phi \right\rangle\right|$ governs how well we can distinguish $\left|\psi\right\rangle$ and $\left|\phi\right\rangle$ by a measurement. The closer this magnitude is to zero, the better we can tell the two states apart. A working cloning machine would improve the distinguishability of non-orthogonal input states by creating outputs that are more nearly orthogonal. But unitary time evolution preserves inner products, and thus cannot increase the distinguishability of the states. No unitary cloning machine can exist.[2]

Notice that the proofs of the no-cloning theorem leave open the possibility of a cloning machine that works for *some* input states. From the second proof, we see that the machine could clone distinct states successfully only if they also happen to be orthogonal. And indeed, it is easy to imagine a way to clone any one of a basis $\{\left|n\right\rangle\}$ of A-states. A basic measurement on A can determine which of the inputs is present, and these data may be used to set up a machine that produces as many copies of the state as we like. (Strictly speaking, this would not be a unitary cloning machine, since it depends on a measurement process. But it is not hard to adapt the essential idea into a unitary scheme.)

It is also possible to imagine an *approximate* cloning machine, which would create imperfect copies, or a *conditional* cloning machine, which would work perfectly with some probability but would otherwise fail (producing an "error" message). More sophisticated analyses are required for these new situations (see Problems 7.1 and 7.2). In general, for non-distinguishable inputs $\left|\phi\right\rangle$ and $\left|\psi\right\rangle$, an approximate cloning machine cannot produce very accurate copies, and a conditional machine cannot succeed with very high probability.

---

[2] We should add that the unitarity of the time evolution is almost as basic as the principle of superposition itself. In Section 5.1 we derived this unitarity from the principle of superposition, plus the simple requirement that state vectors must remain normalized – in effect, that total probability is conserved.

From an information point of view, the no-cloning theorem reveals one of the most significant truths about the quantum world. *Quantum information cannot be copied exactly.* In classical information theory, there is no obstacle to making exact copies of any signal. But if a quantum communication system uses non-orthogonal signal states, no process can perfectly duplicate them. This is the very reason why quantum cryptography is more powerful than classical cryptography.[3] It is also, rather unexpectedly, the principle that guarantees the no-communication theorem and prevents entangled quantum states from providing faster-than-light information transfer.

Exercise 7.5 We previously imagined a machine that, depending on its control setting, could produce arbitrarily many copies of the spin state $|z_+\rangle$ or $|x_+\rangle$. Explain why such a machine is not a violation of the no-cloning theorem.

7.3 Ebits

The *Bell states* are a set of entangled two-qubit states introduced by John Bell. They are

$$\left|\Phi_\pm^{(12)}\right\rangle = \frac{1}{\sqrt{2}}\left(|0,0\rangle \pm |1,1\rangle\right),$$

$$\left|\Psi_\pm^{(12)}\right\rangle = \frac{1}{\sqrt{2}}\left(|0,1\rangle \pm |1,0\rangle\right). \tag{7.13}$$

(Some of these are familiar; see Eq. 6.15 and 6.50.) The four Bell states are orthonormal and thus form a basis for the two-qubit Hilbert space. A basic two-qubit measurement using this basis is called a *Bell measurement*.

The Bell states have the property that any one of them can be transformed into any other by a unitary operation on just *one* of the qubits. For example, the Pauli operator $Z^{(1)}$ is unitary, and

$$Z^{(1)}\left|\Phi_+^{(12)}\right\rangle = \left|\Phi_-^{(12)}\right\rangle. \tag{7.14}$$

Exercise 7.6 Arrange the four Bell states in a square:

$$\left|\Phi_+^{(12)}\right\rangle \left|\Phi_-^{(12)}\right\rangle$$

$$\left|\Psi_+^{(12)}\right\rangle \left|\Psi_-^{(12)}\right\rangle. \tag{7.15}$$

Show that the operator $Z^{(1)}$ exchanges columns in this array, while $X^{(2)}$ exchanges rows. Show that $Z^{(2)}$ and $X^{(1)}$ do almost the same, except for some overall changes of phase.

---

[3] No classical communication system can accomplish the task of secret key distribution between Alice and Bob, since the eavesdropper Eve can make her own copies of all of the exchanged messages. But the BB84 quantum key distribution protocol uses non-orthogonal quantum signal states, and these cannot be cloned by Eve.

To get a physical picture of these mathematical results, suppose Alice and Bob share a pair of qubits in one of the Bell states. Then they can come to share any one of the Bell states simply by performing a local operation on one of the qubits.

Exercise 7.7 Consider the two-qubit state

$$\left|\Gamma^{(12)}\right\rangle = \frac{1}{\sqrt{2}}\left(|0,a\rangle \pm |1,b\rangle\right), \tag{7.16}$$

where $|a\rangle$ and $|b\rangle$ are an orthonormal pair of one-qubit states. Show that $|\Gamma^{(12)}\rangle$ can be converted into a Bell state by means of an operation on *either* qubit.

As we saw in Section 6.2, independent systems cannot become entangled by local operations. However, local operations can transform one entangled state to another. This reminds us of the discussion in Section 1.1 of the transformability of information. A one-bit message can have many different equivalent physical representations, and one representation can be changed to another by a physical operation. This motivates the following definition: When Alice and Bob jointly possess a qubit pair whose state is locally equivalent to a Bell state, then we say that they share an *ebit*, or a "bit of entanglement."

The idea is to think of "ebits" as one kind of resource with which Alice and Bob may perform various tasks. To formalize this notion, we suppose that Alice and Bob each have separate laboratories in which they can perform any local unitary operations and local measurements on the quantum systems in their possession. In addition, they have three types of resource that might be used for information tasks:

- **Bits.** Alice can send an ordinary one-bit message to Bob, or vice versa.
- **Qubits.** Alice can transfer a qubit quantum system to Bob, or vice versa.
- **Ebits.** Alice and Bob can share an entangled qubit pair in a Bell state (or some equivalent state).

Note that the first two types are *directed resources*; it matters whether Alice sends to Bob or Bob sends to Alice. Ebits, however, are not a directed resource.

To compare resources, we will use the symbol "$\succeq$" (pronounced "is at least as strong as"), so that

$$\text{this} \succeq \text{that}, \tag{7.17}$$

means that the resources enumerated in "this" can be used to do any task that can be performed using "that." Since Alice could always use a 10-bit message to send 5 bits of information, we write 10 bits $\succeq$ 5 bits. To take a more interesting example, it is always possible for Alice to encode a one-bit message in the basis states of a qubit, then transfer this qubit to Bob. Bob can recover the message by measuring the qubit in Alice's basis. We represent this as

$$1 \text{ qubit} \succeq 1 \text{ bit.} \tag{7.18}$$

Exercise 7.8 Explain why

$$1 \text{ qubit} \succeq 1 \text{ ebit.} \tag{7.19}$$

Resources are not always comparable, of course. As we have seen, shared entanglement does not in itself allow Alice to send messages to Bob using only local operations. It is also true that exchanging classical messages cannot produce entanglement in otherwise independent systems. This means that, for any N,

$$N \text{ ebits} \not\succeq 1 \text{ bit},$$

$$N \text{ bits} \not\succeq 1 \text{ ebit}. \tag{7.20}$$

Classical communication and entanglement are not comparable resources.

Exercise 7.9 Explain why 1 qubit $\not\succeq$ 2 bits. Is it true that 2 bits $\succeq$ 1 qubit?

So far, we have not said what sort of task is made possible by shared entanglement. This will be the business of the next section. For now, we will simply note the fact (worked out in Problem 6.5) that shared ebits would enable Alice and Bob to create a secure cryptographic key that could be used to communicate secret messages.

7.4 Using entanglement

Superdense coding

The basic decoding theorem (Eq. 4.8) tells us that Alice cannot reliably send two bits of information to Bob via a single qubit, because the number of possible messages ($N = 4$) is larger than the dimension of the Hilbert space ($d = 2$). Suppose, however, that Alice and Bob initially share an ebit. By a local unitary transformation on her own member of the qubit pair, Alice can change the state of the whole system into any of the four Bell states. She can use this to represent two bits of information:

Message	Bell state
00	$|\Phi_+\rangle$
01	$|\Phi_-\rangle$
10	$|\Psi_+\rangle$
11	$|\Psi_-\rangle$

Now Alice transfers her qubit to Bob. Bob, possessing both qubits, performs a Bell measurement to distinguish the four possible states. From the result he can exactly determine Alice's message. Therefore,

$$1 \text{ ebit} + 1 \text{ qubit} \succeq 2 \text{ bits}. \tag{7.21}$$

Entanglement by itself is no help for communication, but it apparently allows us to send two bits via a single qubit! For this reason, the technique just described is sometimes called *superdense coding*.

To appreciate how peculiar this is, consider how Alice and Bob obtained their ebit in the first place. One possibility is that Bob created an entangled qubit pair in his own lab and then transferred one qubit to Alice. The overall process then involves two qubit transfers, one in each direction. We can indicate direction with arrows, and write

$$1 \overrightarrow{\text{ qubit}} + 1 \overleftarrow{\text{ qubit}} \succeq 2 \overrightarrow{\text{ bits}}. \tag{7.22}$$

By sending one qubit in each direction, we can send a two-bit message in one direction. This would be impossible to understand if we regard a qubit merely as an ordinary message-carrier with a one-bit capacity. The relation between bits and qubits is far more subtle than that.

Teleportation

When Alice sends a qubit to Bob, the system is transferred with its state intact, whatever that state may be. This process is similar to the process of sending a message, but there are some essential differences. For instance, when Alice transmits an ordinary message she may retain a copy for herself; but the no-cloning theorem tells us that this cannot occur in a qubit transfer. Nevertheless, it is natural to refer to the transfer of an intact quantum state as the *quantum* communication of *quantum* information, by analogy with the *classical* communication of *classical* information in an ordinary message.

We know that we can accomplish classical communication by means of quantum communication. Can we do the opposite? Can Alice send a qubit to Bob using only bits?

The answer is no, because Alice cannot completely determine the state of the qubit by any measurement process, and therefore cannot provide enough classical information to Bob to construct the correct qubit state. But suppose Alice and Bob also share one or more ebits?

Let us analyze this in a more definite way. At the outset, Alice possesses a qubit (qubit #1, or the "input" qubit) that is in an arbitrary quantum state $|\psi\rangle = \alpha |0\rangle + \beta |1\rangle$. Alice and Bob also share a qubit pair (qubits #2 and #3) that are in the Bell state $\left|\Phi_+^{(23)}\right\rangle$. The overall three-qubit state is

$$\left|\Gamma^{(123)}\right\rangle = \left|\psi^{(1)}\right\rangle \otimes \left|\Phi_+^{(23)}\right\rangle$$
$$= \frac{\alpha}{\sqrt{2}} \left(|0,0,0\rangle + |0,1,1\rangle \right) + \frac{\beta}{\sqrt{2}} \left(|1,0,0\rangle + |1,1,1\rangle \right). \tag{7.23}$$

Now Alice makes a Bell measurement on the qubit pair in her possession (qubits #1 and #2). Each possible outcome of this measurement will have a certain probability and will result in a conditional state for Bob's qubit (qubit #3). These can be computed using the partial inner product. For example,

$$\left\langle \Phi_+^{(12)} \middle| \Gamma^{(123)} \right\rangle = \frac{1}{2} \left(\alpha \left|0^{(3)}\right\rangle + \beta \left|1^{(3)}\right\rangle \right). \tag{7.24}$$

This means that the $|\Phi_+\rangle$ outcome occurs with probability $(\frac{1}{2})^2 = \frac{1}{4}$, and that whenever this outcome occurs the conditional state of qubit #3 is $\alpha |0\rangle + \beta |1\rangle$.

In other words, after Alice performs her measurement and obtains the outcome $|\Phi_+\rangle$, the state of Bob's qubit *is exactly the state* $|\psi\rangle$, the original state of the input qubit. This is remarkable, especially considering the fact that Alice and Bob may be widely separated in space. It is as if the quantum state has instantaneously vanished from one location (due to Alice's measurement) and reappeared someplace else.

Of course, Alice cannot count on obtaining the result $|\Phi_+\rangle$ for the measurement, and Bob cannot know the outcome of Alice's measurement without some further communication. Suppose instead that Alice obtains the result $|\Phi_-\rangle$. Then we find that

$$\langle \Phi_-^{(12)} | \Gamma^{(123)} \rangle = \frac{1}{2} \left(\alpha \, |0^{(3)}\rangle - \beta \, |1^{(3)}\rangle \right), \tag{7.25}$$

indicating that this outcome also occurs with probability $\frac{1}{4}$ and leads to the conditional state $\alpha \, |0\rangle - \beta \, |1\rangle$ for qubit #3. This is almost the same as before, and in fact Bob can turn this conditional state into the input state $|\psi\rangle$ by applying the unitary operator Z on his qubit.

Exercise 7.10 For the Bell measurement outcomes $|\Psi_+\rangle$ and $|\Psi_-\rangle$ on Alice's qubit pair, show that the probability of each outcome is again $\frac{1}{4}$, and that the conditional state of Bob's qubit can be transformed into the input state $|\psi\rangle$ by the application of a simple unitary operator that depends on the measurement outcome but *not* $|\psi\rangle$.

Here is the upshot: Alice has the input qubit and one part of an ebit. She performs a Bell measurement on these qubits and obtains one of four possible results. She then communicates her result to Bob (requiring two bits of ordinary communication), and Bob uses this information to make one of four possible unitary transformations on his own qubit. At the end, Bob's qubit is in exactly the same state as the input qubit was – in effect, one qubit has been transferred from Alice to Bob. This process is called *teleportation*, and it provides another relation between communication resources:

$$1 \text{ ebit} + 2 \text{ bits} \succeq 1 \text{ qubit}. \tag{7.26}$$

Once again, teleportation challenges our intuitions about classical and quantum information. The original ebit shared by Alice and Bob could have arisen from quantum communication from Bob to Alice. This implies

$$1 \overleftarrow{\text{qubit}} + 2 \overrightarrow{\text{bits}} \succeq 1 \overrightarrow{\text{qubit}}. \tag{7.27}$$

If quantum communication is possible from Bob to Alice, and classical communication is possible from Alice to Bob, then teleportation allows quantum communication from Alice to Bob. The "quantum" part of quantum communication can be provided by shared entanglement (ebits), without regard to the process by which this entanglement is initially established.

Exercise 7.11 Comment on the following aphorism: "If entanglement is free, then one qubit is worth exactly two bits."

Teleportation and superdense coding tell us that quantum entanglement is a useful resource for the communication of quantum and classical information. The idea of comparing the resources required to do various quantum and classical tasks has become a common theme in quantum information science. One of its first and most enthusiastic proponents has been Charles Bennett, one of the discoverers of both superdense coding and teleportation. We have exhibited some of the basic relationships between bits, qubits, and ebits, which we summarize for convenience:

$$1 \text{ qubit} \succeq 1 \text{ bit}, \tag{Re 7.18}$$

$$1 \text{ qubit} \succeq 1 \text{ ebit}, \tag{Re 7.19}$$

$$1 \text{ ebit} + 1 \text{ qubit} \succeq 2 \text{ bits}, \tag{Re 7.21}$$

$$1 \text{ ebit} + 2 \text{ bits} \succeq 1 \text{ qubit}. \tag{Re 7.26}$$

These expressions were first written in this form by Bennett. Collectively, we may call them *Bennett's Laws*.

7.5 What is quantum information?

We have discussed the transmission of bits and qubits as if they involved the actual transport of particular physical systems from one point to another. But of course, our experience with classical communication tells us that the matter may be more complicated (see Exercise 1.1!). We say that a classical bit is communicated from Alice to Bob if the final bit possessed by Bob (call it b') is in a 0/1 state identical to that of the initial bit b possessed by Alice, whether or not this bit is represented in the same physical system. Similarly, we say that a qubit is communicated from Alice to Bob if Bob's final qubit Q' has the same state as Alice's initial qubit Q, whether or not they are the same physical system. (Indeed, in teleportation the two systems are never the same.)

The qubit is the one type of resource that appears in each of Bennett's Laws. But careful reflection reveals that this resource is used in two apparently different ways:

- Alice's Q may be initially in a definite pure state $|\psi^{(Q)}\rangle$, which during the communication process becomes the final state $|\psi^{(Q')}\rangle$ of Bob's qubit. This is how things work when we use a qubit for transmitting classical messages (Eq. 7.18), or in teleportation (Eq. 7.26).
- Alice's Q may be entangled with another system R, so that their joint state is $|\Psi^{(RQ)}\rangle$. During the communication process, this entanglement is transferred to Bob's qubit Q', so that the final state of RQ' is $|\Psi^{(RQ')}\rangle$. This is the process used in both sharing entanglement (Eq. 7.18) and superdense coding (Eq. 7.21).

We will call these "Type I" and "Type II" quantum communication, respectively. They embody two distinct answers to the question, "What is quantum information?" For Type I, quantum information lies in the (possibly unknown) state vector of the system Q. This is more analogous to classical communication, in which a definite classical state is faithfully reproduced at the destination. For Type II, quantum information lies in the entangled state of system Q and a "bystander" system R, which does not participate in the communication process but merely serves as an "anchor" for the entanglement carried by Q. This seems to be a very different picture.[4]

The two types, however, are very closely related. We will now argue that the ability to do Type II quantum communication for a *single* entangled input state guarantees the ability to do Type I quantum communication for *any* input state.

Suppose R and Q each have Hilbert space dimension d and start out in a *maximally entangled state*, a uniform, "diagonal" superposition of product basis states:

$$\left|\Phi^{(RQ)}\right\rangle = \frac{1}{\sqrt{d}} \sum_k \left|k^{(R)}, k^{(Q)}\right\rangle. \tag{7.28}$$

(The Bell states are maximally entangled states of two qubits.) If any basic measurement is performed on R, each outcome has probability $1/d$. The outcome associated with the R-basis element $|a^{(R)}\rangle$ leads to the conditional Q-state

$$\left|\psi_a^{(Q)}\right\rangle = \sqrt{d} \left\langle a^{(R)} \middle| \Phi^{(RQ)}\right\rangle, \tag{7.29}$$

(compare Eq. 6.49).

Exercise 7.12 Show that, for any Q-state $|\psi^{(Q)}\rangle$, there is an R-state $|a^{(R)}\rangle$ so that $|\psi^{(Q)}\rangle = |\phi_a^{(Q)}\rangle$. That is, given $|\Phi^{(RQ)}\rangle$, any Q-state could arise as a conditional state for some R-measurement outcome.

Suppose the entanglement of Q with R is transferred to Q′ by a Type II quantum communication process. As in the no-cloning theorem, we assume that the entire system (including all parts of the communication mechanism, which we will collectively call C) can be regarded as informationally isolated during the process. It is also isolated from the bystander system R. The overall time evolution will be described by a unitary operator of the form $\mathbf{1}^{(R)} \otimes U^{(QQ'C)}$.

Now consider the following exercise:

Exercise 7.13 The composite quantum system AB is initially in the state $|\Psi^{(AB)}\rangle$, and B evolves by the unitary operator $V^{(B)}$. Show that, for any A-state $|a^{(A)}\rangle$,

$$\left\langle a^{(A)} \middle| \left(\mathbf{1}^{(A)} \otimes V^{(B)}\right) \middle| \Psi^{(AB)}\right\rangle = V^{(B)} \left\langle a^{(A)} \middle| \Psi^{(AB)}\right\rangle. \tag{7.30}$$

We can interpret this exercise by comparing two processes:

[4] From another point of view, however, Type II quantum communication is not so different from classical communication. If Alice sends a bit b to Bob, what does it mean operationally to say that Bob's received bit b' is correct? This presumes the potential existence of a "reference bit," a fiducial copy of b to which b' may be compared. Alice, for example, could retain her own copy of the transmitted message, and later on this can be compared with Bob's version. The communication is accurate provided the two bits are properly correlated. In quantum communication we cannot make and keep copies, but the bystander system R plays a similar role. The quantum communication from Alice to Bob is successful provided R and Q′ are properly correlated (i.e., in the correct entangled state).

- First we evolve the joint state $|\Psi^{(AB)}\rangle$ according to $\mathbf{1}^{(A)} \otimes V^{(B)}$, then we use the result to find the conditional state for $|a^{(A)}\rangle$.
- First we use $|\Psi^{(AB)}\rangle$ to find the conditional state for $|a^{(A)}\rangle$, then we evolve this state according to $V^{(B)}$.

Both processes lead to exactly the same final state of B. The dynamical evolution of B *commutes* with the measurement process on A.[5]

Let us apply this fact to our communication problem. By assumption, our procedure produces a final RQ$'$ state $|\Phi^{(RQ')}\rangle$ that is the same as the initial state. A subsequent R-measurement with outcome basis state $|a^{(R)}\rangle$ will result in a conditional Q$'$-state $|\phi_a^{(Q')}\rangle$. But the R-measurement must commute with the communication process. If we do the R-measurement first, the pure state $|\phi_a^{(Q)}\rangle$ is the input to the communication process, and this process must lead to the same output state $|\phi_a^{(Q')}\rangle$. Exercise 7.12 tells us that any Q-state could arise in this way. Therefore, a Type II communication procedure for Q that accurately transfers the entangled state $|\Phi^{(RQ)}\rangle$ will also work as a Type I communication procedure for any input Q-state.

It is also true that any perfect Type I quantum communication process will also convey entanglement equally well. (We will defer a general argument for this until Problem 9.3 of Chapter 9, but see Problem 7.5 at the end of this chapter.) The two apparently distinct ways of thinking about the quantum information in Q – as the definite unknown state of Q or as the entanglement between Q and an outside system R – are really equivalent.

So far, we have discussed *perfect* quantum communication. But suppose the process is not perfect? In classical information, we often characterize this by the probability of error P_E, which is the probability that the output of the communication process disagrees with the input. This is, for example, the "figure of merit" we used in the basic decoding theorem of Section 4.1 and the basic distinguishability theorem of Section 4.2. If P_E is small, then the probability of success $P_S = 1 - P_E$ is nearly 1. This would mean that our process is *nearly* perfect, and in many practical situations this will be good enough.

For quantum information, the corresponding "figure of merit" of a communication process is *fidelity*. If the desired state is $|\psi\rangle$ but the actual state is $|\psi'\rangle$, then the fidelity F of the process is

$$F = |\langle \psi | \psi' \rangle|^2. \tag{7.31}$$

The fidelity satisfies $0 \leq F \leq 1$. It is zero when the two states are distinguishable, and is close to 1 when they are nearly indistinguishable. If $F = 1$, the two states must be identical up to overall phase.[6] (We are implicitly supposing that the input and output states are states

---

[5] A variant of this argument is the key step in the proof of the representation theorem for generalized quantum evolution maps, given in Section D.2 of Appendix D.

[6] Some authorities define the fidelity to be $|\langle \psi | \psi' \rangle|$, which we would call $\sqrt{F}$. This alternative definition has some mathematical advantages. We prefer Eq. 7.31, however, because it has a simple interpretation: F is the probability that $|\psi'\rangle$ would pass a basic measurement test designed to determine whether the state is $|\psi\rangle$. The fidelity F will be generalized to mixed states in Problem 8.4.

of the *same* system Q, rather than *isomorphic* systems Q and Q'. This useful simplification will be used for the rest of the discussion.)

The general idea of fidelity is applied in somewhat different ways to Type I and Type II quantum communication schemes. In the Type I situation, we imagine that the input state $|\psi_k\rangle$ is sent with probability p_k, and that it winds up as the output state $|\psi'_k\rangle$. Then it makes sense to compute the *average fidelity* $\bar{F}$:

$$\bar{F} = \sum_k p_k \left|\langle \psi_k | \psi'_k \rangle\right|^2. \tag{7.32}$$

In a Type II situation, an entangled state $|\Psi^{(RQ)}\rangle$ is sent, yielding the output state $|\Psi'^{(RQ)}\rangle$. Then we compute the *entanglement fidelity* F_e:

$$F_e = \left|\langle \Psi^{(RQ)} | \Psi'^{(RQ)} \rangle\right|^2. \tag{7.33}$$

In either case, a fidelity of 1 means that the quantum communication is perfect.

The two fidelities are connected. Suppose that, for some basis states $|k^{(R)}\rangle$ for R, we can write the entangled (Type II) input state as

$$|\Psi^{(RQ)}\rangle = \sum_k \sqrt{p_k} \, |k^{(R)}, \psi_k^{(Q)}\rangle. \tag{7.34}$$

If we make an R-measurement in this basis, we will produce the conditional Q-state $|\psi_k^{(Q)}\rangle$ with probability p_k, exactly as in the Type I problem. Since the R-measurement must commute with the quantum communication process, the final Type II state must be

$$|\Psi'^{(RQ)}\rangle = \sum_k \sqrt{p_k} \, |k^{(R)}, \psi_k'^{(Q)}\rangle, \tag{7.35}$$

for the Type I output states $|\psi_k'^{(Q)}\rangle$.

Exercise 7.14 Explain in detail why this is so.

The entanglement fidelity is

$$F_e = \left|\langle \Psi^{(RQ)} | \Psi'^{(RQ)} \rangle\right|^2 = \left|\sum_{kl} \sqrt{p_k p_l} \, \langle k^{(R)} | l^{(R)} \rangle \langle \psi_k | \psi'_l \rangle\right|^2$$

$$= \left|\sum_k p_k \langle \psi_k | \psi'_k \rangle\right|^2. \tag{7.36}$$

What can we do with this last expression? Let the complex number $\langle \psi_k | \psi'_k \rangle = x_k + iy_k$ for real x_k and y_k. These real numbers are the values of two real random variables X and Y with common probabilities p_k. Then

$$F_e = |\langle X + iY \rangle|^2 = |\langle X \rangle + i \langle Y \rangle|^2$$
$$= \langle X \rangle^2 + \langle Y \rangle^2$$
$$\leq \left\langle X^2 \right\rangle + \left\langle Y^2 \right\rangle$$
$$= \left\langle |X + iY|^2 \right\rangle = \bar{F}. \tag{7.37}$$

Exercise 7.15 Explain each step in this derivation.

Therefore $F_e \leq \bar{F}$. The Type I problem is "no harder" than the related Type II problem, since its fidelity is always at least as large. If the entanglement fidelity approaches 1, then the average fidelity must do so as well. The reverse is not necessarily true; it may be that $\bar{F} = 1$ for some *particular* collection of input states, but that F_e is nevertheless small.

Exercise 7.16 A unitary Z rotation of one qubit can transform the Bell state $|\Phi_+\rangle$ into $|\Phi_-\rangle$. Use this fact to construct an example in which $\bar{F} = 1$ for some set of inputs, but the corresponding entanglement fidelity $F_e = 0$.

How shall we answer the question posed in this section: "What is quantum information?" We have proposed Type I and Type II ideas and argued that they are, in some sense, equivalent. However, two facts suggest that the Type II viewpoint is primary. The exact transfer of a *single* entangled input by a quantum communication process is enough to guarantee that *every* pure state input is transferred correctly. Furthermore, we know that $F_e \leq \bar{F}$ for related Type I and Type II problems; and we can find examples where $F_e = 0$ and $\bar{F} = 1$. This suggests that F_e is a stricter, more comprehensive way to characterize the overall "quantum fidelity" of the process.

For these reasons, we conclude that the deepest answer to the question is that quantum information lies in the entanglement between systems. Quantum communication, in this view, is fundamentally about the transfer of that entanglement from one system to another. The essential role of Bennett's qubit resource is to move ebits around.

Problems

Problem 7.1 Quantum cloning machine I. In this problem and the next, you will try your hand at designing a possible quantum machine that does something akin to cloning two non-orthogonal qubit states $|0\rangle$ and $|+\rangle$. The first is an *approximate* cloning machine, and is required to have the following characteristics:

- The inner workings of the machine are completely unitary.
- The states update according to

$$|0,0\rangle \rightarrow |c_{00}\rangle \qquad |+,0\rangle \rightarrow |c_{++}\rangle, \tag{7.38}$$

where $|c_{00}\rangle$ and $|c_{++}\rangle$ are the "cloned" $|0\rangle$ and $|+\rangle$ states.

- The cloning fidelity $F_c = |\langle 00 | c_{00} \rangle|^2 = |\langle + + | c_{++} \rangle|^2$, equal for the two possible states, is as close to 1 as possible. ($F_c = 1$ would mean perfect cloning.)

Design your machine. How large can you make F_c?

Problem 7.2 Quantum cloning machine II. Here is another approach at almost-cloning for non-orthogonal input states $|0\rangle$ and $|+\rangle$. This second device is a *conditional* cloning machine, and is must have the following characteristics:

- The inner workings of the machine may employ random choices ("coin flips") and measurements.
- The machine does not always succeed in its task. When it fails, it announces "FAIL" and destroys the input qubit state.
- When the machine does succeed, it can produce as many exact duplicates ($F_c = 1$) of the input state as may be required.
- The machine is equally likely to work on either of the two inputs.

Design your machine. What is the probability that it works?

Problem 7.3 Preparation machines. A device P interacts with a quantum system Q as follows: initially, Q is in the standard state $|0\rangle$, while P is initially either in $|s_1\rangle$ or $|s_2\rangle$. The unitary time evolution of PQ maps

$$
\begin{aligned}
|s_1\rangle \otimes |0\rangle &\rightarrow |s_1\rangle \otimes |\psi_1\rangle, \\
|s_2\rangle \otimes |0\rangle &\rightarrow |s_2\rangle \otimes |\psi_2\rangle.
\end{aligned}
$$

In other words, the state of P is unchanged, but Q is changed to one state or another. You should interpret $|s_1\rangle$ and $|s_2\rangle$ as two different "settings" of the device P, and these settings determine which of the two Q-states $|\psi_1\rangle$ or $|\psi_2\rangle$ are prepared. Show that, if $|\langle \psi_1 | \psi_2 \rangle| < 1$, then $|s_1\rangle$ and $|s_2\rangle$ are orthogonal.[7]

Problem 7.4 Bennett also introduced a fourth information resource, the *sbit*. Alice and Bob possess an sbit if they share a random classical bit of data that is secret from everyone else, including any potential eavesdropper. Sbits can be used as the key in a cryptographic scheme, so the problem of key distribution is exactly the problem of distributing sbits. Devise some extensions of Bennett's Laws to show how sbits are related to bits, qubits, and ebits.

Problem 7.5 Suppose in the teleportation protocol, Alice's input qubit is entangled with another qubit, which we will designate #0. The initial state of the four qubits involved is thus

$$
\left| \Gamma^{(0123)} \right\rangle = \left| \Psi^{(01)} \right\rangle \otimes \left| \Phi_+^{(23)} \right\rangle. \tag{7.39}
$$

---

[7] Let us explain this result in more concrete terms. In a laboratory, the equipment configuration that creates $|\psi_1\rangle$ must be entirely distinguishable from that which creates $|\psi_2\rangle$, even if $|\psi_1\rangle$ and $|\psi_2\rangle$ are not very distinguishable as Q-states. Thus, if we need to make a preparation machine P that can create 10^6 different Q-states, then we need dim $\mathcal{H}^{(P)} \geq 10^6$, even if Q itself is just a qubit.

The teleportation process on qubit #1 proceeds just as before. Show that, at the end of the procedure, qubit #0 is entangled with qubit #3 in the state $|\Psi^{(03)}\rangle$. This shows that the teleportation protocol may be used for either Type I or Type II quantum communication.

Problem 7.6 If we have more than two communicators, then the laws governing quantum information resources become less well understood. Suppose Alice, Bob, and Charles (our "players") can share entanglement and send quantum messages to each other. Further suppose that classical communication is so easy that we do not even keep track of it as a resource. Any number of bits may be transmitted for free. (This makes ebits equivalent to qubits.)

Let us keep track of two kinds of resource: ebits (between various pairs of players) and *ghzbits*, in which the three players share a triple of qubits in a GHZ state (see Eq. 6.64). Prove the following two relations:

$$1 \text{ ghzbit} \succeq 1 \text{ ebit}^{(AB)},$$
$$1 \text{ ebit}^{(AB)} + 1 \text{ ebit}^{(BC)} \succeq 1 \text{ ghzbit}. \tag{7.40}$$

Density operators

8.1 Beyond state vectors

We cannot always assign a definite state vector $|\psi\rangle$ to a quantum system Q. It may be that Q is part of a composite system RQ that is in an entangled state $|\Psi^{(RQ)}\rangle$. Or it may be that our knowledge of the preparation of Q is insufficient to determine a particular state $|\psi\rangle$. Consider, for instance, a qubit sent from Alice to Bob during the BB84 key distribution protocol from Section 4.4. The state of this qubit could be $|0\rangle$, $|1\rangle$, $|+\rangle$ or $|-\rangle$, each with equal likelihood. In either case – whether Q is a subsystem of an entangled system, or the state of Q is determined by a probabilistic process – we cannot specify a quantum state vector $|\psi\rangle$ for Q.

Nevertheless, in either case we are in a position to make statistical predictions about the outcomes of measurements on Q. In this chapter we describe the mathematical machinery for doing this.

Mixtures of states

Suppose the state of Q arises by a random process, so that the state $|\psi_\alpha\rangle$ is prepared with probability p_α. The possible states $|\psi_\alpha\rangle$ need not be orthogonal (as you can see in the BB84 example above). We call this situation a *mixture* of the states $|\psi_\alpha\rangle$, or a *mixed state* for short.

One way to interpret a mixed state is to return to the idea of an ensemble of systems, which we introduced in Section 3.3. The individual systems in our ensemble are prepared in one of the various possible states, with a fraction p_α of them prepared in $|\psi_\alpha\rangle$. Now suppose we measure an observable A on this system. For the subset of the ensemble with the state $|\psi_\alpha\rangle$, the average measured value will be $\langle A\rangle_\alpha = \langle\psi_\alpha|A|\psi_\alpha\rangle$. Over the entire ensemble the average will be

$$\langle A\rangle = \sum_\alpha p_\alpha \langle A\rangle_\alpha$$

$$= \sum_\alpha p_\alpha \operatorname{Tr} |\psi_\alpha\rangle\langle\psi_\alpha| A. \tag{8.1}$$

(To remind yourself of the definition and properties of the trace, refer to the discussion that begins with Eq. 3.44.) We now define the *density operator* ρ to be

$$\rho = \sum_\alpha p_\alpha |\psi_\alpha\rangle\langle\psi_\alpha|. \tag{8.2}$$

This operator depends on the possible states and their probabilities, but not on the observable A. Since the trace operation is linear, the expectation is

$$\langle A\rangle = \mathrm{Tr}\,\rho\,\mathsf{A}. \tag{8.3}$$

The density operator ρ therefore lets us compute the (ensemble average) expectation for any observable A.

Exercise 8.1 We have introduced the density operator before, at the end of Section 3.3. The state vector $|\psi\rangle$ was associated with the density operator $\rho = |\psi\rangle\langle\psi|$. Eq. 8.3 here is identical to Eq. 3.72 there. Explain how the previous concept of the density operator is a special case of the present one.

This sort of probabilistic combination of states is called a *mixture* of states, and the resulting situation (described by ρ) is a *mixed state*. A mixture of states is a very different thing from a superposition. A superposition yields a definite state vector, whereas a mixture does not and so must be described by a density operator. The following exercise makes this distinction clearer:

Exercise 8.2 Think of a qubit with basis states $|0\rangle$ and $|1\rangle$. Consider (a) a superposition of $|0\rangle$ and $|1\rangle$ with equal amplitudes, and (b) a mixture of these two states with equal probabilities. Show that these two situations lead to the same value for $\langle Z\rangle$ but different values for $\langle X\rangle$.

Those situations for which a definite state vector exists are also known as *pure states*.

Different mixtures of states can lead to the same density operator. Consider an equiprobable mixture of qubit states $|0\rangle$ and $|1\rangle$. Then

$$\rho = \frac{1}{2}|0\rangle\langle 0| + \frac{1}{2}|1\rangle\langle 1| = \frac{1}{2}\mathbf{1}. \tag{8.4}$$

Alternatively, suppose $|+\rangle$ and $|-\rangle$ appear with equal likelihood:

$$\rho = \frac{1}{2}|+\rangle\langle +| + \frac{1}{2}|-\rangle\langle -| = \frac{1}{2}\mathbf{1}. \tag{8.5}$$

Imagine two ensembles, one composed of equal numbers of $|0\rangle$ and $|1\rangle$ states and the other composed of equal numbers of $|+\rangle$ and $|-\rangle$ states. Since the two mixtures have the same density operator ρ, Eq. 8.3 tells us that the two ensembles have exactly the same expectations $\langle A\rangle$ for any observable. In other words, the two ensembles are statistically indistinguishable.

Exercise 8.3 Devise a mixture of three equally-likely qubit states that also yields the density operator $\rho = \frac{1}{2}\mathbf{1}$.

Subsystems of entangled systems

Now suppose that a composite system RQ is in an entangled state $|\Psi^{(RQ)}\rangle$. As we saw in Section 6.4, a basic measurement on R will lead to a mixture of conditional states on Q. For the R-basis element $|k^{(R)}\rangle$, we use the partial inner product to write:

$$\sqrt{p_k}\,|\psi_k^{(Q)}\rangle = \langle k^{(R)}\,|\Psi^{(RQ)}\rangle, \tag{8.6}$$

where p_k is the probability of the kth measurement result and $|\psi_k^{(Q)}\rangle$ is the corresponding conditional Q-state. The density operator for the mixture of Q-states is

$$\begin{aligned}
\rho^{(Q)} &= \sum_k p_k\,|\psi_k^{(Q)}\rangle\langle\psi_k^{(Q)}| \\
&= \sum_k \langle k^{(R)}\,|\Psi^{(RQ)}\rangle\langle\Psi^{(RQ)}\,|k^{(R)}\rangle. \tag{8.7}
\end{aligned}$$

We can formally write this as

$$\rho^{(Q)} = \sum_k \langle k^{(R)}|\,\rho^{(RQ)}\,|k^{(R)}\rangle, \tag{8.8}$$

where $\rho^{(RQ)} = |\Psi^{(RQ)}\rangle\langle\Psi^{(RQ)}|$.

The expression on the right-hand side of Eq. 8.8 needs a bit of explanation. It resembles the trace operation on $\rho^{(RQ)}$ (compare Eq. 3.46), except that it involves only the basis for $\mathcal{H}^{(R)}$.

We make sense of this by introducing the *partial trace* of an operator on $\mathcal{H}^{(RQ)}$. Suppose we have an operator that is an outer product of product vectors: $G^{(RQ)} = |\alpha^{(R)}, \phi^{(Q)}\rangle\langle\beta^{(R)}, \psi^{(Q)}|$. Then we define

$$\begin{aligned}
\mathrm{Tr}_{(R)} G^{(RQ)} &= \mathrm{Tr}_{(R)}\left(|\alpha^{(R)}, \phi^{(Q)}\rangle\langle\beta^{(R)}, \psi^{(Q)}|\right) \\
&= \mathrm{Tr}_{(R)}\left(|\alpha^{(R)}\rangle\langle\beta^{(R)}| \otimes |\phi^{(Q)}\rangle\langle\psi^{(Q)}|\right) \\
&= \langle\beta^{(R)}\,|\alpha^{(R)}\rangle\,|\phi^{(Q)}\rangle\langle\psi^{(Q)}|. \tag{8.9}
\end{aligned}$$

The partial trace turns the outer product on $\mathcal{H}^{(R)}$ into the corresponding inner product. We can extend this definition to arbitrary operators by requiring that $\mathrm{Tr}_{(R)}$ is linear. In any case, the partial trace $\mathrm{Tr}_{(R)}$ of an operator on $\mathcal{H}^{(RQ)}$ is an operator on $\mathcal{H}^{(Q)}$.

The partial trace can be computed using a basis for $\mathcal{H}^{(R)}$. For the operator $G^{(RQ)}$ above

$$\begin{aligned}
\mathrm{Tr}_{(R)} G^{(RQ)} &= \left(\sum_k \langle\beta^{(R)}\,|k^{(R)}\rangle\langle k^{(R)}\,|\alpha^{(R)}\rangle\right)|\phi^{(Q)}\rangle\langle\psi^{(Q)}| \\
&= \sum_k \langle k^{(R)}\,|\alpha^{(R)}, \phi^{(Q)}\rangle\langle\beta^{(R)}, \psi^{(Q)}\,|k^{(R)}\rangle \\
&= \sum_k \langle k^{(R)}|\,G^{(RQ)}\,|k^{(R)}\rangle. \tag{8.10}
\end{aligned}$$

This result can be extended via linearity to an arbitrary operator on $\mathcal{H}^{(AB)}$. Because our original definition of $\mathrm{Tr}_{(R)}$ did not depend on a choice of basis for $\mathcal{H}^{(R)}$, this expression for the partial trace is basis-independent.

The density operator $\rho^{(Q)}$ from Eq. 8.8 (describing the mixture of conditional states after a measurement on R) is thus

$$\rho^{(Q)} = \mathrm{Tr}_{(R)}\rho^{(RQ)}. \tag{8.11}$$

Because the partial trace is basis-independent, the same density operator would arise for the conditional Q-states for *any* R-measurement. Therefore, the choice of R-measurement cannot have any effect on the ensemble averages of any Q-observable.

Even if we have not made any R-measurement at all, it makes sense to use the density operator $\rho^{(Q)}$ from Eq. 8.11 to describe the state of subsystem Q. For a composite system RQ in the joint state $|\Psi^{(RQ)}\rangle$, the expectation value of an observable $A^{(Q)}$ on Q is

$$
\begin{aligned}
\langle A^{(Q)}\rangle &= \langle \Psi^{(RQ)}|\left(\mathbf{1}^{(R)}\otimes A^{(Q)}\right)|\Psi^{(RQ)}\rangle \\
&= \sum_k \langle \Psi^{(RQ)}|\left(|k^{(R)}\rangle\langle k^{(R)}|\otimes A^{(Q)}\right)|\Psi^{(RQ)}\rangle \\
&= \sum_k \langle \Psi^{(RQ)}|k^{(R)}\rangle A^{(Q)}\langle k^{(R)}|\Psi^{(RQ)}\rangle \\
&= \mathrm{Tr}_{(Q)}\left(\sum_k \langle k^{(R)}|\Psi^{(RQ)}\rangle\langle \Psi^{(RQ)}|k^{(R)}\rangle\right) A^{(Q)}.
\end{aligned}
\tag{8.12}
$$

Therefore,

$$\langle A^{(Q)}\rangle = \mathrm{Tr}_{(Q)}\rho^{(Q)}A^{(Q)}. \tag{8.13}$$

The density operator $\rho^{(Q)} = \mathrm{Tr}_{(R)}|\Psi^{(RQ)}\rangle\langle\Psi^{(RQ)}|$ gives us the expectation value of any observable on Q. If RQ is in an entangled state, we cannot assign a pure state to Q. But when we consider only the observable properties of Q by itself, we can reasonably assign to Q the mixed state $\rho^{(Q)}$ given by the partial trace.

Exercise 8.4 Consider the entangled states $|\Phi_+\rangle$ and $|\Psi_-\rangle$ for a pair of qubits, defined in Eq. 6.15 and 6.50. In each case, find density operators to describe the subsystem states of each qubit. Do the subsystem density operators completely determine the joint state of the composite system?

General considerations

We have used the density operator $\rho^{(Q)}$ to describe two distinct situations:

- The exact preparation of Q is not known, but various possible states $|\psi_\alpha\rangle$ have probabilities p_α. In this case, $\rho^{(Q)}$ describes the statistical properties of the mixture of states.

- Q is part of a composite system RQ that has a definite state vector $|\Psi^{(RQ)}\rangle$; however, only experiments on Q alone are to be performed. In this case, $\rho^{(Q)}$ describes the statistical properties of the subsystem Q.

These two situations are related. Suppose the composite system RQ is in a known joint entangled state $|\Psi^{(RQ)}\rangle$, but R and Q are located in two separated laboratories. Experimenters working with system Q will ascribe a density operator $\rho^{(Q)}$ to this system, since they are in the second situation above. Now word comes that the other laboratory has made a measurement on R (though the measurement result is not announced). The Q-experimenters retain the same density operator $\rho^{(Q)}$, though their interpretation of it is now different. Depending on the result of the R-measurement, the system Q is in one pure state or another, but it is not possible to say which one. In other words, the experimenters now say that $\rho^{(Q)}$ describes a mixture of states, the first situation above.

Exercise 8.5 Imagine that the composite system RQ is itself described by a mixture of joint states. Show that Eq. 8.11 still makes sense to define the state of the subsystem Q.

Here is a trio of important facts, which we will present as exercises:

Exercise 8.6 Show that any density operator ρ must be a positive operator.

Exercise 8.7 Consider the density operator ρ on $\mathcal{H}^{(Q)}$. Show that

$$\mathrm{Tr}\,\rho = 1, \tag{8.14}$$

whether ρ arises from (a) a mixture of Q-states, or (b) a pure (possibly entangled) state of a composite system RQ.

Exercise 8.8 Suppose ρ is a positive operator on $\mathcal{H}^{(Q)}$ whose trace is 1. Show that ρ is a possible density operator arising from either (a) a mixture of Q-states, or (b) a pure state of a composite system RQ.

Together, these establish that the set of possible density operators for Q is exactly the set of positive operators on $\mathcal{H}^{(Q)}$ that have unit trace. A final exercise shows that a probabilistic mixture of mixed states is a possible mixed state:

Exercise 8.9 Suppose ρ_1 and ρ_2 are density operators, and let p_1 and p_2 be non-negative real numbers with $p_1 + p_2 = 1$. Show that

$$\rho = p_1\rho_1 + p_2\rho_2, \tag{8.15}$$

is a possible density operator – i.e. is positive and has unit trace.

8.2 Matrix elements and eigenvalues

Populations and coherences

Let us write the density operator ρ in terms of the basis $\{\,|n\rangle\,\}$:

$$\rho = \sum_{m,n} \rho_{mn}\,|m\rangle\langle n|\,. \tag{8.16}$$

The matrix representation of a density operator is called, reasonably enough, a *density matrix*. The diagonal matrix element $\rho_{nn} = \langle n|\,\rho\,|n\rangle$ is the probability that a basic measurement in this basis will yield the result n. If we imagine performing such a measurement on the members of an ensemble of systems described by ρ, then ρ_{nn} is the fraction of those systems that produce the result n. For this reason, the diagonal elements ρ_{nn} of a density matrix are sometimes called *populations*.

The off-diagonal density matrix elements tell us something different. Suppose we know only that the diagonal matrix elements $\rho_{00} = \rho_{11} = \frac{1}{2}$ for the density operator of a qubit in the standard basis. There are many possible states consistent with this information. For example, an equal mixture of $|0\rangle$ and $|1\rangle$ will have a density matrix

$$(\rho_{mn}) = \frac{1}{2}\begin{pmatrix} 1 & 0 \\ 0 & 1 \end{pmatrix}, \tag{8.17}$$

whose off-diagonal elements are zero. On the other hand, the superposition $|+\rangle$ has the density matrix

$$(\rho_{mn}) = \frac{1}{2}\begin{pmatrix} 1 & 1 \\ 1 & 1 \end{pmatrix}. \tag{8.18}$$

The distinction between a *mixture* and a *superposition* of $|0\rangle$ and $|1\rangle$ can be found in the off-diagonal elements of the density matrix.

These matrix elements encode the relative phases (if any) between the basis elements in the state, as illustrated in the following exercise.

Exercise 8.10 Show that the state $|\psi\rangle = \frac{1}{\sqrt{2}}\left(|0\rangle + e^{i\phi}\,|1\rangle\right)$ has a density matrix in the standard basis

$$(\rho_{mn}) = \frac{1}{2}\begin{pmatrix} 1 & e^{-i\phi} \\ e^{i\phi} & 1 \end{pmatrix}. \tag{8.19}$$

In a superposition of basis states, there is a definite phase relation between the terms of the superposition. We say that a superposition is *coherent*, and the off-diagonal elements of the density matrix are called *coherences*. But suppose that the relative phase ϕ in the state $|\psi\rangle$ above is not known. Instead of a single state $|\psi\rangle$ with a single ϕ, we have an equal mixture of all possible values of ϕ between 0 and 2π. The density matrix will then be an average of matrices like Eq. 8.19 over the possible values of ϕ. The populations (diagonal elements) are unaffected by the average, but the ensemble average $\langle e^{i\phi}\rangle = 0$. The resulting density matrix will look like Eq. 8.17, an "incoherent" mixture of $|0\rangle$ and $|1\rangle$. A mixture of superposition states with random relative phases is thus equivalent to a mixture of the basis states.

The difference between a superposition and a mixture is the possibility of interference effects. Neither $|0\rangle$ nor $|1\rangle$ has a definite value of the observable X, but the coherent superposition $|+\rangle$ does, because the probability amplitudes for the measurement outcomes interfere (constructively for $X = +1$, destructively for $X = -1$). A mixture of $|0\rangle$ and $|1\rangle$ will show no such interference effects; the probabilities of the measurement results ± 1 will be simple mixtures of the probabilities given by the basis states.

Exercise 8.11 Explain the preceding paragraph in terms of the two-beam interferometer from Chapter 2, using the $|u\rangle$ and $|l\rangle$ states as the standard basis. What measurement corresponds to the observable X?

Between the two extremes – between a perfectly coherent superposition and a completely incoherent mixture – there are intermediate cases in which the coherences of the density matrix are smaller in magnitude but not zero. Interference effects are observable for these states, but are weaker.

Exercise 8.12 Consider the qubit state with density matrix

$$(\rho_{mn}) = \frac{1}{2} \begin{pmatrix} 1 & K \\ K & 1 \end{pmatrix}, \tag{8.20}$$

where $0 \leq K \leq 1$. A measurement of X is made on this qubit. Calculate $|P(+1) - P(-1)|$ as a function of K. (In the two-beam interferometer realization, this is sometimes called the *visibility* of the interference effects.)

Any physical process which has the effect of suppressing the coherences of the density matrix (in some particular basis) is called a *decoherence* process. We will discuss such processes in Chapter 9. For now, we simply note that a decoherence process destroys interference effects between probability amplitudes. Problem 8.5 analyzes an explicit example.

Schmidt decomposition

Next, suppose that our basis $\{|k\rangle\}$ is composed of eigenstates of the density operator ρ. The density matrix in this basis is diagonal. The eigenvalue spectrum $\{\lambda_k\}$ is a collection of non-negative real numbers that add up to 1 – in other words, a probability distribution. We can write ρ in its spectral decomposition:

$$\rho = \sum_k \lambda_k |k\rangle\langle k|. \tag{8.21}$$

This looks like a mixture in which the eigenbasis state $|k\rangle$ appears with probability λ_k, and indeed this is one possible mixture that leads to ρ. (As we have seen, it is by no means the only such mixture.)

Imagine that the density operator $\rho^{(R)}$ of system R arises from the pure entangled state $|\Psi^{(RQ)}\rangle$ of the composite system RQ. We can use the eigenbasis of $\rho^{(R)}$ to expand the entangled state:

$$|\Psi^{(RQ)}\rangle = \sum_k |k^{(R)}\rangle \otimes |\psi_k^{(Q)}\rangle. \tag{8.22}$$

The density operator is thus

$$\rho^{(R)} = \text{Tr}_{(Q)} |\Psi^{(RQ)}\rangle\langle\Psi^{(RQ)}| = \sum_{k,k'} \langle\psi_{k'}^{(Q)}|\psi_k^{(Q)}\rangle |k^{(R)}\rangle\langle k'^{(R)}|. \tag{8.23}$$

Comparing this to the spectral decomposition of $\rho^{(R)}$, we conclude that

$$\langle\psi_{k'}^{(Q)}|\psi_k^{(Q)}\rangle = \lambda_k \, \delta_{kk'}. \tag{8.24}$$

The vectors $|\psi_k^{(Q)}\rangle = \langle k^{(R)}|\Psi^{(RQ)}\rangle$ are orthogonal. We can find an orthonormal set of vectors $|k^{(Q)}\rangle$ such that

$$|\psi_k^{(Q)}\rangle = \sqrt{\lambda_k}\,|k^{(Q)}\rangle. \tag{8.25}$$

We can, if need be, extend this set into a basis for $\mathcal{H}^{(Q)}$. Our entangled state is

$$|\Psi^{(RQ)}\rangle = \sum_k \sqrt{\lambda_k}\,|k^{(R)}, k^{(Q)}\rangle. \tag{8.26}$$

This is called the *Schmidt decomposition* of the entangled state $|\Psi^{(RQ)}\rangle$.

Here is what we have proven: Given a joint state vector $|\Psi^{(RQ)}\rangle$ for RQ, there exist bases $\{|k^{(R)}\rangle\}$ for $\mathcal{H}^{(R)}$ and $\{|k^{(Q)}\rangle\}$ for $\mathcal{H}^{(Q)}$ such that Eq. 8.26 holds. This is remarkable! If $\dim\mathcal{H}^{(R)} = \dim\mathcal{H}^{(Q)} = d$, then a joint state vector $|\Psi^{(RQ)}\rangle$ would typically require d^2 complex coefficients when written in terms of a product basis. Equation 8.26 has only d real coefficients $\sqrt{\lambda_k}$. There must be something very special about the Schmidt bases $\{|k^{(R)}\rangle\}$ and $\{|k^{(Q)}\rangle\}$. And indeed there is: these are eigenbases for $\rho^{(R)}$ and $\rho^{(Q)}$, respectively. Choosing these bases gives us a much more compact expression for $|\Psi^{(RQ)}\rangle$.

Exercise 8.13 For a pair of qubits, find the Schmidt decomposition for the state

$$|\Upsilon\rangle = \frac{1}{\sqrt{2}}\left(|0,0\rangle + |1,+\rangle\right). \tag{8.27}$$

The Schmidt decomposition is a powerful tool for analyzing the states of a composite system. Here are some useful results that are easy to prove using the Schmidt decomposition:

Exercise 8.14 Show that if RQ is in a pure state, $\rho^{(R)}$ and $\rho^{(Q)}$ have exactly the same non-zero eigenvalues.

Exercise 8.15 The *Schmidt number* of a joint state $|\Psi^{(RQ)}\rangle$ is the number of non-zero terms in its Schmidt decomposition. What is the Schmidt number of a product state? If $\dim\mathcal{H}^{(R)} = 3$ and $\dim\mathcal{H}^{(Q)} = 5$, what is the maximum possible Schmidt number?

Exercise 8.16 Consider a mixed state $\rho^{(R)}$ of R. A state $|\Psi^{(RQ)}\rangle$ is said to be a *purification* of $\rho^{(R)}$ in Q if $\text{Tr}_{(Q)} |\Psi^{(RQ)}\rangle\langle\Psi^{(RQ)}| = \rho^{(R)}$. Show that any two purifications of $\rho^{(R)}$ in Q are related by a unitary operator acting only on system Q. That is,

$$\left|\Psi_2^{(RQ)}\right\rangle = \left(\mathbf{1}^{(R)} \otimes U^{(Q)}\right)\left|\Psi_1^{(RQ)}\right\rangle,\tag{8.28}$$

for purifications $\left|\Psi_2^{(RQ)}\right\rangle$ and $\left|\Psi_1^{(RQ)}\right\rangle$.

8.3 Distinguishing mixed states

In Sections 4.1 and 4.2 we proved two fundamental results about how well we can distinguish between quantum states. These theorems were proven for pure states described by state vectors; now we need to adapt them to mixed states described by density operators.

First, consider a situation in which Alice wishes to send Bob a message by preparing a quantum system in one of a collection of "signal" states. It might be that these are mixed states of the system described by density operators. Then we need the

Basic decoding theorem for mixed states. If Alice encodes N equally likely messages as mixed states in a quantum system with $\dim \mathcal{H} = d$, and if Bob decodes this by performing a basic measurement (perhaps on a larger Hilbert space) and inferring the message from the result, Bob's probability of error is bounded by

$$P_E \geq 1 - \frac{d}{N}.\tag{Re 4.8}$$

Let us prove this. Each message α is encoded in a state described by the density operator ρ_α, which operates on the Hilbert space $\mathcal{H}$. This space may be a subspace of a larger space. If Π is the projection onto $\mathcal{H}$, then $\rho_\alpha = \Pi\rho_\alpha\Pi$ and $\operatorname{Tr}\Pi = d$.

Now, as in Section 4.1, we suppose that Bob makes a basic measurement using the basis $\{|k\rangle\}$, then groups the various outcomes into disjoint subsets C_α. If the measurement result lies in C_α, then Bob will infer that Alice's message is α. In other words, we can assign to each message α the projection operator

$$\pi_\alpha = \sum_{k \in C_\alpha} |k\rangle\langle k|,\tag{8.29}$$

such that $\sum_\alpha \pi_\alpha = \mathbf{1}$. Then the probability that Bob infers the correct message is

$$\begin{aligned}
P_S &= \sum_\alpha \frac{1}{N}\operatorname{Tr}\rho_\alpha\pi_\alpha \\
&= \frac{1}{N}\sum_\alpha \operatorname{Tr}\Pi\rho_\alpha\Pi\pi_\alpha \\
&= \frac{1}{N}\sum_\alpha \operatorname{Tr}\rho_\alpha\left(\Pi\pi_\alpha\Pi\right).
\end{aligned}\tag{8.30}$$

For any positive operator P and density operator ρ, $\operatorname{Tr} \rho P \leq \operatorname{Tr} P$. Thus,

$$P_S \leq \frac{1}{N} \sum_\alpha \operatorname{Tr} \Pi \pi_\alpha \Pi$$

$$= \frac{1}{N} \operatorname{Tr} \Pi \left(\sum_\alpha \pi_\alpha \right) \Pi$$

$$= \frac{1}{N} \operatorname{Tr} \Pi = \frac{d}{N}. \tag{8.31}$$

The probability of error P_E in this procedure must therefore satisfy Eq. 4.8. In other words, our basic decoding theorem generalizes without change to embrace the possibility of mixed signals.

Exercise 8.17 Compare this proof to the one given in Section 4.1, and fill in any details that have been skimmed over. Is the present argument fundamentally more complicated than the previous one?

The basic decoding theorem led us to say that the information capacity of a system with $\dim \mathcal{H} = d$ is just $\log d$ bits. This point still holds if mixed signals are allowed.

We can actually arrive at a stronger result if we know something about the signal states ρ_α. Suppose that each ρ_α is a uniform density operator on a subspace of dimension s. That is,

$$\rho_\alpha = \frac{1}{s} D_\alpha, \tag{8.32}$$

where D_α is a projection operator. Then we find that

$$P_S = \frac{1}{Ns} \sum_\alpha \operatorname{Tr} D_\alpha \left(\Pi \pi_\alpha \Pi \right)$$

$$\leq \frac{1}{Ns} \operatorname{Tr} \Pi = \frac{d}{Ns}. \tag{8.33}$$

Exercise 8.18 Fill in the gaps in this derivation.

It follows that the error probability is

$$P_E \geq 1 - \frac{d}{Ns}. \tag{8.34}$$

Exercise 8.19 Explain why the information capacity of a system with $\dim \mathcal{H} = d$ using such uniform signal states is $\log d - \log s$. Show that this can be achieved by some code. (The situation in which d is not a multiple of s requires some thinking.)

The basic distinguishability theorem of Section 4.2 can be generalized to mixed states as well, although we need to formulate the new version carefully. We imagine that we have two equally likely mixed states ρ_0 and ρ_1, and we define the operator $\Delta = \rho_0 - \rho_1$. The difference operator Δ is Hermitian and has zero trace. If $\Delta = 0$, then ρ_1 and ρ_2 are the same and make identical predictions for all measurement. This means that no procedure for determining which state is present will succeed with a probability better than mere chance, i.e. $P_S = \frac{1}{2}$. If Δ is non-zero, then we can do better.

Let δ be the sum of the positive eigenvalues of Δ. This is non-zero if and only if the two states are different. Our generalized theorem is this:

Basic distinguishability theorem for mixed states. Suppose a system is prepared in one of two equally likely mixed states ρ_0 and ρ_1, with Δ and δ as defined above. Then the probability of correctly identifying the state by a basic measurement is bounded by

$$P_S \leq \frac{1 + \delta}{2}, \tag{8.35}$$

where equality is achievable by a suitable choice of observable.

The proof is a straightforward adaptation of the one given in Section 4.2. We will therefore leave it as an exercise:

Exercise 8.20 Prove the basic distinguishability theorem for mixed states.

Exercise 8.21 Explain why our new result is really a generalization of the original basic distinguishability theorem.

One final point. If $\rho_1 = \rho_2$, then the two states are entirely indistinguishable. However, if $\rho_1 \neq \rho_2$, then $\delta > 0$ and the best possible $P_S > 1/2$. This is better than guessing. Even if we cannot distinguish the two states perfectly, we can always obtain *some* information about how the system was prepared. Whenever two situations are described by different density operators, they are at least slightly distinguishable.

8.4 The Bloch sphere

There is an elegant way to visualize the set of density operators for a qubit system, which we will now describe.

Given a Hilbert space $\mathcal{H}$ of finite dimension, the set $\mathcal{B}(\mathcal{H})$, the set of all operators on $\mathcal{H}$, is itself a vector space, since the linear combination of two operators is also an operator. Note that $\mathcal{B}(\mathcal{H})$ is in fact a Hilbert space with inner product

$$\langle A, B \rangle = \operatorname{Tr} A^\dagger B. \tag{8.36}$$

Exercise 8.22 Show that $\langle A, B \rangle$ has the required properties of symmetry, linearity, and positive-definiteness.

Exercise 8.23 If $\dim \mathcal{H} = d$, show that $\dim \mathcal{B}(\mathcal{H}) = d^2$. Given an orthonormal basis $\{|n\rangle\}$ for $\mathcal{H}$, show that the operators $|m\rangle\langle n|$ form an orthonormal basis for $\mathcal{B}(()\mathcal{H})$.

The density operator ρ, of course, is an element of $\mathcal{B}(\mathcal{H})$.

For a qubit system, $\dim \mathcal{Q} = 2$. A convenient basis set for $\mathcal{B}(\mathcal{Q})$ is $\{\mathbf{1}, X, Y, Z\}$. This is a set of $2^2 = 4$ operators, but how do we know that it forms a basis? The following exercise does show this:

Exercise 8.24 Show that the operators $1, X, Y,$ and Z are orthogonal to one another in $\mathcal{B}(\mathcal{Q})$.

We call this basis set the *Pauli basis* for $\mathcal{B}(\mathcal{Q})$. Any operator at all can be written as a linear combination of the operators in the Pauli basis.[1] Since the basis operators are Hermitian, any Hermitian operator is a *real* linear combination. We note that $\operatorname{Tr} 1 = 2$ and all of the other basis operators are traceless. It follows that any unit-trace ρ can be written

$$\rho = \frac{1}{2}\left(1 + a_X X + a_Y Y + a_Z Z\right). \tag{8.37}$$

A density operator for a qubit system may be specified by three real numbers a_X, a_Y, and a_Z – the components of a vector $\vec{a}$ in real 3-D space. Formally, we can introduce a "vector" $\vec{\sigma}$ whose components are $X, Y,$ and Z operators, and write

$$\rho = \frac{1}{2}\left(1 + \vec{a} \cdot \vec{\sigma}\right). \tag{8.38}$$

The vector $\vec{a}$ is called the *Bloch vector* for the density operator ρ.

But we are not quite done. Any real Bloch vector $\vec{a}$ defines a trace-1 Hermitian operator ρ, but in order for ρ to be a density operator it must also be positive. Which Bloch vectors yield legitimate density operators? We will first answer this question for pure states of a qubit system, for which the following characterization is useful:

Exercise 8.25 Suppose ρ is a Hermitian operator on a qubit Hilbert space $\mathcal{Q}$. Show that ρ is the density operator for a pure state if and only if $\operatorname{Tr} \rho = 1$ and $\operatorname{Tr} \rho^2 = 1$. (Hint: Consider the eigenvalues of ρ.)

(See Problem 8.2.) In terms of the Bloch vector $\vec{a}$, we have:

$$\operatorname{Tr} \rho^2 = \frac{1}{2}\left(1 + (a_X)^2 + (a_Y)^2 + (a_Z)^2\right). \tag{8.39}$$

Exercise 8.26 Prove this. You will use the fact that $X^2 = Y^2 = Z^2 = 1$, and that the traces of other products of Pauli operators ($XY, YZ,$ etc.) are all zero.

Therefore, the pure states can be identified with the $\vec{a}$ vectors for which

$$\vec{a} \cdot \vec{a} = 1 \quad \text{(pure state)}. \tag{8.40}$$

The Bloch vectors for pure states form a sphere in our real 3-D space. This is called the *Bloch sphere*, and is shown in Fig. 8.1.

The "north and south poles" of the Bloch sphere are the states with Bloch vectors $(0, 0, 1)$ and $(0, 0, -1)$. With respect to the standard basis $\{|0\rangle, |1\rangle\}$ for $\mathcal{H}$ these are

---

[1] Our basis elements are not normalized in $\mathcal{B}(\mathcal{H})$, since $\operatorname{Tr} X^2 = \operatorname{Tr} Y^2 = \operatorname{Tr} Z^2 = \operatorname{Tr} 1 = 2$. We could normalize them by dividing by $\sqrt{2}$, but we will find it a little more convenient here to relax our usual convention and use a non-normalized basis set.

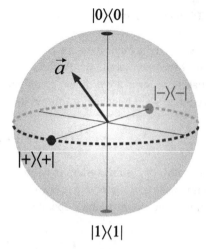

Fig. 8.1 The Bloch sphere describing qubit density operators. The outer surface is made up of pure states, for which $|\vec{a}| = 1$. Some familiar states are shown. Points inside the sphere represent mixed qubit states; the very center is identified with the density operator $\frac{1}{2}\mathbf{1}$, for which $\vec{a} = 0$.

$$\frac{1}{2}(1 + Z) = \frac{1}{2}\Big(|0\rangle\langle 0| + |1\rangle\langle 1| + |0\rangle\langle 0| - |1\rangle\langle 1| \Big)$$

$$= |0\rangle\langle 0|, \tag{8.41}$$

$$\frac{1}{2}(1 - Z) = \frac{1}{2}\Big(|0\rangle\langle 0| + |1\rangle\langle 1| - |0\rangle\langle 0| + |1\rangle\langle 1| \Big)$$

$$= |1\rangle\langle 1|. \tag{8.42}$$

The orthogonal state vectors $|0\rangle$ and $|1\rangle$ yield *antipodal* points on the Bloch sphere. Angles between Bloch vectors are *not* angles in Hilbert space.

Exercise 8.27 What is the Bloch vector associated with the pure state $|+\rangle$? What is the state vector associated with the Bloch vector $(0, 0.6, 0.8)$?

Exercise 8.28 Suppose our qubit is a spin-1/2 system with $|0\rangle = |z_+\rangle$, etc. We make a measurement of S_z on a spin with Bloch vector $\vec{a}$. Show that $\langle S_z \rangle = a_Z \hbar/2$. What are $\langle S_x \rangle$ and $\langle S_y \rangle$?

Exercise 8.29 Given a unit vector $\hat{n}$ with components (n_x, n_y, n_z), we can form the spin component observable S_n in that direction. Show that the eigenstate $|n_+\rangle$ of S_n with eigenvalue $+\hbar/2$ has the Bloch vector $\vec{a} = \hat{n}$.

These exercises show that the Bloch sphere is an easy way to visualize the pure states of a qubit. The representation is particularly straightforward if the qubit happens to be a spin-1/2 system.

What happens when we consider mixtures of states? For example, consider

$$\rho = p\rho_1 + (1-p)\rho_2. \tag{8.43}$$

Let $\vec{a}_1$ and $\vec{a}_2$ be the Bloch vectors for ρ_1 and ρ_2 respectively. Then

$$\rho = \frac{p}{2}\left(1 + \vec{a}_1 \cdot \vec{\sigma}\right) + \frac{1-p}{2}\left(1 + \vec{a}_2 \cdot \vec{\sigma}\right)$$

$$= \frac{1}{2}\left(1 + (p\vec{a}_1 + (1-p)\vec{a}_2) \cdot \vec{\sigma}\right). \tag{8.44}$$

The Bloch vector for ρ is $p\vec{a}_1 + (1-p)\vec{a}_2$, which lies on the line segment between $\vec{a}_1$ and $\vec{a}_2$. This means that the mixed states of a qubit system have Bloch vectors that lie in the interior of the Bloch sphere. Mixing simply "fills in" the sphere, producing the *Bloch ball*.[2] A mixed state has a Bloch vector satisfying

$$\vec{a} \cdot \vec{a} < 1 \quad \text{(mixed state)}. \tag{8.45}$$

Exercise 8.30 What mixed state has $\vec{a} = 0$?

One might hope that there is an equally nice picture of the density operators for other quantum systems. Unfortunately, things are not so simple when $\dim \mathcal{H} > 2$.[3]

We will conclude our introduction to the Bloch sphere with a final exercise, which proves a very useful fact.

Exercise 8.31 Suppose ρ_1 and ρ_2 are qubit density operators with Bloch vectors $\vec{a}_1$ and $\vec{a}_2$. Show that

$$\operatorname{Tr}\rho_1\rho_2 = \frac{1}{2}\left(1 + \vec{a}_1 \cdot \vec{a}_2\right). \tag{8.46}$$

8.5 Time evolution

The state vector $|\psi(t)\rangle$ of an informationally isolated system evolves in time according to the Schrödinger equation

$$H|\psi(t)\rangle = i\hbar \frac{d}{dt}|\psi(t)\rangle. \tag{Re 5.23}$$

We describe the evolution over the interval from time 0 to t by the unitary operator $U(t)$:

$$|\psi(t)\rangle = U(t)|\psi(0)\rangle. \tag{8.47}$$

---

[2] The "filled-in" Bloch sphere is also sometimes called the "Bloch sphere," though mathematically speaking the sphere is only the outer surface.

[3] For $\dim \mathcal{H} = 3$, for instance, the density operators are represented as vectors in a 15-dimensional space, and the outer "boundary" of the set of possible density operators has a more complicated shape.

The density operator corresponding to this pure state is $\rho(t) = |\psi(t)\rangle\langle\psi(t)|$. The unitary time evolution of this operator is thus given by

$$\rho(t) = U(t)\,\rho(0)\,(U(t))^{\dagger}, \tag{8.48}$$

the general rule being, if $|\psi\rangle \to U|\psi\rangle$ then $\rho \to U\rho U^{\dagger}$.

What about the Schrödinger equation? For a tiny interval of time dt, the evolution operator is

$$U(dt) = 1 - \frac{i}{\hbar}H\,dt, \tag{8.49}$$

where we have omitted terms of order $\mathcal{O}\left(dt^2\right)$ and higher. Then the time evolution from t to $t + dt$ yields

$$\rho(t + dt) = \left(1 - \frac{i}{\hbar}H\,dt\right)\rho(t)\left(1 + \frac{i}{\hbar}H\,dt\right)$$

$$= \rho(t) - \frac{i}{\hbar}\left(H\rho(t) - \rho(t)H\right)dt. \tag{8.50}$$

Therefore, at the time t,

$$\frac{d\rho}{dt} = \frac{1}{i\hbar}[H, \rho]. \tag{8.51}$$

This is the Schrödinger equation for the pure-state density operator $\rho = |\psi\rangle\langle\psi|$.

Exercise 8.32 Derive Equation 8.51 in a different way, by taking the time derivative of $|\psi\rangle\langle\psi|$ and using the product rule for the derivative of the outer product.

Equations 8.48 and 8.51 also apply to mixed states. To see this, consider an initial mixture of pure states:

$$\rho(0) = \sum_{\alpha} p_{\alpha}\,|\psi_{\alpha}(0)\rangle\langle\psi_{\alpha}(0)|. \tag{8.52}$$

If the initial state is actually the pure state $|\psi_{\alpha}(0)\rangle$, then it evolves over time to $|\psi_{\alpha}(t)\rangle = U(t)|\psi_{\alpha}(0)\rangle$. The final mixed state is just the corresponding mixture of the possible final pure states

$$\rho(t) = \sum_{\alpha} p_{\alpha}\,|\psi_{\alpha}(t)\rangle\langle\psi_{\alpha}(t)|$$

$$= \sum_{\alpha} p_{\alpha}\,U(t)\,|\psi_{\alpha}(0)\rangle\langle\psi_{\alpha}(0)|\,(U(t))^{\dagger}$$

$$= U(t)\,\rho(0)\,(U(t))^{\dagger}, \tag{8.53}$$

as before. A similar argument generalizes Eq. 8.51. The essential mathematical point is that both Eq. 8.48 and 8.51 are *linear in the density operator*, and that probabilistic mixtures yield density operators that are linear combinations of the states in the mixture. It is also true that the Schrödinger equation, Eq. 8.51, holds for each subsystem of a non-interacting composite system (see Problem 8.8).

Exercise 8.33 Derive the generalization of Eq. 5.27 describing the time evolution of the expectation $\langle A \rangle$ of an observable A.

For qubit density operators, we can visualize the time evolution of the state as a transformation of the Bloch vector of the state. How do Bloch vectors change under unitary evolution? Suppose two different qubit states ρ_1 and ρ_2 evolve via the same unitary U. Then

$$\operatorname{Tr} \rho_1 \rho_2 \rightarrow \operatorname{Tr} U \rho_1 U^\dagger U \rho_2 U^\dagger = \operatorname{Tr} \rho_1 \rho_2. \tag{8.54}$$

From Exercise 8.31, we can conclude that the dot product $\vec{a}_1 \cdot \vec{a}_2$ of the states' Bloch vectors remains unchanged in time. This means that the angle between two pure-state Bloch vectors (each with unit magnitude) is constant under the action of U. Unitary time evolution must therefore correspond to a rigid rotation of the Bloch sphere.

8.6 Uniform density operators

Suppose we know that the state of a quantum system lies in a subspace $\mathcal{T}$ of its Hilbert space $\mathcal{H}$, but we do not have any additional reason to consider one state vector more likely than another. Then the most reasonable density operator to assign is a *uniform* density operator on the subspace $\mathcal{T}$:

$$\rho = \frac{1}{d} \Pi, \tag{8.55}$$

where Π is the projection operator onto $\mathcal{T}$ and $d = \dim \mathcal{T} = \operatorname{Tr} \Pi$.

The uniform density operator is analogous to a uniform probability distribution over a set. And indeed, the two ideas are connected. If we choose a basis $\{ |k\rangle \}$ for $\mathcal{T}$ and assign each basis vector the same probability $1/d$, then the resulting mixture has our uniform density operator. But the same density operator could be produced by other mixtures, some of which do not have equal probabilities.

Exercise 8.34 Suppose a qubit system has a uniform density operator on its entire Hilbert space. Devise a mixture of pure states with unequal probabilities that yields this density operator.[4]

As we saw in Section 1.1, if we have a uniform probability distribution over M possibilities we assign an entropy H according to

$$H = \log M, \tag{Re 1.2}$$

where the logarithm has base 2. By analogy, if we have a uniform density operator on a d-dimensional subspace $\mathcal{T}$, we will assign a *quantum entropy S* according to

$$S = \log d. \tag{8.56}$$

---

[4] We thought of three different easy ways to do this. One was based on the BB84 states from Section 4.4. Another used the geometry of the Bloch ball. A third applied the result of Problem 8.3. The same problem can be attacked from many directions!

A uniform density operator on the Hilbert space of a qubit system has the numerical value $S = \log 2 = 1$; therefore, we say that S is measured in "qubits" (just as H was measured in "bits"). The quantum entropy S is a measure of how "spread out" a uniform density operator is. (Later, we will generalize both H and S to non-uniform situations.)

Exercise 8.35 A system of five qubits is in a mixed state described by a uniform density operator over its Hilbert space. What is the quantum entropy S of this density operator?

A mixture of states has values both of H (for the probability distribution of possible states) and S (for the resulting density operator), but these do not need to agree.

Exercise 8.36 Consider the following mixtures of states for a qubit system.

(a) Suppose we have a mixture of $|0\rangle$ and $|1\rangle$, each with equal likelihood. Show that both H and S have numerical value 1.
(b) Consider the BB84 mixture, an equal mixture of $|0\rangle$, $|1\rangle$, $|+\rangle$, and $|-\rangle$. What is H? What is S?

Consider a situation in which we know various "external parameters" of a quantum system – the number of particles present, the volume of the container enclosing the system, the value of an externally applied magnetic field, etc. These parameters determine the Hilbert space for the system and its Hamiltonian H. We also know one additional piece of information: the energy E of the system. This tells us that the quantum state of the system must lie in the eigenspace of H with eigenvalue E. The projection onto this subspace is denoted $\Pi(E)$. In the absence of further information, we will assign the system a density operator that is uniform on this subspace:

$$\rho(E) = \frac{1}{\Omega(E)} \, \Pi(E), \qquad (8.57)$$

where $\Omega(E) = \mathrm{Tr}\,\Pi(E)$ is the dimension of the eigenspace. In statistical mechanics, this situation is sometimes called the *microcanonical ensemble*, and can be summed up by saying that every state with the known energy E is equally likely.

Alternatively, we might know only the energy to finite precision – that is, we know that the energy is between E and $E + \delta E$. (This is a more realistic assumption for very large systems.) Then $\Pi(E)$ will be the projection onto the subspace spanned by energy eigenstates in the specified range. The microcanonical density operator is then defined exactly as before.

A macroscopic system will have a huge number of energy levels with a lot of degeneracy and near-degeneracy. For such systems, we will assume that the degeneracy function $\Omega(E)$ has the following properties for some range of energies:

- $\Omega(E) \gg 1$;
- $\Omega(E)$ is an increasing function of E;
- $\Omega(E)$ is continuous and differentiable.

The third property cannot be literally true, of course, since $\Omega(E)$ takes only integer values. However, because of the first property, changes in $\Omega(E)$ can be small *compared to* $\Omega(E)$. This fact permits us to treat $\Omega(E)$ as an effectively smooth function.

The quantum entropy of the microcanonical state $\rho(E)$ is

$$S = \log \Omega(E). \tag{8.58}$$

This is related to the thermodynamic entropy S_θ of the macroscopic system:

$$S_\theta = k_B \ln \Omega(E) = k_B \ln 2 \, S, \tag{8.59}$$

where $k_B = 1.38 \times 10^{-23}$ J/K is Boltzmann's constant. (Equation 8.59 is the quantum version of Eq. 1.3.) From thermodynamics, we recall that if heat is added to a system, the changes in the energy E and thermodynamic entropy S_θ are related by

$$dE = T \, dS_\theta, \tag{8.60}$$

where T is the absolute temperature. Therefore,

$$\frac{1}{T} = \frac{dS_\theta}{dE}$$
$$= k_B \frac{d}{dE} \ln \Omega(E). \tag{8.61}$$

Because $\Omega(E)$ is increasing, $T > 0$.

Exercise 8.37 The heat capacity C of a system tells how changes in temperature are related to changes in energy: $dE = C \, dT$. For most systems, an increase in temperature corresponds to an increase in energy, so $C > 0$. For such systems, show that

$$\frac{d^2}{dE^2} \ln \Omega(E) < 0. \tag{8.62}$$

8.7 The canonical ensemble

Consider a system composed of three independent qubits. For each qubit, the energy levels are the standard basis states, with $E_0 = 0$ and $E_1 = \epsilon$. The total energy for the three-qubit system ranges from 0 to 3ϵ. Suppose we know that the total energy is ϵ. The uniform density operator on this eigenspace is

$$\rho^{(123)} = \frac{1}{3} \Big(|0,0,1\rangle\langle 0,0,1| + |0,1,0\rangle\langle 0,1,0| + |1,0,0\rangle\langle 1,0,0| \Big). \tag{8.63}$$

(The degeneracy function $\Omega^{(123)}(\epsilon) = 3$.) Now consider the state of just one of these qubits:

$$\rho^{(1)} = \text{Tr}_{(23)} \rho^{(123)}$$
$$= \frac{2}{3} |0\rangle\langle 0| + \frac{1}{3} |1\rangle\langle 1|. \tag{8.64}$$

A uniform density operator for the global system can yield a non-uniform density operator for a subsystem, and the energy of the subsystem need not have a definite value even though the overall system energy does.

Exercise 8.38 What is the expectation of the energy of qubit #1? Explain why this makes sense.

Exercise 8.39 Suppose we have five independent qubits like the ones above, with a total energy of 2ϵ. What is $\Omega^{(12345)}(2\epsilon)$? Given a uniform density operator for the entire system, what is $\rho^{(1)}$? How about $\rho^{(3)}$?

The microcanonical state of the previous section is an appropriate assumption when a system is isolated. The system cannot exchange energy with its surroundings, and so its energy must remain at one fixed value. But if it can exchange energy with its surroundings, then it may have any one of many different energies, and we do not expect the density operator to be uniform. How can we describe a situation like this?

Imagine a quantum system Q is in contact with another quantum system R. We assume that the systems are nearly non-interacting, so that their energies are additive (see Section 6.2). On the other hand, we do allow the possibility of a small interaction between Q and R so that they can exchange energy over time. We further assume that R is a macroscopic system (as described above), and that it is so much larger than Q that its properties do not change significantly as it gains or loses energy from Q. (We call such a system R a *thermal reservoir*.)

The total energy of QR is E, and we ascribe a microcanonical state to the joint system. What is the resulting density operator? To find this, we need to find the projection $\Pi^{(QR)}(E)$ onto the eigenspace with energy E. Since the systems are non-interacting, we can find a joint eigenbasis of product states of QR. Let $\{\,|n\rangle\langle n|\,\}$ be an energy eigenbasis for Q with corresponding energy eigenvalues E_n. The QR basis elements are of the form $|n\rangle \otimes |\phi_{n,\alpha}\rangle$, where $|\phi_{n,\alpha}\rangle$ is an energy eigenstate for the reservoir R with energy $E - E_n$. Thus

$$\Pi^{(QR)}(E) = \sum_{n,\alpha} |n\rangle\langle n| \otimes |\phi_{n,\alpha}\rangle\langle\phi_{n,\alpha}|$$

$$= \sum_n |n\rangle\langle n| \otimes \Pi^{(R)}(E - E_n). \tag{8.65}$$

The trace of this operator is $\Omega^{(QR)}(E)$, and so

$$\rho^{(QR)} = \frac{1}{\Omega^{(QR)}} \sum_n |n\rangle\langle n| \otimes \Pi^{(R)}(E - E_n). \tag{8.66}$$

The state of the subsystem Q is $\rho^{(Q)} = \mathrm{Tr}_{(R)}\rho^{(QR)}$, which is

$$\rho^{(Q)} = \frac{1}{\Omega^{(QR)}} \sum_n \Omega^{(R)}(E - E_n)\, |n\rangle\langle n|. \tag{8.67}$$

This is not a uniform density operator, since $\Omega^{(R)}$ depends on the energy of R.

Now we will estimate $\Omega^{(R)}(E - E_n)$. The energy E_n is a small fraction of the total energy E, and so we can use the approximation

$$\ln \Omega^{(R)}(E - E_n) = \ln \Omega^{(R)}(E) - E_n \frac{d}{dE} \ln \Omega^{(R)}(E)$$

$$= \ln \Omega^{(R)}(E) - \frac{E_n}{k_B T}, \tag{8.68}$$

where T is the absolute temperature of the reservoir R. Exponentiating this to obtain $\Omega^{(R)}$, we arrive at

$$\rho^{(Q)} = \frac{\Omega^{(R)}(E)}{\Omega^{(QR)}(E)} \sum_k e^{-E_n/k_B T} |k\rangle\langle k| . \tag{8.69}$$

We can conveniently express this in terms of the Hamiltonian operator $H^{(Q)}$ for system Q:

$$\rho^{(Q)} = \frac{1}{\mathcal{Z}} \exp\left(-H^{(Q)}/k_B T\right), \tag{8.70}$$

where $\mathcal{Z} = \mathrm{Tr}\, \exp(-H^{(Q)}/k_B T)$, a normalization factor also called the *partition function*.

Equation 8.70 gives the density operator we would assign to a system that can exchange energy with a thermal reservoir, in the absence of additional information. It depends only on the temperature T and not on any other details of the reservoir. In statistical mechanics, this situation is called the *canonical ensemble*; the canonical state is the equilibrium state of a system that can exchange energy with its surroundings. Both the microcanonical and canonical ensemble are idealizations. The microcanonical ensemble supposes the system to be completely isolated, while the canonical ensemble supposes it to be just a tiny part of a much larger isolated system.

As an example, consider our qubit with energies 0 and ϵ, in equilibrium with a large thermal reservoir at temperature T. The canonical state is

$$\rho = \frac{1}{\mathcal{Z}} \left(|0\rangle\langle 0| + e^{-\epsilon/k_B T} |1\rangle\langle 1| \right). \tag{8.71}$$

Exercise 8.40 Find the partition function $\mathcal{Z}$ for this state.

This looks like a mixture of $|0\rangle$ (with probability $1/\mathcal{Z}$) and $|1\rangle$ (with probability $e^{-\epsilon/k_B T}/\mathcal{Z}$). If we take this view of ρ, then we see that the lower energy state $|0\rangle$ is always more probable than $|1\rangle$, with the probabilities being nearly equal for $k_B T \gg \epsilon$. However, we need not assume that this is the actual probabilistic mixture of states for the qubit. The density operator ρ by itself yields all statistical predictions for the system.

Exercise 8.41 For the qubit state ρ, calculate $\langle E\rangle$, $\langle Z\rangle$, and $\langle X\rangle$.

Here is a related example. Suppose we have a spin-1/2 particle like a proton in an external magnetic field, which we take to lie in the z-direction. Then the Hamiltonian of the spin is $H = -\gamma B S_z$, where γ is the gyromagnetic ratio for the spin. The energy eigenstates for this are $|\downarrow\rangle$ and $|\uparrow\rangle$, with energies $+\frac{\gamma B \hbar}{2}$ and $-\frac{\gamma B \hbar}{2}$, respectively. The canonical density operator for the spin at temperature T is

$$\rho = \frac{1}{\mathcal{Z}} \left(e^{\gamma B \hbar/2 k_B T} |\uparrow\rangle\langle\uparrow| + e^{-\gamma B \hbar/2 k_B T} |\downarrow\rangle\langle\downarrow| \right), \tag{8.72}$$

where $\mathcal{Z} = e^{\gamma B \hbar/2 k_B T} + e^{-\gamma B \hbar/2 k_B T}$.

Exercise 8.42 A spin is in a canonical state inside a magnetic field in the z-direction.

(a) Show that

$$\langle Z \rangle = \tanh\left(\frac{\gamma B \hbar}{2 k_{\mathrm{B}} T}\right), \tag{8.73}$$

where tanh is the hyperbolic tangent function.[5]

(b) Show that, for a given B, $\langle Z \rangle \to 0$ as $T \to \infty$, and $\langle Z \rangle \to 1$ as $T \to 0$.

(c) If the temperature $T \gg \gamma B \hbar / 2k$, derive the approximate expression

$$\langle Z \rangle \approx \frac{\gamma B \hbar}{2 k_{\mathrm{B}} T}. \tag{8.74}$$

Exercise 8.43 A proton has gyromagnetic ratio $\gamma_p = 2.675 \times 10^8 \ \mathrm{s^{-1} T^{-1}}$. Suppose this proton is in a 10.00 T magnetic field and is in equilibrium with its surroundings at room temperature, about $T = 300$ K. Calculate $\langle Z \rangle$ and show that $\langle Z \rangle \ll 1$. How low would we have to make T so that $\langle Z \rangle = 0.5$? How about $\langle Z \rangle = 0.99$?

An external magnetic field causes nuclear spins to tend to "line up" with the field, since this state has a lower energy. However, for typical magnetic fields and ordinary temperatures, this tendency is weak. Nuclear magnetic resonance experiments exploit this slight effect, which works because a macroscopic sample of hydrogen contains a large ensemble of protons. See Chapter 18 for more details.

Exercise 8.44 Both the microcanonical and canonical states of a system are time-independent under the system's internal dynamics. Explain.

Problems

Problem 8.1 Angles between Bloch vectors are not angles in Hilbert space, but the two are related. Suppose pure states $|\psi\rangle$ and $|\phi\rangle$ satisfy $|\langle \psi | \phi \rangle| = \cos\theta$. Show that the Bloch vectors for these states form an angle 2θ.

Problem 8.2 The characterization of pure state density operators in Exercise 8.25 does not extend to higher-dimensional Hilbert spaces.

(a) Construct an explicit counter-example if dim $\mathcal{H} = 3$. That is, find a Hermitian operator ρ such that $\mathrm{Tr}\,\rho = 1$ and $\mathrm{Tr}\,\rho^2 = 1$, but $\rho \neq |\psi\rangle\langle\psi|$ for any state $|\psi\rangle$.

(b) For arbitrary finite dim $\mathcal{H}$, suppose we know that $\mathrm{Tr}\,\rho = 1$, $\mathrm{Tr}\,\rho^2 = 1$, *and* $\mathrm{Tr}\,\rho^3 = 1$. Prove that ρ is the density operator for a pure state.

Problem 8.3 Suppose $\{(p_\alpha, |\psi_\alpha\rangle)\}$ and $\{(q_\beta, |\phi_\beta\rangle)\}$ are two mixtures that yield the same density operator ρ. (The notation $\{(p_\alpha, |\psi_\alpha\rangle)\}$ means that the state $|\psi_\alpha\rangle$ occurs with

---

[5] The hyperbolic sine and cosine functions are $\sinh x = (e^x - e^{-x})/2$ and $\cosh x = (e^x + e^{-x})/2$, respectively. The hyperbolic tangent is the ratio $\sinh x / \cosh x$. If you are unfamiliar with the hyperbolic trigonometric functions, consider it an additional exercise to work out a few of their basic properties, analogous to familiar ones for sin, cos, and tan.

probability p_α.) By including states with probability zero, we may assume that α and β run over the same range. Show that there exists a unitary matrix $U_{\alpha\beta}$ such that

$$\sqrt{p_\alpha} \, |\psi_\alpha\rangle = \sum_\beta U_{\alpha\beta} \sqrt{q_\alpha} \, |\phi_\beta\rangle. \tag{8.75}$$

(Hint: Append an additional system and find purifications that give rise to the mixtures as mixtures of conditional states. Recall that every two purifications are related by a unitary operator on the purifying system.)

Problem 8.4 The fidelity F between two pure states was introduced in Eq. 7.31. This can be generalized to mixed states. Suppose $\rho_1^{(A)}$ and $\rho_2^{(A)}$ are two density operators for system A. Then we define

$$F\left(\rho_1^{(A)}, \rho_2^{(A)}\right) = \max \left|\left\langle \Psi_1^{(AB)} \, \middle| \, \Psi_2^{(AB)} \right\rangle\right|^2, \tag{8.76}$$

where $|\Psi_i^{(AB)}\rangle$ is a purification of $\rho_i^{(AB)}$ in the larger system AB, and the maximum is taken over all choices of such purifications.[6]

(a) Show that, if the two A-states are pure, this definition reduces to Eq. 7.31.
(b) Show that $F = 1$ if and only if the two states are exactly the same.
(c) If $\rho_1 = |\phi_1\rangle\langle\phi_1|$ is pure but ρ_2 is not, show that

$$F = \langle\phi_1| \rho_2 |\phi_2\rangle. \tag{8.77}$$

Hint: You will need the fact that

$$\left|\sum_k \alpha_k \beta_k^*\right|^2 \le \left(\sum_k |\alpha_k|^2\right) \left(\sum_k |\beta_k|^2\right), \tag{8.78}$$

with equality if and only if $\beta_k = C\alpha_k$ for all k. (This is just a form of the Schwartz inequality of Eq. 4.31.)

(d) Consider mixed states $\rho_1^{(AC)}$ and $\rho_2^{(AC)}$ of a composite system AC. Show that

$$F\left(\rho_1^{(AC)}, \rho_2^{(AC)}\right) \le F\left(\rho_1^{(A)}, \rho_2^{(A)}\right). \tag{8.79}$$

The fidelity between the states can only increase if we discard the subsystem C. This property is called the *monotonicity of the fidelity*. Devise a qubit example in which the fidelity of the AC states is 0 but that of the A states is 1.

Problem 8.5 For a quantum system with basis $\{|k\rangle\}$, consider all unitary operators of the form

$$U = \sum_k \alpha_k \, |k\rangle\langle k|, \tag{8.80}$$

where $\alpha_k = \pm 1$. These operators flip some of the phases of the basis states $|k\rangle$ and leave the rest unchanged.

---

[6] It is possible, but much more difficult, to derive an explicit formula for the fidelity F in terms of the two density operators. The interested student is advised to consult a more specialized text in quantum information theory.

There are 2^d such phase-flipping operators, where $d = \dim \mathcal{H}$. We consider a physical process $\mathcal{D}$ that consists of choosing one of these unitary operators at random (with equal probability 2^{-d}) and applying it to an initial state of the system. This process can be viewed as a map on density operators:

$$\rho \to \mathcal{D}(\rho), \tag{8.81}$$

where $\mathcal{D}(\rho)$ is the final density operator, averaged over the random choice of U.

Write the density operator as a density matrix with respect to the basis $\{|k\rangle\}$. Show that the effect of $\mathcal{D}$ is to wipe out the coherences of this matrix while leaving the populations unchanged. In other words, random phase-flipping produces complete decoherence.

Comment on the connection between this problem and Exercise 1.11 in Chapter 1.

Problem 8.6 A qubit system undergoes a decoherence process in the standard $\{|0\rangle, |1\rangle\}$ basis, perhaps by way of the random phase flip in Problem 8.5. Show how this process changes vectors in the Bloch ball.

Problem 8.7 Show that the no-communication theorem of Section 6.4 also holds for mixed states of the composite system AB. (You will need to devise the mixed-state generalization of the idea of conditional states.)

Problem 8.8 Let AB be a composite system with joint state $|\Psi^{(AB)}\rangle$.

(a) For a B-state $|\beta^{(B)}\rangle$ and the operators $G^{(A)}$ and $K^{(B)}$, show that

$$\langle \beta^{(B)}| G^{(A)} \otimes K^{(B)} |\Psi^{(AB)}\rangle = G^{(A)} \left(\langle \beta^{(B)}| K^{(B)} \right) |\Psi^{(AB)}\rangle. \tag{8.82}$$

(b) If the joint state $|\Psi^{(AB)}\rangle$ evolves according to the joint Hamiltonian $H^{(AB)}$, and the subsystems A and B are non-interacting, then show

$$\frac{d\rho^{(A)}}{dt} = \frac{1}{i\hbar}[H^{(A)}, \rho^{(A)}]. \tag{8.83}$$

In other words, the Schrödinger equation (Eq. 8.51) holds for the subsystem A.

Problem 8.9 A *pseudo-pure state* is a mixture between a uniform density operator on a Hilbert space and the density operator for a particular pure state $|\psi\rangle$. It thus has the form

$$\pi = \left(\frac{1-\eta}{d}\right) \mathbf{1} + \eta |\psi\rangle\langle\psi|, \tag{8.84}$$

where $d = \dim \mathcal{H}$. The parameter η is the "purity" of the state, and ranges between 0 and 1.

(a) Under unitary time evolution, show that the pseudo-pure state for $|\psi\rangle$ with purity η evolves into the pseudo-pure state for $U|\psi\rangle$ with the same purity.

(b) Suppose A is an observable with zero trace. Show that for a pseudo-pure state

$$\langle A \rangle = \eta \langle\psi| A |\psi\rangle. \tag{8.85}$$

(c) Show that the qubit canonical state in Eq. 8.71 is pseudo-pure. What is its purity?

For very noisy systems, like nuclear spins at room temperature, pure states may be impractical to make. Pseudo-pure states are a convenient substitute for many purposes.

Problem 8.10 Suppose Π_a and Π_b are projection operators onto subspaces $\mathcal{T}_a$ and $\mathcal{T}_b$, respectively. Show that Π_a and Π_b are orthogonal (in the sense of the operator inner product of Eq. 8.36) if and only if the subspaces $\mathcal{T}_a$ and $\mathcal{T}_b$ are orthogonal in $\mathcal{H}$.

Problem 8.11 A *Werner state* of two qubits has a density operator

$$\rho_\lambda = \lambda \, |\Psi_-\rangle\langle\Psi_-| + \frac{1-\lambda}{3} \left(|\Psi_+\rangle\langle\Psi_+| + |\Phi_+\rangle\langle\Phi_+| + |\Phi_-\rangle\langle\Phi_-| \right), \qquad (8.86)$$

for a parameter $1/4 \le \lambda \le 1$ and Bell states $|\Psi_\pm\rangle$ and $|\Phi_\pm\rangle$ defined in Eq. 7.13. The family of Werner states ranges from a pure entangled state ($\lambda = 1$) to a uniform density operator ($\lambda = 1/4$) on $\mathcal{Q} \otimes \mathcal{Q}$.

(a) Rewrite ρ_λ in terms of $|\Psi_-\rangle\langle\Psi_-|$ and the identity operator $\mathbf{1}$ on $\mathcal{Q} \otimes \mathcal{Q}$.

(b) Suppose each qubit is subjected to the same unitary transformation U. Show that

$$(U \otimes U) \, \rho_\lambda \, (U \otimes U)^\dagger = \rho_\lambda. \qquad (8.87)$$

Since U may represent a rotation of a spin-1/2 state in three dimensions, we may say that the Werner states are "rotationally invariant."

(c) In Section 6.6 we used the Bell state $|\Psi_-\rangle$ (Werner state ρ_1) to prove Bell's theorem. Repeat this analysis for the general Werner state ρ_λ and find an expression analogous to Eq. 6.58. Over what range of λ will the Werner state violate the CHSH inequality (Eq. 6.54) for the spin measurements shown in Fig. 6.2?

Open systems

9.1 Open system dynamics

Any quantum system Q in our laboratory is a subsystem of something bigger. Nevertheless, we can often ignore the rest of the world and consider only states, dynamics, and measurements pertaining to Q itself. When can we do this? That is the question that concerns us in this chapter.

Consider two quantum systems, a system of interest Q and another, external system E. We will suppose that Q can have any initial state, but E starts out in a fixed pure state that we will call $|0\rangle$.[1] Thus, Q and E initially have no correlations with each other. The two systems now interact, evolving via the unitary operator U on $\mathcal{H}^{(QE)}$. In the situation described, we call Q an *open system*, and the external system E is its *environment*. The time evolution of Q cannot in general be described by an "internal" unitary operator – that is, an operator on $\mathcal{H}^{(Q)}$. What sort of description is possible?

If the initial state of Q is $|\phi\rangle$, then the final state of Q will be described by the density operator $\text{Tr}_{(E)} U \left(|\phi\rangle\langle\phi| \otimes |0\rangle\langle 0| \right) U^\dagger$. This is easily generalized to mixed initial states of Q. If Q starts out in the state ρ, then its final state will be

$$\rho' = \text{Tr}_{(E)} U \left(\rho \otimes |0\rangle\langle 0| \right) U^\dagger. \tag{9.1}$$

With both $|0\rangle$ and U fixed, the final Q-state ρ' is a function of the initial Q-state. We can write $\rho' = \mathcal{E}(\rho)$, where $\mathcal{E}$ is a map on operators defined by

$$\mathcal{E}(G) = \text{Tr}_{(E)} U \left(G \otimes |0\rangle\langle 0| \right) U^\dagger. \tag{9.2}$$

Although we have expressed the function $\mathcal{E}$ using the environment E, it is important to remember that $\mathcal{E}$ itself only refers to Q – that is, to the relation between initial and final states of the open system. The function $\mathcal{E}$ provides a way of describing the evolution of the state of Q.

Exercise 9.1 Show that the map $\mathcal{E}$ is linear and trace-preserving. That is, given operators A and B and a scalar c, show that the following are true:

---

[1] The fixed initial state of E means that we are considering fixed "external conditions" for the evolution of Q. The requirement that E is in a pure state is not very restrictive. If E were initially in a mixed state, we could always regard E as a subsystem of some larger E′ in an entangled pure state.

(a) $\mathcal{E}(c\mathsf{A}) = c\mathcal{E}(\mathsf{A})$.

(b) $\mathcal{E}(\mathsf{A} + \mathsf{B}) = \mathcal{E}(\mathsf{A}) + \mathcal{E}(\mathsf{B})$.

(c) $\mathrm{Tr}\,\mathcal{E}(\mathsf{A}) = \mathrm{Tr}\,\mathsf{A}$.

Exercise 9.2 A map on operators is said to be *positive* if positive operators are always mapped to positive operators. Show that $\mathcal{E}$ is a positive map. Why is this important?

The map $\mathcal{E}$ is a linear operator that acts on operators, an object that is sometimes called a *superoperator*.

Let us consider an example. Suppose that Q and E are both qubits and that their interaction is described by the evolution operator

$$U_c = \mathbf{1} \otimes |0\rangle\langle 0| + \mathsf{X} \otimes |1\rangle\langle 1|. \tag{9.3}$$

(This is the CNOT operator of Chapter 18, where Q is the "target" qubit and E is the "control" qubit.) The effect of this interaction on the Q-state will depend on the initial state of the environment E:

- If the initial state of E is $|0\rangle$, then Q evolves unitarily according to the identity operator $\mathbf{1}$.
- If the initial state of E is $|1\rangle$, then Q evolves unitarily according to the operator X.
- For initial E-states that are superpositions of $|0\rangle$ and $|1\rangle$, the evolution of Q is not unitary.

Consider the third case with an initial E-state $|+\rangle = \frac{1}{\sqrt{2}}(|0\rangle + |1\rangle)$. Then the interaction leads to a map $\mathcal{F}$ on Q-operators:

$$\mathcal{F}(\mathsf{G}) = \mathrm{Tr}_{(E)}\mathsf{U}_c \left(\mathsf{G} \otimes |+\rangle\langle +| \right) \mathsf{U}_c^\dagger. \tag{9.4}$$

What is $\mathcal{F}$? Any operator for qubit Q can be written as a combination of the outer products $|0\rangle\langle 0|$, $|1\rangle\langle 1|$, $|0\rangle\langle 1|$, and $|1\rangle\langle 0|$. Thus, we need only to determine how $\mathcal{F}$ acts on these four operators. We can do this in a couple of simple exercises:

Exercise 9.3 Show that

$$\mathsf{U}_c\,|0, +\rangle = \frac{1}{\sqrt{2}} \left(|0, 0\rangle + |1, 1\rangle \right) = |\Phi_+\rangle,$$

$$\mathsf{U}_c\,|1, +\rangle = \frac{1}{\sqrt{2}} \left(|1, 0\rangle + |0, 1\rangle \right) = |\Psi_+\rangle, \tag{9.5}$$

where $|\Phi_+\rangle$ and $|\Psi_+\rangle$ are two of the Bell states from Eq. 7.13.

Exercise 9.4 Use Exercise 9.3 to show that

$$\mathcal{F}(\,|0\rangle\langle 0|\,) = \frac{1}{2}\,\mathbf{1}, \quad \mathcal{F}(\,|1\rangle\langle 1|\,) = \frac{1}{2}\,\mathbf{1},$$

$$\mathcal{F}(\,|0\rangle\langle 1|\,) = \frac{1}{2}\,\mathsf{X}, \quad \mathcal{F}(\,|1\rangle\langle 0|\,) = \frac{1}{2}\,\mathsf{X}, \tag{9.6}$$

where X is the Pauli operator.

This completely defines $\mathcal{F}$. To understand the implications of this, let us take our analysis a further step. Section 8.4 shows how to represent qubit density operators geometrically as vectors in the Bloch sphere. If ρ is represented by the Bloch vector $\vec{a}$, then

$$\rho = \frac{1}{2}\left(\mathbf{1} + a_X X + a_Y Y + a_Z Z\right). \tag{Re 8.37}$$

To understand $\mathcal{F}$, we just need to know how it affects the Pauli operators.

Exercise 9.5 Show that $\mathcal{F}(\mathbf{1}) = \mathbf{1}$, $\mathcal{F}(X) = X$, $\mathcal{F}(Y) = 0$, and $\mathcal{F}(Z) = 0$.

Thus, $\mathcal{F}$ has a very neat description: the Bloch vector (a_X, a_Y, a_Z) maps to $(a_X, 0, 0)$, projecting $\vec{a}$ onto the X-axis.

Exercise 9.6 Repeat this entire analysis if the interaction is

$$U_d = \mathbf{1} \otimes |0\rangle\langle 0| + Z \otimes |1\rangle\langle 1|, \tag{9.7}$$

and E is initially in the state $|+\rangle$. How is the Bloch vector of the initial Q state affected by the resulting map on Q-states?

The map $\mathcal{E}$ defined in Eq. 9.2 takes initial Q-states to final Q-states. Is there a way of representing that map without an explicit reference to the environment E?

There is. Choose a particular basis $\{|e_k\rangle\}$ for the Hilbert space of E. For each k define the operator A_k by

$$A_k |\phi\rangle = \langle e_k| U |\phi, 0\rangle, \tag{9.8}$$

for any $|\phi\rangle$ in $\mathcal{H}^{(Q)}$. Although we have used the environment E and the interaction U in this definition, the A_k operators themselves act on $\mathcal{H}^{(Q)}$ alone. We can use the $\{|e_k\rangle\}$ basis to do the partial trace in Eq. 9.2. If we act on a pure state $|\phi\rangle$, then

$$\mathcal{E}(|\phi\rangle\langle\phi|) = \sum_k \langle e_k| U \left(|\phi\rangle\langle\phi| \otimes |0\rangle\langle 0|\right) U^\dagger |e_k\rangle$$

$$= \sum_k A_k |\phi\rangle\langle\phi| A_k^\dagger. \tag{9.9}$$

Since any density operator for Q is a linear combination of pure state density operators, we have in general that

$$\mathcal{E}(\rho) = \sum_k A_k \rho A_k^\dagger. \tag{9.10}$$

This is called an *operator–sum representation* or *Kraus representation* of the map $\mathcal{E}$, and the operators A_k are called *Kraus operators*.

Exercise 9.7 Pick a basis for the environment E and find an operator–sum representation for the map $\mathcal{F}$ of Eq. 9.4.

The Kraus operators satisfy a normalization condition. If $\rho' = \mathcal{E}(|\phi\rangle\langle\phi|)$ for a normalized pure state $|\phi\rangle$, then

$$\mathrm{Tr}\,\rho' = \sum_k \langle\phi|\,A_k^\dagger A_k\,|\phi\rangle$$

$$= \langle\phi|\left(\sum_k A_k^\dagger A_k\right)|\phi\rangle. \qquad (9.11)$$

We must have $\mathrm{Tr}\,\rho' = 1$ for every normalized $|\phi\rangle$, and thus

$$\sum_k A_k^\dagger A_k = \mathbf{1}. \qquad (9.12)$$

We can describe the map $\mathcal{E}$ entirely in terms of Kraus operators A_k satisfying Eq. 9.12. It is often more convenient to describe $\mathcal{E}$ in this way, without dragging in the whole environment E and its dynamics. This can be a very good thing, since E might be very large and complex. An operator–sum representation is a compact description of how E affects the evolution of the state of Q.

Exercise 9.8 Explain why the unitary evolution of a density operator (Eq. 8.48) is a special case of Eq. 9.10. In the unitary case, what is the significance of the normalization condition in Eq. 9.12?

We have seen how to go from unitary evolution on the system QE to a map on density operators of Q, represented by a set of Kraus operators. Problem 9.1 shows how to do the reverse, starting with a set of Kraus operators on Q and arriving at a unitary evolution on a larger system. In Appendix D, we outline the proof of an even more powerful result, which we can state informally as follows. Suppose $\mathcal{E}$ is a linear map on Q-operators. Then these three conditions are equivalent:

- $\mathcal{E}$ represents a "physically reasonable" evolution for density operators on Q. (For a clear definition of "physically reasonable," see Appendix D.)
- $\mathcal{E}$ is given by unitary evolution on an extended system, as in Eq. 9.2.
- $\mathcal{E}$ has a Kraus representation (Eq. 9.10), where the Kraus operators are normalized according to Eq. 9.12.

In short, the type of quantum dynamics we have described here is the most general type that can be represented by a superoperator $\mathcal{E}$.

9.2 Informationally isolated systems

We can now give a precise mathematical description of a heuristic principle that we introduced as long ago as Section 1.2. There we said that a quantum system evolves in a way that respects the principle of superposition – and thus could exhibit interference phenomena – provided that the system is "informationally isolated." By this phrase we meant that no record is produced anywhere in the Universe of the system's exact state. But what does it mean for a system to be "informationally isolated"?

Once again, we imagine that our system of interest Q is part of a composite system QE, that E is initially in a fixed state $|0\rangle$, and that the overall evolution of QE is described by the unitary operator U. For the input state vector $|\phi\rangle$ for Q, the final joint state for QE is $|\Psi\rangle = U|\phi, 0\rangle$.

We can define an evolution map $\mathcal{E}$ for Q according to Eq. 9.2, as before; but now we are interested also in what happens to the external system E. We say that Q is *informationally isolated* if there is no information transfer from Q to E in this process – that is, if the final state of E is independent of the initial state of Q. That final E-state is

$$\sigma = \text{Tr}_{(Q)}|\Psi\rangle\langle\Psi| = \text{Tr}_{(Q)}U\left(|\phi\rangle\langle\phi| \otimes |0\rangle\langle0|\right)U^\dagger. \tag{9.13}$$

The system Q is informationally isolated if σ is the same for every possible input state of Q.

Assume that Q is informationally isolated. The final density operator σ for E has eigenvalues λ_k and yields a basis of eigenstates $|e_k\rangle$. Since σ is independent of $|\phi\rangle$, so are its eigenvalues and eigenstates. For the initial $|\phi\rangle$, we can construct a Schmidt decomposition (see Eq. 8.26) for the final joint state $|\Psi\rangle$:

$$|\Psi\rangle = \sum_k \sqrt{\lambda_k}|q_k, e_k\rangle, \tag{9.14}$$

where the Q-states $|q_k\rangle$ are orthonormal and are the only parts of the right-hand side that may depend on $|\phi\rangle$.

How do the $|q_k\rangle$ kets depend on $|\phi\rangle$? For every k with $\lambda_k > 0$, define the operator V_k on Q-states by

$$V_k|\phi\rangle = \frac{1}{\sqrt{\lambda_k}}\langle e_k|U|\phi, 0\rangle = \frac{1}{\sqrt{\lambda_k}}\langle e_k|\Psi\rangle = |q_k\rangle. \tag{9.15}$$

Exercise 9.9 Compare the definitions of the Kraus operators A_k in Eq. 9.8 and the V_k operators of Eq. 9.15. How do they differ, and why?

The assumption of informational isolation of Q means that λ_k and $|e_k\rangle$ are independent of $|\phi\rangle$, which shows that V_k acts as a linear operator. Because the $|q_k\rangle$s are normalized, then for any V_k,

$$\langle\phi|V_k^\dagger V_k|\phi\rangle = \langle q_k|q_k\rangle = 1, \tag{9.16}$$

for all Q-states $|\phi\rangle$. Thus, we know that $V_k^\dagger V_k = \mathbf{1}$, and so V_k must be unitary (see Problem 3.7).

Now imagine that we have V_k and V_l with $k \neq l$. The operator $V_k^\dagger V_l$ must be unitary. However, for any state $|\phi\rangle$,

$$\langle\phi|V_k^\dagger V_l|\phi\rangle = \langle q_k|q_l\rangle = 0, \tag{9.17}$$

and so $V_k^\dagger V_l = 0$, which is certainly not unitary. This contradiction tells us that there can be only one V_k operator. The final E density operator σ has only one non-zero eigenvalue

$\lambda = 1$ and so the Schmidt decomposition for $|\Psi\rangle$ has only a single non-zero term $|q, e\rangle$:

$$U |\phi, 0\rangle = |\Psi\rangle = |q, e\rangle = \left(V |\psi\rangle\right) \otimes |e\rangle. \tag{9.18}$$

The state of the system Q has evolved from the pure state $|\phi\rangle$ to the pure state $V |\phi\rangle$, where V is unitary.

We have shown that informational isolation of Q implies that the system Q evolves via a unitary operator V. The converse is also true. Suppose we know that the net effect of the joint evolution U on the initial state $|\phi, 0\rangle$ is to take the Q-state from $|\phi\rangle$ to $V |\phi\rangle$ for some unitary V. The final state of QE must be a purification of $V |\phi\rangle$, which is of the form $V |\phi\rangle \otimes |e_\phi\rangle$. In this notation, we have left open the possibility that $|e_\phi\rangle$ depends on the initial Q-state $|\phi\rangle$ – that is, that Q is not informationally isolated. However, consider the final states for two different inputs $|\phi\rangle$ and $|\phi'\rangle$:

$$\begin{aligned} U |\phi, 0\rangle &= V |\phi\rangle \otimes |e_\phi\rangle, \\ U |\phi', 0\rangle &= V |\phi'\rangle \otimes |e_{\phi'}\rangle. \end{aligned} \tag{9.19}$$

The inner product between these two states is

$$\begin{aligned} \langle \phi, 0| U^\dagger U |\phi', 0\rangle &= \langle \phi| V^\dagger V |\phi\rangle \langle e_\phi |e_{\phi'}\rangle, \\ \langle \phi |\phi'\rangle &= \langle \phi |\phi'\rangle \langle e_\phi |e_{\phi'}\rangle. \end{aligned} \tag{9.20}$$

If $\langle \phi |\phi'\rangle \neq 0$, it follows that $|e_\phi\rangle = |e_{\phi'}\rangle$.

Exercise 9.10 Show that this is also true in the case where $\langle \phi |\phi'\rangle = 0$. (You will need a third Q-state $|\phi''\rangle$ that is not orthogonal to either $|\phi\rangle$ or $|\phi'\rangle$.)

In other words, the final state of E does not depend at all on the initial state of Q, so no information is transferred from Q to E. Thus Q is informationally isolated. We have now established an important result:

> **Isolation theorem.** Suppose the composite system QE evolves according to the unitary operator U, and that E is initially in some fixed pure state $|0\rangle$. Then the evolution of Q itself is unitary if and only if Q is informationally isolated.

This is a deeper fact than may first appear, and it encompasses several of our previous theorems. Here is an example:

Exercise 9.11 Use the isolation theorem to give yet another proof of the no-cloning theorem of Section 7.2.

Exercise 9.12 Consider the interaction U_c between qubits Q and E defined in Eq. 9.3. We have seen that the Q-state evolution is not unitary when the initial state of the E qubit is $\frac{1}{\sqrt{2}}(|0\rangle + |1\rangle)$. In this situation, find an explicit example to show that the final state of E can depend on the initial state of Q – that is, that Q is not informationally isolated.

When Q is informationally isolated, we can describe its time evolution by a simple unitary operator on $\mathcal{H}^{(Q)}$. But even when this is true, the unitary operator may depend on the particular state of the external system E. This is a significant conceptual point. A spin with a magnetic moment is not really dynamically isolated from its surroundings, since it interacts with the external magnetic field. However, for a given external field, the spin may be informationally isolated, with a time evolution described by a unitary operator that depends on the field.

9.3 The Lindblad equation

For the open system Q, the map $\mathcal{E}$ represents the system's evolution over a finite interval of time. As Exercise 9.8 suggests, this is analogous to the unitary evolution of an informationally isolated system. As we have seen, there is also an "instantaneous" way of expressing unitary evolution, via the Schrödinger equation for density operators:

$$\frac{d\rho}{dt} = \frac{1}{i\hbar}[H, \rho].$$
(Re 8.51)

Is there a generalization of this for open systems?

This is a delicate question. Suppose we consider the evolution of Q over a time interval that includes the times $0 < t_1 < t_2$. In the unitary case, the operator $U(t_1, 0)$ describes the evolution from 0 to t_1, while $U(t_2, t_1)$ describes the evolution from t_1 to t_2. The whole interval from 0 to t_2 is described by

$$U(t_2, 0) = U(t_2, t_1)U(t_1, 0),$$
(9.21)

(see Eq. 5.10). In other words, the time evolution over a larger interval of time is just a composition of separate unitary time evolutions over the subintervals composing the larger interval. We arrive at the differential equation in Eq. 8.51 by considering how the state of Q evolves over very short intervals at various times.

What about open systems? The existence of the evolution map $\mathcal{E}$ depends on our assumption that the system E begins in a fixed initial state, independent of the state of Q. But as Q and E interact, they do become entangled. If our assumption about the E-state holds at the time $t = 0$, then we can find a map $\mathcal{E}_1$ that tells how the state of Q evolves from 0 to t_1, and another map $\mathcal{E}_2$ that describes the evolution from 0 to t_2. Since Q and E are most likely in an entangled state at $t = t_1$, however, no such map exists to describe how the Q-state changes from t_1 to t_2. The evolution of an open system is not in general a sequence of separate stages. Therefore, we should not expect to be able to arrive at a differential equation describing that evolution in the manner of Eq. 8.51.

Having explained carefully why a generalization of the Schrödinger equation for open systems is not feasible, we will now proceed to derive such a generalization, called the *Lindblad equation*. How can we possibly justify such a contradictory approach?

To begin with, our derivation will simply establish the necessary *form* of the Lindblad equation, without making any claim that it is exactly applicable to any given system. Moreover, the Lindblad equation may be a good *effective model* for the evolution of the Q-state in some physical situations, even if we cannot take it to be an exact description.

Our equation will be a good model if the environment E has internal dynamics that "hides" any entanglement with Q as quickly as it arises.

Consider the example of an atom Q in the surrounding electromagnetic field E, which is initially in the "vacuum" (zero-photon) state $|0\rangle$. If Q emits a photon into E, it quickly propagates away. Once this has happened, the immediate vicinity of Q again appears to be empty, and any entanglement of the atom's state with the distant photon has no effect on the subsequent behavior of Q. It may be a reasonable approximation to suppose that, at any point in time, the evolution of Q proceeds *as if* the environment E is in a fixed uncorrelated state.

Now we derive the form of the Lindblad equation. Over a short interval of time δt, the state of Q evolves from ρ to $\mathcal{E}(\rho) = \rho + \delta\rho$, where $\delta\rho$ is small.[2] The map $\mathcal{E}$ is given by an operator–sum representation:

$$\rho + \delta\rho = \sum_k A_k \rho A_k^\dagger. \tag{9.22}$$

We assume that one of the Kraus operators (call it A_0) is nearly the identity operator $\mathbf{1}$, while the other terms in Eq. 9.22 are of order $\mathcal{O}(\delta t)$. To this order we write

$$A_0 = \mathbf{1} + (L_0 - ih)\delta t,$$
$$A_k = L_k \sqrt{\delta t}, \tag{9.23}$$

where $k \neq 0$. (From this point on, we will consider the $k = 0$ case separately, and treat k as ranging over all other values.) The operators L_0 and h are Hermitian. The terms in Eq. 9.22 are

$$A_0 \rho A_0^\dagger = \rho + \left(L_0 \rho + \rho L_0 - ih\rho + i\rho h\right)\delta t + \mathcal{O}\left(\delta t^2\right),$$
$$A_k \rho A_k^\dagger = L_k \rho L_k^\dagger \, \delta t. \tag{9.24}$$

To first order in δt, therefore,

$$\delta\rho = \left(\{L_0, \rho\} + \frac{1}{i}[h, \rho] + \sum_k L_k \rho L_k^\dagger\right)\delta t. \tag{9.25}$$

(You should recognize the commutator $[\cdot, \cdot]$ and the anticommutator $\{\cdot, \cdot\}$ of various operators.) Our differential equation for ρ is

$$\frac{d\rho}{dt} = \{L_0, \rho\} + \frac{1}{i}[h, \rho] + \sum_k L_k \rho L_k^\dagger. \tag{9.26}$$

To finish our derivation, we need two more ingredients. First, by comparing our results so far to Eq. 8.51, we recognize that the Hamiltonian of the open system Q must be $H = \hbar h$. This Hamiltonian may include both the internal dynamics of Q and correction terms due to the interaction of Q with its environment. The following exercise completes the analysis:

---

[2] The physical assumptions behind our derivation can be expressed this way: $\delta t \ll T_Q$, the timescale over which changes in the Q-state take place; but $\delta t \gg T_E$, the timescale over which entanglement is "lost" in the dynamics of system E.

Exercise 9.13 Since $\mathrm{Tr}\,\rho = 1$ for all times, the trace of $\frac{d\rho}{dt}$ must be zero whatever ρ may be. Show that this implies that

$$L_0 = -\frac{1}{2}\sum_k L_k^\dagger L_k. \tag{9.27}$$

We arrive at the *Lindblad equation*:

$$\frac{d\rho}{dt} = \frac{1}{i\hbar}[H, \rho] + \sum_k \left(L_k \rho L_k^\dagger - \frac{1}{2}\{L_k^\dagger L_k, \rho\} \right). \tag{9.28}$$

The L_k are called *Lindblad operators*. These operators have units of $(\text{time})^{-1/2}$.

How do we know what Lindblad operators to use for a given physical situation? Since we regard the Lindblad equation merely as a plausible model of open system dynamics, valid only when the entanglement is rapidly "lost" in the environment E, we will not try to derive the Lindblad operators from the overall Hamiltonian evolution of QE. Rather, we will choose the L_k operators in a pragmatic way to describe the dynamical phenomena for Q that we wish to model.

As a first example, suppose Q is a two-level atom with ground state $|0\rangle$ and excited state $|1\rangle$. Over time, the atom decays to its ground state, perhaps by the emission of a photon. This is not a unitary process, since any initial state should eventually approach $|0\rangle$. To model this process with the Lindblad equation, we will need Lindblad operators to produce the transition $|1\rangle \to |0\rangle$. We therefore consider a single Lindblad operator

$$L = \Lambda\,|0\rangle\langle 1|, \tag{9.29}$$

where Λ is real. For simplicity, we shall take the system Hamiltonian $H = 0$. The Lindblad equation with a single Lindblad operator is

$$\frac{d\rho}{dt} = L\rho L^\dagger - \frac{1}{2}\left(L^\dagger L\rho + \rho L^\dagger L \right). \tag{9.30}$$

How does the state of the two-level atom evolve under this equation?

Noting that $L^\dagger L = \Lambda^2\,|1\rangle\langle 1|$, our Lindblad equation becomes

$$\frac{d\rho}{dt} = \Lambda^2 \left(\langle 1|\rho|1\rangle\,|0\rangle\langle 0| - \frac{1}{2}\left(|1\rangle\langle 1|\,\rho + \rho\,|1\rangle\langle 1| \right) \right). \tag{9.31}$$

To solve this, we write the density operator using its matrix elements with respect to the standard basis:

$$\rho = \rho_{00}\,|0\rangle\langle 0| + \rho_{01}\,|0\rangle\langle 1| + \rho_{10}\,|1\rangle\langle 0| + \rho_{11}\,|1\rangle\langle 1|, \tag{9.32}$$

where ρ_{00}, etc., depend on time. From Eq. 9.31 we obtain the system of differential equations

$$\frac{d}{dt}\begin{pmatrix} \rho_{00} & \rho_{01} \\ \rho_{10} & \rho_{11} \end{pmatrix} = \Lambda^2 \begin{pmatrix} \rho_{11} & -\rho_{01}/2 \\ -\rho_{10}/2 & -\rho_{11} \end{pmatrix}. \tag{9.33}$$

Exercise 9.14 Fill in the steps to derive Eq. 9.33 from Eq. 9.31.

Exercise 9.15 Solve the system in Eq. 9.33 and show that

$$
\begin{pmatrix} \rho_{00}(t) & \rho_{01}(t) \\ \rho_{10}(t) & \rho_{11}(t) \end{pmatrix} = \begin{pmatrix} 1 - \rho_{11}(0)e^{-\Lambda^2 t} & \rho_{01}(0)e^{-\Lambda^2 t/2} \\ \rho_{10}(0)e^{-\Lambda^2 t/2} & \rho_{11}(0)e^{-\Lambda^2 t} \end{pmatrix}. \tag{9.34}
$$

Any initial density operator will over time approach the density operator $|0\rangle\langle 0|$. The population ρ_{11}, which is the probability that the atom would be found in the state $|1\rangle$, decays exponentially with time constant $\tau = 1/\Lambda^2$. The off-diagonal elements (coherences) also decay exponentially, with a longer time constant 2τ.

Exercise 9.16 How would this result change if we introduce a Hamiltonian $H = \epsilon\,|1\rangle\langle 1|$ for the atom Q? (This Hamiltonian gives $|0\rangle$ zero energy and $|1\rangle$ an energy ϵ.)

Here is another qubit example. If we randomly flip the relative phase of $|0\rangle$ and $|1\rangle$, then the effect will be to wipe out the coherences in the density operator ρ. This "decoherence" process was analyzed in Problem 8.5. The relative phase flip could be accomplished by the Pauli Z operator. Thus, to model a continuous decoherence process using the Lindblad equation, we again introduce a single Lindblad operator $L = \Lambda Z$. Since $Z^2 = 1$, the Lindblad equation (Eq. 9.30) has a very simple form:

$$
\frac{d\rho}{dt} = \Lambda^2 \left(Z\rho Z - \rho \right). \tag{9.35}
$$

This can be solved as before.

Exercise 9.17 Show that under Eq. 9.35 the density operator populations ρ_{00} and ρ_{11} remain constant, while the coherences ρ_{01} and ρ_{10} decay exponentially with time constant $\tau = 1/(2\Lambda^2)$.

As the time $t \to \infty$, the coherences are suppressed.

Further examples of Lindblad dynamics can be found in the problems at the end of the chapter.

9.4 Heat and work

Our discussion of open system dynamics can shed some light on the basic ideas of thermodynamics. A thermodynamic system can exchange energy with its surroundings in two ways, called *heat* and *work*. The heat dQ transferred to the system represents energy changes in the microscopic degrees of freedom of the system. The work dW done by the system is associated with changes in its external parameters – changes in volume, external magnetic field, etc. The net change in the thermodynamic energy E_θ of the system is

$$
dE_\theta = dQ - dW. \tag{9.36}
$$

This is known as the *First Law of Thermodynamics*. The signs in Eq. 9.36 – heat transferred *to* the system, work done *by* the system – are a customary, if slightly confusing, convention. The symbol d denotes an inexact differential. The heat and work involved in a process generally depend on the exact "path" that the process follows, not just the beginning and ending states of the system. The net change in the energy, on the other hand, depends only on the starting and ending states, so we represent that change by an exact differential dE_θ.

The thermodynamic system is a quantum system whose state is described by a density operator ρ. The thermodynamic energy of the system is the expectation value of its energy: $E_\theta = \langle E \rangle = \text{Tr}\,\rho H$, where H is the system Hamiltonian. Adopting the convenient "dot" notation for time derivatives, the rate of change of E_θ is

$$\dot{E}_\theta = \text{Tr}\,\dot{\rho}H + \text{Tr}\,\rho\dot{H}. \tag{9.37}$$

Heat transfer is associated with changes in the microscopic state of the system, while the external parameter changes in work affect the system by modifying its Hamiltonian. Thus, we can identify the two terms in Eq. 9.37 as the heat transfer rate P_Q and the work rate P_W:

$$P_Q = \text{Tr}\,\dot{\rho}H \qquad \text{and} \qquad P_W = -\text{Tr}\,\rho\dot{H}. \tag{9.38}$$

These rates are defined so that $dQ = P_Q dt$ and $dW = P_W dt$ for a small interval of time dt. They have units of power (J/s).

Suppose our system is informationally isolated during some process. The time evolution will be unitary, described by a (possibly time-dependent) Hamiltonian according to

$$\frac{d\rho}{dt} = \frac{1}{i\hbar}[H, \rho]. \tag{Re 8.51}$$

Then we have

$$\dot{E}_\theta = \frac{1}{i\hbar}\text{Tr}\,(H\rho H - \rho H^2) + \text{Tr}\,\rho\dot{H}. \tag{9.39}$$

But $\text{Tr}\,H\rho H = \text{Tr}\,\rho H^2$, so there is no heat transfer. Thermodynamically, a process in which a system remains informationally isolated is also an *adiabatic* process.[3]

A particularly simple case arises when the Hamiltonian changes suddenly, much more rapidly than ρ changes. No heat is transferred in such a process. If the Hamiltonian changes from H_a to H_b, the net work done by the system is

$$W = -\Delta E_\theta = \text{Tr}\,\rho(H_a - H_b). \tag{9.40}$$

We are sometimes interested in the external work done *on* the system in a transformation. We shall denote this by $\mathbb{W} = -W$. A sudden change in the Hamiltonian requires external work $\mathbb{W} = \Delta E_\theta$.

Imagine that the system evolution can be modeled by the Lindblad equation, Eq. 9.28. The work rate for the system is still $P_W = -\text{Tr}\,\rho\dot{H}$, and the heat rate is given by the Lindblad operators L_k.

---

[3] The term *adiabatic* is also used in quantum theory to describe a gradual change in the Hamiltonian of an informationally isolated system. According to the *adiabatic theorem*, a system in an energy eigenstate will "follow" a slow change in the Hamiltonian of the system, remaining in an energy eigenstate at every stage. This use of the term *adiabatic* is related to, but distinct from, the thermodynamic usage here. We apply the term even to processes involving rapid changes in H, provided there is no heat transfer.

Exercise 9.18 Show that

$$P_Q = \frac{1}{2} \mathrm{Tr}\, \rho \left(\sum_k L_k^\dagger [H, L_k] + [L_k^\dagger, H] L_k \right). \tag{9.41}$$

One special case is of interest. If the L_k operators commute with the Hamiltonian H, then $P_Q = 0$. No heat is transferred, so the process is adiabatic in the thermodynamic sense. However, the system is not at all informationally isolated from its environment. The typical effect of this type of evolution is to destroy coherences between energy eigenstates of H. Here is a simple example.

Exercise 9.19 A qubit evolves by the Lindblad equation with Hamiltonian εZ and a single Lindblad operator ΛZ. Show that the system's evolution is thermodynamically adiabatic, but that the system is not informationally isolated.

A slightly more elaborate example is given in Problem 9.8.

The interaction of a thermodynamic system with a thermal reservoir brings it to thermal equilibrium, leading to the canonical state given in Eq. 8.70. This process could be modeled with an appropriate set of Lindblad operators. Suppose the system continues in contact with a reservoir at fixed T while its Hamiltonian undergoes a very slow change. If this happens slowly enough, the system will essentially remain in a canonical state at any given moment:

$$\rho(t) = \frac{1}{\mathcal{Z}(t)} \exp\left(-H(t)/k_{\mathrm{B}}T\right), \tag{9.42}$$

where we recall the partition function $\mathcal{Z} = \mathrm{Tr}\exp(-H/k_{\mathrm{B}}T)$. If the system continually remains in equilibrium with the reservoir, we say that it undergoes an *isothermal* process. The isothermal work rate is

$$P_W = -\frac{\mathrm{Tr}\,\dot{H}\, e^{-H/k_{\mathrm{B}}T}}{\mathcal{Z}} = k_{\mathrm{B}}T \frac{\dot{\mathcal{Z}}}{\mathcal{Z}} = \frac{d}{dt}\left(k_{\mathrm{B}}T \ln \mathcal{Z}\right). \tag{9.43}$$

We define the *free energy* $F_\theta = -k_{\mathrm{B}}T \ln \mathcal{Z}$. This is a function of the system's (canonical) state, not the process by which it arrived at the state. The work done by the system in a small interval of time is thus $đW = -dF_\theta$. For an isothermal process, the total work done by the system is given by the net change in the free energy function:

$$W = -\Delta F_\theta. \tag{9.44}$$

(Again, the external work required is $\mathbb{W} = -W = \Delta F_\theta$.)

Equation 9.44 is reminiscent of the relation between the work done by forces in a mechanical system and the change in the potential energy function for those forces. For this reason, the free energy F_θ is called a *thermodynamic potential* for isothermal processes. We will return to this subject, armed with more powerful mathematical tools, in Section 19.5.

9.5 Measurements on open systems

A quantum system Q that undergoes a measurement is certainly open. However, we have found that we do not need to provide a detailed description of the measuring apparatus in order to figure out the probabilities and expectations of measurement results. That is, the effect of the measurement can be represented using vectors and operators for Q itself. This is very much in the spirit of our discussion of open system dynamics in the previous sections. The map $\mathcal{E}$ describes how Q's own state is changed by its interaction with the environment, without giving any details about how the environment itself is affected. In this section, we will delve a little deeper into the mathematical description of the measurement process.

Our most elementary type of measurement is the basic measurement, in which each outcome k is associated with an element $|k\rangle$ of an orthonormal basis for the Hilbert space $\mathcal{H}^{(Q)}$. If the system is in the pure state $|\psi\rangle$, then the probability that the outcome k occurs is $p(k) = |\langle k|\psi\rangle|^2$. In terms of the density operator $\rho = |\psi\rangle\langle\psi|$,

$$p(k) = \langle k|\rho|k\rangle = \operatorname{Tr}\rho\,\Pi_k, \tag{9.45}$$

where $\Pi_k = |k\rangle\langle k|$ is the projection onto the basis state $|k\rangle$. In the measurement of a real-valued observable, the basis state $|k\rangle$ is associated with a numerical value A_k. The same basic probability rule also holds for mixed states.

We have also discussed incomplete measurements, which correspond to observables with degeneracy. Each outcome α is associated with a projection operator Π_α. For a real-valued observable, this is the projection onto the eigenspace of eigenvalue A_α. The probability of measurement outcome α is

$$p(\alpha) = \operatorname{Tr}\rho\,\Pi_\alpha, \tag{9.46}$$

(compare Eq. 3.118). The projection operators Π_α map to orthogonal subspaces and satisfy the completeness relation

$$\sum_\alpha \Pi_\alpha = \mathbf{1}. \tag{Re 3.119}$$

The role of the projections Π_α in the measurement process was further explored in Section 4.3.

Again, note that these rules refer to Q itself and not to the details of the external apparatus that performs the measurement.

Now suppose that Q is part of a larger system QE, where the external system E is in the fixed initial state $|0\rangle$. We make a measurement on the joint system QE described by a set of projection operators Π_α. If Q is in the state ρ, the probability of the outcome α is

$$p(\alpha) = \operatorname{Tr}\left(\rho \otimes |0\rangle\langle 0|\right)\Pi_\alpha. \tag{9.47}$$

We can think of this in a different way, as a measurement procedure applied to Q only. The external system E is to be regarded merely as a part of the apparatus for making this measurement. What sort of mathematical description can we give for this? For each α, define the Q-operator E_α by the partial inner product

$$E_\alpha = \langle 0| \, \Pi_\alpha \, |0\rangle . \tag{9.48}$$

It follows that

$$p(\alpha) = \text{Tr}\,\rho E_\alpha . \tag{9.49}$$

This has the same form as Eq. 9.46, but the operators E_α are not necessarily projection operators. The following exercise, however, tells us two important facts:

Exercise 9.20 Use Eq. 9.48 to prove:

(a) E_α is a positive operator.

(b) $\sum_\alpha E_\alpha = \mathbf{1}$. (Compare Eq. 3.119.)

Nothing we have said involves any real extension of the quantum theory. The measurement we perform on QE is one of the usual sort. But if we regard E as part of the apparatus and treat the process as a measurement on Q alone, we are forced to consider it as a new and more general type of measurement. The possible outcomes α of the procedure are associated with positive operators E_α, rather than projections. These outcome operators sum to the identity operator $\mathbf{1}$. The rule for calculating outcome probabilities is given in Eq. 9.49. We call this new type of measurement procedure a *generalized* or *positive operator* measurement, to distinguish it from the *ordinary* or *projection operator* variety.

A generalized measurement can differ from an ordinary one in many ways. For example, suppose that dim $\mathcal{H}^{(Q)} = d$. Then a basic measurement on Q has exactly d possible outcomes, and a projection operator measurement has no more than d. But a generalized measurement is really a measurement on the larger system QE, and so may have more than d possible outcomes. Here is an example of a four-outcome generalized measurement on a qubit:

Exercise 9.21 Suppose Q and E are qubits and E is initially in the state $|+\rangle = \frac{1}{\sqrt{2}}(|0\rangle + |1\rangle)$. We make a basic measurement on QE using the following basis:

$$\begin{array}{ll} |0,0\rangle & |\Psi_+\rangle , \\ |1,1\rangle & |\Psi_-\rangle , \end{array} \tag{9.50}$$

where $|\Psi_\pm\rangle$ are two of the Bell states from Eq. 7.13. Verify that these do indeed form an orthonormal basis, and then determine the outcome operators for the resulting generalized measurement on Q.

Now suppose that we wish to describe the most general possible type of measurement procedure. Such a procedure would be characterized by a set of outcomes α, each with a corresponding rule that takes the quantum state ρ as input and calculates the probability $p(\alpha)$ as output. We could denote this function by $p(\alpha|\rho)$. What sort of probability rule is possible?

We first note that $p(\alpha|\rho)$ should be linear in ρ. Consider an ensemble of systems, a fraction p_1 of them in the state ρ_1 and the remaining p_2 of them in the state ρ_2. The overall ensemble is described by $\rho = p_1\rho_1 + p_2\rho_2$. If the same measurement is made on each member of the ensemble, then for a given outcome α,

$$p(\alpha|p_1\rho_1 + p_2\rho_2) = p_1p(\alpha|\rho_1) + p_2p(\alpha|\rho_2).   \tag{9.51}$$

We conclude that the function $p(\alpha|\rho)$ must be a linear functional on density operators.

Exercise 9.22 Explain this reasoning in more detail.

A linear functional on the operator Hilbert space $\mathcal{B}(\mathcal{H})$ is always given by an operator inner product with some element in $\mathcal{B}(\mathcal{H})$. By Eq. 8.36, we conclude that there must exist operators E_α such that for each α

$$p(\alpha|\rho) = \text{Tr}\,\rho E_\alpha.   \tag{9.52}$$

This is just the same as Eq. 9.49. It remains only to show that the outcome operators E_α satisfy the appropriate properties.

Exercise 9.23 From the fact that probabilities are non-negative and sum to unity, show that the outcome operators E_α must satisfy the two properties in Exercise 9.20.

Although we have derived the positive operator measurements by thinking about projection operator measurements on a larger system, we now see that this generalized framework will suffice to describe any sort of experimental procedure.

9.6 Information and open systems

Let us review our discussion by considering a prototype of a quantum physics experiment. We prepare a system Q in a state $|\phi\rangle$ and then subject it to time evolution described by the unitary operator U, after which the state is $|\phi'\rangle = U|\phi\rangle$. At this point we make a basic measurement using the basis $\{|k\rangle\}$, obtaining the result k with probability $|\langle k|\phi'\rangle|^2$.

The experiment thus consists of three stages: state preparation, dynamical evolution, and measurement. In each, we consider Q "in itself," without reference to any external systems. But an external system E may intrude at any stage, or indeed at all three. How do we handle this? One possibility is simply to include E in the system, and describe the preparation, dynamics, and measurement of QE. In many cases, this may be the best and easiest thing that we can do. But if the external system is only involved in *one* of the three stages, we can get away with something simpler:

- Suppose E is involved only in the preparation stage, so that QE is prepared in some joint state $|\Psi\rangle$. Then Q and E do not interact afterwards, and the final measurement only involves Q. Then we can adopt a generalized description of the initial state of Q, ascribing to it a density operator

$$\rho = \text{Tr}_{(E)}\,|\Psi\rangle\langle\Psi|.   \tag{9.53}$$

This evolves according to $\rho \rightarrow \rho' = U\rho U^\dagger$, and the result k occurs in the final measurement with probability $\text{Tr}\,\rho'\,|k\rangle\langle k|$.

- Suppose E is not involved in the preparation stage, but instead is always in a fixed initial state $|0\rangle$. Systems Q and E do interact during the dynamical evolution; but since our final measurement only involves Q, we need only to know the final state of that subsystem. In this case, we can adopt a generalized description of the dynamics of Q. If the initial state is given by the density operator ρ, we get

$$\rho \longrightarrow \rho' = \mathcal{E}(|\phi\rangle\langle\phi|). \tag{9.54}$$

Again, the final result k occurs with probability $\text{Tr}\,\rho'\,|k\rangle\langle k|$.

- Finally, suppose that E has a fixed initial state and does not interact with Q at all during its time evolution, but that the final measurement is actually a joint measurement on QE. Then we can adopt a generalized description for this measurement process. If the final state of Q is ρ', then the outcome k occurs with probability

$$p(k) = \text{Tr}\,\rho'\,E_k, \tag{9.55}$$

where E_k is a positive outcome operator.

In each case, the fact that E is involved in only one stage of the experiment allows us to arrive at a generalized description of that stage that only refers to Q itself. Density operators, general evolution maps, and positive operator measurements are the mathematical expressions of the preparation, time evolution, and measurement of open quantum systems.

All of this generalization raises important issues. At various places in this book we have derived results about the information properties of quantum systems. These have been important guiding principles for our understanding of quantum states. A list of the key theorems would certainly include:

- the basic decoding theorem of Section 4.1;
- the basic distinguishability theorem of Section 4.2;
- the no-communication theorem of Section 6.4; and
- the no-cloning theorem of Section 7.2.

Our original proofs of these theorems involved pure states, unitary time evolution, and basic measurements. In Section 8.3 and Problem 8.7, we adapted all but the last for mixed states. How well do these results hold up when we consider the generalized forms of time evolution and measurement appropriate for open systems?

All four of our key theorems still hold good in the open system context. A close review of the original proofs will show why. In each case, we could append an external system E in the state $|0\rangle$ and allow any measurement or dynamical evolution to involve E. For the basic decoding and distinguishability theorems, our proofs explicitly allowed that the measurements might take place in a larger Hilbert space. The no-communication theorem is unchanged by the involvement of an external E (as pointed out in Exercise 6.26). The details are left to the following exercises.

Exercise 9.24 Review the proof of the basic decoding theorem in Section 4.1, and show that it still holds if we allow a measurement that involves an external system E. What if we also allow mixed states, as in Section 8.3?

Exercise 9.25 Review the proof of the basic distinguishability theorem in Section 4.2 and show that it still holds if we allow a measurement that involves an external system E. What if we also allow mixed states, as in Section 8.3?

Exercise 9.26 Explain why the availability of an external system E for Alice does not alter the no-communication theorem of Section 6.4. State a general form of the no-communication theorem for open systems.

The no-cloning theorem is even easier to generalize. Since we can always regard an external system E to be part of the cloning machine M, the theorem stands for open systems without any changes.[4]

We have now established our key information theorems for a system Q that interacts in a unitary way with an external system E. But are the resulting generalized dynamical evolution and measurement processes the only possibilities? Or could we imagine *even more generalized* dynamics and measurements, for which our information results could fail? Within quantum theory, the answer is (roughly speaking) that no further generalization is possible; therefore, the various information theorems are secure. The technical details of this answer may be found in Appendix D.

Problems

Problem 9.1 Consider a quantum system Q described by a Hilbert space $\mathcal{H}$.

(a) Suppose we are given a subspace $\mathcal{T}$ of $\mathcal{H}$ and a linear map from kets in $\mathcal{T}$ to others in $\mathcal{H}$: $|\psi\rangle \rightarrow |\psi'\rangle$ (linear). This map preserves inner products of vectors in $\mathcal{T}$, so that $\langle\phi'|\psi'\rangle = \langle\phi|\psi\rangle$. Show that there exists a unitary operator V on $\mathcal{H}$ such that $|\psi'\rangle = V|\psi\rangle$ for any $|\psi\rangle$ in $\mathcal{T}$. Hint: Use the fact that any orthonormal set of vectors may be extended to a complete basis.

This allows us to build up a "unitary-like" map on a subspace into a full unitary operator – a handy fact in some mathematical arguments. For example:

(b) Start with a map $\mathcal{E}$ on density operators for Q that has the operator–sum form (Eq. 9.10) for a set of Kraus operators satisfying Eq. 9.12. Append a system E in state $|0\rangle$ and consider the following linear map on kets:

$$|\psi\rangle \otimes |0\rangle \longrightarrow \sum_k \left(A_k |\psi\rangle \right) \otimes |e_k\rangle, \qquad (9.56)$$

for an orthonormal set of E-states $|e_k\rangle$. Show that there exists a unitary $U^{(QE)}$ that does this.

---

[4] The generalization of the no-cloning theorem to mixed states – called the *no-broadcasting theorem* – is outside the scope of this book.

In other words, every $\mathcal{E}$ on Q that has a normalized operator–sum representation can be realized as the result of unitary evolution on a larger system QE.

Problem 9.2 Given systems Q and E with an interaction U, show that Q is informationally isolated for *every* initial E-state if and only if Q and E are dynamically isolated. Note: "if" is easy, but "only if" is much harder. For each initial E-state $|\chi\rangle$, we have

$$U |\phi, \chi\rangle = V_\chi |\phi\rangle \otimes |e_\chi\rangle, \tag{9.57}$$

where the Q evolution operator V_χ and the final E-state $|e_\chi\rangle$ might depend on $|\chi\rangle$, but not $|\phi\rangle$. The essential trick is to adjust the phases of the $|e_\chi\rangle$ states so that the corresponding V_χ operators are all the same.

Problem 9.3 Suppose Q interacts with an external environment E according to Eq. 9.1, and suppose the net effect of this is to leave every pure initial Q-state fixed: $|\phi\rangle\langle\phi| \rightarrow |\phi\rangle\langle\phi|$. Show that the same dynamics will also preserve any entangled state between Q and a bystander system R. In the language of Section 7.5, an evolution that preserves Type I quantum information will also preserve Type II quantum information.

Problem 9.4 Describe the qubit density operator ρ by its Bloch vector $\vec{a}$. Express the Lindblad equation for a decaying atom (Eq. 9.31) as a system of differential equations for the components of $\vec{a}$, and solve this system. Compare your results to those in Eq. 9.34.

Problem 9.5 Suppose that a two-level system Q is subject to both decay and excitation processes – that is, both $|1\rangle \rightarrow |0\rangle$ and $|0\rangle \rightarrow |1\rangle$ may occur as a result of the interaction of Q with its environment. (This could happen, for example, if the environment of an atom already contained photons that could excite it.) Model this situation using a Lindblad equation with two Lindblad operators:

$$L_- = \Lambda_- |0\rangle\langle 1|, \qquad L_+ = \Lambda_+ |1\rangle\langle 0|. \tag{9.58}$$

For simplicity, take the Hamiltonian $H = 0$.

(a) Write down and solve the Lindblad equation for the evolution of the density operator ρ for Q.
(b) As $t \rightarrow \infty$, show that the density operator ρ must approach a particular *equilibrium* density operator ρ_∞. Show that ρ_∞ is diagonal in the standard basis and find its diagonal elements in terms of the $\Lambda_\pm$ coefficients.

Problem 9.6 Suppose $\mathcal{E}$ is a map describing the evolution of an open quantum system, represented by n Kraus operators A_k.

(a) Let U_{kl} be an $n \times n$ unitary matrix of complex numbers, and define the operators

$$B_k = \sum_l U_{kl} A_l. \tag{9.59}$$

Show that the B_k operators form an operator–sum representation of the same map $\mathcal{E}$.
(b) How is U_{kl} related to a change of basis for the partial trace in Eq. 9.2?

(c) Given a set of Lindblad operators L_k, show that another set of operators

$$K_k = \sum_l U_{kl} L_l, \tag{9.60}$$

gives rise to an equivalent Lindblad equation.

The equivalent representations for the time evolution of an open system are sometimes called different *unravelings* of that evolution.

Problem 9.7 A qubit with a Hamiltonian εZ is in thermal equilibrium with a reservoir at absolute temperature T.

(a) Find the canonical state of the qubit. Express this as a Bloch vector $\vec{a}$.
(b) Suppose the Hamiltonian suddenly changes to εX. How much work is done by the system? (Negative work means that the external agency that changed H had to do work *on* the system.)
(c) Suppose instead that the Hamiltonian is slowly changed from εZ to εX, so that the process is isothermal. How much work is done by the system? Explain the difference between your answers in parts (b) and (c).

Problem 9.8 A system has a fixed Hamiltonian H with energy eigenvalues E_n and eigen-states $|n\rangle$. Its interaction with the environment is modeled by Lindblad operators of the form $\lambda |n\rangle\langle n|$ (one for each $|n\rangle$). Show that the interaction of the system with its environment involves neither work nor heat transfer. Also show that the effect of the Lindblad evolution is to suppress the coherences between the $|n\rangle$ basis states.

Problem 9.9 Even though the operator–sum representation gives an "internal" view of the dynamics of an open system, we can still learn something about the environment from the A_k operators. For a given evolution map $\mathcal{E}$ and an input state ρ, define W_{kl} by

$$W_{kl} = \text{Tr}\, A_k \rho A_l^\dagger. \tag{9.61}$$

Show that these numbers are the matrix elements of the final density operator of the environment E, if the initial state of E is pure (as in Eq. 9.1).

To make this more definite, suppose the Kraus operators for a qubit system are $a\mathbf{1}$ and $b\mathsf{Z}$, where a and b are real parameters such that $a^2 + b^2 = 1$. First, confirm that this pair of operators is normalized according to Eq. 9.12. Then find the eigenvalues of the final *environment* state when the input qubit states are $|0\rangle\langle 0|$, $|+\rangle\langle +|$, and $\frac{1}{2}\mathbf{1}$.

Problem 9.10 Suppose two similar systems R and Q are initially in a maximally entangled state $|\Psi^{(RQ)}\rangle$. System Q undergoes time evolution involving interaction with the environment E, initially in $|0^{(E)}\rangle$. System R, meanwhile, has no external interactions or internal dynamics of its own. Show that the following two conditions are equivalent:

• The final (mixed) state of RE is a product state – that is, there are no correlations between R and E.
• The evolution of Q is unitary.

Comment on the relation, if any, between this result and the isolation theorem of Section 9.2.

Problem 9.11 Return to the two-beam interferometer example of Section 2.1, as extended in Section 4.3. The interferometer has basis states $|0\rangle$ (no photon), $|u\rangle$ (one photon in the upper beam, and $|l\rangle$ (one photon in the lower beam). Suppose a measurement is made on the interferometer system using *non-ideal* photon detectors. These devices are plagued with two imperfections:

- The detectors are not 100% efficient. If a photon is present, there is a probability η that the detector does not "click" at all.
- The detectors are also noisy. Even if a photon is not present, there is a probability ν that the detector will click anyway.

Show how to represent the measurement procedure by a set of positive operators in the cases where (a) only one detector is used, on the upper beam; and (b) a detector is used on each beam. The second case has four possible outcomes, even though $\dim \mathcal{H} = 3$. Why?

A particle in space

10.1 Continuous degrees of freedom

The quantum systems we have discussed so far have been described by finite-dimensional Hilbert spaces. Basic measurements on such systems have a finite number of possible outcomes, and a quantum state predicts a discrete probability distribution over these. Now we wish to extend our theory to handle systems with one or more *continuous* degrees of freedom, such as the position of a particle that can move in one dimension. This will require an extension of our theory to Hilbert spaces of infinite dimension, and to systems with continuous observables.

There is a philosophical issue here. How do we know that there really *are* infinitely many distinct locations for a particle? The short answer is, we don't. It might be that space itself is both discrete (at the tiniest scales) and bounded (at the largest), so that the number of possible locations of a particle is some very large but finite number. If this is the case, then the continuum model for space is nothing more than a convenient approximation. Infinity is just a simplified way of describing a quantity that is immense, but still finite.

In this section, we will adopt this view of infinity. We will imagine that any continuous variable is really an approximation of a "true" discrete variable. This idea will motivate the continuous quantities and operations that we need. But we can always appeal to the underlying finiteness to resolve any mathematical perplexities that arise.

Continuous probability

Suppose we have a discrete probability distribution $P_N(k)$, where k takes on $2N + 1$ distinct integer values from $-N$ to $+N$. The real variable x_k takes on $2N + 1$ values that are symmetric around zero and are evenly spaced by Δx. Thus, $x_k = k\Delta x$. The values of x_k range between $-L$ and $+L$, where $L = N\Delta x$.

We are interested in the situation where the values x_k are very closely spaced and have a wide range – and, in the limit, can take on any real value. We will take this limit in a careful way. Fix a parameter ℓ (with units of length) and let

$$\Delta x = \frac{\ell}{\sqrt{N}}. \tag{10.1}$$

As we take $N \to \infty$, Δx becomes arbitrarily small. The limiting size of the x_k values

$$L = N \frac{\ell}{\sqrt{N}} = \ell\sqrt{N} \to \infty. \tag{10.2}$$

Thus, in the limit $N \to \infty$, we can regard the x_ks as values of a continuous variable x that ranges over the entire real line.

Instead of the Nth probability distribution $P_N(k)$, we will be interested in the Nth density function

$$\mathcal{P}_N(x_k) = \frac{P_N(k)}{\Delta x}. \tag{10.3}$$

We restrict our attention to families of probability assignments $P_N(k)$ for which, as $N \to \infty$, the density functions $\mathcal{P}_N(x_k)$ converge to a continuous real function $\mathcal{P}(x)$. This means that, for large values of N, no particular k can have a large probability. Furthermore, nearby values of k (with nearby values of x_k) must have probabilities that are about equal. The limiting function $\mathcal{P}(x)$ is called a *probability density*.

Each property of the continuous probability density $\mathcal{P}(x)$ can be viewed as the limiting case of the corresponding properties of the discrete distributions $P_N(k)$ that converge to it. For instance, the normalization of probability is

$$1 = \sum_k P_N(k) = \sum_{x_k=-L}^{+L} \mathcal{P}_N(x_k)\Delta x. \tag{10.4}$$

As $N \to \infty$, this sum becomes an integral over x:

$$1 = \int_{-\infty}^{+\infty} \mathcal{P}(x)\, dx. \tag{10.5}$$

Similarly, suppose we wish to find the probability that the value of x_k is in the interval $[a, b]$. We have

$$\Pr(x_k \in [a, b]) = \sum_{x_k \in [a,b]} \mathcal{P}_N(x_k)\Delta x, \tag{10.6}$$

which in the limit becomes

$$\Pr(x \in [a, b]) = \int_a^b \mathcal{P}(x)\, dx. \tag{10.7}$$

Finally, consider some function $F(x_k)$ of the real variable x_k. The expectation value of this function is

$$\langle F \rangle = \sum_k F(x_k)P_N(k) = \sum_{x_k=-L}^{+L} F(x_k)\mathcal{P}_N(x_k)\Delta x. \tag{10.8}$$

This also becomes an integral expression in the limit:

$$\langle F \rangle = \int_{-\infty}^{+\infty} F(x)\mathcal{P}(x)\, dx. \tag{10.9}$$

What we have found is that for "regular" families of distributions – i.e. those that are not too concentrated on particular values and which are not too rapidly varying – we can replace discrete quantities by continuous ones for large N. Finite sums are replaced by integrals. In the limit, the probability that the variable x takes on any specific value is zero. Only intervals of non-zero length can have non-zero probability.

Quantum states and wave functions

We can take the continuous limit for quantum states in a similar way. For each N, we have a quantum system described by the Hilbert space $\mathcal{H}_N$ of dimension $\dim \mathcal{H}_N = 2N + 1$. There is an observable X_N on this system with eigenstates $|k\rangle$ and eigenvalues x_k, where as before k ranges from $-N$ to $+N$ and the values of x_k are evenly spaced by Δx. Once again, we will let $N \to \infty$ and $\Delta x \to 0$ according to Eq. 10.1.

Instead of the orthonormal eigenstates $|k\rangle$, we will find it more convenient to consider the related vectors

$$|x_k\rangle = \frac{|k\rangle}{\sqrt{\Delta x}}. \tag{10.10}$$

These are all orthogonal like the $|k\rangle$s, though their magnitudes diverge as $\Delta x \to 0$. Rather than a sequence of distributions $P_N(k)$, we will consider a sequence of state vectors $|\psi_N\rangle$ in $\mathcal{H}_N$. In the X_N eigenbasis, these have components $\langle k|\psi_N\rangle$. But we are more interested in the functions

$$\psi_N(x_k) = \frac{\langle k|\psi_N\rangle}{\sqrt{\Delta x}} = \langle x_k|\psi_N\rangle. \tag{10.11}$$

We require that the quantum states $|\psi_N\rangle$ yield "well-behaved" functions $\psi_N(x_k)$, so that in the limit $N \to \infty$ we arrive at a continuous limiting function $\psi(x)$. That is, for large N, the probability amplitude $\langle k|\psi_N\rangle$ for any particular $|k\rangle$ is small, and the amplitudes for nearby values of k are nearly equal.

As before, we will rewrite sums over the index k as sums over the eigenvalues x_k, and then interpret these sums as integrals in the continuum limit. That is, our approach can be symbolically written as

$$\sum_k (\cdots) = \sum_{x_k} (\cdots)\,\Delta x \quad \longrightarrow \quad \int_{-\infty}^{+\infty} (\cdots)\, dx. \tag{10.12}$$

However, we must include a caveat. In the large-N limit, the non-normalized eigenvectors $|x_k\rangle$ diverge in magnitude. Thus, when we write $|x_k\rangle \to |x\rangle$, we do not mean $|x\rangle$ to be an actual vector. The improper ket $|x\rangle$ only makes sense inside an integral over the continuous variable x. In that situation, we may suppose that the integral actually represents an approximation of a discrete sum over an immense number of closely-spaced values of x_k.[1]

---

[1] In much the same way, when we write $\Delta x \to dx$, we do not mean that the infinitesimal dx is an actual number. We are simply choosing our notation to remind ourselves how the continuous integral arises from a discrete sum.

For example, we can write the quantum state $|\psi_N\rangle$ as

$$|\psi_N\rangle = \sum_k \langle k|\psi_N\rangle \, |k\rangle = \sum_{x_k=-L}^{+L} \psi_N(x_k) \, |x_k\rangle \, \Delta x, \qquad (10.13)$$

and obtain in the continuum limit an expression for the limiting quantum state $|\psi\rangle$:

$$|\psi\rangle = \int_{-\infty}^{+\infty} \psi(x) \, |x\rangle \, dx, \qquad (10.14)$$

where $\psi(x) = \langle x|\psi\rangle$ is called the continuous *wave function* for the quantum state $|\psi\rangle$. This is a complex function that is the x-representation for the state $|\psi\rangle$. (In this expression, you should note that, even though $|x\rangle$ has a divergent magnitude and thus is not literally a vector, the amplitude $\psi(x) = \langle x|\psi\rangle$ is well-defined in the limit.)

A typical application of the wave function idea is the quantum theory of a particle moving in one dimension. The variable x is then the position of the particle, and the wave function $\psi(x)$ is a full description of the quantum state. If the particle's state $|\psi\rangle$ depends on time, then the wave function $\psi(x, t) = \langle x|\psi(t)\rangle$ is a function of both x and t. (We will ignore for now the time dependence of ψ and return to it when we discuss the dynamics of a particle moving in one dimension.)

The wave function $\psi(x)$ can be used to calculate probabilities. The likelihood that x_k is found to be in an interval $[a, b]$ is

$$\Pr(x_k \in [a, b]) = \sum_{x_k \in [a,b]} |\langle\psi_N|x_k\rangle|^2 \, \Delta x. \qquad (10.15)$$

This becomes

$$\Pr(x \in [a, b]) = \int_a^b |\psi(x)|^2 \, dx. \qquad (10.16)$$

Comparing this to Eq. 10.7, we can identify $\mathcal{P}(x) = |\psi(x)|^2$ as the probability density for x-values in a measurement of the system.

Exercise 10.1 Show that $\langle x\rangle = \int_{-\infty}^{+\infty} x \, |\psi(x)|^2 \, dx.$

Exercise 10.2 Assuming the coordinate x has units of length, what are the physical units of the 1-D wave function $\psi(x, t)$?

It is difficult to draw a graph of $\psi(x)$, since its value at any point is complex. It is much easier to visualize the probability density $|\psi(x)|^2$, see Fig. 10.1.

We can write down various relations of the x-representation in the continuous limit. We shall do that as a set of easy exercises. In each one, you should identify the original finite-dimensional expression from Chapter 3, then show how the appropriate limit is taken. The arguments leading to Eq. 10.14 and 10.16 should give you the general idea.

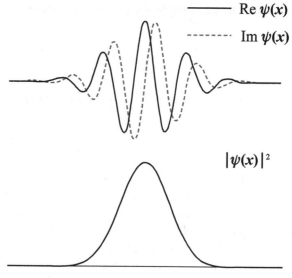

Fig. 10.1 At the top, overlaid graphs of the real and imaginary parts of the wave function $\psi(x)$. At the bottom, a graph of the probability density function $|\psi(x)|^2$. The total area under this curve is 1.

Exercise 10.3 Normalization. Show that

$$\int_{-\infty}^{+\infty} |\psi(x)|^2 \, dx = 1. \tag{10.17}$$

Exercise 10.4 Completeness. Show that

$$\int_{-\infty}^{+\infty} |x\rangle\langle x| \, dx = \mathbf{1}. \tag{10.18}$$

Exercise 10.5 Inner product in x-representation. Show that

$$\langle \phi | \psi \rangle = \int_{-\infty}^{+\infty} \phi^*(x) \, \psi(x) \, dx. \tag{10.19}$$

Each of these should be readily recognizable as a continuous analog to a familiar discrete Hilbert space expression.

One more issue deserves more careful attention: the "orthonormality" of the $|x\rangle$ kets. The underlying discrete basis $\{|k\rangle\}$ for $\mathcal{H}_N$ is orthonormal, so

$$\langle x_k | x_{k'} \rangle = \frac{\delta_{kk'}}{\Delta x}. \tag{10.20}$$

This is zero if $x_k \neq x_{k'}$. Furthermore, the sum

$$\sum_{x_k=-L}^{+L} \langle x_k | x_{k'} \rangle \, \Delta x = 1, \tag{10.21}$$

is independent of N. Therefore, in the continuum limit, we should say that $\langle x \,|\, x' \rangle = 0$ whenever $x \neq x'$, and

$$\int_{-\infty}^{+\infty} \langle x \,|\, x' \rangle \, dx = 1. \tag{10.22}$$

These are the properties of the delta function (see Appendix B). We conclude that

$$\langle x \,|\, x' \rangle = \delta(x' - x). \tag{10.23}$$

This is the continuous version of the orthonormality condition, and based on this we say that the $|x\rangle$s form a *continuous orthonormal basis* for the infinite-dimensional Hilbert space. The delta function is not literally a function, just as the kets $|x\rangle$ are not literally elements of a Hilbert space. Nevertheless, Eq. 10.23 does make sense within integrals, as a continuous approximation of Eq. 10.20.

Using the continuous orthonormality condition, we can see that

$$\mathbf{1} \,|x\rangle = \int_{-\infty}^{+\infty} |x'\rangle \langle x' \,|\, x \rangle \, dx' = \int_{-\infty}^{+\infty} |x'\rangle \, \delta(x - x') \, dx' = |x\rangle, \tag{10.24}$$

as we would expect. The delta function "collapses" the integral, just as the Kronecker delta "collapses" discrete sums.[2]

When we pass to the continuous limit for probability distributions, we make the assumption that the limiting probability density $\mathcal{P}(x)$ exists. This automatically excludes from consideration a great many discrete distributions that are too concentrated or too rapidly varying. These do not have nice continuous approximations, and we ignore them entirely.

In the same way, when we use the continuum approximation in quantum mechanics, we must supplement the mathematical limiting procedure with an additional physical requirement, namely that the "regular" states are, in the limit, *the only physically possible quantum states*. Our physical Hilbert space consists of those states for which a continuous wave function $\psi(x)$ exists. Whether or not we believe that physical space is really at some level discrete and finite – whether or not we believe that other, highly "irregular" and discontinuous states could in principle exist – we readily accept this restriction for everyday use.

10.2 Continuous observables

Position representation of operators

In a Hilbert space of finite dimension d, quantum states and operators are represented by matrices. Given a basis $\{|n\rangle\}$, the state $|\psi\rangle$ is represented by a $d \times 1$ matrix whose elements are

---

[2] Notice that, when we use the completeness relation, we choose a new symbol x' for the integration variable. This is the continuous version of the rule that, when a discrete sum is introduced, the bound summation index should not be the same as any index already being used in the expression.

$$\psi_n = \langle n | \psi \rangle. \tag{10.25}$$

An operator A is given by a $d \times d$ matrix with elements

$$A_{mn} = \langle m | A | n \rangle. \tag{10.26}$$

The operator acts via matrix multiplication. If $|\phi\rangle = A |\psi\rangle$, then

$$\phi_m = \sum_n A_{mn} \psi_n. \tag{10.27}$$

The same sort of machinery can be put in place for a quantum system with a continuous degree of freedom. The wave function values $\psi(x) = \langle x | \psi \rangle$ are the elements of a continuous matrix representing the state $|\psi\rangle$. An operator A has a continuous matrix representation

$$A(x, x') = \langle x | A | x' \rangle. \tag{10.28}$$

If $|\phi\rangle = A |\psi\rangle$, then we can write

$$\phi(x) = \int_{-\infty}^{+\infty} A(x, x') \, \psi(x') \, dx'. \tag{10.29}$$

Exercise 10.6 Justify Eq. 10.29. (You might start with the continuous completeness relation in Eq. 10.18.)

Therefore, the x-representation of an operator is a function of two continuous variables, and the action of the operator is given by the continuous matrix product in Eq. 10.29.

In practice we can often make things much simpler. Consider the operator X, defined by a continuous spectral decomposition

$$X = \int_{-\infty}^{+\infty} x \, |x\rangle \langle x| \, dx. \tag{10.30}$$

This is the "position observable" for a particle moving in one dimension. The operator X has the expected effect on an improper "eigenstate" $|x\rangle$:

Exercise 10.7 Show that $X |x\rangle = x |x\rangle$.

The x-representation of X is

$$X(x, x') = \langle x | X | x' \rangle = x' \langle x | x' \rangle = x' \delta(x' - x). \tag{10.31}$$

If we start with vector $|x\rangle$ having a wave function $\psi(x)$, the vector $X |\psi\rangle$ has a wave function

$$\int_{-\infty}^{+\infty} X(x, x') \, \psi(x') \, dx = \int_{-\infty}^{+\infty} x' \delta(x' - x) \, \psi(x') \, dx = x \, \psi(x). \tag{10.32}$$

Another way to describe the x-representation of the operator X would simply be to say that it affects the wave function by $X : \psi(x) \rightarrow x \psi(x)$. This rule contains exactly the same information as the matrix $X(x, x') = x' \delta(x' - x)$, but in a more intuitive form.

Here is another easy example. The 1-D *parity* operator Π is a reflection about the point $x = 0$, so that $Π |x\rangle = |-x\rangle$ for any x. The operator Π is Hermitian, so we can also write that

$\langle x|\, \mathsf{q} = \langle -x|$. A state $|\psi\rangle$ with wave function $\psi(x)$ is mapped to a state $\mathsf{q}\,|\psi\rangle$ with wave function

$$\langle x|\,\mathsf{q}\,|\psi\rangle = \langle -x|\psi\rangle = \psi(-x). \tag{10.33}$$

The parity operator is expressed by the simple rule that $\mathsf{q} : \psi(x) \to \psi(-x)$.

Exercise 10.8 Find the continuous matrix representation $\mathsf{q}(x, x')$ for the parity operator.

Exercise 10.9 Show that the only possible eigenvalues of q are ± 1. (Hint: What is q^2?) What can we say about the wave functions for the eigenstates of parity?

Parity will be useful in Chapter 15, as we consider the stationary states of particles in 1-D potential wells.

We have shown that, instead of writing down a continuous matrix for an operator, it can be much more convenient to describe the x-representation of the operator by giving a rule about how wave functions are changed by the application of the operator. This is particularly true of the X operator. It is also true for various functions of X. We define $f(\mathsf{X})$, as we defined the operator exponential, by a power series in X:

$$f(\mathsf{X}) = c_0\mathbf{1} + c_1\mathsf{X} + c_2\mathsf{X}^2 + \ldots \tag{10.34}$$

Exercise 10.10 From Equation 10.34, show that

$$f(\mathsf{X})\,|x\rangle = f(x)\,|x\rangle. \tag{10.35}$$

Also show that $\langle x|f(\mathsf{X}) = f(x)\langle x|$. Why does the complex conjugate not appear in this relation?

The vector $f(\mathsf{X})\,|\psi\rangle$ has a wave function

$$\langle x|f(\mathsf{X})\,|\psi\rangle = f(x)\,\psi(x). \tag{10.36}$$

Therefore, any function $f(\mathsf{X})$ of the position operator X can be expressed by the rule $f(\mathsf{X}) : \psi(x) \to f(x)\psi(x)$.

Exercise 10.11 Can the parity operator q be expressed as a function of the X operator?

Finally, we note that the expectation values of functions of X are given by simple integrals:

$$\langle f(x)\rangle = \langle \psi|f(\mathsf{X})\,|\psi\rangle = \int_{-\infty}^{+\infty} \langle \psi|x\rangle\,\langle x|f(\mathsf{X})\,|\psi\rangle\,dx$$
$$= \int_{-\infty}^{+\infty} \psi^*(x)f(x)\,\psi(x)\,dx. \tag{10.37}$$

This last expression is clearly equal to $\int_{-\infty}^{+\infty} f(x)\,|\psi(x)|^2\,dx$, which makes sense when we recall that $|\psi(x)|^2$ is the probability density over possible measured values of x.

Momentum wave functions

Our system is a particle moving in one dimension. The position x is not the only continuous variable that we could imagine measuring for such a particle. There is, for instance, also the momentum p. We can easily write down momentum analogs for the expressions we have already written for the $|x\rangle$ states:

- The momentum operator p has eigenvalues p and eigenstates $|p\rangle$.
- A physical state $|\psi\rangle$ can be represented by a momentum wave function $\bar{\psi}(p) = \langle p|\psi\rangle$.
- The likelihood of obtaining various results in a measurement of p is given by the probability density $|\bar{\psi}(p)|^2$.
- Completeness: $\int_{-\infty}^{+\infty} |p\rangle\langle p|\, dp = \mathbf{1}$.
- Orthonormality: $\langle p'|p\rangle = \delta(p - p')$.

The momentum "eigenstates" $|p\rangle$, like the $|x\rangle$s above, are not literal vectors in the Hilbert space. Nevertheless, as before, we can formally treat them as if they were.

How is the $\{|p\rangle\}$ basis related to the $\{|x\rangle\}$ basis? The Planck–De Broglie relations (Eq. 1.8) tell us that the momentum of the particle is related to the wavelength of the wave function. Thus, a state of definite momentum p must have a wave function with a definite wave vector (a scalar in 1-D) k. We conclude that

$$\langle x|p\rangle = Ce^{ikx}, \tag{10.38}$$

where $\hbar k = p$ and C is a constant. Note that $|\langle x|p\rangle|^2 = |C|^2$, independent of x. This is not really a probability density, since $|p\rangle$ is not an actual physical state. But informally, we can say that, if the particle has a definite momentum p, then it is equally likely to be found at every position x.

What is the constant C? We can choose our momentum eigenstates so that C is real and positive. To determine its magnitude, we consult the continuous orthonormality relation for the $|p\rangle$ states: $\langle p|p'\rangle = \delta(p' - p)$. We find that

$$\langle p|p'\rangle = \int_{-\infty}^{+\infty} \langle p|x\rangle\langle x|p'\rangle\, dx$$

$$= |C|^2 \int_{-\infty}^{+\infty} e^{i(p'-p)x/\hbar}\, dx. \tag{10.39}$$

From the theory of Fourier transforms, we remember that the delta function is given by

$$\delta(z) = \frac{1}{2\pi} \int_{-\infty}^{+\infty} e^{iyz}\, dy. \tag{10.40}$$

We conclude that $C = 1/\sqrt{2\pi\hbar}$, and

$$\langle x|p\rangle = \frac{1}{\sqrt{2\pi\hbar}}\, e^{ipx/\hbar}. \tag{10.41}$$

Now we can work out the relation between the momentum wave function (p-representation) $\bar{\psi}(p)$ and the more familiar position wave function $\psi(x)$.

$$\bar{\psi}(p) = \langle p \,|\, \psi \rangle = \int_{-\infty}^{+\infty} \langle p \,|\, x \rangle \langle x \,|\, \psi \rangle \, dx$$

$$= \frac{1}{\sqrt{2\pi\hbar}} \int_{-\infty}^{+\infty} e^{-ipx/\hbar} \, \psi(x) \, dx. \qquad (10.42)$$

Exercise 10.12 Also show that

$$\psi(x) = \frac{1}{\sqrt{2\pi\hbar}} \int_{-\infty}^{+\infty} e^{ipx/\hbar} \, \bar{\psi}(p) \, dp. \qquad (10.43)$$

In other words, the momentum wave function is essentially the Fourier transform of the position wave function:

$$\bar{\psi}(p) = \frac{1}{\sqrt{\hbar}} \, \tilde{\psi}\left(\frac{p}{\hbar}\right). \qquad (10.44)$$

This is a highly useful result, and it sheds a lot of light on the basic properties of Fourier transforms developed in Appendix B. For instance, consider two quantum states $|\phi\rangle$ and $|\psi\rangle$. We can compute their inner product using either the position or momentum wave functions:

$$\langle \phi \,|\, \psi \rangle = \int_{-\infty}^{+\infty} \phi^*(x)\psi(x) \, dx = \int_{-\infty}^{+\infty} \bar{\phi}^*(p) \, \bar{\psi}(p) \, dp. \qquad (10.45)$$

The fact that these two expressions must be equal is just a restatement of Parseval's theorem for Fourier transforms (Eq. B.26).

The momentum operator

The momentum operator is

$$\mathsf{p} = \int_{-\infty}^{+\infty} p \, |p\rangle\langle p| \, dp. \qquad (10.46)$$

If we represent a state $|\psi\rangle$ using its momentum wave function $\bar{\psi}(p)$, then we expect that the momentum operator has a simple expression.

Exercise 10.13 Show that the momentum wave function of the vector $\mathsf{p}\,|\psi\rangle$ is just $p\,\bar{\psi}(p)$.

Suppose instead we use the position wave function $\psi(x)$ to describe the state of the particle. How does the operator p affect $\psi(x)$?

$$\langle x | \mathsf{p} | \psi \rangle = \int_{-\infty}^{+\infty} p \, \langle x | p \rangle \, \langle p | \psi \rangle \, dp$$

$$= \frac{1}{\sqrt{2\pi\hbar}} \int_{-\infty}^{+\infty} p \, e^{ipx/\hbar} \, \bar{\psi}(p) \, dp$$

$$= \frac{1}{\sqrt{2\pi\hbar}} \int_{-\infty}^{+\infty} -i\hbar \frac{\partial}{\partial x} \left(e^{ipx/\hbar} \right) \bar{\psi}(p) \, dp$$

$$= -i\hbar \frac{\partial}{\partial x} \left(\frac{1}{\sqrt{2\pi\hbar}} \int_{-\infty}^{+\infty} e^{ipx/\hbar} \bar{\psi}(p) \, dp \right). \tag{10.47}$$

Equation 10.43 now tells us that the p operator acts on a state $|\psi\rangle$ so that the position wave function changes according to

$$\mathsf{p} : \psi(x) \to -i\hbar \frac{\partial}{\partial x} \psi(x). \tag{10.48}$$

For a quantum state $|\psi\rangle$, the expectation value of momentum is

$$\langle p \rangle = \langle \psi | \mathsf{p} | \psi \rangle$$

$$= \int_{-\infty}^{+\infty} \langle \psi | x \rangle \, \langle x | \mathsf{p} | \psi \rangle \, dx$$

$$\langle p \rangle = \int_{-\infty}^{+\infty} \psi^*(x) \left(-i\hbar \frac{\partial}{\partial x} \right) \psi(x) \, dx. \tag{10.49}$$

The p operator acts as a derivative operator on position wave functions. We can use this fact to compute the commutator of the x and p operators. Given a vector $|\psi\rangle$, the vector $\mathsf{xp} |\psi\rangle$ has a wave function

$$(x) \left(-i\hbar \frac{\partial}{\partial x} \right) \psi(x) = -i\hbar x \frac{\partial}{\partial x} \psi(x). \tag{10.50}$$

The vector $\mathsf{px} |\psi\rangle$ is not quite the same:

$$-i\hbar \frac{\partial}{\partial x} (x \psi(x)) = -i\hbar x \frac{\partial}{\partial x} \psi(x) - i\hbar \psi(x). \tag{10.51}$$

The commutator $[\mathsf{x}, \mathsf{p}] = \mathsf{xp} - \mathsf{px}$ thus changes the wave function according to the rule

$$[\mathsf{x}, \mathsf{p}] : \psi(x) \to i\hbar \psi(x). \tag{10.52}$$

We conclude that

$$[\mathsf{x}, \mathsf{p}] = i\hbar \mathbf{1}. \tag{10.53}$$

This is called the *canonical commutation relation* for the position and momentum operators. (With a name like that, you may safely infer that Eq. 10.53 is an important result!)

The commutator, of course, tells us a great deal. For instance, the uncertainty relation in Eq. 4.46 has an especially simple form for x and p observables. Since $\langle 1 \rangle = 1$ for any state,

$$\Delta x\, \Delta p \geq \frac{\hbar}{2}. \tag{10.54}$$

This relation was first discovered by Heisenberg (before the more general uncertainty relation of Eq. 4.46), and is often called the *Heisenberg uncertainty principle*. The position and momentum of a particle are complementary observables, and there are inescapable trade-offs in our knowledge of them. The more precisely one is determined, the more uncertain the other must be.

This is sometimes explained by physical arguments involving the measurement processes for position and momentum variables. For instance, if we determine a particle's position by scattering light from it, the best possible resolution of the measurement will be limited by the wavelength λ of the light. A measurement with fine resolution (small λ) must therefore use photons with greater energy and momentum, which necessarily disturb the momentum of the particle.

Interesting as such heuristic arguments may be, they do not get to the heart of the matter. The Heisenberg uncertainty principle does not depend on this or that particular physical realization of a measurement process. Nor does it actually refer to the disturbance produced by a measurement. It is a statement of the basic structure of quantum states of a particle with a continuous degree of freedom. Our simultaneous information about x and p – that is, our ability to predict the results of measurements of these two variables – is limited by Eq. 10.54, *however we come to obtain that information.*

Exercise 10.14 Compare and contrast Eq. 5.33 and 10.54.

The canonical commutation relation between x and p leads to more general commutation relations between functions of these operators. We can work these out through a series of exercises.

Exercise 10.15 For an integer $n > 0$, show that

$$[x^n, p] = i\hbar n\, x^{n-1}. \tag{10.55}$$

Exercise 10.16 Suppose the function f is given by a power series. Show that

$$[f(x), p] = i\hbar f'(x). \tag{10.56}$$

Exercise 10.17 In a similar way, show that

$$[x, g(p)] = i\hbar g'(p). \tag{10.57}$$

10.3 Wave packets

The eigenstates of momentum $|p\rangle$, like the position eigenstates $|x\rangle$, are not actual physical states of a particle moving in one dimension. A plane wave $C\, e^{ikx}$ is not an acceptable wave

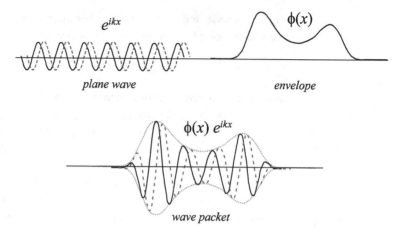

Fig. 10.2 Construction of a wave packet from a plane wave e^{ikx} and an envelope function $\phi(x)$.

function, since it cannot be normalized. Nevertheless, we do observe particles moving with pretty well defined momenta – and pretty well defined positions as well. What sort of wave function would we need to describe such a situation?

Here is a simple model that is reasonably easy to analyze: the *wave packet*. A wave packet is a wave function of the form

$$\psi(x) = \phi(x)\, e^{ikx}, \tag{10.58}$$

where $\phi(x)$ is a well-behaved *envelope function*, which differs significantly from zero only in a bounded region. The envelope $\phi(x)$ thus trims the perfect (but infinite) plane wave down to something that can be normalized. Figure 10.2 gives the general idea. (The wave function sketched in Fig. 10.1 is a wave packet.)

The envelope function $\phi(x)$ determines a lot about the properties of the wave function $\psi(x)$. For example, the probability density for this wave function is

$$\mathcal{P}(x) = |\psi(x)|^2 = |\phi(x)|^2, \tag{10.59}$$

independent of k. The particle is as localized in space as the envelope function $\phi(x)$ makes it, and the expectation $\langle x \rangle$ is the same for both $\phi(x)$ and $\psi(x)$. The momentum wave function corresponding to $\psi(x)$ is given by

$$\bar{\psi}(p) = \bar{\phi}(p - \hbar k), \tag{10.60}$$

where $\bar{\phi}$ would be the momentum wave function if the $\phi(x)$ were the actual position wave function. In other words, a *modulation* of the envelope $\phi(x)$ by e^{ikx} corresponds to a *momentum shift* of the wave function by $\hbar k$.

Exercise 10.18 Use the properties of the Fourier transform discussed in Appendix B to derive Eq. 10.60.

Exercise 10.19 Show that the envelope function $\phi(x)$ determines the uncertainties in both position (Δx) and momentum (Δp) for $\psi(x)$, even if we do not know the value of k.

A useful special case arises when the envelope function is real: $\phi(x)^* = \phi(x)$. Symmetry properties of $\phi(x)$ can sometimes simplify matters further.

Exercise 10.20 Suppose the real envelope $\phi(x)$ is symmetric about some point a, so that $\phi(a - x) = \phi(a + x)$. Show that $\langle x \rangle = a$.

Unless we specify otherwise, we will henceforth consider wave packets with real envelope functions.

The expectation value $\langle p \rangle$ is

$$\langle p \rangle = -i\hbar \int_{-\infty}^{+\infty} \psi^*(x) \frac{\partial}{\partial x} \psi(x) \, dx$$

$$= -i\hbar \int_{-\infty}^{+\infty} \phi(x) \left(\phi'(x) + ik\phi(x) \right) dx$$

$$= -i\hbar \int_{-\infty}^{+\infty} \phi(x)\phi'(x) \, dx + \hbar k \int_{-\infty}^{+\infty} (\phi(x))^2 \, dx. \tag{10.61}$$

Now we take a shortcut. Both integrals in the last equation are real. Since p is Hermitian, we know that $\langle p \rangle$ must be real. Therefore, the first integral can only be zero.

Exercise 10.21 Integrate directly to show that $\int_{-\infty}^{+\infty} \phi(x)\phi'(x) \, dx = 0$.

The remaining part becomes

$$\langle p \rangle = \hbar k \int_{-\infty}^{+\infty} (\phi(x))^2 \, dx = \hbar k. \tag{10.62}$$

So the plane wave part of our wave packet determines the expectation value $\langle p \rangle$ of the momentum.

We can construct wave packets with various specific envelope functions $\phi(x)$. A common choice is the *Gaussian* wave packet, in which the envelope function has a Gaussian form. Since $|\phi(x)|^2$ would then be a Gaussian probability distribution, Equation C.8 from the Appendix tells us that

$$|\phi(x)|^2 = \frac{1}{\sqrt{2\pi\sigma^2}} \exp\left(-\frac{x^2}{2\sigma^2} \right). \tag{10.63}$$

(Here $\langle x \rangle = 0$ and $\langle \Delta x^2 \rangle = \sigma^2$.) The probability density does not determine $\phi(x)$ uniquely, of course. Choosing $\phi(x)$ to be real and positive,

$$\phi(x) = \frac{1}{\sqrt[4]{2\pi\sigma^2}} \exp\left(-\frac{x^2}{4\sigma^2} \right). \tag{10.64}$$

The full wave function is $\psi(x) = \phi(x) e^{ikx}$, as before.

It is a straightforward exercise to determine the momentum wave function from all this, using the Fourier transform.

Exercise 10.22 Do the suggested calculation, arriving at

$$\bar{\psi}(p) = \sqrt[4]{\frac{2\sigma^2}{\pi\hbar^2}} \, \exp\left(-\frac{\sigma^2(p-\hbar k)^2}{\hbar^2}\right). \tag{10.65}$$

The probability density over momentum is $\left|\bar{\psi}(p)\right|^2$, which is also a Gaussian. In fact (by once again comparing to Equation C.8), we can "read off" the variance in p:

$$\left\langle \Delta p^2 \right\rangle = \frac{\hbar^2}{4\sigma^2}. \tag{10.66}$$

Exercise 10.23 Explain where the factor of 4 comes from in this expression.

Therefore, for the Gaussian wave packet,

$$\left\langle \Delta x^2 \right\rangle \left\langle \Delta p^2 \right\rangle = \sigma^2 \frac{\hbar^2}{4\sigma^2},$$

$$\Delta x \, \Delta p = \frac{\hbar}{2}. \tag{10.67}$$

The Heisenberg uncertainty principle tells us that $\Delta x \, \Delta p \geq \frac{\hbar}{2}$, so we can see that a Gaussian wave packet has the smallest joint uncertainty possible. Thus, it is sometimes called the *minimum uncertainty wave packet*.

10.4 Reflection and recoil

Since we can now describe a quantum system with a continuous degree of freedom, we can do an analysis that illuminates some of the important ideas of the last few chapters.

Consider a photon in the two-beam interferometer from Section 2.1, one of our prototype qubit systems. The interferometer apparatus changed the state of the photon in a unitary way. In Section 9.2, we found that unitary evolution occurs only when the system is informationally isolated. But is it reasonable to suppose that the photon is informationally isolated in the interferometer? For example, when the photon encounters a beamsplitter, the light interacts with the material of the mirror. Is it reasonable to suppose that this interaction leaves *no trace whatsoever* behind in the mirror? Or could a detailed examination of the mirror afterwards determine whether the photon had been reflected or transmitted – and hence determine the photon's state?

If a record were made of the photon's state ($|u\rangle$ or $|l\rangle$) during its interaction with the beamsplitter, then it would not be possible to observe interference effects in the resulting beams. In the laboratory, however, such effects are not hard to produce. Our job is to reconcile this experimental fact with our theoretical description of the situation. We want to show that it is reasonable to treat the photon as informationally isolated, despite its strong interaction with the mirror.

Suppose the photon approaches the beamsplitter in the upper beam. In our previous treatment, the beamsplitter changes the photon state according to

$$|u\rangle \longrightarrow \frac{1}{\sqrt{2}} \left(|u\rangle + |l\rangle \right).$$ (10.68)

Now we imagine that the mirror is free to move perpendicular to its face. Its position is given by a continuous degree of freedom x, and its initial state is $|\psi_0\rangle$ with wave function $\psi_0(x) = \langle x | \psi_0 \rangle$. It is reasonable to assume that $\psi_0(x)$ for the mirror is fairly well localized, so that it differs from zero only in the close vicinity of $x = 0$.

The interaction of the photon with the beamsplitter affects the mirror's state. For simplicity, we assume that photon transmission leaves the mirror in state $|\psi_0\rangle$; but if the photon is reflected, then the mirror recoils. The effect of this recoil is described by an operator R on the beamsplitter state. The joint state of the photon and the beamsplitter evolves according to

$$|u\rangle \otimes |\psi_0\rangle \longrightarrow \frac{1}{\sqrt{2}} \left(|u\rangle \otimes (R\,|\psi_0\rangle) + |l\rangle \otimes |\psi_0\rangle \right).$$ (10.69)

To what degree is the final beamsplitter state a physical record of whether the photon has been transmitted? If we examine only the beamsplitter, how well can we determine whether the photon is finally in state $|u\rangle$ or $|l\rangle$? This is the problem of distinguishing the beamsplitter conditional states $|\psi_R\rangle = R\,|\psi_0\rangle$ and $|\psi_0\rangle$, each of which is equally likely. The basic distinguishability theorem of Section 4.2 (generalized in Section 9.6) gives us the answer. If we define θ so that $|\langle \psi_0 | \psi_R \rangle| = \cos\theta$, then the probability P_S of successfully distinguishing between the states is bounded by

$$P_S \le \frac{1}{2} \left(1 + \sin\theta \right).$$ (Re 4.20)

If $|\langle \psi_0 | \psi_R \rangle| \approx 1$, then $P_S \approx 1/2$, which is what pure guessing allows. In this case, an examination of the beamsplitter would provide almost no information at all about the photon's path, and so the photon would be approximately informationally isolated.

If the photon reflects from the mirror, it transfers some momentum q to the mirror. Thus, the effect of R is to shift the mirror's momentum wave function by q. If $\tilde{\psi}_0(p) = \langle p | \psi_0 \rangle$ is the initial momentum wave function, then the final one is $\tilde{\psi}_R(p) = \langle p | \psi_R \rangle = \tilde{\psi}_0(p - q)$. We know from Eq. 10.60 (and Exercise 10.18) that a shift in $\tilde{\psi}$ corresponds to a modulation of ψ. Thus, the spatial wave function for the recoiling mirror is related to its initial wave function by

$$\psi_R(x) = e^{iqx/\hbar}\psi_0(x).$$ (10.70)

Therefore, the inner product is

$$\langle \psi_0 | \psi_R \rangle = \int_{-\infty}^{+\infty} e^{iqx/\hbar} |\psi_0(x)|^2 \, dx.$$ (10.71)

The mirror's initial wave function $\psi_0(x)$ is pretty well localized in space, so we can see how this works out in two extreme cases:

- If q is very small, then the function $e^{iqx/\hbar}$ varies much more slowly than $|\psi_0(x)|^2$. This means that $e^{iqx/\hbar} \approx 1$ in the bounded region around $x = 0$ where $|\psi_0(x)|^2$ is non-negligible. In this situation, $|\langle \psi_0 | \psi_R \rangle| \approx 1$, and the two states are practically indistinguishable.
- On the other hand, if the momentum transfer q is very large, then the oscillating factor $e^{iqx/\hbar}$ varies much more rapidly than $|\psi_0(x)|^2$. In this case, $|\langle \psi_0 | \psi_R \rangle| \approx 0$, and the two states are almost perfectly distinguishable ($P_S \approx 1$).

That is, the distinguishability of the mirror states depends on how much momentum q is transferred and on the spread of the undisturbed wave function $\psi_0(x)$.

To make this argument more definite, suppose that the initial mirror state is given by a Gaussian wave packet with position uncertainty Δx. Recalling Eq. 10.64,

$$\langle \psi_0 | \psi_R \rangle = \frac{1}{\sqrt{2\pi(\Delta x)^2}} \int_{-\infty}^{+\infty} e^{iqx/\hbar} \exp\left(-\frac{x^2}{2(\Delta x)^2}\right) dx. \tag{10.72}$$

This is almost the Fourier transform of a Gaussian function, which is also Gaussian.

Exercise 10.24 Consult Eq. C.9 and show that

$$\langle \psi_0 | \psi_R \rangle = \exp\left(-\frac{q^2(\Delta x)^2}{2\hbar^2}\right). \tag{10.73}$$

Thus, if $q\Delta x \ll \hbar$, then the two mirror states are practically indistinguishable. If $q\Delta x \gg \hbar$, then they are almost perfectly distinguishable.

Consider now the mirror in the interferometer. For a definite phase relationship to be maintained between the beams in the apparatus, the mirror must be localized to a small fraction of the wavelength λ of the light used. If $k = 2\pi/\lambda$, we require that $k\Delta x \ll 1$ for the mirror.[3] The wavelength of the light is also related to the momentum carried by the photons, which is $\hbar k$. In the geometry of the Mach–Zehnder interferometer, this means that the momentum transfer to the mirror is $\sqrt{2}\,\hbar k$. From this it follows that $q\Delta x \ll \sqrt{2}\,\hbar$, and so Eq. 10.73 tells us that $\langle \psi_0 | \psi_R \rangle \approx 1$. The unrecoiling and recoiling mirror states are almost indistinguishable, and the photon is effectively informationally isolated, just as originally supposed.

Exercise 10.25 Recast this argument in terms of the momentum uncertainty Δp of the mirror. What can we say about the distinguishability of the two mirror states when $q \ll \Delta p$? When $q \gg \Delta p$?

10.5 More dimensions of space

What if a particle moves, not simply along a one-dimensional line, but in a space of two or three dimensions? We can easily generalize our discussion to describe this.

---

[3] This is easy to achieve for a macroscopic mirror made of trillions upon trillions of atoms.

The position of the particle in space is described by Cartesian coordinates x, y, and z, which are the components of a 3-D position vector $\vec{r}$. The quantum particle has improper position eigenstates $|x, y, z\rangle = |\vec{r}\rangle$, and its state is represented by a wave function

$$\psi(\vec{r}) = \langle \vec{r} | \psi \rangle. \tag{10.74}$$

The squared magnitude $|\psi(\vec{r})|^2$ is a probability density for the position of the particle in an experiment. Thus, the normalization condition for the wave function is

$$\iiint |\psi(\vec{r})|^2 \, d^3r = 1, \tag{10.75}$$

where the integral ranges over all space, i.e. all values of $\vec{r}$. (Our 3-D integrals will be assumed to extend over all space unless otherwise specified.)

The position eigenstates $|\vec{r}\rangle$ have continuous completeness and orthonormality relations. First, completeness:

$$\iiint |\vec{r}\rangle\langle \vec{r}| \, d^3r = \mathbf{1}. \tag{10.76}$$

The $|\vec{r}\rangle$ improper vectors also satisfy

$$\langle \vec{r} | \vec{r}' \rangle = \delta^3(\vec{r}' - \vec{r}), \tag{10.77}$$

where $\delta^3(\vec{r}' - \vec{r})$ is a 3-D delta function, which can be written in terms of the Cartesian coordinates of the vectors:

$$\delta^3(\vec{r}' - \vec{r}) = \delta(x' - x)\delta(y' - y)\delta(z' - z). \tag{10.78}$$

Exercise 10.26 Suppose $|\phi\rangle$ and $|\psi\rangle$ are two states of a particle moving in space. Write down an integral expression for the inner product $\langle \phi | \psi \rangle$.

Exercise 10.27 Consider the 3-D wave function $\psi(\vec{r}) = Ae^{-ar^2}$, where $a > 0$ and $r^2 = x^2 + y^2 + z^2$. Find the normalization constant A (which may be assumed real and positive). Also calculate $\langle \vec{r} \rangle$ and $\langle r^2 \rangle$ for this state.[4]

There is also a momentum representation for the quantum state of a particle in three dimensions. The momentum eigenstates $|\vec{p}\rangle$ correspond to plane waves with a wave vector $\vec{k}$ that satisfies $\vec{p} = \hbar \vec{k}$. We can express this as follows:

$$\langle \vec{r} | \vec{p} \rangle = \frac{1}{(2\pi\hbar)^{3/2}} e^{i\vec{p}\cdot\vec{r}/\hbar}, \tag{10.79}$$

compare Eq. 10.41. The momentum wave function $\bar{\psi}(\vec{p}) = \langle \vec{p} | \psi \rangle$ for the state $|\psi\rangle$ is related to the spatial wave function via a 2-D Fourier transform.

Exercise 10.28 Show that

$$\bar{\psi}(\vec{p}) = \frac{1}{(2\pi\hbar)^{3/2}} \iiint e^{-i\vec{p}\cdot\vec{r}/\hbar} \psi(\vec{r}) \, d^3r. \tag{10.80}$$

---

[4] The expression $\langle \vec{r} \rangle$ indicates a vector whose components are $\langle x \rangle$, $\langle y \rangle$, and $\langle z \rangle$.

What about the 3-D vector "momentum operator" $\vec{p}$? This is simply a heuristic way of considering three component operators at once. Each component operator p_x, p_y, and p_z is exactly like the 1-D momentum operator for the corresponding coordinate. That is, we can write

$$\vec{p} : \psi(\vec{r}) \rightarrow -i\hbar\vec{\nabla}\psi(\vec{r}). \tag{10.81}$$

The gradient $\vec{\nabla}$ has "components" $\dfrac{\partial}{\partial x}$, $\dfrac{\partial}{\partial y}$, and $\dfrac{\partial}{\partial z}$.

Exercise 10.29 Consider the commutation relations among the x, y, z, p_x, p_y, and p_z. Show that all of the commutators are zero except

$$[x, p_x] = i\hbar\mathbf{1}, \qquad [y, p_y] = i\hbar\mathbf{1}, \qquad [z, p_z] = i\hbar\mathbf{1}. \tag{10.82}$$

10.6 How not to think about ψ

A particle moving in space is described by a wave function $\psi(\vec{r}, t)$ that evolves in time according to the Schrödinger equation (Eq. 11.26), which is a linear partial differential equation. The function $\psi(\vec{r}, t)$ is the "wave" part of the wave–particle duality discussed in Section 1.2.

This invites analogies to other wave systems – the vibrations of a stretched wire or membrane, sound waves in the air, or electromagnetic waves. Each of these is a disturbance in a field described by a function of space and time variables. For instance, a sound wave traveling in space can be described by the function $P(\vec{r}, t)$, the air pressure as it depends on $\vec{r}$ and t. This function is governed by a partial differential equation, and the mathematical machinery used in its analysis contains many elements also found in the analysis of quantum wave functions: initial-value problems, boundary conditions, separation of variables, vector space methods, Fourier transforms, etc.

Nevertheless, the mathematical analogy can be misleading. The wave function $\psi(\vec{r}, t)$ is *not* a physical field like the air pressure $P(\vec{r}, t)$.

To see why, consider a situation in which two sound wave pulses move in opposite directions. The pressure field function $P(\vec{r}, t)$ consists of two separate disturbances moving to the left and the right. After a long time, the waves may impinge on two widely separated sound detectors. These detectors will behave completely independently.

This is true whether the detectors have a deterministic response to sound or they act in a noisy, probabilistic way. Because the physics of sound and sound-detectors is entirely local, the left-hand detector only "sees" the left-hand part of the sound wave, and its response only depends on that part of the wave, no matter what is happening on the right side. Knowing the experimental set-up, we gain no information about the response of the left-hand detector from knowledge of the response of the right-hand detector. Or, to put it another way, the local values $P(\vec{r}_1, t)$ and $P(\vec{r}_2, t)$ of the field at two different locations have *independent* experimental meaning, in terms of the responses of local sound-detectors.

The situation is not the same for a quantum particle described by a wave function $\psi(\vec{r}, t)$. We can prepare this system in a state that is a superposition of two localized wave packets moving in opposite directions: $\psi(\vec{r}, t) = \psi_L(\vec{r}, t) + \psi_R(\vec{r}, t)$. Yet here we must be careful in our thinking. There may be two wave packets in $\psi(\vec{r}, t)$, but the wave function still represents the state of a *single* particle. If a measuring device on the right-hand side detects the presence of the particle, we know for certain that a detector on the left-hand side will not. The detection probabilities P_L and P_R are *parts of the same probability distribution, not probabilities for independent events*. Unlike the local values of a physical field, the local values of the wave function do not have an independent meaning.

A second (and even more compelling) reason why the wave function cannot be considered as a physical field arises whenever the system includes more than one particle. If ψ were a field, then we would expect the two-particle situation to be described by a ψ with greater magnitude (e.g. a normalization of 2), or perhaps by two distinct fields ψ_1 and ψ_2 for the two particles. But, as we will see in Chapter 14, neither of these is correct. The two-particle system is described by a single wave function that depends on both particle coordinates and time: $\psi(\vec{r}_1, \vec{r}_2, t)$. In other words, the wave function is now a function of a *six*-dimensional space. For more particles, things get even worse. The wave function $\psi(\vec{r}_1, \ldots, \vec{r}_N, t)$ for an N-particle system in one spatial dimension depends on $3N$ spatial coordinates (and time).

In this way, the quantum wave function $\psi(\vec{r}, t)$ seems more like a probability distribution than a physical field. Imagine a classical particle that moves in space, described at any time t by its position $\vec{r}$ and momentum $\vec{p}$. We may not have exact knowledge of x and p, so we will describe the particle by a probability distribution $\mathcal{P}(\vec{r}, \vec{p}, t)$ that depends on time. Newtonian motion of the particle produces a time evolution of $\mathcal{P}$ that can be expressed as a partial differential equation. Furthermore, if we have more than one particle, the probability distribution will be a function of all of the variables describing the classical state of the system. For two particles, the joint distribution function will have to be written $\mathcal{P}(\vec{r}_1, \vec{p}_1, \vec{r}_2, \vec{p}_2, t)$.

Yet this analogy, too, can be misleading, as the phenomenon of interference shows. The quantum wave function $\psi(\vec{r}, t)$ is neither a field nor a probability distribution. It is a creature of a new and different sort, and we must regard all such analogies with an appropriate suspicion.

Problems

Problem 10.1 The continuous collection of "eigenstates" $|x\rangle$ for our regular Hilbert space is a linearly independent set, although the exact formulation of this idea is a bit tricky. After all, the kets $|x\rangle$ are not themselves regular states. Here is a simple way to express it: For any continuous function $c(x)$ such that

$$\int_{-\infty}^{+\infty} c(x) |x\rangle \, dx = 0, \tag{10.83}$$

then $c(x) = 0$ for all x. (You should compare this definition with the usual condition of linear independence in a finite-dimensional space.) Use the fact that the $|x\rangle$ are linearly independent to show that the wave function for a given state $|\psi\rangle$ is unique.

Problem 10.2 A quantum wave function is given by

$$\psi(x) = \begin{cases} A\left(a^2 - x^2\right)^2 & |x| < a \\ 0 & |x| \geq a \end{cases}, \tag{10.84}$$

where A is a constant chosen to normalize $\psi(x)$.

(a) Show that $\psi(x)$ and its derivative $\psi'(x)$ are both continuous, and find a suitable constant A. What are the units of A?
(b) Calculate $\langle x \rangle$, Δx, $\langle p \rangle$, and Δp for this wave function. Verify the Heisenberg uncertainty principle (Eq. 10.54).

Problem 10.3 Repeat all parts of Problem 10.2 for the slightly modified wave function

$$\phi(x) = \begin{cases} Bx\left(a^2 - x^2\right)^2 & |x| < a \\ 0 & |x| \geq a \end{cases}. \tag{10.85}$$

Are the quantum states in this problem and that one orthogonal to one another?

Problem 10.4 A wave packet is a finite train of plane waves with an envelope function that is a square "pulse" over the interval $[a, b]$,

$$\psi(x) = \begin{cases} Ce^{ikx} & a \leq x \leq b \\ 0 & \text{otherwise} \end{cases}. \tag{10.86}$$

(This $\psi(x)$ is not continuous, but for now ignore that difficulty.) First, find the (real) normalization constant C. Then calculate $\langle x \rangle$, $\left(\Delta x^2\right)$, $\bar{\psi}(p)$, and $\langle p \rangle$. Also show that $\left(\Delta p^2\right)$ is infinite.

Problem 10.5 A wave function is a superposition of two wave packets moving in opposite directions: $\phi_1(x)e^{ikx}$ and $\phi_2(x)e^{-ikx}$, for real envelope functions ϕ_1 and ϕ_2. Is it possible that the two wave packets might exactly cancel out, so that $\psi(x) = 0$ everywhere?

Problem 10.6 Suppose Q and P are Hermitian operators on a Hilbert space of finite dimension dim $\mathcal{H} = d$. Show that these operators cannot possibly have the canonical commutation relation $[Q, P] = i\hbar \mathbf{1}$.

Problem 10.7 The remarkable *Wigner function* was introduced by Eugene Wigner in 1932. It is a useful alternative representation of quantum states. Suppose a particle is moving in 1-D and is described by a wave function $\psi(x)$. Then the Wigner function is

$$\mathcal{P}(x,p) = \frac{1}{2\pi\hbar} \int_{-\infty}^{+\infty} \psi^*(x+y/2)\psi(x-y/2)e^{ipy/\hbar}\, dy. \tag{10.87}$$

The Wigner function has many of the characteristics of a joint probability distribution over x and p. Prove the following properties of $\mathcal{P}(x,p)$:

(a) $P(x,p)$ is real.

(b) $\int_{-\infty}^{+\infty} P(x,p)\,dp = |\psi(x)|^2$.

(c) $\int_{-\infty}^{+\infty} P(x,p)\,dx = |\bar{\psi}(p)|^2$.

(d) For two states $|\psi_1\rangle$ and $|\psi_2\rangle$ with Wigner functions P_1 and P_2,

$$|\langle \psi_1 | \psi_2 \rangle|^2 = 2\pi\hbar \int_{-\infty}^{+\infty} \int_{-\infty}^{+\infty} P_1(x,p)P_2(x,p)\,dx\,dp. \qquad (10.88)$$

If we integrate over either x or p, we obtain the correct quantum probability distributions over p and x. However, the Wigner function cannot be properly regarded as a joint probability distribution. To prove this, construct an example in which $P(0,0) < 0$. (Hint: Consider an odd, real wave function.)

Problem 10.8 The state of a particle moving in 1-D may be described by a density operator ρ. In the position representation, this will be described by a function $\rho(x,x')$.

(a) What is the normalization condition (Eq. 8.14) for $\rho(x,x')$?

(b) Denote by $|G_\sigma(p)\rangle$ the state described by a Gaussian wave packet with an envelope function $\phi(x)$ from Eq. 10.64 and a plane wave e^{ikx} with $\hbar k = p$. Find the position representation $\rho(x,x')$ for the density operator $\rho = |G_\sigma(p)\rangle\langle G_\sigma(p)|$.

(c) Suppose that the particle is in a mixture of Gaussian wave packet states with a continuous range of momenta near zero:

$$\rho = \frac{1}{2q} \int_{-q}^{q} |G_\sigma(q)\rangle\langle G_\sigma(q)|\,dq. \qquad (10.89)$$

Show that the coherences of the density operator are suppressed – that is, show that $\rho(x,x')$ is closer to zero than in part (c), if x and x' are different enough. (Part of your job is to pin down the phrase "different enough.")

11 Dynamics of a free particle

11.1 Dynamics in 1-D

The 1-D Hamiltonian

It is time to think about time. The quantum state $|\psi(t)\rangle$ of a particle in 1-D depends on time t, and so do the position and momentum wave functions:

$$\psi(x,t) = \langle x|\psi(t)\rangle \quad \text{and} \quad \bar\psi(p,t) = \langle p|\psi(t)\rangle. \tag{11.1}$$

All of our previous results about these wave functions and their relationship to one another still hold true. The improper $|x\rangle$ and $|p\rangle$ eigenstates are not time dependent; nor are the X and p observables.

The time dependence of the state is governed by the Schrödinger equation:

$$\mathsf{H}|\psi(t)\rangle = i\hbar\frac{d}{dt}|\psi(t)\rangle. \tag{Re 5.23}$$

But what Hamiltonian operator are we to use? Our answer is based on two ideas. First, the Hamiltonian operator is the energy operator for the particle. Second, the particle's energy is the sum of its kinetic and potential energies, and these are functions of the momentum and position of the particle.

Thus, the kinetic energy K for a non-relativistic particle moving in one dimension is related to the momentum p by

$$K = \frac{p^2}{2\mu}, \tag{11.2}$$

where μ is the particle's mass. The potential energy is a function of the position x only:

$$U = U(x). \tag{11.3}$$

The total energy of the system is thus

$$E = K + U = \frac{p^2}{2\mu} + U(x). \tag{11.4}$$

The Hamiltonian operator should be related to the operators X and p by

$$\mathsf{H} = \frac{1}{2\mu}\mathsf{p}^2 + U(\mathsf{x}). \tag{11.5}$$

This is the operator that governs the time evolution of the quantum state via Eq. 5.23, as follows:

$$\left(\frac{1}{2\mu}\mathsf{p}^2 + U(\mathsf{x})\right) |\psi(t)\rangle = i\hbar\frac{d}{dt}|\psi(t)\rangle. \tag{11.6}$$

Notice that our procedure here is to begin with a prior understanding of a classical system (a particle moving in one dimension) that is analogous to the quantum system of interest. We adapt classical relations between position, momentum, and energy variables into quantum relations between operators, from which we derive the dynamics of the quantum system. This procedure is called *quantization*. This is a good strategy for guessing the Hamiltonian operator governing a quantum system of a given sort. But it is *not* the way that Nature works. Quantization is a heuristic procedure, not a physical process. Yet the analogy between classical and quantum systems, so useful to us here, is by no means an accident. The real connection runs in the opposite direction, from the underlying quantum physics to the approximate, effectively classical picture that emerges from it.

How can this "emergence" happen? This is a difficult question, some aspects of which are still poorly understood. However, we can give a partial answer. Recall that, for a time-independent observable A, the expectation value $\langle A \rangle$ evolves according to

$$\frac{d}{dt}\langle A \rangle = \frac{1}{i\hbar}\langle \psi | [\mathsf{A}, \mathsf{H}] |\psi\rangle. \tag{Re 5.27}$$

We can use this to determine how $\langle x \rangle$ and $\langle p \rangle$ change over time, relying on the canonical commutation relation and its corollaries. For instance,

$$\frac{d}{dt}\langle x \rangle = \frac{1}{i\hbar}\langle \psi | \left[\mathsf{x}, \left(\frac{\mathsf{p}^2}{2\mu} + U(\mathsf{x})\right)\right] |\psi\rangle. \tag{11.7}$$

Since x commutes with any function of x, this reduces to

$$\frac{d}{dt}\langle x \rangle = \frac{1}{2i\mu\hbar}\langle \psi | [\mathsf{x}, \mathsf{p}^2] |\psi\rangle. \tag{11.8}$$

Equation 10.57 tells us that $[\mathsf{x}, \mathsf{p}^2] = 2i\hbar\mathsf{p}$. Therefore,

$$\frac{d}{dt}\langle x \rangle = \frac{\langle p \rangle}{\mu}. \tag{11.9}$$

This has a familiar look to it! The average position $\langle x \rangle$ of the wave function moves with a velocity equal to the average momentum $\langle p \rangle$ divided by μ. The classical connection between momentum and velocity holds good in the quantum realm — at least in expectation values.

In a similar way, we can examine how $\langle p \rangle$ evolves:

$$\frac{d}{dt}\langle p \rangle = \frac{1}{i\hbar}\langle \psi | \left[\mathsf{p}, \left(\frac{\mathsf{p}^2}{2\mu} + U(\mathsf{x})\right)\right] |\psi\rangle$$

$$= -\frac{1}{i\hbar}\langle \psi | [U(\mathsf{x}), \mathsf{p}] |\psi\rangle. \tag{11.10}$$

Recall that $F(x) = -\dfrac{\partial U}{\partial x}$ is the classical force acting on a particle subject to a potential $U(x)$. Then Eq. 10.56 lets us write that

$$\frac{d}{dt}\langle p \rangle = \langle F(x) \rangle. \tag{11.11}$$

Once again, we have recovered an expression that strongly resembles a classical law – in this case, the Newtonian equation of motion of a particle in one dimension.

Equations 11.9 and 11.11 are together known as *Ehrenfest's theorem*. Let us compare these results explicitly to the classical equations of motion for a particle in one dimension, with x_c and p_c representing the classical position and momentum variables.

Classical	Quantum
$\dfrac{dx_c}{dt} = \dfrac{p_c}{\mu}$	$\dfrac{d}{dt}\langle x \rangle = \dfrac{\langle p \rangle}{\mu}$
$\dfrac{dp_c}{dt} = F(x_c)$	$\dfrac{d}{dt}\langle p \rangle = \langle F(x) \rangle.$

$$\tag{11.12}$$

If we replace x_c and p_c by the quantum expectations $\langle x \rangle$ and $\langle p \rangle$, the two sides are almost in agreement: almost, but not quite.

Exercise 11.1 What is the difference between $F(\langle x \rangle)$ and $\langle F(x) \rangle$? Construct an example – that is, describe a function $F(x)$ and a wave function ψ – for which these are quite different. Under what circumstances should these two be about the same?

A particle is in a "quasi-classical" state if it is described by a wave packet that is pretty well localized in both position and momentum (subject to the requirements of the Heisenberg uncertainty relation, of course). From the macroscopic point of view, the expectations $\langle x \rangle$ and $\langle p \rangle$ are the actual effective position and momentum of the particle. In this case, the classical equations of motion will work pretty well.

The Schrödinger equation for wave functions

Since we have been describing our state via its wave function $\psi(x, t)$, it makes sense to find the position representation for the Schrödinger equation – that is, a representation as an equation governing $\psi(x, t)$. We take each part of Eq. 11.6 in turn:

$$\frac{1}{2\mu}\langle x | \mathsf{p}^2 | \psi(t) \rangle = -\frac{\hbar}{2\mu}\frac{\partial^2}{\partial x^2}\psi(x, t),$$

$$\langle x | U(\mathsf{x}) | \psi(t) \rangle = U(x)\,\psi(x, t),$$

$$i\hbar\langle x | \left(\frac{d}{dt}|\psi(t)\rangle\right) = i\hbar\frac{\partial}{\partial t}\psi(x, t). \tag{11.13}$$

We arrive at the following equation for the wave function:

$$-\frac{\hbar^2}{2\mu}\frac{\partial^2}{\partial x^2}\,\psi(x,t) + U(x)\,\psi(x,t) = i\hbar\frac{\partial}{\partial t}\,\psi(x,t). \tag{11.14}$$

The Schrödinger equation thus becomes a partial differential equation governing the wave function $\psi(x,t)$.

The x-representation of the 1-D Schrödinger equation contains an enormous amount of physics in highly compressed form. For instance, consider what it says about stationary states of a particle moving in one dimension. Such a state evolves in time according to

$$|\psi(t)\rangle = e^{-iEt/\hbar}\,|\psi(0)\rangle. \tag{11.15}$$

The wave function is therefore

$$\psi(x,t) = \langle x\,|\psi(t)\rangle = \psi(x)\,e^{-iEt/\hbar}, \tag{11.16}$$

where $\psi(x) = \langle x\,|\psi(0)\rangle$, a function that depends only on the position x. Notice that we have "separated the variables" x and t, writing $\psi(x,t)$ as the product of a function $\psi(x)$ of x and a function $e^{-iEt/\hbar}$ of t.

The state $|\psi(0)\rangle$ is an eigenstate of the Hamiltonian operator, so $\psi(x)$ must satisfy the position representation of the energy eigenvalue equation:

$$-\frac{\hbar^2}{2\mu}\frac{d^2}{dx^2}\,\psi(x) + U(x)\,\psi(x) = E\psi(x). \tag{11.17}$$

(Note that the derivatives are now total derivatives, since the wave function $\psi(x) = \langle x\,|\psi(0)\rangle$ only depends on x.) We can also get this from Eq. 11.14 and 11.16 directly.

Exercise 11.2 Substitute Eq. 11.16 into Eq. 11.14 and obtain the "time-independent Schrödinger equation" (Eq. 11.17).

We have already noted the importance of finding energy eigenstates, so solving Eq. 11.17 for E and $\psi(x)$ will deserve some considerable attention. The whole of Chapter 15, in fact, is devoted to that problem.

The flow of probability

As the quantum state evolves and the wave function $\psi(x,t)$ changes, the probabilities for finding the particle in various locations will also change. The Schrödinger equation can tell us something about how this happens. At any given point, the probability density changes by

$$\frac{\partial}{\partial t}(\psi^*\psi) = \frac{\partial\psi^*}{\partial t}\,\psi + \psi^*\,\frac{\partial\psi}{\partial t}$$

$$= \frac{1}{-i\hbar}\left(-\frac{\hbar^2}{2\mu}\frac{\partial^2\psi^*}{\partial x^2} + U(x)\psi^*\right)\psi$$

$$+ \frac{1}{i\hbar}\psi^*\left(-\frac{\hbar^2}{2\mu}\frac{\partial^2\psi}{\partial x^2} + U(x)\psi\right). \tag{11.18}$$

(Note that we have used both Eq. 11.14 and its complex conjugate.) This looks like a mess; but the terms involving the potential $U(x)$ cancel. After some simplification, we get

$$\frac{\partial}{\partial t}(\psi^*\psi) = -\frac{i\hbar}{2\mu}\left(\frac{\partial^2\psi^*}{\partial x^2}\psi - \psi^*\frac{\partial^2\psi}{\partial x^2}\right). \tag{11.19}$$

Now we define the *probability flux J* to be

$$J(x,t) = \frac{i\hbar}{2\mu}\left(\frac{\partial\psi^*}{\partial x}\psi - \psi^*\frac{\partial\psi}{\partial x}\right). \tag{11.20}$$

Then

$$\frac{\partial}{\partial t}(\psi^*\psi) = -\frac{\partial J}{\partial x}. \tag{11.21}$$

Our claim is that J represents the "flow of probability" along the x-axis. This is a slightly weird idea, but it does make sense. Consider, for instance, a particular interval from a to b. The total probability that the particle is found in that interval is

$$P = \int_a^b \psi^*\psi \, dx. \tag{11.22}$$

As the wave function changes, this probability changes. How? Like this:

$$\begin{aligned}\frac{dP}{dt} &= \int_a^b \frac{\partial}{\partial t}(\psi^*\psi) \, dx \\ &= -\int_a^b \frac{\partial J}{\partial x} \, dx \\ &= J(a,t) - J(b,t).\end{aligned} \tag{11.23}$$

In other words, the probability P that the particle is in $[a, b]$ only changes due to probability flows across the boundaries of the interval at the points a and b. Probability within $[a, b]$ cannot just appear or disappear.

Imagine that the wave function $\psi(x,t)$ is only appreciably non-zero in two disjoint regions on the x-axis. This might represent a particle that is confined to a pair of boxes, but is completely excluded from the region between the boxes. The wave function ψ allows us to calculate the probability of finding the particle in one box or the other. But if the particle can never be found between the boxes, then $\psi = 0$ there, and so $J = 0$ as well. In this case, *the probability for each box must remain constant over time.*

Exercise 11.3 If x has units of length, what are the units of J?

Exercise 11.4 Suppose at a given moment, the wave function is a wave packet $\phi(x)e^{ikx}$, where the envelope function ϕ is real. Show that the probability flow J satisfies

$$J = (\text{probability density}) \times (\text{classical velocity}). \tag{11.24}$$

Which classical velocity do we mean?

Wave functions in three dimensions

All of this can easily be generalized to 3-D space. The Hamiltonian for a particle of mass μ moving in 3-D space typically has the form

$$H = \frac{1}{2\mu}p^2 + U(\vec{r}), \tag{11.25}$$

where $p^2 = p_x^2 + p_y^2 + p_z^2$, and U is the potential energy, a function of position. From this, we can derive the Schrödinger equation for the time-dependent wave function $\psi(\vec{r}, t) = \langle \vec{r}|\psi(t)\rangle$:

$$-\frac{\hbar^2}{2\mu}\nabla^2\psi + U(\vec{r})\psi = i\hbar\frac{\partial\psi}{\partial t}. \tag{11.26}$$

You should recall that in Cartesian coordinates the Laplacian ∇^2 is given by

$$\nabla^2\psi = \frac{\partial^2\psi}{\partial x^2} + \frac{\partial^2\psi}{\partial y^2} + \frac{\partial^2\psi}{\partial z^2}. \tag{11.27}$$

Expressions for ∇^2 in other coordinates are also very useful, as we will see.

If the particle is in an energy eigenstate with energy E, then its wave function must satisfy

$$-\frac{\hbar^2}{2\mu}\nabla^2\psi + U(\vec{r})\psi = E\psi. \tag{11.28}$$

The time-dependent wave function for a stationary state is $\psi(\vec{r}, t) = \psi(\vec{r}, 0)e^{-iEt/\hbar}$.

Finally, in three dimensions the probability flux is a vector:

$$\vec{J} = \frac{i\hbar}{2\mu}\left(\psi\vec{\nabla}\psi^* - \psi^*\vec{\nabla}\psi\right). \tag{11.29}$$

The connection between $\vec{J}$ and changes in probability density is worked out in Problem 11.13.

Exercise 11.5 This is the generalization of Exercise 11.4 to three dimensions. If $\psi(\vec{r}) = \phi(\vec{r})e^{i\vec{k}\cdot\vec{r}}$, where $\phi(\vec{r})$ is real, then show that

$$\vec{J} = |\phi|^2\frac{\hbar\vec{k}}{\mu}. \tag{11.30}$$

11.2 Free particles in 1-D

Time evolution

A particle is a *free particle* if the potential function $U(x) = 0$. Then the Hamiltonian operator for the system is

$$H = \frac{p^2}{2\mu}. \tag{11.31}$$

Clearly, the momentum eigenstates $|p\rangle$ are also energy eigenstates for the system:

$$H |p\rangle = \frac{p^2}{2\mu} |p\rangle . \tag{11.32}$$

Because of this, the time evolution of a quantum state $|\psi(t)\rangle$ of a free particle is particularly simple in the momentum representation. That is,

$$i\hbar \frac{\partial}{\partial t} \bar{\psi}(p, t) = i\hbar \langle p| \left(\frac{d}{dt} |\psi(t)\rangle \right)$$

$$= \langle p| H |\psi(t)\rangle$$

$$= \frac{p^2}{2\mu} \langle p |\psi(t)\rangle$$

$$i\hbar \frac{\partial}{\partial t} \bar{\psi}(p, t) = \frac{p^2}{2\mu} \bar{\psi}(p, t). \tag{11.33}$$

We can solve this to find $\bar{\psi}(p, t)$ in terms of the initial momentum wave function $\bar{\psi}(p, 0)$:

$$\bar{\psi}(p, t) = e^{-ip^2 t/2\mu\hbar} \, \bar{\psi}(p, 0). \tag{11.34}$$

Alternatively, we can use the momentum representation to write down the time-evolution operator $U(t)$:

$$U(t) = e^{-iHt/\hbar} = \int_{-\infty}^{+\infty} e^{-ip^2 t/2\mu\hbar} |p\rangle \langle p| \, dp. \tag{11.35}$$

Exercise 11.6 Show by evaluating $\langle p |\psi(t)\rangle = \langle p| U(t) |\psi(0)\rangle$ that Eq. 11.35 leads to the solution given by Eq. 11.34.

We have now completely solved the Schrödinger equation for a free particle moving in one dimension! The fact that our solution is given in terms of the momentum wave function is a slight inconvenience, of course, since we are often more interested in the x-representation of the quantum state. Nevertheless, if we know the initial position wave function $\psi(x, 0)$, we can find $\psi(x, t)$ at any later time t by a three-step procedure:

- We first find $\bar{\psi}(p, 0)$ via the Fourier transform of $\psi(x, 0)$.
- We then find $\bar{\psi}(p, t)$ from Eq. 11.34.
- Finally, we use the inverse Fourier transform to obtain $\psi(x, t)$.

An alternative, somewhat more direct method is suggested in Problem 11.5.

Moving wave packets

Can we get more insight than this into the behavior of a free particle wave function $\psi(x, t)$? We can – indeed, we are able to draw surprisingly general conclusions about this time evolution. Let $\langle \cdots \rangle_0$ represent the initial expectation value of a quantity, taken with respect to the initial state $|\psi(0)\rangle$. Since p commutes with the Hamiltonian H, the particle's momentum is conserved, and so $\langle p \rangle = \langle p \rangle_0$ at all times. The first half of Ehrenfest's theorem, Eq. 11.9, now tells us that

$$\frac{d}{dt}\langle x \rangle = \frac{\langle p \rangle_0}{\mu}. \tag{11.36}$$

The expectation value $\langle x \rangle$ moves uniformly with time, according to

$$\langle x \rangle = \frac{\langle p \rangle_0}{\mu} t + \langle x \rangle_0. \tag{11.37}$$

The uniform motion of $\langle x \rangle$ and the value of $\langle p \rangle_0$ do not tell the whole story. The wave function $\psi(x,t)$ is spread out around $\langle x \rangle$ in some way, which we measure by the uncertainty Δx. In a similar way, the uncertainty Δp measures how spread out the momentum wave function is. Initially, these have values Δx_0 and Δp_0. How do the uncertainties change over time?

It is mathematically easier to deal with the variances $\langle \Delta x^2 \rangle$ and $\langle \Delta p^2 \rangle$, which are the squares of the uncertainties Δx and Δp. Consider first the momentum. Since $\langle \Delta p^2 \rangle = \langle p^2 \rangle - \langle p \rangle^2$, and since both p and p^2 commute with H, the variance in momentum remains at the constant value $\langle \Delta p^2 \rangle_0$.

Exercise 11.7 Show that $\langle E \rangle = \dfrac{\langle p \rangle_0^2 + \langle \Delta p^2 \rangle_0}{2\mu}$, which stays constant.

To find how $\langle \Delta x^2 \rangle$ changes, we will need two more commutation relations.

Exercise 11.8 Starting with the canonical commutation relation $[x, p] = i\hbar$, show that

(a) $\left[x^2, p^2 \right] = 2i\hbar(xp + px)$.

(b) $\left[(xp + px), p^2 \right] = 4i\hbar p^2$.

Then

$$\frac{d}{dt}\langle x^2 \rangle = \frac{1}{i\hbar}\left\langle \left[x^2, \frac{p^2}{2\mu} \right] \right\rangle = \frac{1}{\mu}\langle xp + px \rangle. \tag{11.38}$$

This expression is not constant, but the second derivative

$$\frac{d^2}{dt^2}\langle x^2 \rangle = \frac{1}{\mu}\frac{d}{dt}\langle xp + px \rangle$$

$$= \frac{1}{2im^2\hbar}\left\langle \left[(xp + px), p^2 \right] \right\rangle$$

$$= \frac{2}{\mu^2}\langle p^2 \rangle = \frac{2}{\mu^2}\left(\langle \Delta p^2 \rangle_0 + \langle p \rangle_0^2 \right), \tag{11.39}$$

is constant. Therefore, $\langle x^2 \rangle$ can be written

$$\langle x^2 \rangle = \left(\frac{\langle \Delta p^2 \rangle_0 + \langle p \rangle_0^2}{\mu^2} \right) t^2 + Ct + D, \tag{11.40}$$

where C and D are constants. Clearly, $D = \langle x^2 \rangle_0$, and

$$C = \left(\frac{d}{dt} \langle x^2 \rangle \right)_{t=0}$$

$$= \frac{1}{\mu} \langle \mathsf{x}\mathsf{p} + \mathsf{p}\mathsf{x} \rangle_0 . \tag{11.41}$$

Exercise 11.9 From the above, show that

$$\langle \Delta x^2 \rangle = \left(\frac{\langle \Delta p^2 \rangle_0}{\mu^2} \right) t^2 + \frac{1}{\mu} \left(\langle \mathsf{x}\mathsf{p} + \mathsf{p}\mathsf{x} \rangle_0 - 2 \langle p \rangle_0 \langle x \rangle_0 \right) t + \langle \Delta x^2 \rangle_0 . \tag{11.42}$$

The variance $\langle \Delta x^2 \rangle$ of an arbitrary free particle wave function $\psi(x, t)$ depends on time according to the quadratic expression in Eq. 11.42.

Things become even simpler when the initial wave function $\psi(x, 0)$ is a wave packet $\phi(x)e^{ikx}$ with a real envelope function $\phi(x)$. The minimum uncertainty (Gaussian) wave packet of Eq. 10.64 is an example of this type. In this case, the linear term in Eq. 11.42 can be computed explicitly.

First, we recall that $\langle p \rangle_0 = \hbar k$ for the wave packet. Also, the canonical commutation relation implies that

$$\mathsf{x}\mathsf{p} + \mathsf{p}\mathsf{x} = i\hbar \mathbf{1} + 2\mathsf{x}\mathsf{p}. \tag{11.43}$$

Thus,

$$\langle \mathsf{x}\mathsf{p} + \mathsf{p}\mathsf{x} \rangle_0 = \int_{-\infty}^{+\infty} \phi(x)e^{-ikx}$$

$$\times \left(i\hbar\phi(x)e^{ikx} - 2i\hbar x \frac{\partial}{\partial x} \left(\phi(x)e^{ikx} \right) \right) dx$$

$$= i\hbar \int_{-\infty}^{+\infty} \phi^2 \, dx - 2i\hbar \int_{-\infty}^{+\infty} x\phi\phi' \, dx$$

$$+ 2\hbar k \int_{-\infty}^{+\infty} x\phi^2 \, dx. \tag{11.44}$$

Now we take a shortcut, the same one we used to derive Eq. 10.62 above. All of the integrals in the last equation are real. Because the operator $\mathsf{x}\mathsf{p} + \mathsf{p}\mathsf{x}$ is Hermitian, its expectation must be real. Thus, the integrals that are multiplied by i must add up to zero. The remaining term yields

$$\langle \mathsf{x}\mathsf{p} + \mathsf{p}\mathsf{x} \rangle_0 = 2 \langle p \rangle_0 \langle x \rangle_0 . \tag{11.45}$$

This means that the the linear coefficient of t in Eq. 11.42 exactly vanishes. If $\psi(x, 0)$ is a wave packet with a real envelope function $\phi(x)$, the variance of x varies with time according to

$$\langle \Delta x^2 \rangle = \left(\frac{\langle \Delta p^2 \rangle_0}{\mu^2} \right) t^2 + \langle \Delta x^2 \rangle_0 . \tag{11.46}$$

The initial value $\langle \Delta x^2 \rangle_0$ of the variance is also its minimum value. With time it grows quadratically, depending on the value of the momentum variance $\langle \Delta p^2 \rangle_0$.

Actually, as Problem 11.6 shows, Eq. 11.46 has nothing distinctively "quantum" about it. Any uncertainty we have about the initial velocity of a particle will contribute to our uncertainty about its later position. What is distinctively "quantum" is that the uncertainties in position and momentum are related by the Heisenberg uncertainty principle. That is,

$$\left\langle \Delta p^2 \right\rangle_0 \geq \frac{\hbar^2}{4 \left\langle \Delta x^2 \right\rangle_0}. \tag{11.47}$$

At a later time,

$$\left\langle \Delta x^2 \right\rangle \geq \left(\frac{\hbar^2}{4 \mu^2 \left\langle \Delta x^2 \right\rangle_0} \right) t^2 + \left\langle \Delta x^2 \right\rangle_0. \tag{11.48}$$

Equality holds in both expressions if $\psi(x, 0)$ is a Gaussian wave packet. The quantum trade-off between uncertainties in x and p means that, the narrower we make the initial wave packet, the more rapidly the wave function will spread out over time.

Exercise 11.10 Let us put some rough numbers to this. We can estimate $\hbar \approx 10^{-34}$ J s, and the mass of an electron is about 10^{-30} kg. A free electron moves along the x-axis. We prepare the electron in an initial Gaussian wave packet. Find the uncertainty Δx in the electron's position after 1 ns if the initial uncertainty is (a) 10 nm, (b) 1 nm, and (c) 0.1 nm. (See if you can do this without touching a calculator.)

It is not difficult to write down the analogous development for a particle moving in 2-D or 3-D. We can analyze the dynamics of 2-D and 3-D wave functions, including the motion and spread of wave packets, as we have already done for 1-D wave functions. In fact, the multidimensional analysis adds almost nothing, as the following exercise shows.

Exercise 11.11 For a free particle moving in 3-D, show that the three Cartesian coordinates x, y, and z behave as three *dynamically independent* 1-D particles.

11.3 Particle on a circle

So far, we have considered a free quantum particle moving in one dimension along an infinite line. It may be, however, that the space in which the particle moves is bounded in some way.

For example, consider a particle that is constrained to move on a circle of finite circumference L, rather than a straight line. We still describe the particle's position by the coordinate x, but now this coordinate is wrapped around the physical space in a periodic way. The physical location designated by coordinate x is the same as the location designated by $x + L$, $x - L$, etc. It follows that the improper position eigenstates satisfy $|x\rangle = |x + L\rangle$ for any x. For the wave function,

$$\psi(x + L) = \langle x + L | \psi \rangle = \langle x | \psi \rangle = \psi(x). \tag{11.49}$$

Thus $\psi(x)$ is a periodic continuous function with period L.

We say that this quantum system is subject to *periodic boundary conditions*. The mathematical development proceeds just as before, except that all integrals over the variable x are restricted to one period of length L, which we can take to be the interval from 0 to L. The completeness relation becomes

$$\int_0^L |x\rangle\langle x| \, dx = \mathbf{1}. \tag{11.50}$$

Exercise 11.12 From the completeness relation, show that the inner product for the particle on the circle is given by

$$\langle \phi | \psi \rangle = \int_0^L \phi^*(x) \, \psi(x) \, dx. \tag{11.51}$$

Compare Equation 10.19.

Now consider the momentum eigenstate $|p\rangle$. As before, we suppose that this has a wave function of the form

$$\langle x | p \rangle = C e^{ikx}, \tag{Re 10.38}$$

where $\hbar k = p$ and C is a constant. However, there are a couple of new wrinkles. Since $\langle x | p \rangle = \langle x + L | p \rangle = e^{ikL} \langle x | p \rangle$, the possible values of k are restricted so that kL is a multiple of 2π. In other words,

$$p = \hbar k = \frac{2\pi \hbar}{L} n, \qquad n = \ldots, -1, 0, 1, 2, \ldots \tag{11.52}$$

The discrete index n is called a *quantum number*. The momentum p is restricted to discrete values related to n, which we may conveniently call p_n. Since the physical space on the circle is bounded, the momentum states $|p\rangle$ are actually normalizable quantum states. This allows us to fix the value of the constant C in Eq. 10.38:

$$\langle x | p_n \rangle = \frac{e^{i p_n x / \hbar}}{\sqrt{L}}. \tag{11.53}$$

Exercise 11.13 Show that the momentum states $|p_n\rangle$ are orthonormal.

The completeness relation for the momentum states is discrete:

$$\sum_n |p_n\rangle\langle p_n| = \mathbf{1}. \tag{11.54}$$

For a particle moving along an unbounded line, the momentum wave function $\bar{\psi}(p)$ is essentially the Fourier transform of the position wave function $\psi(x)$ (see Eq. 10.42). With periodic boundary conditions, the momentum representation of a state $|\psi\rangle$ is the infinite discrete set of Fourier coefficients

$$\bar{\psi}_n = \langle p_n | \psi \rangle = \frac{1}{\sqrt{L}} \int_0^L e^{i 2\pi n x / L} \psi(x) \, dx. \tag{11.55}$$

A free particle moving on a circle has the familiar Hamiltonian

$$\mathsf{H} = \frac{\mathsf{p}^2}{2\mu}. \tag{Re 11.31}$$

Because of the periodic boundary conditions, the energy eigenvalues are now discrete:

$$E_n = \left(\frac{2\pi^2 \hbar^2}{\mu L^2}\right) n^2, \tag{11.56}$$

where n is an integer and μ is the particle mass. The stationary states can be taken to be the momentum states $|p_n\rangle$.

Exercise 11.14 Describe the degeneracies of the energy eigenvalues for the free particle on a circle.

Periodic boundary conditions imply many convenient properties, such as the normalizability of the momentum states and the discreteness of the energy spectrum. However, there are some other properties that are much less easy to understand. For instance, because the coordinate x is not a unique function of position (x is the same point as $x + L$ and so on) the operator X is not well defined. We cannot calculate the expectation $\langle x \rangle$ in an unambiguous way. Momentum, on the other hand, is a different story. We can still define an operator p so that

$$p : \psi \rightarrow -i\hbar \frac{\partial \psi}{\partial x}, \tag{11.57}$$

as before. The expectation $\langle p \rangle$ has a clear meaning and can be calculated using either the position or momentum representation of the state.

Exercise 11.15 Even though we cannot give an unambiguous meaning to the operator X or the expectation $\langle x \rangle$ for the particle on a circle, explain why we can nevertheless define the expectation $\langle \cos(2\pi x/L) \rangle$.

Exercise 11.16 The momentum eigenstates $|p_n\rangle$ are also energy eigenstates for the free particle on the circle. All of these have a probability density $|\langle x | p_n \rangle|^2$ for position that is completely uniform over the circle. Construct other energy eigenstates whose probability density is not uniform – the particle is more likely to be found in one place than another.

11.4 Particle in a box

A second type of boundary condition arises when a particle moving in one dimension is confined to some limited region of the line. For example, we can imagine that the particle's position is required to be on the positive half-line. The Hilbert space describing such a particle is a subspace of the Hilbert space describing an unlimited particle. The subspace is "spanned" by the improper position eigenstates $|x\rangle$ with $x > 0$. The wave function $\psi(x) = \langle x | \psi \rangle$ for states in our subspace must satisfy $\psi(x) = 0$ for $x \leq 0$. Since the wave function is continuous, it must be that $\psi(x) \rightarrow 0$ as $x \rightarrow 0$.

Physically, we can interpret this situation as motion in the presence of a perfectly impenetrable wall at $x = 0$. A particle initially located on the positive side of this wall can never later be found at any point beyond it, no matter what its energy. (As we will see in

Chapter 15, we can regard this ideal type of barrier as a limiting case of a more realistic, hard-to-penetrate barrier.)

In the presence of the wall boundary condition, the position observable X poses no difficulties, since the relation between physical location and coordinate x is one-to-one. Momentum, however, is now problematic. The improper momentum eigenstates have wave functions Ce^{ikx} that extend over the whole line, so they cannot be approximated by the physical states in our subspace.

We may also consider a particle confined between two impenetrable walls. This is sometimes called the *particle in a box*. One wall is at $x = 0$ and the other is at $x = L$, where L is the length of the box. The position eigenstates $|x\rangle$ with $0 < x < L$ satisfy a completeness relation (on the limited subspace)[1]

$$\int_0^L |x\rangle\langle x| \, dx = \mathbf{1}. \tag{11.58}$$

Wave functions for the particle in a box must satisfy $\psi(0) = \psi(L) = 0$.

Again, the momentum observable p is problematic, since momentum eigenstate wave functions do not vanish at the walls. However, the squared momentum p^2 is more promising. Consider the wave function

$$\psi(x) = \begin{cases} A\sin(kx) & 0 < x < L \\ 0 & \text{elsewhere} \end{cases}. \tag{11.59}$$

This satisfies the wall boundary condition provided $kL = \pi n$ for $n = 1, 2, \ldots$ In the interior of the box, the squared momentum acts on this state according to

$$p^2 : A\sin(kx) \rightarrow -\hbar^2 \frac{\partial^2}{\partial x^2} A\sin(kx) = \hbar^2 k^2 A\sin(kx). \tag{11.60}$$

The only difficulty lies at the wall, where the wave function has a step discontinuity in its first derivative (and thus a delta function in the second derivative). A careful resolution of this apparent problem will be given in Chapter 15. Meanwhile, we will "resolve" it simply by agreeing to apply operators *only in the interior of the box*, in the open interval $(0, L)$. The walls themselves are excluded from consideration.

Since p^2 makes sense, we can consider a free particle moving between walls, with the free-particle Hamiltonian (Eq. 11.31) in the interval $(0, L)$. The energy eigenstates $|n\rangle$ have wave functions

$$\psi_n(x) = \sqrt{\frac{2}{L}} \sin(k_n x), \tag{11.61}$$

where $k_n = \pi n/L$ and $n = 1, 2, \ldots$ Again, we have a discrete quantum number n labeling the possible states. They have energies

$$E_n = \frac{\hbar^2 k_n^2}{2\mu} = \left(\frac{\pi^2 \hbar^2}{2\mu L^2}\right) n^2. \tag{11.62}$$

---

[1] This of course looks just like Equation 11.50. The subtle difference lies in how we treat the boundary points $x = 0$ and $x = L$. On the circle, these are the same location; in the box, they are both excluded from the set of possible locations.

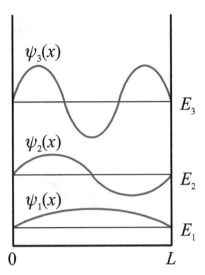

Fig. 11.1 Energy levels and stationary state wave functions for the particle in a 1-D box. It is a common convention for the wave functions to be drawn on the corresponding energy level.

The first three energy eigenstates and eigenvalues are shown in Fig. 11.1.

Exercise 11.17 (a) Verify the normalization of the wave functions $\psi_n(x)$. (b) Verify the energy eigenvalues E_n.

Exercise 11.18 The energy eigenvalues E_n for the 1-D particle in a box appear similar to those for a particle moving on a circle (Eq. 11.56). Both can be written $E_n = n^2 E_1$. Identify two crucial differences in the two expressions. Can you give an intuitive explanation for the differences?

Exercise 11.19 Do the following rough calculation without touching a calculator. An electron with a mass of about 10^{-30} kg moves in a 1-D box of size $L = 1$ nm. It is convenient to approximate $\hbar \approx 10^{-34}$ J s and $2\pi^2 \approx 20$. Find the ground state energy E_1 for the electron, and compare it to the electron-volt (1 eV $= 1.6 \times 10^{-19}$ J), a convenient unit of energy for atomic-scale problems.

How should we understand the stationary states of the particle in a box? We can get an insight into their physical meaning from the identity

$$\sin(kx) = \frac{1}{2i}\left(e^{ikx} - e^{-ikx}\right). \tag{11.63}$$

Within the box, it appears that the energy eigenstates for the particle are superposition of rightward and leftward moving waves. This suggests that we should imagine our particle to be bouncing back and forth between the two walls.

This intuitive picture of a bouncing particle in a box has obvious limitations for describing a *stationary* state of the system! So let us consider time-dependent states. Since we lack a nicely behaved momentum operator p, the approach we followed for a free unlimited particle (via commutation relations and Ehrenfest's theorem) will not work. Instead, we

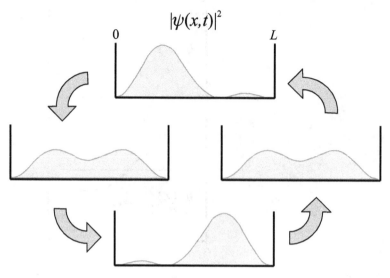

Time-dependent probability density $|\psi(x,t)|^2$ for a particle in a box. At any time, the particle is most likely to be found in a region that moves back and forth between the walls.

suppose that at time $t = 0$ we have a simple superposition of the two lowest eigenstates. That is,

$$|\psi(0)\rangle = \frac{1}{\sqrt{2}}\left(|1\rangle + |2\rangle\right),$$

$$\psi(x, 0) = \frac{1}{\sqrt{L}}\left(\sin(k_1 x) + \sin(k_2 x)\right), \tag{11.64}$$

with k_1 and k_2 defined as above. At a later time t, the wave function will be

$$\psi(x, t) = \frac{1}{\sqrt{L}}\left(e^{-i\omega_1 t}\sin(k_1 x) + e^{-i\omega_2 t}\sin(k_2 x)\right), \tag{11.65}$$

where $\hbar\omega_n = E_n$. We are interested in $|\psi(x, t)|^2$, the time-dependent probability density over the location of the particle. This is

$$
\begin{aligned}
|\psi(x, t)|^2 &= \psi^*(x, t)\psi(x, t) \\
&= \frac{1}{L}\left(\sin^2(k_1 x) + \sin^2(k_2 x)\right. \\
&\quad \left. + 2\cos((\omega_2 - \omega_1)t)\,\sin(k_1 x)\sin(k_2 x)\right).
\end{aligned}
\tag{11.66}
$$

We sketch the result in Fig. 11.2. Over time, $|\psi(x, t)|^2$ sloshes back and forth in a way that suggests the bouncing of a classical particle.

Exercise 11.20 Confirm Eq. 11.66 and explain how it leads to the behavior in Fig. 11.2. What time is required for $|\psi(x, t)|^2$ to return to its initial configuration?

11.5 Quantum billiards

A rectangular box

We may also consider 2-D and 3-D problems with boundary conditions. Suppose a quantum particle moves inside a 2-D rectangular box. The allowed region is aligned with our coordinate axes – or, more precisely, we have chosen our (mathematical) coordinates to line up with the (physical) box – and it ranges from 0 to L_x on the x-axis and from 0 to L_y on the y-axis. The wave function of the particle is zero outside this region.

We regarded the 1-D particle in a box as a particle bouncing back and forth between two walls. Our 2-D system is therefore something like a particle bouncing around inside a rectangular enclosure, like a ball on a billiard table. Quantum particles in 2-D enclosures are in fact called *quantum billiard* systems, and have been extensively studied in connection with quantum chaos. (Our aims in this section, however, are less ambitious.)

What are the stationary states of the particle in a 2-D rectangular box? Inside the box, they have wave functions of the form

$$\psi_{n_x n_y}(\vec{r}) = A \, \sin\left(\frac{n_x \pi}{L_x} x\right) \sin\left(\frac{n_y \pi}{L_y} y\right), \qquad (11.67)$$

where n_x and n_y are two independent quantum numbers that can take on values $1, 2, \ldots$

Exercise 11.21 Show that the wave function $\psi_{n_x n_y}(\vec{r})$ satisfies the boundary condition at the walls of the box. Also evaluate the normalization constant A.

Exercise 11.22 Verify that $\psi_{n_x n_y}(\vec{r})$ represents a stationary state of the 2-D free particle with energy

$$E_{n_x n_y} = \frac{2\pi^2 \hbar^2}{\mu} \left(\frac{n_x^2}{L_x^2} + \frac{n_y^2}{L_y^2}\right). \qquad (11.68)$$

The rectangular box is easy to analyze because there are independent boundary conditions for x ($\psi = 0$ except between 0 and L_x) and y ($\psi = 0$ except between 0 and L_y). Also, the free particle Hamiltonian is

$$\mathsf{H} = \frac{1}{2\mu}\mathsf{p}_x^2 + \frac{1}{2\mu}\mathsf{p}_y^2, \qquad (11.69)$$

the sum of x and y terms. As in Exercise 11.11, the x and y degrees of freedom for the particle amount to *non-interacting subsystems* of the particle. The stationary states are product states with wave functions of the form $\psi(x)\chi(y)$, and their energies are the sum of two independent terms.

Unfortunately, things are almost never this simple. For any other shape of box, the boundary conditions will not be expressible as independent conditions on x and y. In situations with a high degree of symmetry, however, the problem may still be tractable.

A circular box

Now imagine that a free 2-D quantum particle is inside a circular box of radius a, at whose center we place our coordinate origin. In terms of the Cartesian coordinates x and y, our boundary condition is that the wave function $\psi(x, y) = 0$ when $x^2 + y^2 = a^2$. We certainly do not have separate boundary conditions for x and y.

On the other hand, the problem is much simpler if we re-cast it in plane polar coordinates r and ϕ. These are related to the Cartesian coordinates by

$$x = r \cos \phi \quad \text{and} \quad y = r \sin \phi. \tag{11.70}$$

The wave function now depends on r and ϕ. There are two boundary conditions:

- $\psi(a, \phi) = 0$ for any ϕ; and
- $\psi(r, \phi + 2\pi) = \psi(r, \phi)$ for any r and ϕ.

There are separate boundary conditions for the two coordinates. (The second condition arises because the angular coordinate ϕ wraps around the plane with period 2π.)

To find the stationary states of the particle, we need to solve the energy eigenvalue equation, which in the coordinate representation is

$$-\frac{\hbar^2}{2\mu} \nabla^2 \psi = E\psi, \tag{11.71}$$

subject to our boundary conditions. To do this, we need a polar coordinate form for the Laplacian. This is derived in Problem 11.11. The equation we wish to solve is

$$\frac{\partial^2 \psi}{\partial r^2} + \frac{1}{r}\frac{\partial \psi}{\partial r} + \frac{1}{r^2}\frac{\partial^2 \psi}{\partial \phi^2} = -\frac{2\mu E}{\hbar^2}\psi. \tag{11.72}$$

Now we try a product form for the wave function, $\psi(r, \phi) = R(r)\,\Phi(\phi)$. We are motivated to do this because our boundary conditions affect R and Φ separately: $R(a) = 0$ and $\Phi(\phi + 2\pi) = \Phi(\phi)$. Our equation becomes

$$\Phi R'' + \frac{\Phi}{r}R' + \frac{R}{r^2}\Phi'' = -\frac{2\mu E}{\hbar^2}R\Phi. \tag{11.73}$$

Here R', R'', and Φ'' represent derivatives of these functions with respect to their arguments.

There is a trick that makes this equation much simpler. If we multiply by $r^2/R\Phi$ and rearrange terms, we have

$$\frac{r^2}{R}R'' + \frac{r}{R}R' + \frac{2\mu E}{\hbar^2}r^2 = -\frac{1}{\Phi}\Phi''. \tag{11.74}$$

The left-hand side of this equation is a function only of the radial coordinate r, while the right-hand side depends only on ϕ. The two sides can be equal only if both of them equal the same constant K. We therefore arrive at two separate ordinary differential equations for R and Φ:

$$\Phi'' = -K\,\Phi, \tag{11.75}$$

$$r^2 R'' + rR' + \frac{2mE}{\hbar^2}r^2 R = KR. \tag{11.76}$$

The same K appears in both equations, but otherwise they are independent. This general trick is called *separation of variables*, and the linking constant K is called a *separation constant*.

The angular equation (Eq. 11.75) is easy to solve:

$$\Phi(\phi) = e^{im\phi}, \tag{11.77}$$

where $m^2 = K$. The quantum number m must be a positive or negative integer to satisfy the boundary condition $\Phi(\phi + 2\pi) = \Phi(\phi)$.

What is the meaning of the angular quantum number m? Recall first that the chain rule for partial derivatives is

$$\frac{\partial}{\partial \phi} = \left(\frac{\partial x}{\partial \phi}\right)\frac{\partial}{\partial x} + \left(\frac{\partial y}{\partial \phi}\right)\frac{\partial}{\partial y}. \tag{11.78}$$

From this, it is easy to work out the following exercise:

Exercise 11.23 For a particle moving in 2-D, consider the operator $\mathsf{x}\mathsf{p}_y - \mathsf{y}\mathsf{p}_x$. Show that this affects the wave function by

$$(\mathsf{x}\mathsf{p}_y - \mathsf{y}\mathsf{p}_x) : \psi \to -i\hbar\frac{\partial\psi}{\partial\phi}. \tag{11.79}$$

The combination $\mathsf{x}\mathsf{p}_y - \mathsf{y}\mathsf{p}_x$, of course, is just the *angular momentum* of the particle about the origin, which we denote L_z. (The z label reminds us that the 3-D angular momentum vector is perpendicular to the plane of the particle's motion.) For 2-D wave functions with an angular quantum number m, the angular momentum operator acts according to

$$\mathsf{L}_z : R(r)\,e^{im\phi} \to m\hbar\,R(r)\,e^{im\phi}. \tag{11.80}$$

States with angular quantum number m are *angular momentum eigenstates* with eigenvalues $m\hbar$.

We now turn to the radial part of the problem, Eq. 11.76. The combination $\dfrac{2\mu E}{\hbar^2}$ has units of (length)$^{-2}$. If we define $k = \sqrt{2\mu E/\hbar^2}$, then we can replace the radial coordinate r by the dimensionless coordinate $\rho = kr$. The radial function $R(r)$ will be

$$R(r) = C\,J(kr), \tag{11.81}$$

where the constant C takes care of any necessary wave function normalization. The function $J(\rho)$ satisfies

$$\rho^2 J'' + \rho J' + (\rho^2 - m^2)J = 0, \tag{11.82}$$

where m has a fixed integer value corresponding to some angular momentum.

Equation 11.82 is known as *Bessel's equation*. Its solutions, known as *Bessel functions*, are well understood and discussed in detail in any text on mathematical methods of physics.[2]

---

[2] See, for instance, *Mathematical Methods for Physicists (6th edn)* by G. B. Arfken and H. J. Weber, Chapter 11 (Burlington, MA: Elsevier Academic Press, 2005).

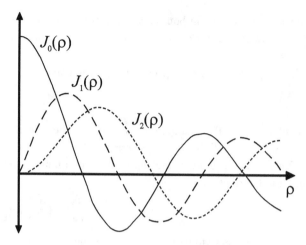

Fig. 11.3   **Graphs of the first three Bessel functions.**

We denote by J_m the Bessel function for a given non-negative integer value of m. This can be expressed as a power series:

$$J_m(\rho) = \frac{\rho^m}{2^m} \sum_{\nu=0}^{\infty} \frac{(-1)^\nu}{2^\nu \nu!(\nu + m)!} \rho^{2\nu}. \tag{11.83}$$

Exercise 11.24 Write down the first five terms of the power series for J_0. Show that J_0 solves Eq. 11.82.

What about a negative index? The Bessel function J_{-m} solves the same equation as J_m. By convention, we let $J_{-m} = (-1)^m J_m$.

A few Bessel functions are plotted in Fig. 11.3. Notice that, for $\rho \ll 1$, the function J_m is about proportional to ρ^m. Thus, for $m \neq 0$, the wave function of the particle vanishes at $r = 0$. If the particle has non-zero angular momentum, it cannot be found exactly at the center of the circular box.

Of particular interest to us are the roots of the Bessel functions, the points where $J_m = 0$. Figure 11.3 shows that each Bessel function has a whole sequence of distinct roots. Define

$$\rho_{nm} = \text{the } n\text{th non-zero root of } J_m, \tag{11.84}$$

for $n = 1, 2, \ldots$ Here are a few of these roots:

	J_0	J_1	J_2	J_3
$n = 1$	2.4048	3.8317	5.1356	6.3802
$n = 2$	5.5201	7.0156	8.4172	9.7610
$n = 3$	8.6537	10.1735	11.6198	13.0152

Exercise 11.25 Seven of these roots appear on the graphs in Fig. 11.3. Identify the ones shown.

The boundary condition at the wall of the box is that $R(a) = C J_m(ka) = 0$, from which we conclude that $ka = \rho_{nm}$ for some value of n. This condition tells us the possible energies of the particle:

$$E_{nm} = \frac{\hbar^2}{2\mu a^2} (\rho_{nm})^2. \tag{11.85}$$

As in the rectangular box, the energy depends on two quantum numbers – in this case, a radial quantum number n and an angular quantum number m.

Exercise 11.26 Derive Eq. 11.85. Comment on the degeneracy of the energy levels, explaining any assumptions you make.

Exercise 11.27 Our stationary state wave functions are

$$\psi_{nm}(r, \phi) = C J_m(k_{nm}r) e^{im\phi}, \tag{11.86}$$

where $k_{nm}a = \rho_{nm}$. For all of these, the probability density $|\psi_{nm}|^2$ depends only on r, not on ϕ. The distributions are *radially symmetric*. Verify this. Then construct some stationary state wave functions for the circular quantum billiard system whose probability densities are not radially symmetric.

Compare the circular quantum billiard system to the 1-D situations we considered at the beginning of this chapter. For a particle with periodic boundary conditions, we could not really define a position observable X, owing to the way the coordinate wrapped around the physical space. We did, however, have a perfectly sensible momentum operator p, which had a discrete spectrum of eigenvalues. For the circular billiard system, we cannot really define an angular position operator ϕ, but the angular momentum L_z exists and has a discrete spectrum.

For the 1-D particle in a box, the wall boundary conditions posed no difficulties for the position operator X. However, the momentum operator became problematic. The radial wall boundary condition in the quantum circular billiard system works the same way. We can define a radial position operator r, calculating expectations like $\langle r \rangle$, etc., but there is no radial momentum operator. Nevertheless, in either case the energy operator of the system (which is quadratic in the momenta) is well defined and has a discrete spectrum.

Problems

Problem 11.1 The kinetic energy part of the Hamiltonian is $K = \frac{1}{2m}p^2$.

(a) Show that the kinetic energy satisfies

$$\langle K \rangle \geq \frac{\langle \Delta p^2 \rangle}{2m}. \tag{11.87}$$

(b) Calculate the minimum $\langle K \rangle$ for an electron whose position uncertainty $\Delta x = 0.1$ nm, the size of a typical atom. Express your answer in electron-volts (1 eV $= 1.6 \times 10^{-19}$ J).

Problem 11.2 We can write any wave function $\psi(x) = R(x)e^{i\alpha(x)}$, where R and α are real functions. Write the probability density and probability flux for this wave function in terms of R and α.

Problem 11.3 We can devise an analog to the probability flux $J(x, t)$ for systems in a finite dimensional Hilbert space. Given a basis $\{|n\rangle\}$ for the space, the probability amplitudes are $\psi_n = \langle n|\psi\rangle$. We wish to know how the probabilities $P_n = |\psi_n|^2$ change with time. Given the Hamiltonian operator H, define

$$J_{nm} = \frac{1}{i\hbar}\left(\psi_n^* H_{nm}\psi_m - \psi_m^* H_{mn}\psi_n\right).\tag{11.88}$$

We interpret J_{nm} as the probability flux from state m to state n. Show that

$$\frac{dP_n}{dt} = \sum_m J_{nm}.\tag{11.89}$$

Explain in words the meaning of this equation. If $\langle m|$ H $|n\rangle = 0$, what can we say about the flow of probability from state m to state n?

Problem 11.4 A particle is moving in one dimension subject to some potential function U. At the instant $t = 0$, the wave function of the system happens to be completely real. Show that the wave function at all times satisfies

$$\psi(x, t) = \psi^*(x, -t).\tag{11.90}$$

Comment on the following: *If the wave function is ever real, the future of the system is a perfect reflection of its past.*

Problem 11.5 For a free particle, find the position representation of the time evolution operator:

$$U(x, x'; t) = \langle x| U(t) |x'\rangle.\tag{11.91}$$

(To evaluate this explicitly, use the fact that the Fourier transform of a Gaussian function in Equation C.9 also applies when the constant a is imaginary.) Write down the wave function $\psi(x, t)$ as an integral expression involving $\psi(x, 0)$.

Problem 11.6 Suppose a random variable Z is the sum of two independent random variables $Z = X + Y$. Show that the variance of Z is

$$\left\langle \Delta Z^2 \right\rangle = \left\langle \Delta X^2 \right\rangle + \left\langle \Delta Y^2 \right\rangle.\tag{11.92}$$

(Note that, if X and Y are independent, then $\langle XY\rangle = \langle X\rangle\langle Y\rangle$.)

Imagine a classical free particle of mass m moving in one dimension. Its position and momentum are not exactly known, and so are described by a probability distribution. At $t = 0$, the variables x and p are independently distributed with variances $\langle \Delta x^2\rangle_0$ and $\langle \Delta p^2\rangle_0$. Show that, at a later time, the variance of the particle's position is

$$\left\langle \Delta x^2 \right\rangle = \left(\frac{\langle \Delta p^2\rangle_0}{\mu^2}\right) t^2 + \left\langle \Delta x^2 \right\rangle_0,\tag{Re 11.46}$$

exactly as we found for a quantum wave packet. (Here, of course, the uncertainties arise from our ignorance of x and p, not any quantum indeterminacy.) In this classical problem, are x and p independent variables at time $t \neq 0$?

Problem 11.7 At $t = 0$, we prepare a free particle moving in one dimension. Our aim is to minimize the uncertainty Δx of the particle at a later time t. We can create an initial wave packet with a real envelope function of any shape. Notice that Δx at time t will have to be large if Δx_0 is either very large or very small. The optimum initial Δx_0 is therefore somewhere in between. Find this optimum in terms of $\hbar$, m, and t. Using the approximate values in Exercise 11.10, find minimum uncertainty Δx for a free electron after 1 ns.

Problem 11.8 For a particle moving on a circle of circumference L, let T_s be the shift operator defined by

$$\mathsf{T}_s |x\rangle = |x + s\rangle. \tag{11.93}$$

Devise an expression for T_s in terms of the momentum states $|p_n\rangle$, and prove that your expression has the correct effect on wave functions.

Problem 11.9 For a particle on a circle, define the operator $\boldsymbol{\Phi}$ by

$$\boldsymbol{\Phi} |x\rangle = R\, e^{-ix/R} |x\rangle, \tag{11.94}$$

where $R = L/2\pi$, the radius of the circle. First, verify that $\boldsymbol{\Phi}$ is well defined despite the many-valued nature of the x-coordinate. Then answer the following:

(a) How does $\boldsymbol{\Phi}$ affect the wave $\psi(x)$ of a particle on the circle?
(b) Find $\boldsymbol{\Phi}^\dagger$. (Note that this operator is not Hermitian!)
(c) What is $\boldsymbol{\Phi} |p_n\rangle$ for the momentum state $|p_n\rangle$?
(d) Compute the commutator $[\boldsymbol{\Phi}, \mathsf{p}]$.

Problem 11.10 This problem investigates connections between the classical and the quantum particle in a box.

(a) Imagine a classical free particle of mass μ and energy E is bouncing back and forth between walls separated by a distance L apart. Show that the period of this motion is

$$\tau = \frac{2L}{\sqrt{2\mu E}}. \tag{11.95}$$

(b) A quantum particle is in a superposition of the ψ_n and ψ_{n+1}. Initially, the state is

$$\psi(x, 0) = \frac{1}{\sqrt{2}} \left(\psi_n(x) + \psi_{n+1}(x) \right). \tag{11.96}$$

Show that the probability distribution $|\psi|^2$ changes over time. Find $\langle x \rangle$ as a function of time.

(c) The probability $|\psi|^2$ returns to its original configuration after a time τ. Compute τ in terms of L, μ, and E_n. How does this compare with the classical expression? How do they compare when $n \gg 1$?

Problem 11.11 Prove that, in polar coordinates, the Laplacian operator is

$$\nabla^2 f = \frac{\partial^2 f}{\partial r^2} + \frac{1}{r}\frac{\partial f}{\partial r} + \frac{1}{r^2}\frac{\partial^2 f}{\partial \phi^2}, \tag{11.97}$$

where $f = f(r, \phi)$.

Problem 11.12 Consider a wave function for the circular billiard system that at $t = 0$ is a superposition of the two lowest stationary states:

$$\psi(r, \phi, 0) = aJ_0(k_{10}r) + bJ_1(k_{11}r)e^{i\phi}. \tag{11.98}$$

For simplicity, suppose that both a and b are real, positive, and non-zero.

(a) Show that the initial probability distribution $|\psi|^2$ is not circularly symmetric. In which half of the circle is the particle most likely to be found?

(b) Show that over time $|\psi|^2$ moves around the circle. How long does it take to return to the original probability distribution?

If you have access to the necessary computer software tools, create an animation of the dynamical behavior of $|\psi|^2$ over time, given $a = b$.

Problem 11.13 The *divergence* of a vector field is, in Cartesian coordinates,

$$\vec{\nabla} \cdot \vec{F} = \frac{\partial F_x}{\partial x} + \frac{\partial F_y}{\partial y} + \frac{\partial F_z}{\partial z}. \tag{11.99}$$

According to the divergence theorem, the integral of $\vec{\nabla} \cdot \vec{F}$ over a bounded volume V is related to the integral of $\vec{F}$ over the closed surface Σ enclosing the volume:

$$\iiint_V \vec{\nabla} \cdot \vec{F}\, d^3r = \iint_\Sigma \vec{F} \cdot \hat{n}\, dA, \tag{11.100}$$

where $\hat{n}$ is an outward-pointing unit vector perpendicular to the surface Σ.

(a) Show that, if the wave function $\psi(\vec{r}, t)$ for a particle in three dimensions satisfies Eq. 11.26, then it follows that

$$\frac{\partial |\psi|^2}{\partial t} + \vec{\nabla} \cdot \vec{J} = 0. \tag{11.101}$$

(b) Let P_V be the total probability that a quantum particle is found in V. Show that

$$\frac{dP_V}{dt} = -\iint_\Sigma \vec{J} \cdot \hat{n}\, dA. \tag{11.102}$$

12 Spin and rotation

12.1 Spin-s systems

Angular momentum is one of the fundamental quantities of Newtonian physics, and in quantum physics its importance is at least as great. In quantum mechanics we often distinguish between two types of angular momentum: *orbital angular momentum*, which a system of particles possesses due to particle motion through space; and *spin angular momentum*, which is an intrinsic property of a particle.[1] The distinction will be important later, but for now we will ignore it. We will here refer to angular momentum of any sort as "spin" and develop general-purpose mathematical tools for its description.

We have already dealt with spin systems, particularly the example of a spin-1/2 particle. Our approach began with the empirical observation that a measurement of any spin component of a spin-1/2 particle could yield only the results $+\hbar/2$ or $-\hbar/2$. We introduced the basis states $|z_{\pm}\rangle$ for the two-dimensional Hilbert space $\mathcal{H}$. We also gave other basis states such as $\{|x_{\pm}\rangle\}$ and $\{|y_{\pm}\rangle\}$ in terms of the $|z_{\pm}\rangle$ states. From basis states and measurement values we constructed operators for the spin components S_x, S_y, and S_z. With the operators in hand, we could then examine the algebraic relations between them (such as the commutation relation in Exercise 3.56).

Our job here is to generalize our analysis to systems of arbitrary spin. To do this, we will reverse our chain of logic. We now *begin* with spin component operators that are assumed to satisfy the same commutation relations we obtained for the spin-1/2 operators. Amazingly, this will be a sufficient foundation to derive everything – the eigenvalues, eigenvectors, and matrix representations for all of the spin operators. Later, in Section 12.3, we will see how the commutation relations themselves follow naturally from the geometry of 3-D rotations.

Spin commutation relations

Here is our basic postulate (which previously appeared in Problem 5.2):

$$[S_x, S_y] = i\hbar S_z, \quad [S_y, S_z] = i\hbar S_x, \quad [S_z, S_x] = i\hbar S_y. \tag{12.1}$$

---

[1] The terminology calls to mind a planetary analogy. The Earth has angular momentum due to its orbital motion about the Sun and also due to its rotation about its axis. But this analogy, like most analogies in quantum physics, is dangerous if pressed too far. Both orbital motion and rotation of a planet involve masses moving in space, and so yield angular momentum of the "orbital" type. Intrinsic spin is a property peculiar to quantum systems.

A cyclic permutation of the x, y, and z axes gives an equivalent right-handed spatial coordinate system, so it is no surprise to find that our three basic commutation relations in Eq. 12.1 are connected by just such a permutation.

We sometimes speak (and are tempted to think) of S_x, S_y, and S_z as the components of a vector quantity $\vec{S}$, the "spin vector." We did this in Section 10.5 for the momentum vector $\vec{p}$ of a particle moving in three dimensions. The spin operator vector $\vec{S}$ is more problematic than that example, however, because the components do not commute with each other. For this reason we usually cannot have eigenstates of $\vec{S}$ – states with "definite spin vectors." Nevertheless, the formal vector $\vec{S}$ is valuable as a *heuristic*, a way of suggesting what calculations are useful and meaningful. For instance, we can combine the component operators to determine the magnitude $S^2 = \vec{S} \cdot \vec{S}$:

$$S^2 = S_x^2 + S_y^2 + S_z^2. \tag{12.2}$$

Despite our qualms about $\vec{S}$, the definition of S^2 is straightforward and accords with our classical understanding of angular momentum as a spatial vector quantity.

Each of the squared component operators S_x^2, etc., is positive, so their sum S^2 is also positive. This means that the eigenvalues of S^2 are real and non-negative. Since S^2 has the units of the square of angular momentum, we can write any such eigenvalue in the form $s(s+1)\hbar^2$, where the dimensionless parameter $s \geq 0$, though it is at the moment otherwise unrestricted.[2]

What commutation relations does S^2 have with the spin component operators S_x, S_y, and S_z? This is easy to answer if we first establish the following handy fact about commutators:

Exercise 12.1 Given three operators A, B, and C, show that

$$[A, BC] = [A, B]C + B[A, C]. \tag{12.3}$$

Now we find that

$$[S_z, S^2] = [S_z, S_x^2] + [S_z, S_y^2]$$
$$= [S_z, S_x]S_x + S_x[S_z, S_x] + [S_z, S_y]S_y + S_y[S_z, S_y]$$
$$= i\hbar S_y S_x + i\hbar S_x S_y - i\hbar S_x S_y - i\hbar S_y S_x$$
$$= 0. \tag{12.4}$$

The spin operator S^2 commutes with S_z, and by a similar argument with the other spin components as well.

To simplify things a bit, we will focus our attention on $\mathcal{H}_s$, the eigenspace of S^2 with some particular eigenvalue $s(s+1)\hbar^2$. If the state of the system is restricted to such a space,

---

[2] Why do we write $s(s+1)$ instead of s^2 or even just s? This is one of those places where we are anticipating how things work out later on. If we write the eigenvalues of S^2 in this way, later expressions become simpler.

At any point in a mathematical derivation, we generally have two questions in mind. First, we ask whether the step is logically sound. The answer should be clear – and had better be "yes"! Second, we ask *why* the step is taken. The exact reason for the step might be hard to see right away, but usually becomes apparent later on. To give an analogy: as we observe a game of chess, we can tell at once whether or not a given move is legal, but we sometimes do not understand the player's strategy until the game develops further.

then we say that it is a *spin-s system*. Because the component operators all commute with S^2, they will act as operators within $\mathcal{H}_s$.

Exercise 12.2 Prove that S_x, S_y, and S_z must map vectors in $\mathcal{H}_s$ to vectors in $\mathcal{H}_s$.

What do the states of a spin-s system look like? We can choose one component, S_z for instance, and find its eigenstates in $\mathcal{H}_s$. These can be written $|m, \alpha\rangle$. The number m indicates the S_z eigenvalue by

$$S_z |m, \alpha\rangle = m\hbar |m, \alpha\rangle . \tag{12.5}$$

The additional label α allows us to identify different eigenstates with the same S_z eigenvalue. At this point we know nothing about the possible values of m except that they, like s, are dimensionless.

Raising and lowering operators

We now define a pair of operators

$$S_\pm = S_x \pm iS_y. \tag{12.6}$$

Note that $S_+^\dagger = S_-$. At first, the significance of these operators may not be apparent. We start to uncover that significance by evaluating the commutator of S_+ with S_z:

$$[S_z, S_+] = [S_z, S_x] + i[S_z, S_y]$$
$$= i\hbar S_y + \hbar S_x$$
$$= \hbar S_+. \tag{12.7}$$

From this, we know that $S_z S_+ = S_+ S_z + \hbar S_+$.

Exercise 12.3 Also prove the corresponding result[3] for S_-:

$$[S_z, S_-] = -\hbar S_-. \tag{12.8}$$

Therefore $S_z S_- = S_- S_z - \hbar S_+$.

The operator S_+ acts on an S_z eigenstate $|m, \alpha\rangle$ to produce a vector $S_+ |m, \alpha\rangle$. If we operate on the new vector with S_z, we obtain

$$S_z \left(S_+ |m, \alpha\rangle\right) = (S_+ S_z + \hbar S_+)\hbar |m, \alpha\rangle = (m + 1)\hbar \left(S_+ |m, \alpha\rangle\right). \tag{12.9}$$

In other words, the vector $S_+ |m, \alpha\rangle$ acts like an eigenvector of S_z with eigenvalue $(m+1)\hbar$. We can thus say that $S_+ |m, \alpha\rangle = K |m + 1, \alpha'\rangle$ for some scalar factor K.

Exercise 12.4 Show that $S_- |m, \alpha\rangle$ acts like an eigenvector of S_z with eigenvalue $(m - 1)\hbar$.

---

[3] Be ready for several similar exercises in the pages ahead. We will show something for S_+ and let you work out the parallel fact about S_-.

Why do we say "*acts like* an eigenvector" instead of "*is* an eigenvector"? It is because the resulting vectors might be null, and only non-null vectors can be eigenvectors. A careful statement of our conclusions about $S_\pm$ would be:

- $S_+ |m, \alpha\rangle$ is either an eigenvector of S_z with eigenvalue $(m + 1)\hbar$, or else null.
- $S_- |m, \alpha\rangle$ is either an eigenvector of S_z with eigenvalue $(m - 1)\hbar$, or else null.

The S_+ and S_- operators are thus "raising" and "lowering" operators, incrementing the value of m by $+1$ or -1 respectively. They link together the different S_z eigenstates in $\mathcal{H}_s$. We can use them to prove a variety of results about the spin component operators. The next exercise gives an example.

Exercise 12.5 Show that $S_x = \dfrac{1}{2}(S_+ + S_-)$. Use this to prove that $\langle m, \alpha| S_x |m', \beta\rangle = 0$ unless $m - m' = \pm 1$.

The vectors $S_+ |m, \alpha\rangle$ and $S_- |m, \alpha\rangle$ will not themselves be physical quantum states of our spin-s system, since they are not normalized. Indeed, it is possible that one or both of these vectors is the null vector. We now investigate the magnitudes of the raised and lowered vectors.

To do this, we will need to know something about the operator product of S_+ and S_-. First, note that

$$S_+ S_- = S_x^2 + S_y^2 + i \left(S_y S_x - S_x S_y \right)$$
$$= S^2 - S_z^2 + \hbar S_z. \tag{12.10}$$

This is an especially convenient relation for us, since the $|m, \alpha\rangle$ states are eigenstates of both S^2 and S_z.

Exercise 12.6 In a similar way, show that

$$S_- S_+ = S^2 - S_z^2 - \hbar S_z. \tag{12.11}$$

Now consider the vector $S_+ |m, \alpha\rangle$. The squared magnitude of this vector is of course non-negative:

$$0 \le \langle m, \alpha| S_- S_+ |m, \alpha\rangle$$
$$= \langle m, \alpha| \left(S^2 - S_z^2 - \hbar S_z \right) |m, \alpha\rangle$$
$$0 \le (s(s + 1) - m(m + 1)) \hbar^2. \tag{12.12}$$

We have arrived at an important connection between the values of m and s, namely that $m \le s$. The non-negative parameter s acts as a "ceiling" for the possible values of m. Also, we note that when $m = s$, the vector $S_+ |m, \alpha\rangle = 0$.

But this suggests a paradox! How do we reconcile the limit $m \le s$ with the action of S_+ to increase the value of m by 1? If we begin with an eigenvector $|m, \alpha\rangle$ and successively apply S_+, we can increase m to $m + 1$, then $m + 2$, and so on, without changing the value of s. We climb up the S_z "ladder" higher and higher. Can we not apply S_+ enough times to climb beyond the $m \le s$ ceiling, and arrive at a contradiction?

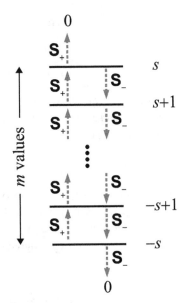

The "ladder" of possible m values for a spin-s system. The rungs are connected by the $S_{\pm}\phi$ operators and have unit separation.

The only hope lies in the observation that, when $m = s$ exactly, the result of applying S_+ is the null vector. Here the upward ladder stops, and further applications of S_+ lead nowhere. We therefore conclude that successive applications of S_+ must eventually lead us to an eigenvector with $m = s$. That means that every possible value for m must differ from s by *exactly an integer*. Otherwise, successive applications of the raising operator, each one increasing m by 1, would jump past the $m = s$ ceiling.

In short, the only possible values of m are of the form s, $s - 1$, $s - 2$, and so on. The m-values form a discrete ladder, separated by unit steps, up to the maximum value $m = s$. Naturally, there is a similar argument for the lowering operator S_-.

Exercise 12.7 By considering the squared magnitude of $S_- |m, \alpha\rangle$, show that

$$0 \geq s(s + 1) - m(m - 1). \tag{12.13}$$

From this, show how to draw the conclusion that $m \geq -s$.

Exercise 12.8 Successive applications of the lowering operator S_- on $|m, \alpha\rangle$ lead to values $m - 1, m - 2$, etc. Show that this downward sequence can only terminate at $m = -s$, where the vector $S_- |m, \alpha\rangle = 0$.

Therefore, the only possible values of m must be of the form $-s$, $-s + 1$, $-s + 2$, etc. Otherwise, successive applications of S_- would miss the unique "floor" value of $m = -s$.

The general situation is illustrated in Fig. 12.1. We know that the top of the ladder must be at $m = s$, that the bottom of the ladder is at $m = -s$, and that adjacent values of m are separated by 1. It follows that *the distance between $-s$ and s must be an integer*. That is, $2s = 0, 1, 2, \ldots,$ or

$$s = 0, \frac{1}{2}, 1, \frac{3}{2}, \ldots \tag{12.14}$$

Only integer or half-integer values for s are allowed. We can have spin-0 systems, spin-1/2 systems, spin-1 systems, and so on; but no other values of s are possible. For a given s, the S_z quantum number m can take on $2s + 1$ different values ranging from $-s$ to s.

Exercise 12.9 Under what circumstances would the value $m = 0$ be possible?

Pause to reflect on our achievement so far. Beginning only with the fundamental commutation relations in Eq. 12.1 (which are in fact all the same, up to a cyclic permutation of $x, y,$ and z), we have deduced the spectra of possible eigenvalues for S^2 and S_z. This seems a remarkable product from such meager ingredients. But we are not done yet!

Simple spin systems

Now that we know all about the possible values of s and m, we turn to the task of figuring out the structure of the Hilbert space $\mathcal{H}_s$. Each rung of the ladder we have constructed represents an S_z eigenvalue of $m\hbar$, which corresponds to a subspace containing all the states of the form $|m, \alpha\rangle$. Let d_m be the dimension of this subspace. What are the dimensions d_m?

In fact, each of these subspaces must have the *same* dimension. Here is why. Suppose $|m, \alpha\rangle$ and $|m, \beta\rangle$ are two orthogonal states in the m eigenspace. The raising operator S_+ maps these to vectors that are still orthogonal:

$$\langle m, \alpha| \, S_- S_+ \, |m, \beta\rangle = \langle m, \alpha| \left(S^2 - S_z^2 - \hbar S_z \right) |m, \beta\rangle$$
$$= (s(s+1) - m(m+1)) \, \hbar^2 \, \langle m, \alpha \, |m, \beta\rangle$$
$$= 0. \tag{12.15}$$

If $m = s$, then S_+ maps everything to the null vector. For $m < s$, an orthonormal basis for the m eigenspace (containing d_m elements) will be mapped to an orthogonal set in the $(m + 1)$ eigenspace. The dimension d_{m+1} of this eigenspace must be large enough to accommodate this set. Therefore, $d_{m+1} \geq d_m$. As usual, a similar argument can be made by going down the ladder using S_-:

Exercise 12.10 By applying S_- to orthogonal states in the $(m + 1)$ eigenspace, show that $d_{m+1} \leq d_m$. Conclude that $d_{m+1} = d_m$.

It follows that all of the d_ms are the same: $d_m = d_m$. Since there are $2s + 1$ different possible values of m, the dimension of the overall space $\mathcal{H}_s$ is $(2s + 1)d_m$. This suggests that we should regard our system as a composite of two subsystems, a "spin" subsystem (with Hilbert space dimension $2s + 1$) and an "other" subsystem (with Hilbert space dimension d_m). The overall Hilbert space is just the tensor product of those for the subsystems.

A *simple spin-s system* is one for which $d_m = 1$ for all m. This means that the spin degree of freedom is the only degree of freedom in the system, and $d = 2s + 1$. A familiar example of this idea is the spin-1/2 particle we analyzed as a part of qubit quantum mechanics. For that system we ignored the position and other degrees of freedom of the particle, considering only its spin. The dimension of its Hilbert space was therefore $2\left(\frac{1}{2}\right) + 1 = 2$.

For a simple spin-s system, we do not need the additional label α for the S_z eigenstates. Instead, we can construct a standard basis of these eigenstates using the following procedure:

- First, choose a normalized state with $m = -s$ to be the state $|-s\rangle$.
- For each $m < s$, define

$$|m + 1\rangle = \frac{1}{\hbar\sqrt{s(s+1) - m(m+1)}} S_+ |m\rangle. \tag{12.16}$$

This provides an inductive definition of all of the states $|m\rangle$, where m ranges over the $2s + 1$ values from $-s$ to s.

- The $|m\rangle$ states are orthogonal to one another, since they are eigenstates of S_z with different eigenvalues. The prefactor in the inductive definition is carefully chosen so that the $|m\rangle$ states are also normalized. Since the dimension of $\mathcal{H}_s$ is $2s + 1$, the $|m\rangle$ states form an orthonormal basis for it.

Building upon this construction, we can express all of our spin operators in the standard basis.

Exercise 12.11 Show that

$$S_+ = \sum_m \hbar\sqrt{s(s+1) - m(m+1)} \, |m+1\rangle\langle m|, \tag{12.17}$$

where the sum ranges over all the values of m from $-s$ to s. (The $m = s$ term is zero, of course.) Use this to write down the matrix representation for S_+ for a spin-1 system.

Exercise 12.12 Show that, for all m,

$$S_- |m\rangle = \hbar\sqrt{s(s+1) - m(m-1)} \, |m-1\rangle. \tag{12.18}$$

(When $m = -s$, this simply means that $S_- |-s\rangle = 0$.)

See also Problem 12.2, where explicit matrix representations for S_x and S_y are derived.

We have learned that the whole quantum theory of systems of arbitrary spin is somehow "encoded" in the component commutation relations postulated in Eq. 12.1. Admittedly, we did have to go to considerable effort to "decode" the commutation relations and draw the appropriate conclusions; but we succeeded in building up the entire mathematical structure. The key step in our effort was the definition of the raising and lowering operators $S_\pm$, whose properties allowed us to construct the ladder of S_z eigenstates. We will see this general strategy again, when we analyze an infinite "ladder system" in Chapter 13. Meanwhile, the commutator has once again proven to be a powerful tool for extracting physical meaning from the abstract operators of quantum mechanics. In the next two sections, we will get some idea where the basic spin commutation relations come from.

12.2 Orbital angular momentum

Suppose a particle moves in three dimensions. Classically, the angular momentum of the particle is $\vec{L} = \vec{r} \times \vec{p}$, the cross product of the position and momentum vectors of the particle. We can extract three component expressions from this definition to motivate the definitions of the three orbital angular momentum components for the quantum particle:

$$L_x = yp_z - zp_y, \qquad L_y = zp_x - xp_z, \qquad L_z = xp_y - yp_x. \qquad (12.19)$$

Given the canonical commutation relations between the position and momentum components (Eq. 10.82), we can establish the commutation relations between the orbital angular momentum components.

Exercise 12.13 Show that

$$[L_x, L_y] = i\hbar L_z. \qquad (12.20)$$

Write down the other two commutators by cyclically permuting the coordinates.

Thus, the components of the orbital angular momentum satisfy the basic commutation relations for spin components (Eq. 12.1). Orbital angular momentum is a kind of "spin." The general analysis of Section 12.1 immediately tells us that we can find simultaneous eigenstates of L^2 and L_z with quantum numbers l and m. The eigenvalues are $l(l+1)\hbar^2$ and $m\hbar$, respectively.

To get a bit more detail, let us explore how angular momentum works in the position representation. The particle is described by a wave function $\psi(\vec{r})$. Though we are interested in the Cartesian components L_x, L_y, and L_z, it is in fact easiest to regard ψ as a function of spherical coordinates r, θ, and ϕ. The relation between spherical and Cartesian coordinates is

$$x = r \sin\theta \, \cos\phi, \qquad y = r \sin\theta \, \sin\phi, \qquad z = r \cos\theta, \qquad (12.21)$$

or equivalently,

$$r = \sqrt{x^2 + y^2 + z^2},$$

$$\theta = \cos^{-1}\left(\frac{z}{\sqrt{x^2 + y^2 + z^2}}\right),$$

$$\phi = \tan^{-1}\left(\frac{y}{x}\right). \qquad (12.22)$$

From the chain rule for partial derivatives (Eq. 11.78 and its generalizations) we can work out how the L_x, L_y, and L_z operators affect the wave function $\psi(r, \theta, \phi)$. The following two exercises show how this is done:

Exercise 12.14 Here are the partial derivatives of the spherical coordinates (r, θ, ϕ) with respect to the Cartesian ones (x, y, z).

$$\frac{\partial r}{\partial x} = \sin\theta \, \cos\phi, \qquad \frac{\partial \theta}{\partial x} = \frac{1}{r}\cos\theta \, \cos\phi, \qquad \frac{\partial \phi}{\partial x} = -\frac{1}{r}\frac{\sin\phi}{\sin\theta},$$

$$\frac{\partial r}{\partial y} = \sin\theta \, \sin\phi, \qquad \frac{\partial \theta}{\partial y} = \frac{1}{r}\cos\theta \, \sin\phi, \qquad \frac{\partial \phi}{\partial y} = \frac{1}{r}\frac{\cos\phi}{\sin\theta},$$

$$\frac{\partial r}{\partial z} = \cos\theta, \qquad \frac{\partial \theta}{\partial z} = -\frac{1}{r}\sin\theta, \qquad \frac{\partial \phi}{\partial z} = 0. \qquad (12.23)$$

Verify as many of these as you need to be convinced of the whole set. You should work out at least one in each column and one in each row. (Not fair picking the one in the lower right!)

Exercise 12.15 From the results in Eq. 12.23, show that

$$y\frac{\partial}{\partial z} - z\frac{\partial}{\partial y} = -\sin\phi\frac{\partial}{\partial\theta} - \cot\theta \, \cos\phi\frac{\partial}{\partial\phi}, \qquad (12.24)$$

$$z\frac{\partial}{\partial x} - x\frac{\partial}{\partial z} = \cos\phi\frac{\partial}{\partial\theta} - \cot\theta \, \sin\phi\frac{\partial}{\partial\phi}, \qquad (12.25)$$

$$x\frac{\partial}{\partial y} - y\frac{\partial}{\partial x} = \frac{\partial}{\partial\phi}.$$

It follows that

$$L_x : \psi \to i\hbar\left(\sin\phi\,\frac{\partial\psi}{\partial\theta} + \cot\theta\,\cos\phi\,\frac{\partial\psi}{\partial\phi}\right),$$

$$L_y : \psi \to i\hbar\left(-\cos\phi\,\frac{\partial\phi}{\partial\theta} + \cot\theta\,\sin\phi\,\frac{\partial\psi}{\partial\phi}\right),$$

$$L_z : \psi \to -i\hbar\left(\frac{\partial\psi}{\partial\phi}\right), \qquad (12.26)$$

and thus the orbital ladder operators $L_\pm = L_x \pm iL_y$ are

$$L_\pm : \psi \to \hbar e^{\pm i\phi}\left(\pm\frac{\partial\psi}{\partial\theta} + i\cot\theta\,\frac{\partial\psi}{\partial\phi}\right). \qquad (12.27)$$

Exercise 12.16 From Eq. 12.26, show that

$$L^2 : \psi \to -\hbar^2\left(\frac{1}{\sin\theta}\frac{\partial}{\partial\theta}\left(\sin\theta\,\frac{\partial\psi}{\partial\theta}\right) + \frac{1}{\sin^2\theta}\frac{\partial^2\psi}{\partial\phi^2}\right). \qquad (12.28)$$

Notice that derivatives with respect to r do not appear in any of these expressions.

Let us suppose that the wave function ψ_{lm} represents an eigenstate of both L^2 and L_z with quantum numbers l and m. The L_z eigenvalue equation yields

$$-i\hbar\frac{\partial\psi_{lm}}{\partial\phi} = m\hbar\psi_{lm}. \qquad (12.29)$$

The solution to this differential equation is

$$\psi_{lm} = f_{lm}(r, \theta)e^{im\phi}, \tag{12.30}$$

for some function f_{lm}. Since the wave function is single-valued, the value of m must be an integer. This tells us that orbital angular momentum quantum numbers must be integers:

$$l = 0, 1, 2, \ldots \quad \text{and} \quad m = -l, \ldots, 0, \ldots, +l. \tag{12.31}$$

What about the function $f_{lm}(r, \theta)$? If $m = -l$, then the state represented by ψ_{lm} is on the bottom rung of the "ladder" of angular momentum states for a given l, so that $\mathsf{L}_- |\psi_{l-l}\rangle = 0$. This leads to

$$-\frac{\partial f_{l-l}}{\partial \theta} + l \cot \theta \, f_{l-l} = 0, \tag{12.32}$$

with solution

$$f_{l-l} = R(r) \sin^l \theta. \tag{12.33}$$

Exercise 12.17 Derive Eq. 12.32 and its solution Eq. 12.33. (For the solution, the substitution $u = \sin \theta$ is handy.)

Thus, $\psi_{l-l} = R(r) \sin^l \theta \, e^{-il\phi}$.

The function $R(r)$ is arbitrary and does not affect the angular momentum of the state. Furthermore, all of the angular momentum eigenstates can be generated from these by application of the L_+ ladder operator, as in Eq. 12.16, which we here rewrite:

$$|\psi_{l\,m+1}\rangle = \frac{1}{\hbar\sqrt{l(l+1) - m(m+1)}} \mathsf{L}_+ |\psi_{lm}\rangle. \tag{12.34}$$

None of these operations will affect $R(r)$ in the wave function at all.

Therefore, we conclude that every eigenstate of both L^2 and L_z can be written

$$\psi_{lm}(r, \theta, \phi) = R(r) \, Y_{lm}(\theta, \phi), \tag{12.35}$$

where the functions Y_{lm} are called *spherical harmonics*.[4] These have the form

$$Y_{lm}(\theta, \phi) = P_{lm}(\theta) \, e^{im\phi}, \tag{12.36}$$

where $P_{l-l}(\theta) = K_l \sin^l \theta$ and

$$P_{l\,m+1}(\theta) = \frac{1}{\sqrt{l(l+1) - m(m+1)}} \left(\frac{dP_{lm}}{d\theta} - m \cot \theta \, P_{lm} \right). \tag{12.37}$$

The constant K_l is determined by normalization. We choose to normalize the spherical harmonics so that $|Y_{lm}|^2$ integrates to unity over the (θ, ϕ) sphere. With a standard (but arbitrary) choice of sign, we obtain

$$K_l = \frac{(-1)^l}{2^l l!} \sqrt{\frac{(2l+1)!}{4\pi}}. \tag{12.38}$$

---

[4] So called because the functions Y_{lm} arise in a great many wave problems with spherical symmetry – for instance, in the "ringing modes" of a spherical bell.

The spherical harmonics are among the most useful functions in mathematical physics. For $l = 0$ and $l = 1$ these are

$$Y_{0\,0} = \frac{1}{\sqrt{4\pi}},$$

$$Y_{1\,\pm 1} = \mp \sqrt{\frac{3}{8\pi}}\, \sin\theta\, e^{\pm i\phi},$$

$$Y_{1\,0} = \sqrt{\frac{3}{4\pi}}\, \cos\theta. \tag{12.39}$$

Exercise 12.18 Given that $Y_{2\,0} = \sqrt{\frac{5}{16\pi}}\,(3\cos^2\theta - 1)$, find $Y_{2\,\pm 1}$ and $Y_{2\,\pm 2}$.

Exercise 12.19 Rewrite each of the spherical harmonics with $l \leq 2$ in the following form:

$$Y_{lm} = r^{-l} f(x, y, z). \tag{12.40}$$

12.3 Rotation

In Section 5.4 we discussed how the state of a spin-1/2 system changes under rotations. We found that, if we rotate the spin by an angle α about the z-axis, then its state is changed by the unitary operator

$$R_z(\alpha) = e^{-iS_z\alpha/\hbar}, \tag{12.41}$$

(compare Eq. 5.56). We said that the spin operator S_z is the "generator" of rotations about the z-axis.

We can now see that a similar relation holds for orbital angular momentum. A rotation by α about the z-axis changes the spherical harmonic function by

$$R_z(\alpha) : Y_{lm}(\theta, \phi) \rightarrow Y_{lm}(\theta, \phi - \alpha) = e^{-im\alpha}\, Y_{lm}(\theta, \phi), \tag{12.42}$$

which allows us to write

$$R_z(\alpha)\,|\psi\rangle = e^{-iL_z\alpha/\hbar}\,|\psi\rangle, \tag{12.43}$$

for a particle moving in 3-D. Again, the z-component of angular momentum is the generator of rotations about the z-axis. And in fact, this is a general relationship. To see why, we will take a closer look at the geometry of rotations in 3-D space.

A rotation in 3-D space is represented by a 3×3 matrix, which tells how the Cartesian coordinates transform:

$$\begin{pmatrix} x' \\ y' \\ z' \end{pmatrix} = \begin{pmatrix} R_{xx} & R_{xy} & R_{xz} \\ R_{yx} & R_{yy} & R_{yz} \\ R_{zx} & R_{zy} & R_{zz} \end{pmatrix} \begin{pmatrix} x \\ y \\ z \end{pmatrix}. \tag{12.44}$$

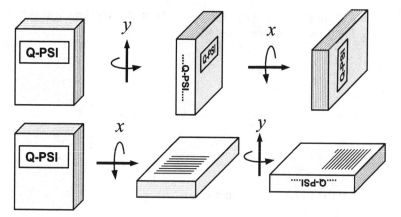

Fig. 12.2 Rotations of a book about the y- and x-axes lead to different results depending on the order of the operations.

It is easy to find matrices describing rotations by α about the x-, y-, and z-axes:

$$\mathbf{R}_x(\alpha) = \begin{pmatrix} 1 & 0 & 0 \\ 0 & \cos\alpha & -\sin\alpha \\ 0 & \sin\alpha & \cos\alpha \end{pmatrix}$$

$$\mathbf{R}_y(\alpha) = \begin{pmatrix} \cos\alpha & 0 & \sin\alpha \\ 0 & 1 & 0 \\ -\sin\alpha & 0 & \cos\alpha \end{pmatrix}$$

$$\mathbf{R}_z(\alpha) = \begin{pmatrix} \cos\alpha & -\sin\alpha & 0 \\ \sin\alpha & \cos\alpha & 0 \\ 0 & 0 & 1 \end{pmatrix}. \tag{12.45}$$

We can consider rotations by a small angle ϵ by expanding the cosine and sine functions in a power series to a specified order.

Exercise 12.20 To order $\mathcal{O}\left(\epsilon^2\right)$, show that

$$\mathbf{R}_x(\epsilon) = \begin{pmatrix} 1 & 0 & 0 \\ 0 & (1-\epsilon^2/2) & -\epsilon \\ 0 & \epsilon & (1-\epsilon^2/2) \end{pmatrix}. \tag{12.46}$$

Also write down $\mathcal{O}\left(\epsilon^2\right)$ expressions for $\mathbf{R}_y(\epsilon)$ and $\mathbf{R}_z(\epsilon)$.

Spatial rotations do not commute with each other. A rotation about the y-axis followed by one about the x-axis produces a different result than the same two rotations performed in the opposite order (see Fig. 12.2). Consider now rotations by a small angle ϵ. We imagine making rotations about the y- and x-axes by ϵ, then about the y- and x-axes by $-\epsilon$. If rotations were commutative, this would result in zero net rotation. However, because they are non-commuting, there is some residual rotation "left over" afterwards. Working to $\mathcal{O}\left(\epsilon^2\right)$ we find

$$\mathbf{R}_x(-\epsilon)\mathbf{R}_y(-\epsilon)\mathbf{R}_x(\epsilon)\mathbf{R}_y(\epsilon) = \begin{pmatrix} 1 & -\epsilon^2 & 0 \\ \epsilon^2 & 1 & 0 \\ 0 & 0 & 1 \end{pmatrix}. \tag{12.47}$$

Exercise 12.21 Confirm this.

Note that the residual rotation is, to $\mathcal{O}\left(\epsilon^2\right)$, a rotation about the z-axis. Thus

$$\mathbf{R}_x(-\epsilon)\mathbf{R}_y(-\epsilon)\mathbf{R}_x(\epsilon)\mathbf{R}_y(\epsilon) = \mathbf{R}_z(\epsilon^2). \tag{12.48}$$

So far, we have only been considering the geometry of rotations in 3-D space. What does this imply about rotation of a quantum system?

Rotation about a fixed axis is a continuous uniform family of unitary operators. Thus, we write the rotation operators

$$R_x(\alpha) = e^{-i\alpha G_x}, \qquad R_y(\alpha) = e^{-i\alpha G_y}, \qquad R_z(\alpha) = e^{-i\alpha G_z}, \tag{12.49}$$

for generators G_x, G_y, and G_z. For rotations by a small angle ϵ, we expand the operator exponentials to $\mathcal{O}\left(\epsilon^2\right)$ and get

$$R_x(\epsilon) = 1 - i\epsilon G_x - \frac{\epsilon^2}{2} G_x^2, \tag{12.50}$$

and so on.

Exercise 12.22 Show that, to $\mathcal{O}\left(\epsilon^2\right)$,

$$R_x(-\epsilon)R_y(-\epsilon)R_x(\epsilon)R_y(\epsilon) = 1 - (G_xG_y - G_yG_x)\epsilon^2. \tag{12.51}$$

But our previous geometrical analysis tells us that this composition of rotations is, to the same order $\mathcal{O}\left(\epsilon^2\right)$, $R_z(\epsilon^2) = 1 - i\epsilon^2 G_z$. Furthermore, all of our relationships should remain the same if x, y, and z are cyclically permuted (amounting to a mere relabeling of a right-handed Cartesian coordinate system). Thus, the generator operators must satisfy

$$[G_x, G_y] = iG_z, \qquad [G_y, G_z] = iG_x, \qquad [G_z, G_x] = iG_y. \tag{12.52}$$

If we now *define* spin component operators $S_x = \hbar G_x$ and so on, then:

- The rotation operators $R_x(\alpha) = e^{-i\alpha S_x/\hbar}$, etc.; and
- The spin components automatically satisfy the fundamental commutation relations of Eq. 12.1.

We might sum this up by saying that angular momentum is related to rotation in a *transformational* rather than a *dynamical* way. Any quantum system whose state is affected by an overall spatial rotation has angular momentum observables that generate the rotation operators. Because of the geometry of 3-D rotations, these observables must have the spin commutation relations. Since a particle's spatial wave function is affected by rotation, it has orbital angular momentum (related to position and momentum variables in the familiar way). If its internal state is also affected, the particle also has an internal spin. These two types of angular momentum, though quite different in physical significance, have just the same general mathematical structure.

12.4 Adding spins

A quantum system may contain more than one angular momentum variable. An electron has orbital angular momentum due to its motion through space and also internal spin. The total angular momentum of the electron system is the sum of these. The addition of quantum angular momenta requires some careful analysis, which is the business of this section.

Consider a composite system made up of two simple spin systems. The total angular momentum is $\vec{S} = \vec{S}^{(1)} + \vec{S}^{(2)}$. This stands for three component operator relations:

$$S_x = S_x^{(1)} + S_x^{(2)}, \qquad S_y = S_y^{(1)} + S_y^{(2)}, \qquad S_z = S_z^{(1)} + S_z^{(2)}. \tag{12.53}$$

Exercise 12.23 Show that the total spin component operators S_x, etc., satisfy the basic spin commutation relations of Eq. 12.1.

Therefore, *the vector sum of two angular momenta is an angular momentum*. We can treat a composite of two spins as a spin system.

Exercise 12.24 Consider a general linear combination of two spins: $\vec{T} = a_1 \vec{S}^{(1)} + a_2 \vec{S}^{(2)}$. If the components of $\vec{T}$ must satisfy the basic spin commutation relations, show that the only non-trivial coefficients are $a_1 = a_2 = 1$. Thus, the vector sum $\vec{S}^{(1)} + \vec{S}^{(2)}$ is an angular momentum, but a more general linear combination is not.

To get a better idea of how this works, we consider a pair of spin-1/2 systems. Each subsystem is described by a Hilbert space of dimension 2, so the joint system has

$$\dim \mathcal{H}^{(12)} = \dim \left(\mathcal{Q}^{(1)} \otimes \mathcal{Q}^{(2)} \right) = 2 \times 2 = 4. \tag{12.54}$$

Eigenstates of S_z can be constructed from the single-spin eigenstates $|\uparrow\rangle$ and $|\downarrow\rangle$ for the individual spins:

$$\begin{aligned} S_z |\uparrow, \uparrow\rangle &= +\hbar |\uparrow, \uparrow\rangle, & S_z |\uparrow, \downarrow\rangle &= 0, \\ S_z |\downarrow, \downarrow\rangle &= -\hbar |\downarrow, \downarrow\rangle, & S_z |\downarrow, \uparrow\rangle &= 0. \end{aligned} \tag{12.55}$$

Exercise 12.25 Verify Eq. 12.55.

Note that the eigenvalues of S_z are $+\hbar$, 0, and $-\hbar$. This is what we would expect for $s = 1$, which would lead us to the comfortable conclusion that two spin-1/2 systems combine to make a spin-1 system. But there are *two* orthogonal eigenstates with eigenstate 0, so this cannot be the whole story. We need a more careful analysis.

We can construct ladder operators for the joint spin system

$$S_\pm = S_x \pm i S_y = S_\pm^{(1)} + S_\pm^{(2)}. \tag{12.56}$$

The state $|\downarrow, \downarrow\rangle$ is the eigenstate of lowest S_z value, so it must be the "bottom rung" of a spin-state ladder with $s = 1$. We write

$$|s, m\rangle = |1, -1\rangle = |\downarrow, \downarrow\rangle. \tag{12.57}$$

We apply S_+ to obtain the higher rungs of the spin-state ladder.

Exercise 12.26 Use Eq. 12.16 to show that

$$|1, 0\rangle = \frac{1}{\hbar\sqrt{2}} S_+ |1, -1\rangle = \frac{1}{\sqrt{2}} \left(|\uparrow, \downarrow\rangle + |\downarrow, \uparrow\rangle \right),$$

$$|1, +1\rangle = \frac{1}{\hbar\sqrt{2}} S_+ |1, 0\rangle = |\uparrow, \uparrow\rangle. \tag{12.58}$$

We have constructed the expected three-rung spin-state ladder for a spin-1 system. What have we missed?

Exercise 12.27 Consider the state $|0, 0\rangle = \frac{1}{\sqrt{2}}(|\uparrow, \downarrow\rangle - |\downarrow, \uparrow\rangle)$ of a pair of spin-1/2 systems. Show that

$$S_+ |0, 0\rangle = S_- |0, 0\rangle = 0, \tag{12.59}$$

so that $|0, 0\rangle$ is the only rung on a spin-state ladder for spin-0.

To summarize, a pair of spin-1/2 systems can be treated as a single spin system with spin number $s = 1$ or $s = 0$. The three $s = 1$ states (with $m = +1, 0, -1$) are called "triplet" states, and the $s = 0$ state is called the "singlet" state of the pair of spins.[5]

In general, the composite of simple spin systems with s_1 and s_2 will have a Hilbert space of dimension $(2s_1 + 1)(2s_2 + 1)$. We can of course describe this Hilbert space using the product basis vectors $|s_1, m_1; s_2, m_2\rangle$ for the individual spins. But we can also form a basis of eigenstates of the total spin observables S^2 and S_z. These are the $|s, m\rangle$ states, and for each possible s value there will be a "ladder" of $2s + 1$ values of m.

We would like to write the $|s, m\rangle$ states in terms of the product basis states $|s_1, m_1; s_2, m_2\rangle$:

$$|s, m\rangle = \sum_{m_1, m_2} |s_1, m_1; s_2, m_2\rangle \langle s_1, m_1; s_2, m_2 |s, m\rangle. \tag{12.60}$$

The components $\langle s_1, m_1; s_2, m_2 |s, m\rangle$ are called *Clebsch–Gordon coefficients*, and they arise in many different quantum mechanical situations. We have already calculated several

---

[5] The singlet state was already introduced for abstract qubits in Eq. 6.50, and was used extensively in our discussion of entanglement, the EPR argument, and Bell's theorem.

Clebsch–Gordon coefficients:

$$\left\langle \frac{1}{2}, +\frac{1}{2}; \frac{1}{2}, +\frac{1}{2} \,\middle|\, 1, +1 \right\rangle = 1,$$

$$\left\langle \frac{1}{2}, +\frac{1}{2}; \frac{1}{2}, -\frac{1}{2} \,\middle|\, 1, 0 \right\rangle = \frac{1}{\sqrt{2}}, \qquad \left\langle \frac{1}{2}, -\frac{1}{2}; \frac{1}{2}, +\frac{1}{2} \,\middle|\, 1, 0 \right\rangle = \frac{1}{\sqrt{2}},$$

$$\left\langle \frac{1}{2}, -\frac{1}{2}; \frac{1}{2}, -\frac{1}{2} \,\middle|\, 1, -1 \right\rangle = 1,$$

$$\left\langle \frac{1}{2}, +\frac{1}{2}; \frac{1}{2}, -\frac{1}{2} \,\middle|\, 0, 0 \right\rangle = \frac{1}{\sqrt{2}}, \qquad \left\langle \frac{1}{2}, -\frac{1}{2}; \frac{1}{2}, +\frac{1}{2} \,\middle|\, 0, 0 \right\rangle = -\frac{1}{\sqrt{2}}. \tag{12.61}$$

All other Clebsch–Gordon coefficients with $s_1 = s_2 = 1/2$ are zero.[6]

Exercise 12.28 Verify the coefficients in Eq. 12.61.

Exercise 12.29 Show that $\langle s_1, m_1; s_2, m_2 \,|\, s, m \rangle = 0$ unless $m = m_1 + m_2$. Hint: Consider the eigenvalue of S_z.

In practice, one usually looks up the Clebsch–Gordon coefficients in a reference book; but they can be computed from first principles, as we have done for the pair of spin-1/2 systems. To illustrate how this works, we will consider a composite of systems with spin 1 and 1/2. The dimension of the composite Hilbert space is $2 \times 3 = 6$. We wish to find a basis of $|s, m\rangle$ states for the total angular momentum of the two spins.

What values of s and m are possible? Since $m = m_1 + m_2$, we can deduce the possible values of m and the corresponding degeneracies (i.e. the dimensions of the S_z eigenspaces).

Exercise 12.30 By considering product basis states $|s_1, m_1; s_2, m_2\rangle$, show that the degeneracies for the m values are

m value	$-\dfrac{3}{2}$	$-\dfrac{1}{2}$	$+\dfrac{1}{2}$	$+\dfrac{3}{2}$,
degeneracy	1	2	2	1.

There is only one way to arrange six such m-eigenstates into angular momentum ladders – that is, into sequences that run from $-s$ to s for one or more values of s. This way is shown in Fig. 12.3. The two possible values of s are $\frac{3}{2}$ and $\frac{1}{2}$. Since $m = -\frac{3}{2}$ has degeneracy 1, we know one of the $|s, m\rangle$ states to be

$$\left| \frac{3}{2}, -\frac{3}{2} \right\rangle = \left| 1, -1; \frac{1}{2}, -\frac{1}{2} \right\rangle, \tag{12.62}$$

so that $\left\langle 1, -1; \frac{1}{2}, -\frac{1}{2} \,\middle|\, \frac{3}{2}, -\frac{3}{2} \right\rangle = 1$. By applying $S_+ = S_+^{(1)} + S_+^{(2)}$ to this "bottom rung" state, we can construct the remaining states in the $s = \frac{3}{2}$ ladder.

---

[6] Since the overall phases of the $|s, m\rangle$ states are arbitrary, we could construct Clebsch–Gordon coefficients with different complex phases. In general the phases are chosen so that the coefficients are all real, as they are here.

Ladders of $|s, m\rangle$ basis states for the combination of spin-1 and spin-1/2 systems.

Exercise 12.31 Show that

$$\left|\frac{3}{2}, -\frac{1}{2}\right\rangle = \sqrt{\frac{2}{3}}\left|1, 0; \frac{1}{2}, -\frac{1}{2}\right\rangle + \frac{1}{\sqrt{3}}\left|1, -1; \frac{1}{2}, +\frac{1}{2}\right\rangle,$$

$$\left|\frac{3}{2}, +\frac{1}{2}\right\rangle = \frac{1}{\sqrt{3}}\left|1, 1; \frac{1}{2}, -\frac{1}{2}\right\rangle + \sqrt{\frac{2}{3}}\left|1, 0; \frac{1}{2}, +\frac{1}{2}\right\rangle,$$

$$\left|\frac{3}{2}, +\frac{3}{2}\right\rangle = \left|1, +1; \frac{1}{2}, +\frac{1}{2}\right\rangle. \tag{12.63}$$

Write down the corresponding Clebsch–Gordon coefficients.

Now we need to write down the basis states in the other, $s = \frac{1}{2}$ ladder. To do this, we note that the "bottom rung" state $\left|\frac{1}{2}, -\frac{1}{2}\right\rangle$ must be a linear combination of the $\left|1, 0; \frac{1}{2}, -\frac{1}{2}\right\rangle$ and $\left|1, -1; \frac{1}{2}, +\frac{1}{2}\right\rangle$ product basis states, and that

$$S_-\left|\frac{1}{2}, -\frac{1}{2}\right\rangle = 0. \tag{12.64}$$

After we solve Eq. 12.64, we can apply S_+ to obtain the remaining ladder state $\left|\frac{1}{2}, +\frac{1}{2}\right\rangle$.

Exercise 12.32 Show that

$$\left|\frac{1}{2}, -\frac{1}{2}\right\rangle = \frac{1}{\sqrt{3}}\left|1, 0; \frac{1}{2}, -\frac{1}{2}\right\rangle - \sqrt{\frac{2}{3}}\left|1, -1; \frac{1}{2}, +\frac{1}{2}\right\rangle,$$

$$\left|\frac{1}{2}, +\frac{1}{2}\right\rangle = \sqrt{\frac{2}{3}}\left|1, 1; \frac{1}{2}, -\frac{1}{2}\right\rangle - \frac{1}{\sqrt{3}}\left|1, 0; \frac{1}{2}, +\frac{1}{2}\right\rangle. \tag{12.65}$$

Write down the corresponding Clebsch–Gordon coefficients.

Exercise 12.33 Suppose we make a composite system from two simple spin-1 systems. Draw a diagram analogous to Fig. 12.3 showing the possible values for the total spin quantum numbers s and m. (Problem 12.4 continues the calculation of the Clebsch–Gordon coefficients in this case.)

In general, if we combine systems with spins s_1 and s_2, the total angular momentum s will take on values between $|s_1 - s_2|$ and $s_1 + s_2$ separated by integers. For a given s, the quantum number m will range from $-s$ to s, also in integer steps. The Clebsch–Gordon coefficients specify how the $|s, m\rangle$ basis states are related to the $|s_1, m_1; s_2, m_2\rangle$ product basis states.

We can apply these ideas to our original example of an electron moving in 3-D space. The orbital angular momentum has quantum numbers l and m_l for observables L^2 and L_z, and the spin angular momentum has quantum numbers $s = 1/2$ and m_s for S^2 and S_z. The total angular momentum $\vec{J} = \vec{L} + \vec{S}$ has quantum numbers j and m_j for J^2 and J_z. These have values

$$j = l \pm \frac{1}{2} \quad \text{and} \quad m_j = -j, \ldots, +j, \qquad (12.66)$$

for $l \neq 0$. (For $l = 0$, only $j = 1/2$ is possible.) The angular momentum state of the electron is characterized by four quantum numbers. These can be chosen to be l, m_l, $s = 1/2$, and m_s, since orbital and spin degrees of freedom are in effect different "subsystems" of the electron. However, we could also choose to describe the electron by l, $s = 1/2$, j, and m_j, and in many problems of atomic physics this choice is better.

Exercise 12.34 Write J^2 in terms of the orbital and spin angular momentum components. What are the possible eigenvalues of the operator $\vec{L} \cdot \vec{S}$? Work out specific examples for $l = 1$ and $l = 2$. (In atomic physics, the *spin–orbit interaction* for an electron involves just such a dot product.)

12.5 Isospin

The properties of angular momentum operators lead us to quantum systems that are naturally described by finite ladders of eigenstates. It is sometimes useful to reverse this logic. In quantum systems with finite-dimensional Hilbert spaces, it can be useful to define operators that have the same properties as angular momentum. These operators, however, need have nothing at all to do with rotation in space.

One famous example of this arises in the theory of nuclear reactions. It is observed that nuclear particles occur in families, the members of which are similar in mass and nuclear interactions, but different in other characteristics such as electric charge. For instance, the family of nucleons includes the proton and the neutron.

Heisenberg suggested that such families be regarded as a single type of particle, differing in an internal spin-like property called *isospin*. The nucleon, for instance, is an isospin-1/2 particle, so it has two distinguishable states: the proton state $|p\rangle$ and the neutron state $|n\rangle$. These form a basis for the internal isospin degree of freedom of the nucleon.

The isospin operators I_1, I_2, and I_3 do not correspond to spatial directions, but they are assumed to satisfy spin-like commutation relations:

$$[I_1, I_2] = iI_3, \qquad [I_2, I_3] = iI_1, \qquad [I_3, I_1] = iI_2. \tag{12.67}$$

Except for factors of $\hbar$, all of the general results about spin systems also hold for isospin. The eigenvalues of the isospin magnitude operator I^2 are $I(I+1)$, where $I = 0, \frac{1}{2}, 1, \ldots$; and the eigenvalues I_3 of I_3 range from $-I$ to I and are separated by 1.

The isospin states $|I, I_3\rangle$ of the proton and neutron are taken to be

$$|p\rangle = \left|\frac{1}{2}, +\frac{1}{2}\right\rangle, \qquad |n\rangle = \left|\frac{1}{2}, -\frac{1}{2}\right\rangle. \tag{12.68}$$

Other families of particles have other values of isospin. For instance, we observe that there are three types of pi meson (or *pion*), denoted π^+, π^0, and π^-. These have nearly the same mass and almost exactly the same properties under the strong nuclear force. We regard these as distinct isospin states of a generic pion, which is an isospin-1 system. The states are

$$\left|\pi^+\right\rangle = |1, +1\rangle, \qquad \left|\pi^0\right\rangle = |1, 0\rangle, \qquad \left|\pi^-\right\rangle = |1, -1\rangle. \tag{12.69}$$

Heisenberg's key insight was that the isospin operators commute with the part of the system Hamiltonian involving the strong nuclear force. In other words, isospin is conserved by strong nuclear interactions.

This idea has many consequences, one of which we will describe here. There is a very short-lived particle called the $\Delta(1232)$. (The "1232" designation refers to the rest energy of this particle, which is around 1232 MeV, but we shall omit this label for the remainder of our discussion.) In fact this is a family of four closely-related particles: Δ^{++}, Δ^+, Δ^0, and Δ^-. We therefore consider these as states of a single generic Δ having isospin 3/2:

$$\left|\Delta^{++}\right\rangle = \left|\frac{3}{2}, +\frac{3}{2}\right\rangle, \quad \left|\Delta^+\right\rangle = \left|\frac{3}{2}, +\frac{1}{2}\right\rangle,$$

$$\left|\Delta^0\right\rangle = \left|\frac{3}{2}, -\frac{1}{2}\right\rangle, \quad \left|\Delta^-\right\rangle = \left|\frac{3}{2}, -\frac{3}{2}\right\rangle. \tag{12.70}$$

The strong nuclear interaction causes the Δ particle to decay rapidly into a pion and a nucleon: $\Delta \rightarrow \pi + N$. To work out the details of this process, we first write down all of the possible decay processes that conserve electric charge (which is conserved in all interactions):

$$\Delta^{++} \longrightarrow \pi^+ + p,$$

$$\Delta^+ \longrightarrow \pi^0 + p \quad \text{or} \quad \pi^+ + n,$$

$$\Delta^0 \longrightarrow \pi^- + p \quad \text{or} \quad \pi^0 + n,$$

$$\Delta^- \longrightarrow \pi^- + n. \tag{12.71}$$

The Δ^{++} and Δ^- each have only one "mode" of decay. The Δ^+, on the other hand, might decay into either $\pi^0 + p$ or $\pi^+ + n$. What is the relative likelihood of these two outcomes?

The answer is given by isospin. In the decay, the isospin-3/2 Δ system turns into a composite of an isospin-1 pion and an isospin-1/2 nucleon, with isospin conserved throughout. We therefore want to write the $\left|\frac{3}{2}, +\frac{1}{2}\right\rangle$ isospin state as a superposition of $\left|\pi^0; p\right\rangle = \left|1, 0; \frac{1}{2}, +\frac{1}{2}\right\rangle$ and $\left|\pi^+; n\right\rangle = \left|1, +1; \frac{1}{2}, -\frac{1}{2}\right\rangle$; the probabilities of each decay mode will be given by the probability amplitudes for these terms. But these amplitudes are exactly the Clebsch–Gordon coefficients calculated in Eq. 12.63:

$$\left\langle 1, 0; \frac{1}{2}, +\frac{1}{2} \middle| \frac{3}{2}, +\frac{1}{2} \right\rangle = \sqrt{\frac{2}{3}},$$

$$\left\langle 1, +1; \frac{1}{2}, -\frac{1}{2} \middle| \frac{3}{2}, +\frac{1}{2} \right\rangle = \frac{1}{\sqrt{3}}. \tag{12.72}$$

We conclude that the decay products $\pi^0 + p$ should occur 2/3 of the time, and $\pi^+ + n$ should occur 1/3 of the time. Experiment confirms this prediction.

Exercise 12.35 What are the relative likelihoods of the two decay modes for the Δ^0 particle?

We have throughout this book emphasized the isomorphism between quantum systems. Quite distinct physical situations nevertheless have deep analogies based on a common mathematical framework. The physics of isospin is a very striking illustration of this general idea. Even though the distinction between proton and neutron has nothing to do with anything spinning in space, we can nevertheless use the methods and ideas developed for angular momentum to help us understand how these particles behave in nuclear reactions.

Problems

Problem 12.1 A spin system can have either $s = 0$ or $s = 1$, but is otherwise a simple spin system. Show that the observable S_z has some degeneracy, and describe its eigenspaces. Also describe how $S_\pm$ act on the Hilbert space of this system.

Problem 12.2 Suppose we have a simple spin-s system with standard basis $|m\rangle$.

(a) Find the matrix representation for the operator S_z in this basis.
(b) Recall that $S_x = \frac{1}{2}(S_+ + S_-)$. Use this fact to find the matrix representation for S_x in the $|m\rangle$ basis.
(c) In a similar way, find the matrix representation for S_y.
(d) Work out the results from parts (b) and (c) for the special case of $s = 1/2$. Do you get the matrices that you expect?

Problem 12.3 The 3-D parity operator $\mathcal{P}$ acts on position eigenstates by $\mathcal{P}\,|\vec{r}\rangle = |-\vec{r}\rangle$. The parity has possible eigenvalues $+1$ (even parity) and -1 (odd parity). What is the parity of an eigenstate of L^2 and L_z described by the wave function

$$\psi_{lm}(r,\theta,\phi) = R(r)\,Y_{lm}(\theta,\phi),$$ (Re 12.35)

for a given l and m?

Problem 12.4 Determine all of the non-zero Clebsch–Gordon coefficients $\langle s_1, m_1; s_2, m_2 | s, m \rangle$ with $s_1 = s_2 = 1$. (Choose your phases so that the coefficients are real.)

Problem 12.5 Three spin-1/2 particles interact, with a Hamiltonian

$$H = g\,\vec{S}^{(1)} \cdot \vec{S}^{(2)} + g\,\vec{S}^{(2)} \cdot \vec{S}^{(3)} + g\,\vec{S}^{(3)} \cdot \vec{S}^{(1)}.$$ (12.73)

Find the energy levels and corresponding degeneracies for this system. Hint: Write down the total angular momentum operator J^2.

Problem 12.6 An electron has both orbital and spin-1/2 angular momentum. Suppose that the Hamiltonian of the electron has a rotational symmetry about the z-axis, so that $R_z(\theta)$ commutes with H.

(a) Write down the rotation operator $R_z(\theta)$ in terms of the $\vec{L}$ and $\vec{S}$ operators.
(b) Show that a term in the Hamiltonian of the form $g(r)\,\vec{L} \cdot \vec{S}$ does commute with the rotation operator. (Hence such a term might be present.)
(c) Which of the following must be conserved for the electron: L_z, S_z, and/or $L_z + S_z$? Carefully explain your answer.

Problem 12.7 A deuteron d is a combination of two nucleons in an isospin-0 state. (Think: How is this different from saying that a deuteron is made up of one proton and one neutron?) Two deuterons collide with each other at high speed and interact via the strong nuclear force. Here is the net reaction:

$$d + d \rightarrow d + d + \text{pions}.$$

The pions produced, of course, must have zero total electric charge.

(a) Is it possible to produce a single pion? Why or why not?
(b) The reaction might produce two π^0s or a $\pi^+\pi^-$ pair. Find the ratio of the probabilities for these two processes.

Ladder systems

13.1 Raising and lowering operators

In Section 12.1 we considered a quantum system of arbitrary spin s. For a given value of s, we constructed a "ladder" of S_x eigenstates. These states were linked by "raising" and "lowering" operators $S_\pm$. The properties of the raising and lowering operators (which followed from the commutation relations between the spin components) completely determined the structure of the Hilbert space $\mathcal{H}_s$ for the spin-s system.

In this chapter, we investigate a similar type of quantum system that we will call a *ladder system*. The ladder system is described by an infinite-dimensional Hilbert space with a basis of states $|0\rangle$, $|1\rangle$, $|2\rangle$, ... These form the "rungs" of a ladder of basis states, which has a bottom rung $|0\rangle$ but extends infinitely upwards. As we will see, many real quantum systems are isomorphic to this. The ladder system is as useful a generalization as the "qubit" of Chapter 2. We will adopt the special symbol $\mathcal{J}$ to denote the Hilbert space of a ladder system.[1]

Our ladder of basis states comes equipped with a lowering operator a. This has the property that, for any n,

$$\mathsf{a}\,|n\rangle = \sqrt{f(n)}\,|n-1\rangle. \tag{13.1}$$

The function $f(n)$ is

$$f(n) = \langle n|\,\mathsf{a}^\dagger \mathsf{a}\,|n\rangle, \tag{13.2}$$

which is the squared norm of the vector $\mathsf{a}\,|n\rangle$. Thus $f(n) \geq 0$. We have chosen the phases of the basis states so that the coefficient $\sqrt{f(n)}$ in Eq. 13.1 is real and non-negative.

The adjoint operator $\mathsf{a}^\dagger$ acts as a raising operator. To see this, note that

$$\langle n|\,\mathsf{a}\,|m\rangle = \sqrt{f(m)}\,\delta_{n,m-1} = \sqrt{f(n+1)}\,\delta_{n+1,m}. \tag{13.3}$$

The vector $\mathsf{a}^\dagger\,|n\rangle$ has components $c_m = \langle m|\,\mathsf{a}^\dagger\,|n\rangle = \langle n|\,\mathsf{a}\,|m\rangle^*$, so we can see that

$$\mathsf{a}^\dagger\,|n\rangle = \sqrt{f(n+1)}\,|n+1\rangle. \tag{13.4}$$

Given the function $f(n)$, Eq. 13.1 and 13.4 define the properties of the raising and lowering operators a and $\mathsf{a}^\dagger$.

---

[1] Why $\mathcal{J}$? Our notation is suggested by Jacob's ladder, as described in the Book of Genesis: *[Jacob] had a dream, and behold, a ladder was set on the earth with its top reaching to heaven; and behold, the angels of God were ascending and descending on it.* It would, of course, be too fanciful to refer to the ascending and descending operators $\mathsf{a}^\dagger$ and a as "angel" operators!

But what function should we choose for $f(n)$? We wish this to be as simple as possible. We require $f(0) = 0$, so that the application of a to the "bottom rung" state $|0\rangle$ yields the null vector. However, we want $f(n) > 0$ for $n > 0$. The most straightforward choice is to let $f(n) = n$, so that

$$a|n\rangle = \sqrt{n}\,|n-1\rangle\,,$$

$$a^\dagger|n\rangle = \sqrt{n+1}\,|n+1\rangle\,. \tag{13.5}$$

From here on, we will adopt Eq. 13.5 as the defining properties for a and $a^\dagger$. (See Problem 13.1 for a parallel development for operators in a finite ladder of states.)

Exercise 13.1 Write the operator a as a sum of outer products of the basis states $|n\rangle$.

From the raising and lowering operators, we can define other operators of interest for our ladder system. For example, we may define the *number operator* n by $n = a^\dagger a$.

Exercise 13.2 Show that n is Hermitian, and

$$n|n\rangle = n|n\rangle\,, \tag{13.6}$$

for all basis states $|n\rangle$.

The basis states $|0\rangle$, $|1\rangle$, ... are exactly the *number eigenstates* of the system, and we can write

$$n = \sum_n n|n\rangle\langle n|\,. \tag{13.7}$$

Any number eigenstate for the ladder system can be constructed from the $|0\rangle$ state by successive applications of the raising operator $a^\dagger$:

$$|1\rangle = a^\dagger|0\rangle\,,$$

$$|2\rangle = \frac{a^\dagger}{\sqrt{2}}|1\rangle = \frac{\left(a^\dagger\right)^2}{\sqrt{2}}|0\rangle\,,$$

$$|3\rangle = \frac{a^\dagger}{\sqrt{3}}|2\rangle = \frac{\left(a^\dagger\right)^3}{\sqrt{2\cdot 3}}|0\rangle\,, \tag{13.8}$$

and so on. In general,

$$|n\rangle = \frac{\left(a^\dagger\right)^n}{\sqrt{n!}}|0\rangle\,. \tag{13.9}$$

As we will see, this is quite useful. It is also useful to examine the commutator of the raising and lowering operators. For any number eigenstate $|n\rangle$,

$$\left[a, a^\dagger\right]|n\rangle = aa^\dagger|n\rangle - a^\dagger a|n\rangle$$

$$= (n+1)|n\rangle - n|n\rangle = |n\rangle\,. \tag{13.10}$$

From this we conclude that

$$\left[a, a^\dagger\right] = \mathbf{1}. \tag{13.11}$$

We have seen that commutation relations express a lot of mathematical information in highly compressed form. In fact, all of the properties of a and $a^\dagger$ can be derived from their commutator, as is demonstrated in the following sequence of exercises. In each exercise, assume only that a and $a^\dagger$ are operators on the Hilbert space $\mathcal{J}$ satisfying Eq. 13.11.

Exercise 13.3 Define the operator $n = a^\dagger a$. Show that n is Hermitian. Let $|\nu\rangle$ be an eigenstate of n with real eigenvalue ν. Show that $\nu \geq 0$.

Exercise 13.4 Begin with the eigenstate $|\nu\rangle$ and show that the vector $a|\nu\rangle$ has squared norm ν. For $\nu > 0$, show that $a|\nu\rangle$ is an eigenvector of n with eigenvalue $\nu - 1$.

Exercise 13.5 Show that the only possible eigenvalues of n are $0, 1, 2, \ldots$

Exercise 13.6 Assume that the spectrum of n is non-degenerate. (This means we have a "simple" ladder system, analogous to the simple spin system of Section 12.1.) Let $|0\rangle$ be the eigenstate with eigenvalue 0, and define $|n\rangle$ according to Eq. 13.9. Show that the $|n\rangle$ states form an eigenbasis for n, and derive Eq. 13.5.

13.2 Oscillators

Ladder systems are not just a pleasant mathematical exercise. They turn out to be closely related to quantum particles moving in one dimension.

To see this, define

$$Q = \frac{1}{\sqrt{2}}\left(a + a^\dagger\right),$$

$$P = \frac{1}{i\sqrt{2}}\left(a - a^\dagger\right). \tag{13.12}$$

The Q and P operators are Hermitian operators such that

$$a = \frac{1}{\sqrt{2}}\left(Q + iP\right),$$

$$a^\dagger = \frac{1}{\sqrt{2}}\left(Q - iP\right). \tag{13.13}$$

In other words, Q and iP are $\sqrt{2}$ times the Hermitian and anti-Hermitian parts of a. The commutator between Q and P is

$$[Q, P] = \frac{1}{2i}\left[a + a^\dagger, a - a^\dagger\right]$$

$$= \frac{1}{2i}\left(-\left[a, a^\dagger\right] + \left[a^\dagger, a\right]\right)$$

$$[Q, P] = i\mathbf{1}. \tag{13.14}$$

This reminds us strongly of the commutation relation between position and momentum operators X and p for a particle moving in one dimension. In fact, we can make the resemblance even closer. Pick a constant α that has units of length. If we define

$$X = \alpha \, Q,$$
$$p = \frac{\hbar}{\alpha} \, P, \tag{13.15}$$

then X and p will have the canonical commutation relation in Eq. 10.53.

Exercise 13.7 Show that, for X and p defined in this way,

$$[X, p] = i\hbar \mathbf{1}. \tag{Re 10.53}$$

Beginning with the raising and lowering operators a and $a^\dagger$, we can first define the dimensionless pair Q and P, and then construct X and p satisfying the canonical commutation relation. Symbolically, our procedure is:

$$\left(\begin{array}{c} a, a^\dagger \\ {[a, a^\dagger]} = \mathbf{1} \end{array} \right) \Longrightarrow \left(\begin{array}{c} Q, P \\ {[Q, P]} = i\mathbf{1} \end{array} \right) \overset{\alpha}{\Longrightarrow} \left(\begin{array}{c} X, p \\ {[X, p]} = i\hbar\mathbf{1} \end{array} \right),$$

(where the second arrow reminds us that we must fix the length scale parameter α).

We could also reverse this procedure. For any particle moving in one dimension, the position and momentum observables can be used to construct dimensionless Q and P operators, from which we can create a and $a^\dagger$ operators satisfying Eq. 13.11.

Exercise 13.8 Write down the operators a and $a^\dagger$ in terms of X, p, $\hbar$, and the scale parameter α. Show that the canonical commutation relation in Eq. 10.53 implies Eq. 13.11.

In other words, every 1-D particle system has a ladder structure built into it. Of course, the number eigenstates $|n\rangle$ do not necessarily correspond to physically important states of the particle system – to energy eigenstates, for example. But as we will see next, in one important special case they do.

A 1-D *harmonic oscillator* has a Hamiltonian

$$H = \frac{1}{2\mu}p^2 + \frac{1}{2}kx^2, \tag{13.16}$$

where k is a constant that determines the "stiffness" of the elastic force experienced by the oscillator. A classical system with this Hamiltonian exhibits periodic motion with a classical frequency

$$\omega = \sqrt{\frac{k}{\mu}}, \tag{13.17}$$

independent of the amplitude of the oscillation. We can rewrite the Hamiltonian in terms of μ and ω:

$$H = \frac{1}{2\mu}p^2 + \frac{\mu\omega^2}{2}x^2. \tag{13.18}$$

We define the quantum length scale α for the oscillator system according to

$$\alpha = \sqrt{\frac{\hbar}{\mu\omega}}. \tag{13.19}$$

Exercise 13.9 Verify that α has units of length.

This lets us define dimensionless operators Q and P according to Eq. 13.15, from which we obtain raising and lowering operators

$$a = \frac{1}{\sqrt{2}}\left(\frac{X}{\alpha} + i\frac{\alpha}{\hbar}P\right),$$

$$a^\dagger = \frac{1}{\sqrt{2}}\left(\frac{X}{\alpha} - i\frac{\alpha}{\hbar}P\right). \tag{13.20}$$

The Hamiltonian operator can be written

$$H = \frac{\hbar\omega}{2}\left(P^2 + Q^2\right). \tag{13.21}$$

In terms of the raising and lowering operators, this has a very nice form.

Exercise 13.10 Show that

$$H = \hbar\omega\left(n + \frac{1}{2}1\right), \tag{13.22}$$

where $n = a^\dagger a$.

The Hamiltonian, in other words, is closely related to the number operator n. This has many useful consequences:

- The energy eigenstates of a harmonic oscillator are exactly the number eigenstates $|n\rangle$ of the ladder system.
- These energy eigenstates have eigenvalues given by

$$H|n\rangle = E_n|n\rangle = \left(n + \frac{1}{2}\right)\hbar\omega|n\rangle. \tag{13.23}$$

- The energy of the ground state $|0\rangle$ is $\hbar\omega/2$, and adjacent energy levels are separated by $\hbar\omega$ (see Fig. 13.1).

What are the wave functions for the stationary states of the oscillator? To find $\psi_n(x)$, we must solve Eq. 15.4, which in this case is

$$\psi_n''(x) = -\frac{2\mu}{\hbar^2}\left(E_n - \frac{\mu\omega^2}{2}x^2\right)\psi_n(x). \tag{13.24}$$

The good news is that we already know $E_n = \left(n + \frac{1}{2}\right)\hbar\omega$. The bad news is that Eq. 13.24 is laborious to solve directly (see Problem 13.3). Luckily, there is a better way.

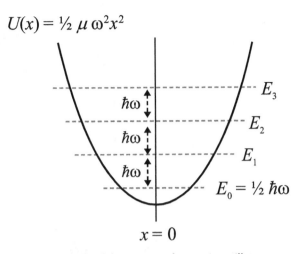

$$U(x) = \tfrac{1}{2}\,\mu\,\omega^2 x^2$$

Fig. 13.1   Energy levels of the quantum harmonic oscillator.

From our discussion in Section 10.2, we know that the position and momentum operators x and p affect the wave function by

$$\mathsf{x} : \psi \to x\psi(x) \quad \text{and} \quad \mathsf{p} : \psi \to -i\hbar\frac{\partial \psi}{\partial x} \ . \tag{13.25}$$

Equation 13.20, therefore, tells us how the lowering operator a changes the wave function:

$$\mathsf{a} : \psi \to \frac{1}{\sqrt{2}}\left(\frac{x}{\alpha}\psi + \alpha\frac{\partial \psi}{\partial x}\right). \tag{13.26}$$

(The raising operator $\mathsf{a}^\dagger$ acts in a similar way, with the "+" changed to "−.")

The ground state satisfies $\mathsf{a}\,|0\rangle = 0$. Thus, the wave function $\psi_0(x)$ for that ground state must satisfy

$$\psi_0'(x) = -\frac{x}{\alpha^2}\,\psi_0(x). \tag{13.27}$$

This is easily solved. We find that

$$\psi_0(x) = Ae^{-x^2/2\alpha^2}, \tag{13.28}$$

where A is determined by normalization. The ground state wave function is a Gaussian. By comparing this to our discussion of the Gaussian wave packet in Section 10.3 (especially around Eq. 10.64), we find that the constant $A = (\pi\alpha^2)^{-1/4}$ and the variance in position $\langle \Delta x^2\rangle = \alpha^2/2$.

Equation 13.28 gives us the wave function $\psi_0(x)$ for the ground state. What about the wave functions $\psi_1(x)$, $\psi_2(x)$, etc.? These are easily obtained by applying the raising operator $\mathsf{a}^\dagger$ to $\psi_0(x)$. The state $|1\rangle = \mathsf{a}^\dagger\,|0\rangle$, so

$$\psi_1 = \frac{1}{\sqrt{2}}\left(\frac{x}{\alpha}\psi_0 - \alpha\frac{\partial \psi_0}{\partial x}\right). \tag{13.29}$$

Exercise 13.11 Show that $\psi_1(x) = Bx\,e^{-x^2/2\alpha^2}$. Find B.

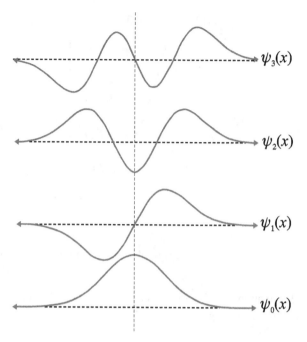

Fig. 13.2   **Wave functions for the first four stationary states of the quantum harmonic oscillator.**

We can of course iterate this idea to find the wave functions for $|2\rangle$, $|3\rangle$, etc.

Exercise 13.12 Find $\psi_2(x)$.

Exercise 13.13 Write $\psi_n(x) = F_n\left(\dfrac{x}{\alpha}\right) e^{-x^2/2\alpha^2}$. Show by mathematical induction that F_n must be an nth order polynomial.

The first few wave functions for the harmonic oscillator stationary states are sketched in Fig. 13.2.

Once we have the wave functions for the stationary states of the oscillator, we can find probability distributions over position and momentum variables. From these we can calculate expectation values $\langle x \rangle$, $\langle x^2 \rangle$, $\langle p \rangle$, and so on. However, it is much easier to calculate these expectations using a and $a^\dagger$. As an example, we will find $\langle \Delta x^2 \rangle = \langle x^2 \rangle - \langle x \rangle^2$ for the stationary state $|n\rangle$.

The observables X and X^2 are

$$X = \frac{\alpha}{\sqrt{2}}\left(a + a^\dagger\right),$$

$$X^2 = \frac{\alpha^2}{2}\left(a + a^\dagger\right)^2$$

$$= \frac{\alpha^2}{2}\left(a^2 + aa^\dagger + a^\dagger a + (a^\dagger)^2\right). \tag{13.30}$$

The expectation $\langle x \rangle$ is

$$\langle x \rangle = \frac{\alpha}{\sqrt{2}} \langle n| \left(a + a^\dagger \right) |n\rangle$$

$$= \frac{\alpha}{\sqrt{2}} \left(\sqrt{n} \langle n|n-1\rangle + \sqrt{n+1} \langle n|n+1\rangle \right)$$

$$= 0. \tag{13.31}$$

We next compute the expectation $\langle x^2 \rangle$. To do this, we note that $\langle n| a^2 |n\rangle = 0$ since $\langle n|n-2\rangle = 0$. In a similar way, $\langle n| (a^\dagger)^2 |n\rangle = 0$. This leaves

$$\langle x^2 \rangle = \frac{\alpha^2}{2} \langle n| \left(aa^\dagger + a^\dagger a \right) |n\rangle$$

$$= \frac{\alpha^2}{2} \langle n| \left(n \langle n|n\rangle + (n+1) \langle n|n\rangle \right)$$

$$\langle x^2 \rangle = \frac{(2n+1)\,\alpha^2}{2}. \tag{13.32}$$

This is also $\langle \Delta x^2 \rangle$, since $\langle x \rangle = 0$. The uncertainty in position for a stationary state of a harmonic oscillator is

$$\Delta x = \sqrt{\frac{(2n+1)\alpha^2}{2}} = \sqrt{\left(n + \frac{1}{2} \right) \frac{\hbar}{2\mu\omega}}. \tag{13.33}$$

Exercise 13.14 Note that Δx is larger for higher values of n. Explain why this makes sense.

Exercise 13.15 Calculate Δp for the energy eigenstate $|n\rangle$, and find the product $\Delta x \, \Delta p$.

13.3 Coherent states

In Section 10.3 we considered wave packets, free particle wave functions with pretty well-defined positions and momenta. Such states also exist for a harmonic oscillator system. Of particular interest and usefulness are the *coherent states*, which we will describe in this section.

The coherent states of an oscillator are eigenstates of a, the lowering operator. Because a is not Hermitian (or even normal), these eigenstates do not form an orthonormal basis. We have already met one of the eigenstates, the ground state $|0\rangle$ of the oscillator:

$$a |0\rangle = 0, \tag{13.34}$$

so its eigenvalue is zero. The other coherent states are labeled by their complex eigenvalue z:

$$a |z\rangle = z |z\rangle. \tag{13.35}$$

We can write the coherent state $|z\rangle$ in terms of the number state basis:

$$|z\rangle = \sum_n c_n |n\rangle. \tag{13.36}$$

Exercise 13.16 Substitute Eq. 13.36 into the definition in Eq. 13.35 and show that

$$c_n = \frac{z^n}{\sqrt{n!}} c_0. \tag{13.37}$$

Exercise 13.17 The value of c_0 is chosen to normalize the state $|z\rangle$. Show that $c_0 = e^{-|z|^2/2}$ works. This will be our standard choice.

Exercise 13.18 Show that the coherent state $|z\rangle$ is not an eigenstate of the raising operator $a^\dagger$. Show, in fact, that $a^\dagger$ has *no* eigenstates.[2]

Since the coherent states are defined in terms of the lowering operator a, it is easy to calculate various observable quantities. For example, suppose we measure the number observable n on an oscillator in a coherent state z. Then the expectation value is

$$\langle n \rangle = \langle z| n |z\rangle = \langle z| a^\dagger a |z\rangle = z^* z = |z|^2. \tag{13.38}$$

Exercise 13.19 What is the expectation $\langle E \rangle$ of the energy of an oscillator in the state $|z\rangle$?

We can find the wave functions $\psi_z(x)$ for the coherent state $|z\rangle$ in the same way we derived Eq. 13.28. The wave function must satisfy

$$\psi_z'(x) = \left(\frac{\sqrt{2}z}{\alpha} - \frac{x}{\alpha^2} \right) \psi_z(x). \tag{13.39}$$

Exercise 13.20 Show that the solution to this equation is

$$\psi_z(x) = A_z \, e^{-(x-b)^2/2\alpha^2} \, e^{ikx}, \tag{13.40}$$

where $b = \sqrt{2}\alpha \, \mathrm{Re}(z)$ and $k = (\sqrt{2}/\alpha)\mathrm{Im}(z)$.

We recognize this at once as a Gaussian wave packet – the ground state wave function $\psi_0(x)$, in fact, shifted by b along the x-axis and modulated by the plane wave e^{ikx}. Even the same normalization factor will work, at least up to an overall phase: $|A_z| = (\pi\alpha^2)^{-1/4}$. (The phase must be chosen to stay consistent with the phase already chosen for the coefficients in Eq. 13.36.)

Several facts immediately follow. For a coherent state $|z\rangle$,

$$\langle x \rangle = b = \sqrt{2}\alpha \, \mathrm{Re}(z), \quad \Delta x = \frac{\alpha}{\sqrt{2}},$$

$$\langle p \rangle = \hbar k = \frac{\sqrt{2}\,\hbar}{\alpha} \, \mathrm{Im}(z), \quad \Delta p = \frac{\hbar}{\sqrt{2}\,\alpha}. \tag{13.41}$$

---

[2] This does not contradict the argument in Section 3.5 establishing the existence of eigenvalues and eigenvectors, because that argument only applied to finite-dimensional Hilbert spaces. The space $\mathcal{J}$ for a ladder system is infinite-dimensional.

(Review the discussion of Gaussian wave packets in Section 10.3 for details.) Every coherent state is a minimum uncertainty wave packet for the oscillator.

We now see that the real and imaginary parts of z are related to expectation values of x and p – that is, to points in the *classical phase space* for a particle in one dimension. We might even say that, for the oscillator system, the coherent states *are* the quantum version of phase space points. Two nearby points, however, are not completely distinguishable.

Exercise 13.21 For coherent states $|z\rangle$ and $|z'\rangle$, show that

$$\left| \langle z' | z \rangle \right|^2 = e^{-|z-z'|^2}. \tag{13.42}$$

No two coherent states are perfectly distinguishable by any measurement, but if the value of $|z - z'|$ is large enough they can be distinguished pretty well.

Exercise 13.22 Use the basic distinguishability theorem of Section 4.2 to find the minimum value of $|z - z'|$ for which two coherent states can be distinguished with success probability $P_S = 0.95$. Repeat this calculation for $P_S = 0.9999$.

A close connection between coherent states of a quantum oscillator and the phase space points of a classical oscillator can be seen in the time evolution of a coherent state. Suppose the initial state of the oscillator is a coherent state $|\psi(0)\rangle = |z\rangle$. How does this state evolve under the oscillator Hamiltonian? The number states $|n\rangle$ are energy eigenstates with energy $E_n = (n + 1/2)\hbar\omega$, so the time evolution operator $U(t)$ acts by

$$U(t) |z\rangle = e^{-|z|^2/2} \sum_n \frac{z^n}{\sqrt{n!}} U(t) |n\rangle$$

$$= e^{-|z|^2/2} \sum_n \frac{z^n}{\sqrt{n!}} e^{-i(n+1/2)\omega t} |n\rangle. \tag{13.43}$$

But $e^{-in\omega t} = \left(e^{-i\omega t}\right)^n$, and so

$$U(t) |z\rangle = e^{-i\omega t/2} \left| e^{-i\omega t} z \right\rangle, \tag{13.44}$$

where $\left| e^{-i\omega t} z \right\rangle$ is the coherent state with eigenvalue $e^{-i\omega t} z$. The overall phase factor $e^{-i\omega t/2}$ is unimportant. The main point is that *coherent states evolve into coherent states*, tracing out circular trajectories in the complex z-plane.

Exercise 13.23 Show that the coherent state trajectory in (x, p) phase space satisfies the harmonic oscillator equation of motion.

Finally, consider the expansion of a coherent state in terms of the number states $|n\rangle$. Each of these states can be built from $|0\rangle$ by applying the raising operator $a^\dagger$, as in Eq. 13.9. We have

$$|z\rangle = e^{-|z|^2/2} \sum_n \frac{z^n}{\sqrt{n!}} |n\rangle = e^{-|z|^2/2} \sum_n \frac{z^n}{n!} \left(a^\dagger\right)^n |0\rangle. \tag{13.45}$$

We recognize an operator power series here, which allows us to write

$$|z\rangle = e^{-|z|^2/2} \exp(za^\dagger) |0\rangle . \tag{13.46}$$

The operator $\exp(za^\dagger)$ shifts the coherent state $|0\rangle$ to a vector parallel to $|z\rangle$. In fact, this is a special case of a more general "shift" property, stated in the next exercise.

Exercise 13.24 Show that

$$\exp(z'a^\dagger) |z\rangle = C |z + z'\rangle , \tag{13.47}$$

and evaluate the scalar C. To do this, you will want to recall that $\exp(A) \exp(B) = \exp(A + B)$, provided that A and B commute.

13.4 Thermal states of a ladder system

A ladder system might be in a mixed state described by a density operator ρ. A particularly useful example of this is the canonical state of Eq. 8.70. Given the Hamiltonian $H = (n + 1/2)\hbar\omega$, this is

$$\rho = \frac{1}{\mathcal{Z}} e^{-H/kT} = \frac{1}{\mathcal{Z}} \sum_n e^{-(n+1/2)\hbar\omega/kT} |n\rangle \langle n| . \tag{13.48}$$

This is the mixed state of an oscillator that is in thermal equilibrium with a reservoir system at absolute temperature T.

Exercise 13.25 Show that the partition function $\mathcal{Z} = \mathrm{Tr}\, e^{-H/kT}$ in the canonical state is

$$\mathcal{Z} = \frac{e^{-\hbar\omega/2kT}}{1 - e^{-\hbar\omega/kT}} . \tag{13.49}$$

This relies on the sum of the geometric series:

$$\sum_{n=0}^{\infty} x^n = \frac{1}{1 - x} , \tag{13.50}$$

for $|x| < 1$. What plays the role of x here?

We wish to calculate expectations of the form $\langle A \rangle = \mathrm{Tr}\, \rho A$. Since the canonical density operator in Eq. 13.48 is diagonal in the number state basis, this is the most convenient basis to use for evaluating the trace. For instance, the average energy of the oscillator in this mixed state is

$$\langle E \rangle = \mathrm{Tr}\, \rho H = \frac{1}{\mathcal{Z}} \sum_{n=0}^{\infty} (n + 1/2)\hbar\omega\, e^{-(n+1/2)\hbar\omega/kT}$$

$$= \left(\frac{\hbar\omega}{e^{\hbar\omega/kT} - 1} \right) + \frac{\hbar\omega}{2} . \tag{13.51}$$

Exercise 13.26 Verify Eq. 13.51. You will find the following fact useful:

$$\sum_{n=0}^{\infty} n x^n = x \frac{d}{dx}\left(\sum_{n=0}^{\infty} x^n\right) = \frac{x}{(1-x)^2}. \tag{13.52}$$

(This can be derived from Eq. 13.50.)

At very low temperatures ($kT \ll \hbar\omega$) the first term in Eq. 13.51 is negligible, and so $\langle E \rangle \approx \hbar\omega/2$, the energy of the ground state. At very high temperatures ($kT \gg \hbar\omega$) we can approximate the exponential by $e^{\hbar\omega/kT} \approx 1 + \hbar\omega/kT$ and obtain $\langle E \rangle \approx kT$.

This last result is in accordance with the equipartition theorem of classical statistical mechanics, which states that each quadratic term in the Hamiltonian of a system in equilibrium should have an average energy of $kT/2$. Since the harmonic oscillator has two quadratic terms, the average oscillator energy should be kT.

The *heat capacity* C of a system measures how the average energy changes with changes in temperature. For a single quantum oscillator, a short calculation finds

$$C = \frac{d\langle E \rangle}{dT} = \frac{e^{\hbar\omega/kT}}{(e^{\hbar\omega/kT}-1)^2} \frac{(\hbar\omega)^2}{kT^2}. \tag{13.53}$$

Again, we can examine the behavior of this at high and low temperatures:

- If $kT \ll \hbar\omega$, then $C \approx e^{-\hbar\omega/kT} \dfrac{(\hbar\omega)^2}{kT^2}$, and so $C \to 0$ as $T \to 0$.
- If $kT \gg \hbar\omega$, then $C \approx k$, a constant.

The quantum heat capacity of an oscillator played an important role in the early history of quantum theory. In the 19th Century, classical statistical mechanics was applied to find the heat capacities of crystalline solids. The atoms in such a solid should behave roughly as 3-D harmonic oscillators. By the equipartition theorem, each atom should thus have an average energy of $3kT$. An assembly of N atoms should have a heat capacity of $3Nk$. One mole has Avogadro's number $N_A = 6.02 \times 10^{23}$ atoms, so the molar heat capacity of a crystalline solid should be

$$3N_A k = 24.9 \text{ J / K mol}. \tag{13.54}$$

And indeed, the molar heat capacity of many solids is close to this value. This fact is called the *Dulong–Petit Law*, discovered in 1819. Other solids, however, have a much smaller heat capacity at room temperature.

In 1907, Einstein explained this behavior by applying early quantum ideas to the oscillating atoms in the solid. In our terminology, he regarded one mole of a solid to be composed of $3N_A$ identical quantum oscillators. The characteristic frequency ω of the oscillators fixed an *Einstein temperature* $T_E = \hbar\omega/k$. The heat capacity of one mole is

$$C = 3N_A k \frac{e^{T_E/T}}{(e^{T_E/T}-1)^2} \left(\frac{T_E}{T}\right)^2. \tag{13.55}$$

Any solid should thus have a heat capacity that agrees with the classical predication for $T \gg T_E$ but approaches zero for $T \ll T_E$. Different solids, however, differ in their values for T_E.

Einstein's theory marked the first time that quantum ideas had been applied to matter by itself. (Previous quantum work, such as Planck's explanation of the blackbody spectrum and Einstein's explanation of the photoelectric effect, concerned the interaction of light with matter.) The theory qualitatively explained the behavior of the heat capacities of solids over a wide range of temperatures. A later refinement of Einstein's ideas by Peter Debye made the agreement quantitatively accurate as well.

Problems

Problem 13.1 Suppose we are constructing a *finite* ladder of basis states $|0\rangle, \ldots, |d-1\rangle$. We need to make a different choice for the function $f(n)$ in Eq. 13.1 and 13.4. To avoid "stepping off the ladder" at the top and bottom, we will require that $f(0) = f(d) = 0$, but that $f(n) > 0$ for $0 < n < d$. The simplest algebraic choice is

$$f(n) = n(d - n). \tag{13.56}$$

Now consider the ladder of S_z eigenstates for a simple spin-s system, discussed in Section 12.1. There are $d = 2s + 1$ states that are indexed by m-values ranging from $m = -s$ to $m = +s$. How are the spin raising and lowering operators $S_\pm$ related to the a and $a^\dagger$ operators defined by Eq. 13.1 and 13.4? (Here is an example where the "simplest choice" for $f(n)$ gives a good result!)

Problem 13.2 The position operator X for the harmonic oscillator is

$$X = \frac{\alpha}{\sqrt{2}} \left(a + a^\dagger \right). \tag{13.57}$$

Use this fact to calculate $\langle x^3 \rangle$ and $\langle x^4 \rangle$ for the energy eigenstate $|n\rangle$.

Problem 13.3 Suppose $\psi(x)$ satisfies

$$\psi''(x) = -\frac{2\mu}{\hbar^2} \left(E - \frac{\mu\omega^2}{2} x^2 \right) \psi(x), \tag{13.58}$$

for some E. Let $\psi(x) = F\left(\frac{x}{\alpha}\right) e^{-x^2/2\alpha^2}$, where α is defined as in Eq. 13.19.

(a) Find a differential equation for the function F.
(b) Assume that F is a polynomial function that is either even or odd. Find the corresponding values E_n, where n is the degree of the polynomial.
(c) Find the first four polynomials F_0, F_1, F_2, and F_3. (Do not worry about normalization.)
(d) (Optional) By considering a general series solution to your differential equation from part (a), show that the condition in part (b) must hold for the resulting wave function $\psi(x)$ to be normalizable.

Problem 13.4 Suppose at $t = 0$ an oscillator is in the state $|\psi(0)\rangle = \sum_n c_n |n\rangle$. Show that, after a time $T = 2\pi/\omega$ (the classical oscillation period) the state $|\psi(T)\rangle$ is indistinguishable from the original state $|\psi(0)\rangle$.

Problem 13.5 Suppose the number observable n is measured on an oscillator system in a coherent state $|z\rangle$. We have already found the expectation $\langle n \rangle = |z|^2$. Find the probability $P(n)$ that the result n is obtained. Also calculate $\langle n^2 \rangle$ and Δn. (Hint: Recall that $\mathsf{aa}^\dagger = \mathsf{a}^\dagger \mathsf{a} + 1$, which is just Eq. 13.11.)

Problem 13.6 Define the displacement operator $\mathsf{D}(z)$ by

$$\mathsf{D}(z) = \exp(z\mathsf{a}^\dagger - z^*\mathsf{a}). \tag{13.59}$$

(a) Show that $\mathsf{D}(z)$ is unitary.
(b) Show that the coherent state $|z\rangle = \mathsf{D}(z)|0\rangle$. (You will need to use the Baker–Campbell–Hausdorff identity from Problem 5.8.)
(c) Evaluate $\mathsf{D}(z')|z\rangle$ for the coherent state $|z\rangle$.

This version of the displacement operator has somewhat more elegant properties than the simpler one mentioned in Exercise 13.24.

Problem 13.7 A ladder system subject to spontaneous decay can be modeled with a Lindblad equation (see Section 9.3). This involves a single Lindblad operator $\Lambda\mathsf{a}$, where Λ is a constant with units $(\text{time})^{-1/2}$. The equation can be written

$$\frac{d}{dt}\rho = \frac{1}{i\hbar}[\mathsf{H}, \rho] + \Lambda^2 \mathsf{a}\rho\mathsf{a}^\dagger - \frac{\Lambda^2}{2}\left(\mathsf{a}^\dagger \mathsf{a}\rho + \rho\mathsf{a}^\dagger \mathsf{a}\right). \tag{13.60}$$

Take the Hamiltonian operator to be $\mathsf{H} = \hbar\omega\mathsf{a}^\dagger\mathsf{a}$. Suppose that the state of the system at any time is diagonal in the number basis:

$$\rho(t) = \sum_n p_n(t)|n\rangle\langle n|. \tag{13.61}$$

Derive rate equations for the time-dependent eigenvalues $p_n(t)$, and show that the expectation $\langle n \rangle$ decays exponentially with time.

Problem 13.8 This is a variation on Problem 13.7. The ladder system with Hamiltonian $\hbar\omega\mathsf{a}^\dagger\mathsf{a}$ undergoes Lindblad-type evolution with two Lindblad operators, $\Lambda\mathsf{a}$ and $\kappa\Lambda\mathsf{a}^\dagger$, with $0 < \kappa < 1$. Thus, the environment can both raise and lower the ladder state, though the lowering process is "stronger." Show that a canonical state $\rho = \mathcal{Z}^{-1}\exp(-\mathsf{H}/k_\mathrm{B}T)$ can be a constant under this evolution, for a particular relation between κ and T.

Many particles

14.1 Two-particle wave functions

In this chapter we will consider quantum systems composed of two or more particles. To make our analysis look as simple as possible, we will suppose that these particles each move in one dimension. However, as the exercises will make clear, our results can be easily generalized to particles moving in 3-D space.

To begin, suppose a system has just two particles, designated #1 and #2. The Hilbert space for this composite system is $\mathcal{H}^{(1)} \otimes \mathcal{H}^{(2)}$, the tensor product of the Hilbert spaces for each particle. Let $|x_1, x_2\rangle = |x_1\rangle \otimes |x_2\rangle$ be the product of improper position eigenstates for the two particles.[1] The wave function for the system in state $|\Psi\rangle$

$$\Psi(x_1, x_2) = \langle x_1, x_2 | \Psi \rangle. \tag{14.1}$$

Just as the composite system has only one state vector $|\Psi\rangle$, it also has only one wave function $\Psi(x_1, x_2)$, which is a function of both particle coordinates.[2] The square of the magnitude of Ψ gives the joint probability density over x_1 and x_2:

$$\mathcal{P}(x_1, x_2) = |\Psi(x_1, x_2)|^2. \tag{14.2}$$

This means that the probability that $x_1 \in [a_1, b_1]$ and $x_2 \in [a_2, b_2]$ is

$$P(x_1 \in [a_1, b_1], x_2 \in [a_2, b_2]) = \int_{a_1}^{b_1} \int_{a_2}^{b_2} |\Psi(x_1, x_2)|^2 \, dx_1 dx_2. \tag{14.3}$$

Exercise 14.1 What is the normalization condition for the joint wave function $\Psi(x_1, x_2)$?

Exercise 14.2 Suppose we have two particles moving in three dimensions. The overall wave function Ψ will be a function of how many Cartesian coordinates?

---

[1] To avoid some later confusion with exponentiation, we will use subscripts to designate the coordinates for different particles. However, we will continue to use our regular superscript notation in other contexts where this confusion is impossible. This is why we write $|x_1, x_2\rangle$ rather than $|x^{(1)}, x^{(2)}\rangle$, but still write $\mathcal{H}^{(1)} \otimes \mathcal{H}^{(2)}$. We will switch from superscripts to subscripts as needed without further comment, trusting that the reader will be able to sort things out without too much trouble.

[2] Note the similarity between this two-particle situation and a single particle moving in two dimensions. From the standpoint of the abstract theory, both are examples of a quantum system with two continuous degrees of freedom.

A product state $|\psi,\phi\rangle$ yields a wave function that is a product of one-particle wave functions:

$$\Psi(x_1,x_2) = \langle x_1,x_2 \,|\, \psi,\phi \rangle = \psi(x_1)\phi(x_2). \tag{14.4}$$

This is the only case in which it is possible to ascribe a definite wave function to each particle separately.

Exercise 14.3 Show that a product state of two 1-D particles leads to a probability distribution in which the position variables x_1 and x_2 are independent.

Exercise 14.4 Given a joint probability distribution $\mathcal{P}(x_1,x_2)$, we can find a distribution over x_1 simply by integrating over x_2. That is,

$$\mathcal{P}(x_1) = \int_{-\infty}^{+\infty} \mathcal{P}(x_1,x_2)\,dx_2. \tag{14.5}$$

Why does this not work with wave functions? What is wrong with defining $\psi(x_1) = \int_{-\infty}^{+\infty} \Psi(x_1,x_2)\,dx_2$?

There are position operators X_1 and X_2 for each particle, and corresponding momentum operators p_1 and p_2. These satisfy the canonical commutation relations

$$[\mathsf{X}_1,\mathsf{p}_1] = i\hbar\mathbf{1} \quad \text{and} \quad [\mathsf{X}_2,\mathsf{p}_2] = i\hbar\mathbf{1}. \tag{14.6}$$

All other pairs of these operators commute. The momentum operators affect the wave function as follows:

$$\mathsf{p}_1 : \Psi \longrightarrow -i\hbar\frac{\partial}{\partial x_1}\Psi \quad \text{and} \quad \mathsf{p}_2 : \Psi \longrightarrow -i\hbar\frac{\partial}{\partial x_2}\Psi. \tag{14.7}$$

Expectation values of observables are calculated from the wave function in the usual way.

The Hamiltonian of a two-particle system typically consists of separate kinetic energy terms for each particle and a potential energy that may depend on both x_1 and x_2. That is,

$$\mathsf{H} = \frac{1}{2\mu_1}\mathsf{p}_1^2 + \frac{1}{2\mu_1}\mathsf{p}_1^2 + U(\mathsf{X}_1,\mathsf{X}_2), \tag{14.8}$$

where μ_1 and μ_2 are the masses of the two particles. The Hamiltonian steers the time evolution of the system via the Schrödinger equation. In terms of the time-dependent wave function $\Psi(x_1,x_2,t)$, this is

$$-\frac{\hbar^2}{2\mu_1}\frac{\partial^2\Psi}{\partial x_1^2} - \frac{\hbar^2}{2\mu_2}\frac{\partial^2\Psi}{\partial x_2^2} + U(x_1,x_2)\,\Psi = i\hbar\frac{\partial\Psi}{\partial t}. \tag{14.9}$$

This single partial differential equation governs the time evolution of the joint wave function Ψ, and therefore the behavior of the entire system.

Exercise 14.5 Write down the Schrödinger equation for the wave function of two particles in three dimensions. Compare Eq. 11.26. (Here is a useful piece of notation: $\vec{\nabla}_1$ is the gradient with respect to the coordinates in $\vec{r}_1$, and ∇_1^2 is the corresponding Laplacian. We define $\vec{\nabla}_2$ and ∇_2^2 in the same way.)

Exercise 14.6 Suppose two 1-D particles are non-interacting, so that the potential energy $U(x_1, x_2) = U_1(x_1) + U_2(x_2)$. Let $\psi_1(x_1, t)$ and $\psi_2(x_2, t)$ be solutions to the separate single-particle Schrödinger equations. Show that the product

$$\Psi(x_1, x_2, t) = \psi_1(x_1, t)\psi_2(x_2, t), \qquad (14.10)$$

is a solution to Eq. 14.9.

Exercise 14.6 makes an important conceptual point. In classical mechanics, a system of many particles has dynamics described by a system of many equations of motion, which are ordinary differential equations in the variable t. To analyze the dynamics of an isolated subsystem, we need only consider the equations of motion for the subsystem's own degrees of freedom – equations which will be uncoupled from the rest of the system.

However, in quantum mechanics, even a system of many particles has only *one* equation of motion, the Schrödinger equation, which is a partial differential equation involving all particle coordinates and t. Exercise 14.6 nevertheless tells us that, if the particles are dynamically isolated and the system is in a product state, we can effectively analyze the dynamics of the individual particles via separate one-particle "Schrödinger equations."[3] Since, as a practical matter, we are always applying quantum mechanics to small pieces of the entire world, it is a very good thing that such effective descriptions exist!

14.2 Center of mass and relative coordinates

It may happen that our two-particle system is subject only to internal forces, so that the potential energy depends only on the relative position of the particles:

$$U(x_1, x_2) = U(x_2 - x_1). \qquad (14.11)$$

This means that, if we translate both particles by exactly the same amount, the potential is unchanged. In fact, such a joint translation is a dynamical symmetry, as defined in Section 5.4.

Exercise 14.7 Let $\mathsf{T}(s)$ be the unitary operator that translates both particles by s. That is,

$$\mathsf{T}(s) : \Psi(x_1, x_2) \longrightarrow \Psi(x_1 - s, x_2 - s). \qquad (14.12)$$

(Why the minus signs?) If Eq. 14.11 holds, show that

$$[\mathsf{T}(s), \mathsf{H}] = 0, \qquad (14.13)$$

for the two-particle Hamiltonian of Eq. 14.8.[4]

---

[3] This is also implied by the general discussion in Section 6.2, which you might wish to review at this time.
[4] This is the definition of a dynamical symmetry given in Eq. 5.61. As an additional exercise, find the Hermitian observable that generates the one-parameter family $\mathsf{T}(s)$.

We can adopt an alternative description of our two-particle system that exploits this symmetry. Define the operators

$$\mathsf{X} = \frac{1}{M}(\mu_1 \mathsf{x}_1 + \mu_2 \mathsf{x}_2) \quad \text{and} \quad \mathsf{x} = \mathsf{x}_2 - \mathsf{x}_1, \tag{14.14}$$

where $M = \mu_1 + \mu_2$. We recognize X as the operator corresponding to the system's center of mass, while x is the relative position of particle #2 with respect to particle #1. Now our potential is $U(\mathsf{x})$, a function of x only.

A joint position eigenstate $|x_1, x_2\rangle$ is also an eigenstate of X and x, so we can also write it as $|X, x\rangle$, where X and x are the corresponding eigenvalues.

Exercise 14.8 What is the X eigenvalue X for the (improper) joint position eigenstate $|x_1, x_2\rangle$? What about the x eigenvalue?

Given a two-particle quantum state $|\Psi\rangle$, we can write down its wave function $\Psi(X, x) = \langle X, x | \Psi \rangle$ in terms of variables X and x.

Our system has been introduced as a composite of two 1-D particles – that is, as two quantum systems, each with a single continuous degree of freedom. The overall Hilbert space is $\mathcal{H} = \mathcal{H}^{(1)} \otimes \mathcal{H}^{(2)}$. Now we wish to regard the same system in a different way, as a composite of a center-of-mass degree of freedom (associated with coordinate operator X) and a relative position degree of freedom (associated with x). The same Hilbert space $\mathcal{H}$ will be decomposed as $\mathcal{H}^{(\text{cm})} \otimes \mathcal{H}^{(\text{rel})}$. To do this, we must find momentum operators P and p corresponding to the new position operators X and x. These momentum operators should affect $\Psi(X, x)$ in the expected way:

$$\mathsf{P} : \Psi \longrightarrow -i\hbar \frac{\partial}{\partial X} \Psi \quad \text{and} \quad \mathsf{p} : \Psi \longrightarrow -i\hbar \frac{\partial}{\partial x} \Psi. \tag{14.15}$$

These X and x derivatives are related to the x_1 and x_2 derivatives by the chain rule:

$$\frac{\partial}{\partial X} = \left(\frac{\partial x_1}{\partial X} \right) \frac{\partial}{\partial x_1} + \left(\frac{\partial x_2}{\partial X} \right) \frac{\partial}{\partial x_2} = \frac{\partial}{\partial x_1} + \frac{\partial}{\partial x_2},$$

$$\frac{\partial}{\partial x} = \left(\frac{\partial x_1}{\partial x} \right) \frac{\partial}{\partial x_1} + \left(\frac{\partial x_2}{\partial x} \right) \frac{\partial}{\partial x_2} = -\frac{\mu_2}{M} \frac{\partial}{\partial x_1} + \frac{\mu_1}{M} \frac{\partial}{\partial x_2}. \tag{14.16}$$

Therefore, we define

$$\mathsf{P} = \mathsf{p}_1 + \mathsf{p}_2 \quad \text{and} \quad \mathsf{p} = \frac{1}{M}(\mu_1 \mathsf{p}_2 - \mu_2 \mathsf{p}_1). \tag{14.17}$$

So the momentum P associated with the center of mass X is exactly the total momentum of the system. The other new momentum p associated with the relative coordinate x is less easily recognized. If we define the *reduced mass* μ for the system by

$$\frac{1}{\mu} = \frac{1}{\mu_1} + \frac{1}{\mu_2} = \frac{M}{\mu_1 \mu_2}, \tag{14.18}$$

then the definition of p becomes

$$\mathsf{p} = \mu \left(\frac{\mathsf{p}_2}{\mu_2} - \frac{\mathsf{p}_1}{\mu_1} \right). \tag{14.19}$$

Heuristically, the relative momentum p is just the reduced mass μ times the relative velocity of the particles.

Exercise 14.9 Show that the reduced mass μ is less than either μ_1 or μ_2. What is μ if $\mu_2 = \mu_1$? If $\mu_2 \ll \mu_1$?

Exercise 14.10 Show that the operators X, P, x, and p satisfy the canonical commutation relations

$$[X, P] = i\hbar\mathbf{1} \quad \text{and} \quad [x, p] = i\hbar\mathbf{1}, \tag{14.20}$$

with all other commutators equal to zero.

Exercise 14.11 Show that

$$p_1 = \frac{\mu_1}{M}P - p, \qquad p_2 = \frac{\mu_2}{M}P + p. \tag{14.21}$$

This last exercise leads to a very significant result. We can write the kinetic energy terms in the two-particle Hamiltonian (Eq. 14.8) using P and p. When we include the potential, the complete Hamiltonian is

$$H = \frac{1}{2M}P^2 + \frac{1}{2\mu}p^2 + U(x). \tag{14.22}$$

Exercise 14.12 Verify Eq. 14.22.

Notice that we can write this Hamiltonian as $H^{(cm)} + H^{(rel)}$, where

$$H^{(cm)} = \frac{1}{2M}P^2, \qquad H^{(rel)} = \frac{1}{2\mu}p^2 + U(x). \tag{14.23}$$

In other words, the center of mass and the relative position are *completely non-interacting* quantum systems. The center of mass behaves as a free particle of mass M, while the relative position behaves independently as a particle of mass μ moving in an external potential $U(x)$.

The same system can be regarded as a composite of two particles (with coordinate operators x_1 and x_2, or as a composite of a center of mass (X) and a relative position (x). So which decomposition is the "right" one? The answer, of course, is that both are equally "correct." Nothing in quantum theory requires us to adopt one or another way of dividing an overall system into subsystems. For a given physical situation, our choice of decomposition will be made on purely practical grounds. For instance, if our measurement devices allow us to observe the two particles individually, it may be easier to think of x_1 and x_2 as observables for separate subsystems. On the other hand, when external forces are absent, the Hamiltonian in Eq. 14.22 suggests that the X, x view may have significant advantages.

Our discussion so far easily generalizes to a pair of particles moving in three dimensions. Our particles have positions $\vec{r}_1$ and $\vec{r}_2$. As before, we define center of mass position $\vec{R}$ and relative position $\vec{r}$, and write the wave function in terms of these coordinates. We leave the details for:

Exercise 14.13 Start at the beginning of this section and derive all of the analogous results for two particles moving in 3-D. For most steps, this is no more complicated than writing down a vector version of the 1-D equation, but in a couple of places (Eq. 14.17, for instance) you may need more of a derivation.

Exercise 14.14 A two-particle system in 3-D is subject only to internal forces, and we write its time-dependent wave function $\Psi(\vec{R},\vec{r},t)$ in terms of center-of-mass and position coordinates. Show that the Schrödinger equation for Ψ can be written

$$-\frac{\hbar^2}{2M}\nabla_R^2\Psi - \frac{\hbar^2}{2\mu}\nabla_r^2\Psi + U(\vec{r})\,\Psi = i\hbar\frac{\partial}{\partial t}\,\Psi, \tag{14.24}$$

where ∇_R^2 and ∇_r^2 are the Laplacians for the $\vec{R}$ and $\vec{r}$ coordinates, respectively.

Everything can also be generalized to quantum systems having more than just two particles. For N particles moving in 1-D, the Hilbert space is $\mathcal{H}^{(1)} \otimes \cdots \otimes \mathcal{H}^{(N)}$. The wave function Ψ is a function of all N coordinates:

$$\Psi(x_1,\dots,x_N) = \langle x_1,\dots,x_N \,|\,\Psi\rangle. \tag{14.25}$$

Thus the spatial wave function for a system of many particles is a function of the entire *configuration* of the system. The evolution of the joint wave function Ψ is described by a single Schrödinger equation.

We can devise center-of-mass and relative coordinates for systems of many particles. For instance, consider three particles numbered #0, #1, and #2. Define

$$X = \frac{1}{M}\left(\mu_0 x_0 + \mu_1 x_1 + \mu_2 x_2\right), \qquad x_1' = x_1 - x_0, \qquad x_2' = x_2 - x_0, \tag{14.26}$$

where $M = \mu_0 + \mu_1 + \mu_2$. (In this scheme, we have singled out particle #0 for a special role as the reference point for the relative coordinates.) If there are only internal forces in this system, then the potential energy is a function only of the relative coordinates: $U(x_1', x_2')$.

Exercise 14.15 Explain why this is true, even if there are forces between each pair of particles, so that the potential energy depends on all three relative positions: $x_1 - x_0$, $x_2 - x_0$, and $x_2 - x_1$.

As before, we define momentum operators P, p_1', and p_2' so that the wave function $\Psi(X, x_1', x_2')$ is affected by the expected derivatives. This enables us to write the three-particle Hamiltonian in terms of the center-of-mass and relative variables.

Exercise 14.16 Derive the operators P, p_1', and p_2' in terms of the single-particle momenta p_0, p_1, and p_2. Show that the Hamiltonian of the three-particle system is

$$H = \frac{1}{2M}P^2 + \frac{1}{2\mu_1'}{p_1'}^2 + \frac{1}{2\mu_2'}{p_2'}^2 + U(x_1', x_2'), \tag{14.27}$$

where the reduced masses μ_k' ($k = 1, 2$) are

$$\frac{1}{\mu_k'} = \frac{1}{\mu_0} + \frac{1}{\mu_k}. \tag{14.28}$$

It is straightforward to extend this scheme to three dimensions and to more than three particles.

An atom is a composite system containing a postively-charged nucleus (particle #0) and N electrons (particles #1 through #N). The atom is held together by internal forces. Thus, it is most convenient to consider the atom in terms of center-of-mass and relative coordinates. Simplifying to 1-D, the center of mass X gives the position of the "whole atom," which moves through space as a free particle. The relative coordinates $x'_1, \ldots, x'_N$ of the various electrons with respect to the nucleus are internal degrees of freedom of the atom. These are dynamically isolated from the center-of-mass motion and thus may be considered separately from it.

Exercise 14.17 Consider the hydrogen atom, which consists of one proton and one electron. By what percentage does the reduced mass μ of their relative position differ from the electron's mass? Repeat the same calculation for positronium (whose "nucleus" is a positron having the same mass as the electron). Also repeat it for the relative position of an electron and the nucleus of a gold atom.

These ideas provide an answer to a question that may have occurred to the reader back in Chapter 1. In Problem 1.4 we described an experiment demonstrating interference effects in a beam of C_{60} molecules, known as "buckyballs." The wavelength we assumed for these molecules was 3 pm. Yet this is hundreds of times smaller than the diameter of a buckyball, which is about 1 nm!

No real paradox is involved, however. The wavelength is associated with the motion of the molecule's center of mass, which acts like a *point particle* moving freely through space. The molecule's diameter depends on all the relative positions of the constituent carbon atoms. The center of mass and relative positions are completely independent and non-interacting "subsystems" of the buckyball.[5]

14.3 Identical particles

According to proverb, no two snowflakes are identical. We may generalize the principle: in fact, no two macroscopic objects of any sort are identical. This is because macroscopic objects have vast numbers of microscopic degrees of freedom – the exact atom-by-atom composition, the spatial arrangement of quadrillions of molecules, and so forth. The likelihood that even two grains of sand on a beach have exactly the same internal configuration is effectively zero. Thus, we could use the internal configuration of a sand grain as a sort of label, a unique physical "serial number" that identifies that grain and no other.

---

[5] The dynamical isolation of center of mass and relative position also helps to explain why the complicated buckyballs in the experiment can be informationally isolated enough to exhibit interference effects. Even though the carbon atoms have very strong interactions with each other, these interactions have nothing to do with the location of the center of mass.

Elementary quantum systems, however, are radically simpler than macroscopic objects. All electrons have exactly the same charge, mass, and so forth. They behave exactly the same way in all experiments. Therefore, if we exchange two electrons in a system, we arrive at a physical situation that is exactly equivalent in every respect to the original one. No experimental test, however refined, would allow us to distinguish one situation from the other.

In other words, electrons are truly *identical particles*. They do not possess unique physical "serial numbers" of any kind. Furthermore, the same is true of other types of particles such as protons, neutrons, photons, etc. This fact has far-reaching implications for the quantum mechanics of many-particle systems.

Suppose the state of a single particle is described by a state vector in the Hilbert space $\mathcal{H}$. We shall denote these one-particle states by $|\alpha\rangle$, $|\beta\rangle$, and so on. According to our previous work on composite systems, the state of two similar particles should be a vector in the tensor product space $\mathcal{H} \otimes \mathcal{H}$. The state might either be a product state $|\alpha, \beta\rangle$ or an entangled state.

If the particles are identical, though, difficulties arise. Suppose the one-particle states $|\alpha\rangle$ and $|\beta\rangle$ are orthogonal (and thus distinguishable). Under the usual tensor product rule, the two-particle system will have two distinguishable states $|\alpha, \beta\rangle$ and $|\beta, \alpha\rangle$. But for truly identical particles, these state vectors should actually correspond to *indistinguishable* physical situations.

The problem is that $\mathcal{H} \otimes \mathcal{H}$ is too large to describe the states of a pair of identical particles. For this reason, we take the physical Hilbert space for the pair to be a proper subspace of $\mathcal{H} \otimes \mathcal{H}$. But which subspace? To find the right one, we introduce the *particle exchange* operator $\mathbb{X}$, which acts on $\mathcal{H} \otimes \mathcal{H}$. On product states, this has the effect

$$\mathbb{X}|\alpha, \beta\rangle = |\beta, \alpha\rangle. \tag{14.29}$$

The particle exchange operator has several key properties.

Exercise 14.18 Show that the particle exchange $\mathbb{X}$ is both unitary and Hermitian, that $\mathbb{X}^2 = \mathbf{1}$, and that the only possible eigenvalues for $\mathbb{X}$ are ± 1.

If the particles are really identical, then the energy of the system is always unchanged if we exchange the particles. This means that $\mathbb{X}$ and the Hamiltonian H must commute:

Exercise 14.19 Suppose that, for all $|\psi\rangle$ in $\mathcal{H} \otimes \mathcal{H}$, $|\psi\rangle$ and $\mathbb{X}|\psi\rangle$ have exactly the same expectation value for H. Show that $[\mathbb{X}, H] = 0$. (You may wish to recall the useful facts from Problem 3.7.)

Thus $\mathbb{X}$ is a dynamical symmetry of the two-particle system. Considered as an observable, $\mathbb{X}$ is always exactly conserved.

We postulate that the physical Hilbert space for the two-particle system must be an eigenspace of the exchange operator $\mathbb{X}$. This is an attractive notion because this exchange property (the eigenvalue of $\mathbb{X}$) must remain constant over time. There are two possibilities:

- The eigenvalue may be $+1$, so that $\mathbb{X}|\psi\rangle = +|\psi\rangle$. We denote this Hilbert space by $\mathcal{H}_+^{[2]}$. In this case, we say that the particles are *Bose–Einstein particles*, or just *bosons*.
- The eigenvalue may be -1, so that $\mathbb{X}|\psi\rangle = -|\psi\rangle$. We denote this Hilbert space by $\mathcal{H}_-^{[2]}$. In this case, we say that the particles are *Fermi–Dirac particles*, or just *fermions*.

In quantum theory, there are two possible ways that particles can be identical. Both ways occur in nature. Electrons, protons, and neutrons are all fermions. Photons are bosons; entire atoms like ^{4}He can behave as bosons as well.

Suppose the particles are bosons. Then the physical Hilbert space $\mathcal{H}_+^{[2]}$ contains all product states of the form $|\phi, \phi\rangle$, in which both particles are in the same one-particle state. Given orthogonal one-particle states $|\alpha\rangle$ and $|\beta\rangle$, neither $|\alpha, \beta\rangle$ nor $|\beta, \alpha\rangle$ is present, but we do have the symmetric combination

$$\frac{1}{\sqrt{2}}\left(|\alpha, \beta\rangle + |\beta, \alpha\rangle\right). \tag{14.30}$$

Now suppose the particles are fermions. The physical Hilbert space $\mathcal{H}_-^{[2]}$ contains antisymmetric combinations like

$$\frac{1}{\sqrt{2}}\left(|\alpha, \beta\rangle - |\beta, \alpha\rangle\right). \tag{14.31}$$

However, $\mathcal{H}_-^{[2]}$ contains no state of the form $|\phi, \phi\rangle$, in which the two particles are in the same state. Hence, two identical fermions cannot be in the same quantum state. This is the famous *Pauli exclusion principle*.

Exercise 14.20 For any one-particle state $|\phi\rangle$ and any fermion state $|\Psi\rangle$ in $\mathcal{H}_-^{[2]}$, show that $\langle \phi, \phi | \Psi \rangle = 0$. Thus, the probability that a measurement finds two identical fermions in the same state is always zero. (How is this statement of the Pauli exclusion principle different from the one we just gave?)

Exercise 14.21 Prove that every state of two identical fermions is entangled.

Exercise 14.22 If $\dim \mathcal{H} = d$, find $\dim \mathcal{H}_+^{[2]}$ and $\mathcal{H}_-^{[2]}$.

Boson and fermion states are said to be *symmetric* and *antisymmetric* under particle exchange, respectively. What about systems of more than two identical particles? Again, the physical Hilbert space is a subspace of the tensor product space $\mathcal{H}^{\otimes N}$. Here the exchange symmetries are more complicated, since every pair of particles has an exchange operator: $\mathbb{X}_{12}$ exchanges particles #1 and #2, $\mathbb{X}_{23}$ exchanges #2 and #3, etc. Once again we postulate that there are just two types of exchange symmetry. For a collection of N identical bosons, the quantum state is an eigenstate of *all* of the exchange operators, with eigenvalue $+1$. This specifies the completely symmetric subspace $\mathcal{H}_+^{[N]}$. For a collection of identical fermions, the quantum state is an eigenstate of all of the exchange operators, with eigenvalue -1. The state resides in the completely antisymmetric subspace $\mathcal{H}_-^{[N]}$.

Exercise 14.23 Show that the Pauli exclusion principle must hold for any state of N identical fermions. Hint: If particles #m and #n are in the same state, then $\mathbb{X}_{mn}|\Psi\rangle = -|\Psi\rangle$.

To put it more generally, for a collection of N identical particles we have $N!$ *permutation operators* $\mathbb{P}$. Each permutation can be written as a product of the two-particle exchange operators $\mathbb{X}_{12}$, etc. It turns out that permutations fall into two distinct types: *even* permutations can only be written as products of an even number of pairwise exchanges, and *odd* permutations are always products of an odd number of exchanges.

Exercise 14.24 Write down all six permutations of the three letters ABC. (Do not forget the *identity* permutation ABC.) Which permutations are even and which are odd?

A system of bosons must be in an eigenstate of all of the particle permutation operators, each of them having eigenvalue $+1$. That is, $\mathbb{P}|\Psi\rangle = |\Psi\rangle$ for any $\mathbb{P}$. A system of fermions is also in an eigenstate of all of the permutations, but its eigenvalues are $+1$ for the even permutations and -1 for the odd. This may be written

$$\mathbb{P}|\Psi\rangle = (-1)^{\pi(\mathbb{P})}|\Psi\rangle, \tag{14.32}$$

where $\pi(\mathbb{P})$ is the *parity* of the permutation $\mathbb{P}$, and equals 0 for even and 1 for odd.

But how do we find such symmetric and antisymmetric states? Suppose we choose a basis $\{|a\rangle, |b\rangle, \ldots\}$ of one-particle states. Then we proceed as follows:

- Write down an N-particle product state $|a, b, \ldots\rangle$. For symmetric states, the same one-particle state may appear more than once; for antisymmetric states, all of the one-particle basis states must be distinct.
- Write down all $N!$ permutations of the product state. (For the symmetric case with a repeated basis state, some of these may be the same.)
- Construct a superposition of the permuted states. In the symmetric case, each term will appear with a $+$ sign. In the antisymmetric case, even permutation terms will appear with a $+$ sign and odd permutations will appear with a $-$ sign.
- Normalize the resulting state.

Symbolically, we write

$$|\mathbb{S}(a, b, \ldots)\rangle = \mathcal{N}\sum_{\mathbb{P}}\mathbb{P}|a, b, \ldots\rangle,$$

$$|\mathbb{A}(a, b, \ldots)\rangle = \mathcal{N}\sum_{\mathbb{P}}(-1)^{\pi(\mathbb{P})}\mathbb{P}|a, b, \ldots\rangle. \tag{14.33}$$

For the antisymmetric state $|\mathbb{A}(a, b, \ldots)\rangle$, all of the terms in the superposition are orthogonal to each other, so the normalization constant may be chosen to be $1/\sqrt{N!}$. In the symmetric case, the normalization may vary depending on whether any one-particle state is repeated.

It remains to show that the symmetric and antisymmetric states we have constructed satisfy the requirements of boson and fermion states. This is done in general in Problem 14.2. Here, we consider a few examples.

First, consider some antisymmetric states. For the one-particle states $|a\rangle$ and $|b\rangle$ for two particles, our procedure yields

$$| \mathbb{A}\,(a,b)\rangle = \frac{1}{\sqrt{2}} \left(|a,b\rangle - |b,a\rangle \right),\qquad (14.34)$$

which is a correct quantum state for two identical fermions, one in state $|a\rangle$ and one in $|b\rangle$ (see Eq. 14.31). For three particles, we need three one-particle states $|a\rangle$, $|b\rangle$, and $|c\rangle$. Then

$$| \mathbb{A}\,(a,b,c)\rangle = \frac{1}{\sqrt{6}} \Big(|a,b,c\rangle - |b,a,c\rangle + |b,c,a\rangle$$

$$- |c,b,a\rangle + |c,a,b\rangle - |a,c,b\rangle \Big).\qquad (14.35)$$

Exercise 14.25 Show that $| \mathbb{A}\,(a,b,c)\rangle$ is an eigenstate of any pairwise exchange of particles with eigenvalue -1. That is, $| \mathbb{A}\,(a,b,c)\rangle$ is a correct quantum state for three identical fermions in the states $|a\rangle$, $|b\rangle$, and $|c\rangle$.

For the symmetric states $| \mathbb{S}\,(a,b)\rangle$ and $| \mathbb{S}\,(a,b,c)\rangle$, we obtain similar expressions in which only $+$ signs appear.

Exercise 14.26 Show that $| \mathbb{S}\,(a,b)\rangle$ and $| \mathbb{S}\,(a,b,c)\rangle$ are correct quantum states for two or three identical bosons.

Exercise 14.27 Write down the state $| \mathbb{S}\,(a,a,b)\rangle$ and show that it is a correct boson state.

From state vectors, we can derive wave functions. Suppose two identical particles move in one dimension.[6] The particle exchange operator affects the position basis states in this way:

$$\mathbb{X}\,|x_1,x_2\rangle = \pm\,|x_2,x_1\rangle,\qquad (14.36)$$

with $+$ for bosons and $-$ for fermions. Thus, the wave function for the system will have the property

$$\Psi(x_1,x_2) = \pm\Psi(x_2,x_1).\qquad (14.37)$$

As before, we can construct symmetric and antisymmetric wave functions for systems of two or more bosons or fermions. The fermion case is particularly interesting. Suppose $\psi_1(x)$, $\psi_2(x)$, etc., are the one-particle wave functions for orthogonal states of a particle, and our system consists of N identical fermions. Then an antisymmetric wave function can be formally written as a matrix determinant:

$$\Psi(x_1,\ldots,x_N) = \frac{1}{\sqrt{N!}} \begin{vmatrix} \psi_1(x_1) & \cdots & \psi_N(x_1) \\ \vdots & \ddots & \vdots \\ \psi_1(x_N) & \cdots & \psi_N(x_N) \end{vmatrix}.\qquad (14.38)$$

---

[6] We are assuming for the moment that each particle is completely characterized by its position and has no other degrees of freedom such as spin.

This construction is called a *Slater determinant*. Exchange of two particles amounts to exchanging two rows in the matrix, which multiplies the determinant by -1, consistent with Eq. 14.37.

Exercise 14.28 Write a wave function for Eq. 14.35 as a Slater determinant.

14.4 Energy levels

Next we discuss the energy spectrum of a system composed of two or more non-interacting particles. If the particles are identical, then their exchange symmetry will have important effects on this spectrum.

Each particle has a Hamiltonian H. The stationary states (energy levels) are denoted by $|0\rangle, |1\rangle, \ldots$, with corresponding energies E_0, E_1, and so on. These energies may or may not be degenerate. For simplicity, we here assume that the Hilbert space $\mathcal{H}$ for one particle has finite dimension $\dim \mathcal{H} = d$; but many of our results also hold for an infinite-dimensional Hilbert space.

Suppose now that we have two such particles. Because the particles are non-interacting, the total Hamiltonian is $H^{(1)} + H^{(2)}$. The particles are similar enough so that the two Hamiltonians are the same, but are otherwise distinguishable. The energy level states of the pair can be taken to be products of the one-particle states: $|0,0\rangle, |0,1\rangle, |1,0\rangle, |1,1\rangle$, etc. The energy of the $|m,n\rangle$ state is just $E_{m,n} = E_m + E_n$. Some of these levels are necessarily degenerate; for instance, $E_{4,2} = E_{2,4}$. Other "accidental" degeneracies might also occur, as the following exercise shows:

Exercise 14.29 Consider two distinguishable similar particles. Each particle has three energy levels with energies $E_1 = 1$ eV, $E_2 = 2$ eV, and $E_3 = 3$ eV. Write down the nine energy levels for the composite system, together with their energies. Draw an energy level diagram for the system, showing all the degeneracies.

Exercise 14.30 In the previous exercise, suppose that the energy E_2 is increased by a slight amount, say 0.1 eV. Draw an energy level diagram to show how the various levels of the composite system change. How have the degeneracies changed?

If the particles are identical, then the energy levels are not quite the same. Identical boson energy levels are of the symmetric form $|\mathbb{S}(m,n)\rangle$, while those of identical fermions are the antisymmetric states $|\mathbb{A}(m,n)\rangle$. Each of these will have the energy $E_m + E_n$. However, the energy level diagram for each situation will be different. The boson diagram will have less degeneracy, since for $m \neq n$ there will be only one state, $|\mathbb{S}(m,n)\rangle$ rather than two degenerate states $|m,n\rangle$ and $|n,m\rangle$. The fermion diagram will have less degeneracy as well; also, certain energies will be missing entirely, since there are no states of the form $|\mathbb{A}(m,m)\rangle$.

Exercise 14.31 Suppose a single particle has four energy levels with energies of 1 eV, 2 eV, 4 eV, and 7 eV. Draw side-by-side energy level diagrams for (a) two similar but distinguishable particles, (b) two identical bosons, and (c) two identical fermions. How many energy level states are there in each case?

It is also straightforward to work out the energy levels for three or more particles.

Exercise 14.32 Given the same single-particle energy levels as the previous exercise, find all the possible energy levels (and their degeneracies) for three identical fermions.

The effect of exchange symmetry on the energy level structure of a multiparticle system can have far-reaching consequences. To get some inkling of this, consider the following toy model.[7] We imagine N particles of a type that has only two energy level states, which we can label by energy: $|0\rangle$ and $|\varepsilon\rangle$. The N-particle system is in equilibrium with a thermal reservoir at absolute temperature T. Thus, by Eq. 8.70, the density operator for the system is

$$\rho = \sum_s \frac{e^{-E_s/kT}}{\mathcal{Z}} |s\rangle\langle s|, \tag{14.39}$$

where k is Boltzmann's constant, the $|s\rangle$ states are the stationary states of the system, and $\mathcal{Z} = \sum_s e^{-E_s/kT}$. We wish to calculate the probability P_0 that the entire system will be found in its ground state – that is, that all N particles would be found in the state $|0\rangle$. This will be

$$P_0 = \mathrm{Tr}\,\rho\,|0\cdots 0\rangle\langle 0\cdots 0| = \frac{1}{\mathcal{Z}}. \tag{14.40}$$

The energy $E_s = n\varepsilon$, where n is the number of excited particles in the state $|s\rangle$. If the particles are distinguishable, the binomial coefficient $\binom{N}{n}$ gives the number of orthogonal energy eigenstates with this energy. This yields

$$\mathcal{Z} = \sum_{n=0}^{N} \binom{N}{n} e^{-n\varepsilon/kT} = \left(1 + e^{-\varepsilon/kT}\right)^N. \tag{14.41}$$

The probability that all particles are found in the ground state is thus

$$P_0 = \left(1 + e^{-\varepsilon/kT}\right)^{-N}. \tag{14.42}$$

In a macroscopic system, the number of particles N is extremely large. If we take $N \to \infty$, then we find that $P_0 \to 0$. The likelihood that all of the N particles will be found in the ground state is negligible.

---

[7] *Toy model* is a term for an example that is far too simplified to be useful for calculating real-world effects, but nevertheless does illustrate some feature of the basic physics of a situation.

On the other hand, suppose that the particles are bosons. Then the energy level states are symmetric states:

$$|\phi_0\rangle = |\mathbb{S}(0,0,\ldots,0)\rangle,$$

$$|\phi_1\rangle = |\mathbb{S}(\varepsilon,0,\ldots,0)\rangle,$$

$$\vdots$$

$$|\phi_N\rangle = |\mathbb{S}(\varepsilon,\ldots,\varepsilon)\rangle. \tag{14.43}$$

For any n between 0 and N, there is only one symmetric state $|\phi_n\rangle$ with energy $n\varepsilon$. The spectrum of the N boson system has no degeneracy at all. Thus, we find that

$$\mathcal{Z} = \sum_{n=0}^{N} e^{-n\varepsilon/kT} = \frac{1 - e^{-(N+1)\varepsilon/kT}}{1 - e^{-\varepsilon/kT}}, \tag{14.44}$$

and so

$$P_0 = \frac{1 - e^{-\varepsilon/kT}}{1 - e^{-(N+1)\varepsilon/kT}}. \tag{14.45}$$

In the macroscopic limit $N \to \infty$, we find that $P_0 \to 1 - e^{-\varepsilon/kT}$, which is not zero. Indeed, for $kT \ll \varepsilon$, this might be close to unity, and it would be quite likely to find all N identical bosons in exactly the same quantum state.

Although our model is too simplified to be taken very seriously, it does suggest the real phenomenon of *Bose condensation*, which can occur in systems of identical bosons at low temperatures. In Bose condensation, a macroscopically significant number of the particles are in fact in the same quantum state. The exchange symmetry of the particles can have a drastic effect on the system's low-temperature behavior.

Exercise 14.33 Why did we not consider a system of N two-state fermions?

14.5 Exchange effects

Suppose we have two particles (#1 and #2) moving in one dimension. Two particles occupy orthogonal quantum states $|a\rangle$ and $|b\rangle$ with wave functions $\psi_a(x)$ and $\psi_b(x)$. We will consider three possible states of the composite system: the product state $|a,b\rangle$ (appropriate for distinguishable particles) and the symmetric and antisymmetric states $|\mathbb{S}(a,b)\rangle$ and $|\mathbb{A}(a,b)\rangle$.

The particles have position observables X_1 and X_2, and the joint wave function $\Psi(x_1, x_2)$ yields a joint probability distribution over the particle positions.

Exercise 14.34 Evaluate the probability density $\mathcal{P}(x, x)$ where the two positions are equal for (a) the product state, (b) the symmetric state, and (c) the antisymmetric state. Summarize your results as a qualitative statement about the likelihood of finding the particles close together in the three cases.

As this exercise suggests, the symmetry or antisymmetry of the two-particle state influences how probable it is to find the particles near to one another.

Let us approach this question in a slightly different way. Let $s = x_2 - x_1$ be the relative position of the two particles. We wish to consider s^2, the particles' squared separation. This is

$$s^2 = x_1^2 + x_2^2 - 2x_1 x_2. \qquad (14.46)$$

The expectation $\langle s^2 \rangle \geq 0$. The larger the value of $\langle s^2 \rangle$, the more widely separated we expect to find the two particles. For the simple product state,

$$\langle s^2 \rangle = \langle a, b | s^2 | a, b \rangle$$

$$= \langle x^2 \rangle_a + \langle x^2 \rangle_b - 2 \langle x \rangle_a \langle x \rangle_b, \qquad (14.47)$$

where $\langle \cdots \rangle_a$ and $\langle \cdots \rangle_b$ denote single-particle expectations for states $|a\rangle$ and $|b\rangle$, respectively.

Exercise 14.35 Fill in the derivation of Eq. 14.47. Note that the single-particle expectations do not depend on which particle we are talking about, but only on its state. Thus $\langle x_1 \rangle_a = \langle x_2 \rangle_a = \langle x \rangle_a$ and so on.

We designate this expectation $\langle s^2 \rangle_{\text{dist}}$, where "dist" stands for "distinguishable."

How are things different if the particles are in a symmetric or an antisymmetric state? These two states will differ only by a sign; thus, we will combine our calculations to find $\langle s^2 \rangle_{\pm}$:

$$\langle s^2 \rangle_{\pm} = \frac{1}{2} \left(\langle a, b | \pm \langle b, a | \right) s^2 \left(|a, b\rangle \pm |b, a\rangle \right). \qquad (14.48)$$

When we substitute Eq. 14.46 and evaluate this expectation, we obtain all of the terms we had before plus a few more. Some of the additional terms are zero. For instance,

$$\langle a, b | x_1^2 | b, a \rangle = \langle a | x^2 | b \rangle \langle b | a \rangle = 0. \qquad (14.49)$$

Other terms are not necessarily zero,

$$\langle a, b | x_1 x_2 | b, a \rangle = \langle a | x | b \rangle \langle b | x | a \rangle = |x_{ab}|^2, \qquad (14.50)$$

where $x_{ab} = \langle a | x | b \rangle$, the off-diagonal matrix element for the single particle operator x. Putting everything together, we find

$$\langle s^2 \rangle_{\pm} = \langle s^2 \rangle_{\text{dist}} \mp |x_{ab}|^2. \qquad (14.51)$$

Exercise 14.36 Explain why the $\pm$ gets "inverted" to $\mp$ in Eq. 14.51.

Thus, $|x_{ab}|^2$ determines how large an effect the exchange symmetry of the two-particle state has on the separation of the particles. Particles in a symmetric state tend to be closer together, while those in an antisymmetric state tend to be farther apart. This tendency is sometimes called an *exchange force*, since the particles behave as if they were subject to a peculiar attraction (symmetric) or repulsion (antisymmetric), depending on their exchange properties. But no force is present, for the effect has nothing to do with dynamics. It is simply a consequence of the structure of the symmetric and antisymmetric two-particle Hilbert spaces $\mathcal{H}_\pm^{[2]}$.

In terms of the wave functions ψ_a and ψ_b for the two states, the matrix element x_{ab} is

$$x_{ab} = \int_{-\infty}^{\infty} x\, \psi_a^*(x)\, \psi_b(x)\, dx. \tag{14.52}$$

This has an interesting physical consequence. Suppose that ψ_a and ψ_b have little or no overlap – that is, the wave functions are significantly different from zero only in disjoint regions of space. (We might imagine two well-separated wave packets, for instance.) In this case the matrix element x_{ab} is zero, and there are no exchange effects in $\langle s^2 \rangle$. Exchange effects can only be important when the wave functions overlap significantly.

The "exchange force" is not a real force, but it can have an indirect effect on the energy of the system if the particles are interacting in some way. Suppose the Hamiltonian of the system includes an interaction potential term $U_{int}(x_1, x_2)$. If the particles repel one another, for example, then U_{int} is greater when the particles are closer together. Thus, we expect that $\langle U_{int} \rangle$ would be larger for a symmetric two-particle state and smaller for an antisymmetric one.

Exercise 14.37 For two particles moving in 3-D, we define the squared separation operator to be

$$s^2 = (x_2 - x_1)^2 + (y_2 - y_1)^2 + (z_2 - z_1)^2. \tag{14.53}$$

Derive a version of Eq. 14.51 for symmetric and antisymmetric states of two particles in 3-D.

We have been careful so far in this section to refer to "symmetric" and "antisymmetric" spatial wave functions, without mentioning "bosons" and "fermions." The reason is that real particles are often characterized by more than just their position degree of freedom. They may also have spin.

Consider electrons, which are spin-1/2 fermions. As we mentioned in Section 6.3, the Hilbert space for a single electron can be written

$$\mathcal{H} = \mathcal{H}_{space} \otimes \mathcal{Q}_{spin}. \tag{Re 6.28}$$

Any quantum state of a pair of electrons is an antisymmetric state in the Hilbert space[8]

$$\left(\mathcal{H}_{space}^{(1)} \otimes \mathcal{H}_{space}^{(2)} \right) \otimes \left(\mathcal{Q}_{spin}^{(1)} \otimes \mathcal{Q}_{spin}^{(2)} \right). \tag{14.54}$$

Suppose that the two-electron state is a product of spatial and spin degrees of freedom:

$$\left| \Psi^{(12)} \right\rangle = \left| \psi_{space}^{(12)} \right\rangle \otimes \left| \chi_{spin}^{(12)} \right\rangle. \tag{14.55}$$

---

[8] For convenience, we have rearranged and grouped the terms in this tensor product. This has no effect on the structure of the Hilbert space.

Then the electrons have both a spatial wave function (in 1-D) $\psi(x_1, x_2) = \langle x_1, x_2 \,|\, \psi_{\text{space}}^{(12)} \rangle$ and a spin state $\big| \chi_{\text{spin}}^{(12)} \big\rangle$. The electron exchange operator $\mathbb{X}$ exchanges both the spatial and spin degrees of freedom. That is,

$$\mathbb{X} = \mathbb{X}_{\text{space}} \otimes \mathbb{X}_{\text{spin}}. \qquad (14.56)$$

We know that $\mathbb{X} | \Psi^{(12)} \rangle = - | \Psi^{(12)} \rangle$. This can happen in two ways:

- The spin state $\big| \chi_{\text{spin}}^{(12)} \big\rangle$ is antisymmetric under $\mathbb{X}_{\text{spin}}$, but the spatial wave function $\psi(x_1, x_2)$ is symmetric.
- The spin state $\big| \chi_{\text{spin}}^{(12)} \big\rangle$ is symmetric under $\mathbb{X}_{\text{spin}}$, but the spatial wave function $\psi(x_1, x_2)$ is antisymmetric.

The spatial wave function of a pair of electrons can be either symmetric or antisymmetric, depending on the spin state. The spin state symmetry is related to the total angular momentum:

Exercise 14.38 Review Section 12.4 on the addition of spins. Show that:

(a) $\mathbb{X}_{\text{spin}} \big| \chi_{\text{spin}}^{(12)} \big\rangle = - \big| \chi_{\text{spin}}^{(12)} \big\rangle$ if and only if the pair of spins is in a spin-0 ("singlet") state.
(b) $\mathbb{X}_{\text{spin}} \big| \chi_{\text{spin}}^{(12)} \big\rangle = + \big| \chi_{\text{spin}}^{(12)} \big\rangle$ if and only if the pair of spins is in a spin-1 ("triplet") state.

If an electron pair has total spin 0, we may say informally that the electrons have "opposite" spins. In this case, their spatial wave function is symmetric – and indeed, the two electrons might be in the same spatial state. If the spins are not opposite, so that the total spin is 1, then the spatial wave function must be antisymmetric. The exchange effect in Eq. 14.51 is thus spin-dependent.

Finally, notice that nothing in our derivation of Eq. 14.51 relied on the fact that x_1 and x_2 were position observables. Particles tend to be "clumped together" in a symmetric state and "spread apart" in an antisymmetric state, regardless of the variable we are considering.

Exercise 14.39 For a pair of spin-1/2 systems, consider three states: $|\uparrow\downarrow\rangle$, the symmetric (spin 1) combination $\frac{1}{\sqrt{2}}(|\uparrow\downarrow\rangle + |\downarrow\uparrow\rangle)$, and the antisymmetric (spin 0) combination $\frac{1}{\sqrt{2}}(|\uparrow\downarrow\rangle - |\downarrow\uparrow\rangle)$. We are interested how far apart the x-components of the spins are likely to be. Use Eq. 14.51 to find $\langle (S_x^{(1)} - S_x^{(2)})^2 \rangle$ for each state.

14.6 Occupation numbers

Consider a composite quantum system composed of two subsystems, each described by the Hilbert space $\mathcal{H}$. According to Section 6.1, the joint system must be described by the tensor product space $\mathcal{H} \otimes \mathcal{H}$. As we have seen in this chapter, however, this rule does not hold for a pair of identical particles. The states of such a pair are restricted to lie in the symmetric or antisymmetric subspace $\mathcal{H}_{\pm}^{[2]}$ of $\mathcal{H} \otimes \mathcal{H}$. The unrestricted tensor product rule is thus not valid for systems containing identical particles.

In this section, we sketch a very different point of view. We will argue that the symmetric and antisymmetric states introduced so far, though useful for calculations, disguise a more fundamental point. The unrestricted tensor product rule is in fact always valid in quantum mechanics, even for systems containing identical particles. The essential difficulty is that *particles are not subsystems*.

First, we introduce the idea of the *direct sum* of two or more Hilbert spaces. Given $\mathcal{H}_1$ and $\mathcal{H}_2$, then the direct sum $\mathcal{H}_1 \oplus \mathcal{H}_2$ is the smallest Hilbert space in which $\mathcal{H}_1$ and $\mathcal{H}_2$ are orthogonal subspaces.[9] Any vector in the direct sum can be written

$$|\psi\rangle = |\psi_1\rangle + |\psi_2\rangle, \tag{14.57}$$

for $|\psi_1\rangle$ in $\mathcal{H}_1$ and $|\psi_2\rangle$ in $\mathcal{H}_2$.

Exercise 14.40 (a) Carefully explain the mathematical difference between $\mathcal{H}_1 \otimes \mathcal{H}_2$ and $\mathcal{H}_1 \oplus \mathcal{H}_2$. (b) How are these different as Hilbert spaces of quantum systems? (c) If $\dim \mathcal{H}_1 = d_1$ and $\dim \mathcal{H}_1 = d_2$, find the dimension of each larger space.

The direct sum Hilbert space has a basis $\{ |j_1\rangle, |k_2\rangle \}$, including all the elements of the $\{ |j_1\rangle \}$ basis for $\mathcal{H}_1$ and the $\{ |k_2\rangle \}$ basis for $\mathcal{H}_2$.

Now consider a system of identical fermions, each described by the space $\mathcal{H}$ with dimension d. For n such fermions, we have described the system by the antisymmetric subspace $\mathcal{H}_{-}^{[n]}$ of $\mathcal{H}^{\otimes n}$. We can also consider the possibility that there are no fermions at all, in which case there can only be a single possible physical state, the *vacuum* state $|\text{vac}\rangle$. We summarize all this in the following table:

0 particles	$\mathcal{H}_{-}^{[0]}$, which has $\dim \mathcal{H}_{-}^{[0]} = 1$
1 particle	$\mathcal{H}_{-}^{[1]} = \mathcal{H}$
2 particles	$\mathcal{H}_{-}^{[2]}$, a subspace of $\mathcal{H} \otimes \mathcal{H}$
3 particles	$\mathcal{H}_{-}^{[3]}$, a subspace of $\mathcal{H} \otimes \mathcal{H} \otimes \mathcal{H}$

and so on. Since $\dim \mathcal{H} = d$, we can have no more than d identical fermions and still respect the Pauli exclusion principle. The Hilbert space that encompasses all of these possibilities – and also superpositions of them (i.e. states of *indeterminate* particle number) – would be the direct sum

$$\mathcal{H}_{-}^{[0]} \oplus \mathcal{H}_{-}^{[1]} \oplus \mathcal{H}_{-}^{[2]} \oplus \cdots \oplus \mathcal{H}_{-}^{[d]} = \bigoplus_{n=0}^{d} \mathcal{H}_{-}^{[n]}. \tag{14.58}$$

The various subspaces $\mathcal{H}_{-}^{[n]}$ are the eigenspaces of a "total particle number" observable N with eigenvalues n.

Exercise 14.41 Work out the dimension of this Hilbert space if $d = \dim \mathcal{H}$ has the values 2, 3, or 4. Conjecture a general formula for arbitrary d. (See Problem 14.5.)

---

[9] In the language of Problem 3.2, $\mathcal{H}_1 = (\mathcal{H}_2)^{\perp}$ in this larger space.

Given a basis $\{|k\rangle\}$ of one-particle states in $\mathcal{H}$, we can write down a basis for the whole space:

$$\{|\mathrm{vac}\rangle, |k\rangle, |\mathbb{A}(j,k)\rangle, |\mathbb{A}(j,k,l)\rangle, \ldots\}. \tag{14.59}$$

We can label these basis states in another way. The basis state $|\mathbb{A}(j,k,\ldots)\rangle$ is identified by specifying which states $j, k, \ldots$ have a particle in them; from this we can construct the appropriate antisymmetric combination of product states. Thus, we could equally well specify the state simply by listing all of the single particle basis states and specifying how many particles (0 or 1) are present in them. This idea is the kernel of the *occupation number representation* for a system of fermions. It works like this:

- Find a basis $\{|k\rangle\}$ for one-particle states of the system.
- For each k, introduce a qubit system $\mathcal{Q}^{(k)}$. The operator $\mathsf{n}^{(k)} = |1^{(k)}\rangle\langle 1^{(k)}|$ is the *occupation number observable* for the state k. It has eigenvalues 0 (no particle in state k) and 1 (one particle in state k).
- The Hilbert space for the entire system is the tensor product $\mathcal{H} = \bigotimes_k \mathcal{Q}^{(k)}$.

In other words, the d one-particle basis states have become d independent subsystems, each one a qubit that stores the information about whether or not the state is occupied. The correspondence between our previous antisymmetric basis states and the new occupation number states is

$$|\mathbb{A}(j,k,\ldots)\rangle \leftrightarrow \left|0^{(1)}, \ldots, 1^{(j)}, \ldots, 1^{(k)}, \ldots, 0^{(d)}\right\rangle. \tag{14.60}$$

Exercise 14.42 Explain why this is a one-to-one correspondence – that is, why there is exactly one occupation number basis state for each antisymmetric basis state, and vice versa.

Arbitrary superpositions of these basis states are also possible.

The total particle number operator is $\mathsf{N} = \sum_k \mathsf{n}^{(k)}$. If we know that we have a given number N of particles, then the system's state is restricted to the eigenspace of N with eigenvalue N. This subspace is the same as the antisymmetric space $\mathcal{H}_-^{[N]}$. If total particle number N is conserved (as often happens), then we may be able simply to restrict our attention to one such subspace. But the occupation number approach is also applicable to situations in which particles are created or destroyed.

For identical bosons, things work out in a similar way, with the following difference. The number of particles occupying a given one-particle state can be 0, 1, 2, ... The independent subsystems assigned to the one-particle states will not be qubits, but the ladder systems of Chapter 13. The procedure is thus:

- Find a basis $\{|k\rangle\}$ for one-particle states of the system.
- For each k, introduce a ladder system $\mathcal{J}^{(k)}$. The operator $\mathsf{n}_k = \mathsf{a}_k^\dagger \mathsf{a}_k$ is the *occupation number observable* for the state k. It has eigenvalues 0 (no particle in state k) and 1 (one particle in state k).
- The Hilbert space for the entire system is the tensor product $\bigotimes_k \mathcal{J}^{(k)}$.

Again, there is a correspondence between symmetric basis states and the new occupation number states. Any given $|\mathbb{S}(j, k, \ldots)\rangle$ can be uniquely identified by how many particles are in each basis state.

To see these ideas in action, recall the two-beam interferometer of Section 2.1. Photons are identical bosons. The one-photon basis states $|u\rangle$ and $|l\rangle$ (for the photon in the upper or lower beams, respectively) are now identified with two distinct ladder systems. The occupation number state $|m, n\rangle$ describes a situation in which there are m photons in the upper beam and n photons in the lower beam. So far, we have considered the states $|u\rangle = |1, 0\rangle$ and $|l\rangle = |0, 1\rangle$ – and, in Section 4.3, the "no photon" state $|0\rangle = |0, 0\rangle$. Now we can describe states containing two, three or more photons.

Each mode has its own photon annihilation and creation operators. We denote these by a and $a^\dagger$ for the upper beam and b and $b^\dagger$ for the lower beam. Then

$$a\,|m, n\rangle = \sqrt{m}\;|m-1, n\rangle\,, \quad a^\dagger\,|m, n\rangle = \sqrt{m+1}\;|m+1, n\rangle\,,$$

$$b\,|m, n\rangle = \sqrt{n}\;|m, n-1\rangle\,, \quad b^\dagger\,|m, n\rangle = \sqrt{n+1}\;|m, n+1\rangle\,. \tag{14.61}$$

Exercise 14.43 Write down all of the commutation relations between the operators a, $a^\dagger$, b, and $b^\dagger$.

The total photon number operator is $N = a^\dagger a + b^\dagger b$, the sum of the number operators for the two modes.

Exercise 14.44 Show that $N\,|m, n\rangle = (m + n)\,|m, n\rangle$.

Photon number states can be constructed from the vacuum state by the application of photon creation operators:

$$|m, n\rangle = \frac{1}{\sqrt{m!\,n!}}\left(a^\dagger\right)^m \left(b^\dagger\right)^n |0, 0\rangle\,. \tag{14.62}$$

Any quantum state of the interferometer system is a superposition of these basis states. Thus, any quantum state of the system can be written in the form $|\text{state}\rangle = (\text{operator})\,|\text{vacuum}\rangle$.

If the system evolves according to the unitary operator U, then the photon number state $|m, n\rangle$ evolves to

$$U\,|m, n\rangle = \frac{1}{\sqrt{m!\,n!}}\,U\left(a^\dagger\right)^m \left(b^\dagger\right)^n |0, 0\rangle$$

$$= \frac{1}{\sqrt{m!\,n!}}\left(U a^\dagger U^\dagger\right)^m \left(U b^\dagger U^\dagger\right)^n U\,|0, 0\rangle\,. \tag{14.63}$$

Exercise 14.45 Where did all these extra U and $U^\dagger$ operators come from?

Thus, in order to specify how U acts on the basis state $|m, n\rangle$, we need to specify:

- $U\,|0, 0\rangle$, the effect of U on the vacuum state, and
- $U a^\dagger U^\dagger$ and $U b^\dagger U^\dagger$, the effect of U on the two photon creation operators.

Recall that the interferometer contains only linear optical devices. The unitary U describing such a device has the following properties. First, the vacuum state is unaffected by the device, so that $U|0,0\rangle = |0,0\rangle$. (This makes sense because the devices do not emit photons.) The second property is that U transforms each photon creation operator into a *linear combination* of both creation operators. That is,

$$U a^\dagger U^\dagger = w\, a^\dagger + x\, b^\dagger,$$
$$U b^\dagger U^\dagger = y\, a^\dagger + z\, b^\dagger, \qquad\qquad (14.64)$$

where w, x, y, and z are complex scalars. Formally, we may write this as

$$U \begin{pmatrix} a^\dagger \\ b^\dagger \end{pmatrix} U^\dagger = \begin{pmatrix} w & x \\ y & z \end{pmatrix} \begin{pmatrix} a^\dagger \\ b^\dagger \end{pmatrix}. \qquad\qquad (14.65)$$

We can relate this to the analysis in Section 2.1 by considering how U affects single-photon states. We find that

$$U|1,0\rangle = U a^\dagger U^\dagger U |0,0\rangle$$
$$= (w a^\dagger + x b^\dagger)|0,0\rangle = w|1,0\rangle + x|0,1\rangle. \qquad\qquad (14.66)$$

In a similar way, $U|0,1\rangle = y|1,0\rangle + z|0,1\rangle$.

Exercise 14.46 Suppose that the initial state of the interferometer is $|\psi\rangle = \alpha|1,0\rangle + \beta|0,1\rangle$, which evolves into $U|\psi\rangle = \alpha'|1,0\rangle + \beta'|0,1\rangle$. Show that

$$\begin{pmatrix} \alpha \\ \beta \end{pmatrix} \longrightarrow \begin{pmatrix} \alpha' \\ \beta' \end{pmatrix} = \begin{pmatrix} w & y \\ x & z \end{pmatrix} \begin{pmatrix} \alpha \\ \beta \end{pmatrix}. \qquad\qquad \text{(Re 2.7)}$$

Notice that we have met this exact equation before in our analysis of the single-photon interferometer!

Exercise 14.47 Show that unitarity of U implies that $\begin{pmatrix} w & x \\ y & z \end{pmatrix}$ is a unitary matrix.

When we apply these ideas to multiphoton states, we arrive at some fascinating results. Suppose our linear device is a balanced beam-splitter B_l, described by the coefficient matrix in Eq. 2.10. If we send in an input state $|1,1\rangle$, what output state emerges?

Exercise 14.48 Show that

$$B_l |1,1\rangle = \frac{1}{\sqrt{2}}\left(|2,0\rangle - |0,2\rangle\right). \qquad\qquad (14.67)$$

In other words, if one photon goes into the beamsplitter along each beam, then the output photons will always emerge together.[10]

Further developments are found in the problems at the end of this chapter.

Consider a system of two electrons, designated #1 and #2. If we describe its state by an antisymmetric state vector, we are faced with a conceptual puzzle. No product state is antisymmetric; thus, it appears that every physical state of the electron pair must

---

[10] This example is the simplest instance of an observable quantum effect that is called *photon bunching*.

be entangled. Since any electron anywhere can be considered one member of a pair of electrons, it appears that no electron could ever be found in a pure state.

Yet such a conclusion seems incorrect. If a single electron resides in a lab in Cambridge, surely it may be prepared in a pure state. The mere existence of a second electron in a distant lab – in Pasadena, say – should have no effect on this fact. Where does our argument go wrong?

To resolve this puzzle, let us first consider the Hilbert space of a single electron that might be in *either* Cambridge or Pasadena. This electron's state lies in a direct sum Hilbert space $\mathcal{H}^{(1)}_{\text{Cam}} \oplus \mathcal{H}^{(1)}_{\text{Pas}}$. This has a basis $\{ |k_{\text{Cam}}\rangle , |n_{\text{Pas}}\rangle \}$ that includes all the elements of the $\{ |k_{\text{Cam}}\rangle \}$ basis for $\mathcal{H}^{(1)}_{\text{Cam}}$ and the $\{ |n_{\text{Pas}}\rangle \}$ basis for $\mathcal{H}^{(1)}_{\text{Pas}}$.

Now consider any number of electrons. In the occupation number representation, we identify one-particle basis states with distinct subsystems, each described by a qubit Hilbert space $\mathcal{Q}$. The overall Hilbert space for any number of electrons in either Cambridge or Pasadena would then be

$$\mathcal{H} = \left(\bigotimes_{k_{\text{Cam}}} \mathcal{Q}_{k_{\text{Cam}}} \right) \otimes \left(\bigotimes_{n_{\text{Pas}}} \mathcal{Q}_{n_{\text{Pas}}} \right) = \mathcal{H}_{\text{Cam}} \otimes \mathcal{H}_{\text{Pas}}. \qquad (14.68)$$

In the occupation number representation, therefore, *the two laboratories are the distinct subsystems*, described by $\mathcal{H}_{\text{Cam}}$ and $\mathcal{H}_{\text{Pas}}$. The single-electron subspaces mentioned above may be regarded as $N = 1$ subspaces of these two.

Exercise 14.49 Explain how $\mathcal{H}^{(1)}_{\text{Cam}}$ can be regarded as a subspace of $\mathcal{H}_{\text{Cam}}$. To do this, write each of the basis vectors $\{ |k_{\text{Cam}}\rangle \}$ for $\mathcal{H}^{(1)}_{\text{Cam}}$ in the appropriate occupation number representation.

A situation with one electron in each laboratory can therefore be described as an element of the tensor product space $\mathcal{H}^{(1)}_{\text{Cam}} \otimes \mathcal{H}^{(1)}_{\text{Pas}}$. Product states are perfectly possible.

In other words, we have "permission" to think of two identical particles as distinct subsystems, provided that there is some physical observable (location, for instance) that can be used as an effective label. We choose a single-particle basis to be compatible with this label observable. In the occupation number representation, the different label values then correspond to distinct subsystems, with a tensor product Hilbert space.

Let us return to our original puzzle, in which identical electrons #1 and #2 are represented by an antisymmetric state. It is true that every antisymmetric state of the two electrons must be entangled; but this does not mean that any correlations are observable. The subsystem labels #1 and #2 do not correspond to any physical difference between the particles, and so we cannot construct measuring devices that observe them individually. No "correlation" between #1 and #2 is directly observable.

But the labs in Cambridge and Pasadena can examine their electrons individually. This fact is more clearly expressed in the occupation number representation, which is mathematically equivalent to the antisymmetric state approach. In this representation, measurement results in the two labs may or may not be correlated, depending on the quantum state.

Exercise 14.50 Are exchange effects (as described in Section 14.5) observable for the electrons in Cambridge and Pasadena?

Exercise 14.51 We have framed our discussion for identical fermions in the two laboratories. What, if anything, would work out differently if the particles were bosons?

Problems

Problem 14.1 From the Hamiltonian in Equation 14.8, derive a two-particle version of Ehrenfest's theorem:

$$\frac{d}{dt}\langle x_1\rangle = \frac{\langle p_1\rangle}{\mu_1}, \qquad \frac{d}{dt}\langle x_2\rangle = \frac{\langle p_2\rangle}{\mu_2},$$
$$\frac{d}{dt}\langle p_1\rangle = \left\langle -\frac{\partial U}{\partial x_1}\right\rangle, \qquad \frac{d}{dt}\langle p_2\rangle = \left\langle -\frac{\partial U}{\partial x_2}\right\rangle. \tag{14.69}$$

(See Eq. 11.9 and 11.11.)

Problem 14.2
(a) Suppose $\{\mathbb{P}\}$ is the set of all possible permutations of N particles. Suppose we multiply each permutation by the particle exchange $\mathbb{X}_{mn}$. Show that the resulting set $\{\mathbb{X}_{mn}\mathbb{P}\}$ is still the set of all possible permutations of the N particles.
(b) Show that the symmetric and antisymmetric states defined in Eq. 14.33 satisfy the postulates for boson and fermion states of N particles.

Problem 14.3 Two non-interacting particles move in the 1-D box of length L described in Section 11.4. The particles are in the $n = 1$ and $n = 2$ stationary states. Calculate the expected square separation $\langle s^2\rangle$ of the two particles if they are in (a) a product state, (b) a symmetric state, and (c) an antisymmetric state. Your answers should all be of the form cL^2, where c is a numerical factor.

Problem 14.4 Show that

$$\mathcal{H}_A \otimes (\mathcal{H}_B \oplus \mathcal{H}_C) = (\mathcal{H}_A \otimes \mathcal{H}_B) \oplus (\mathcal{H}_A \otimes \mathcal{H}_C). \tag{14.70}$$

That is, tensor products distribute over direct sums of Hilbert spaces. (Mathematicians may wish to ponder what the "=" sign means!)

Problem 14.5 Suppose the one-fermion Hilbert space has dimension $\dim \mathcal{H} = d$ and we have n identical fermions.

(a) Show that $\dim \mathcal{H}_{-}^{[n]} = \binom{d}{n}$, the binomial coefficient.

(b) Show that $\dim \left(\bigoplus_{n=0}^{d} \mathcal{H}_{-}^{[n]} \right) = 2^d$. How does this fact relate to the occupation number representation for the system?

Problem 14.6 Consider a system of identical bosons. The one-particle Hilbert space has $\dim \mathcal{H} = 3$, with basis states $|a\rangle$, $|b\rangle$, and $|c\rangle$. The occupation number basis states can be written $|n_a, n_b, n_c\rangle$. Write down the symmetric basis state forms of the occupation number basis states $|1,0,0\rangle$, $|2,0,0\rangle$, $|1,0,1\rangle$, $|1,1,1\rangle$, and $|2,1,0\rangle$. What about $|0,0,0\rangle$?

Problem 14.7 Consider a two-beam interferometer for fermions. The ladder operators for the qubit Hilbert space Q are $a = |0\rangle\langle 1|$ and $a^\dagger = |1\rangle\langle 0|$. A linear element of an interferometer alters these operators just as in Eq. 14.64. Suppose the two-fermion state $|1, 1\rangle$ is acted upon by a balanced fermion beamsplitter. What happens?

Problem 14.8 Quasi-classical light. Photon number eigenstates are quite difficult to produce in the laboratory. The light produced by a laser is actually in a coherent state $|z\rangle$ of that beam. Suppose a two-beam interferometer has an input state $|z_u, z_l\rangle$, a product of two coherent beam states. Show that any linear optical element will yield an output state that is also a product of two coherent beam states. Show also that the complex coherent state parameters z can exhibit constructive and destructive interference in an interferometer. (Hint: Represent the coherent states using Eq. 13.46.)

Stationary states in 1-D

15.1 Wave functions and potentials

In this chapter we will continue our discussion of the dynamics of a particle of mass μ moving in one dimension, which we began back in Chapters 10 and 11. Our focus here will be on the stationary states (energy eigenstates) and associated energies of such a particle. As we have seen, a knowledge of the stationary states and their energies is enough to determine the time evolution of an arbitrary, non-stationary state as well.

We have already studied stationary states of 1-D particles in some important special cases:

- In Chapter 11 we found the energy eigenstates for free particles. These are "plane waves" with any $E \geq 0$ for a particle that moves along an unbounded line. When there are special boundary conditions, the situation may change. For example, if the particle is constrained to move between impenetrable walls separated by a distance L, only discrete energies are possible

$$E_n^{(\text{box})} = \left(\frac{\pi^2 \hbar^2}{2\mu L^2} \right) n^2, \tag{15.1}$$

for $n = 1, 2, \ldots$ (see Eq. 11.62).

- In Chapter 13 we saw that a particle moving in a harmonic oscillator potential $U(x) = \mu \omega^2 x^2 / 2$ has an infinite ladder of stationary states, with

$$E_n^{(\text{osc})} = \left(n + \frac{1}{2} \right) \hbar \omega, \tag{15.2}$$

for $n = 0, 1, 2, \ldots$ (see Eq. 13.23).

These two special examples are well worth keeping in mind.

The purpose of this chapter will be to study the energies and stationary states of more general 1-D systems. Each of the examples just given involves particles that are spatially "trapped" (by impenetrable walls or the harmonic oscillator force) and have a discrete energy spectrum. As we will see, other types of energy eigenstate are possible in the more general case.

The energy eigenvalue equation for a particle moving in one dimension is

$$\mathsf{H} |\psi\rangle = \left(\frac{1}{2\mu} \mathsf{p}^2 + U(\mathsf{x}) \right) |\psi\rangle = E |\psi\rangle. \tag{15.3}$$

As we have seen, this may be written in terms of the wave function $\psi(x)$ for the energy eigenstate $|\psi\rangle$ as follows:

$$-\frac{\hbar^2}{2\mu}\frac{d^2}{dx^2}\psi(x) + U(x)\psi(x) = E\psi(x). \qquad \text{(Re 11.17)}$$

Thus $\psi(x)$ satisfies

$$\psi''(x) = -\frac{2\mu}{\hbar^2}(E - U(x))\psi(x). \qquad (15.4)$$

Typically, we know the potential $U(x)$ and wish to find the various possible eigenvalues of the energy E together with the corresponding wave functions $\psi(x)$.

In Section 10.1, we required the wave function $\psi(x)$ to be continuous. For energy eigenstates, Eq. 15.4 gives us some additional restrictions. For instance, if the potential function $U(x)$ is everywhere continuous, then the second derivative $\psi''(x)$ must also exist and be continuous for all x. This obviously implies that the first derivative $\psi'(x)$ also exists and is continuous.

What if the potential $U(x)$ is *not* continuous? Then we note that, for a reference point x_0,

$$\psi'(x) = \psi'(x_0) - \frac{2\mu}{\hbar^2}\int_{x_0}^{x}(E - U(x'))\psi(x')\,dx'. \qquad (15.5)$$

The potential $U(x)$ might be discontinuous at some value of x but still bounded – for instance, there might be a "step" discontinuity in $U(x)$ there. In this case, the integral expression above will still be continuous at x. We draw the following general conclusions about the wave function $\psi(x)$ for an energy eigenstate:

- $\psi(x)$ is continuous for all x.
- If the potential $U(x)$ is bounded in a region, then the wave function has a continuous derivative $\psi'(x)$ in that region.
- If the potential is continuous in a region, then the second derivative $\psi''(x)$ of the wave function must also be continuous.

If the potential has a vertical asymptote or actually "becomes infinite" at some point x, then the derivative $\psi'(x)$ may be discontinuous there.

Exercise 15.1 Suppose $U(x) = -\epsilon\,\delta(x)$, which is singular at $x = 0$. For an energy eigenstate find the difference between

$$\lim_{x\to 0^-}\psi'(x) \quad \text{and} \quad \lim_{x\to 0^+}\psi'(x),$$

in terms of ϵ, $\hbar$, μ, and $\psi(0)$.

Except for idealized limiting cases like this one, we will suppose that the potential $U(x)$ is finite everywhere.

Equation 15.4 is a second-order ordinary differential equation. For a given E, we can uniquely determine a particular solution to this equation by choosing a point x_0 and specifying the values of $\psi(x_0)$ and $\psi'(x_0)$ there. This has an interesting consequence. Can we ever have a stationary state wave function with both $\psi(x_0) = 0$ and $\psi'(x_0) = 0$ at some point? No, because these data uniquely specify the mathematical solution $\psi(x) = 0$ to Eq. 15.4, which is not a physical wave function.

What energies E are possible for a given potential function? Let us suppose that the potential is bounded below, so that $U(x) \geq U_{min}$. Then for any state $|\psi\rangle$, the expectation value of the energy is

$$\langle E\rangle = \langle\psi|\,H\,|\psi\rangle$$

$$= \frac{1}{2\mu}\,\langle\psi|\,p^2\,|\psi\rangle + \langle\psi|\,U(x)\,|\psi\rangle. \tag{15.6}$$

The "kinetic energy" term $\dfrac{1}{2\mu}\,\langle\psi|\,p^2\,|\psi\rangle \geq 0$. That means that

$$\langle E\rangle \quad \geq \quad \langle\psi|\,U(x)\,|\psi\rangle = \int_{-\infty}^{+\infty}\psi^*(x)U(x)\psi(x)\,dx$$

$$= \int_{-\infty}^{+\infty}\mathcal{P}(x)U(x)\,dx, \tag{15.7}$$

where $\mathcal{P}(x) = |\psi(x)|^2$, the probability density over the position x. Let U_{min} be the minimum value of the potential function. Since $U(x) \geq U_{min}$, the expectation $\langle E\rangle$ must satisfy

$$\langle E\rangle \geq U_{min}. \tag{15.8}$$

If the state $|\psi\rangle$ is a stationary state, then its energy $E \geq U_{min}$. Thus, U_{min} is a lower bound for the spectrum of possible energy values of the system.

Exercise 15.2 Show that this result also holds for a classical particle moving in one dimension.

Solutions or states?

Yet there is a difficulty here. Look back at Eq. 15.4. This is a second-order linear ordinary differential equation. We can find a mathematical solution $\psi(x)$ for this equation for *any* value E of the energy, including values for which $E < U_{min}$. How does this square with our conclusion that $E \geq U_{min}$?

Consider once again a free particle, for which $U(x) = 0$ everywhere. We found that the stationary states of a free particle are momentum eigenstates $|p\rangle$. For these,

$$H\,|p\rangle = \frac{p^2}{2\mu}\,|p\rangle, \tag{15.9}$$

so that $E = \dfrac{p^2}{2\mu} \geq 0$.

The wave function for a $|p\rangle$ state is $\psi(x) = (2\pi\hbar)^{-1/2}\,e^{ipx/\hbar}$, a plane wave. This wave function is not properly normalized, however, reflecting the fact that $|p\rangle$ is not a proper physical state in the Hilbert space $\mathcal{H}$.

Nevertheless, there are physical states represented by wave packets $\phi(x)e^{ipx/\hbar}$ (as described in Section 10.3) that are *approximate* momentum and energy eigenstates.

Exercise 15.3 For a free particle wave packet $\phi(x)e^{ipx/\hbar}$ with a real envelope function $\phi(x)$, show that:

(a) $\langle p \rangle = p$.

(b) $\langle E \rangle = \dfrac{p^2}{2\mu} + \dfrac{\langle \Delta p^2 \rangle}{2\mu}$.

(c) If the wave packet is Gaussian with variance σ^2, as in Eq. 10.64, that $\Delta p \to 0$ and $\Delta E \to 0$ as $\sigma^2 \to \infty$.

For a very wide, flat envelope function $\phi(x)$, the wave function closely resembles a plane wave over a wide region, and Exercise 15.3 shows that this leads to a physical state that has *almost* definite values for momentum and energy. Even though the improper eigenstates $|p\rangle$ are not physical states, we can use them as a convenient shorthand for these very wide wave packet states.[1]

Now suppose we try to construct a state with negative energy. The solutions to Eq. 15.4 with $E < 0$ are of the form e^{bx} and e^{-bx}, where b is a real parameter such that $b^2 = -2\mu E/\hbar^2$. Like the plane wave solutions, these exponential functions cannot be normalized, and so they cannot represent physical states. On the other hand, inspired by the wave packet analysis above, we might introduce an envelope function $\phi(x)$. Does this lead us to an *approximate* eigenstate with negative energy?

In fact, it does not. The function $\phi(x)e^{\pm bx}$ is a normalized wave function, so that Eq. 15.8 applies; thus $\langle E \rangle \geq 0$ no matter how wide and flat the envelope function might be. Therefore, the negative-energy exponential solutions $e^{\pm bx}$ cannot represent a limit of physically meaningful quantum states.

To sum up, solutions to the differential equation in Eq. 15.4 come in three varieties:

- Some solutions represent normalized energy eigenstates.
- Some solutions, like the plane wave solutions for a free particle, are not normalizable, but can be approximated by normalized wave functions.
- Some solutions, like the negative energy exponential solutions for the free particle, can be neither normalized nor approximated by normalized wave functions.

In either of the first two cases, we accept the solution together with its corresponding energy value E as a physically meaningful situation. These are our "stationary states." Solutions of the third type have no physical meaning, and must be discarded even as idealized descriptions of a physical state.

The interaction zone

Suppose a particle moves in one dimension and interacts with its surroundings. The particle remains informationally isolated, so that no "record" is created in the environment of the

---

[1] This is much like we use Dirac's delta function – as a notational shortcut that stands for a sophisticated limiting argument.

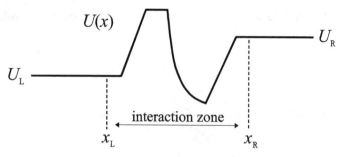

Fig. 15.1   The potential $U(x)$ is constant on each side of a bounded interaction zone.

particle's position. However, the interaction does determine the potential function $U(x)$ and the consequent forces experienced by the particle.

For the next few sections (through Section 15.5), we will consider the situation in which these forces are localized in space. We will assume that the potential $U(x)$ is constant everywhere except within a region called the *interaction zone* lying between x_L and x_R. This is shown in Fig. 15.1. To the left of the interaction zone (for $x < x_L$) the potential $U(x) = U_L$, and to the right of it (for $x > x_R$) the potential $U(x) = U_R$. The constants U_L and U_R may be, but need not be, equal.

Exercise 15.4 Does the free particle fall into this general type of problem? How about the 1-D harmonic oscillator?

In any region where $U(x) = U_0$ (a constant), the mathematical solutions to Eq. 15.4 fall into two categories:

- If $E > U_0$, then the general solution is of the form

$$\psi(x) = Ae^{ikx} + Be^{-ikx}, \tag{15.10}$$

where $k = \sqrt{(2\mu/\hbar^2)(E - U_0)}$. That is, the general solution is a superposition of rightward and leftward moving plane waves.
- If $E < U_0$, then the general solution is of the form

$$\psi(x) = Ae^{bx} + Be^{-bx}, \tag{15.11}$$

where $b = \sqrt{(2\mu/\hbar^2)(U_0 - E)}$. This is a superposition of increasing and decreasing exponential functions.

(The special case $E = U_0$ can be viewed as a limit of either of these.) One or the other of these situations must obtain in the regions outside the interaction zone.

What sort of stationary states are possible for this type of localized potential? There are three distinct cases, depending on the energy E. This energy might be larger than both U_L and U_R, it might be less than both, or it might be larger than one and less than the other (see Fig. 15.2). These three cases lead to three different types of state.

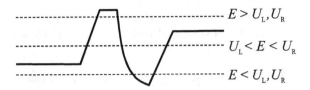

Fig. 15.2 Three possible relations between the energy eigenvalue E and the constants U_L and U_R.

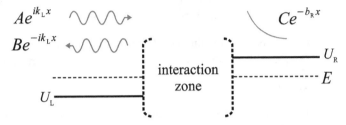

Fig. 15.3 General form of a reflecting state.

15.2 Reflecting, scattering, and bound states

Reflecting states

Let us suppose that $U_L < U_R$, and that E lies between them: $U_L < E < U_R$. In this case we call the stationary state a *reflecting state*. On the left-hand side of the interaction zone ($x < x_L$), the wave function must be a superposition of plane waves $e^{ik_L x}$ and $e^{-ik_L x}$, where $k_L = \sqrt{(2\mu/\hbar^2)(E - U_L)}$. On the right-hand side ($x > x_R$), the wave function is a superposition of exponentials $e^{b_R x}$ and $e^{-b_R x}$, where $b_R = \sqrt{(2\mu/\hbar^2)(U_R - E)}$. However, as we have seen, the exponentially increasing solution $e^{b_R x}$ cannot be interpreted as a physical wave function. The exponentially decreasing solution is okay because it applies only in the region $x > x_R$, where the exponential remains bounded. We therefore conclude that

$$\psi(x) = \begin{cases} Ae^{ik_L x} + Be^{-ik_L x} & x < x_L \\ Ce^{-b_R x} & x > x_R \end{cases}, \tag{15.12}$$

and something else between x_L and x_R, in the interaction zone. Any solution of this form describes a physically meaningful situation for the particle.

Even without a detailed knowledge of the potential $U(x)$ within the interaction zone, we can say more about the relations between the coefficients A, B, and C. Suppose we choose some particular value of C. This will determine both $\psi(x_0)$ and $\psi'(x_0)$ for a point $x_0 > x_R$. These completely determine the solution to Eq. 15.4 at all points, including the coefficients A and B. This means that we can find physically meaningful improper states for any value of E between U_L and U_R. The spectrum of possible energies of reflecting states is continuous.

We have shown that, if we specify a value of C, the values of both A and B are determined. Suppose instead that we specify A. Does this also determine the solution everywhere? It does:

Exercise 15.5 Suppose we have two reflecting solutions to Eq. 15.4 with the same energy E but different coefficients A and A'. Show that these two solutions cannot have the same C coefficients. (It might be useful to consider the *difference* between the solutions, which is also a solution.) Thus, the relationship between A and C is one-to-one, and a solution may be uniquely specified by giving the value for either C or A.

For a stationary state, the probability density $|\psi(x)|^2$ is independent of time. This means that the probability flux J defined in Eq. 11.20 is independent of x.

Exercise 15.6 Explain why.

Let us evaluate J on both sides of the interaction zone for a reflecting state described by Eq. 15.12.

$$J = \frac{\hbar k_L}{\mu}\left(|A|^2 - |B|^2\right) \qquad x < x_L,$$

$$J = 0 \qquad\qquad\qquad\qquad x > x_R. \qquad (15.13)$$

Exercise 15.7 Verify the calculations of J in Eq. 15.13.

Since J must be the same for all x, we conclude that $|A|^2 = |B|^2$ for a reflecting state.

This has a simple interpretation. The flux J represents the "flow of probability" in the positive direction. On the left-hand side of the interaction zone ($x < x_L$), the terms Ae^{ik_Lx} and Be^{-ik_Lx} (rightward and leftward moving plane waves) make positive and negative contributions to this flow. Since the two contributions cancel, we conclude that particles in a reflecting state "bounce back" from the potential with probability 1. In the region where $x > x_R$, even though the wave function is non-zero there is no rightward flow of probability.

In classical mechanics, a particle moving from the left with energy $E < U_R$ does not have sufficient energy to reach the right-hand side of the interaction zone. At some point it must turn around and travel back to the left. So far, the classical and quantum theories are in accord. However, in the quantum case the wave function $\psi(x) \neq 0$ for $x > x_R$, so there is a non-zero probability for finding the quantum particle on the "wrong" side of the interaction zone – a region it is classically forbidden to reach.

Our analysis is easily adapted to the case where $U_R < E < U_L$. In this situation, particles impinge on the interaction from the right and are reflected, with an exponential "tail" of the wave function found on the left-hand side.

Scattering states

Suppose that E is greater than both U_L and U_R. The wave function $\psi(x)$ for a stationary state involves plane waves on either side of the interaction zone. We will consider the special situation in which the wave function in the region $x > x_R$ contains no leftward moving plane wave. That is, the "incoming" probability comes entirely from the left-hand side of

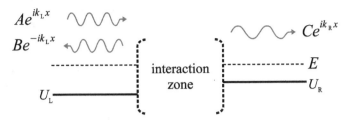

Fig. 15.4   General form of a scattering state.

the interaction zone. This models the situation in which a particle comes from the left and "scatters" from the potential, either passing through the interaction zone or reflecting from it. We call this a *scattering state*.

Outside the interaction zone, the solution to Eq. 15.4 for a scattering state has the form

$$\psi(x) = \begin{cases} Ae^{ik_L x} + Be^{-ik_L x} & x < x_L \\ Ce^{ik_R x} & x > x_R \end{cases}, \tag{15.14}$$

where $k_{L,R} = \sqrt{(2\mu/\hbar^2)(E - U_{L,R})}$. As before, if we specify any one of the coefficients A, B, and C, the other two are determined. This means that any energy E greater than both U_L and U_R is a possible energy for a scattering state. Scattering states have a continuous range of energy.

Exercise 15.8 Write down the form of a scattering state analogous to Eq. 15.14, except that the particle is coming in from the right. Show that the general scattering state is a superposition of "in-from-the-left" and "in-from-the-right" states.

The probability flux J is

$$J = \frac{\hbar k_L}{\mu}\left(|A|^2 - |B|^2\right) \qquad x < x_L,$$

$$J = \frac{\hbar k_R}{\mu}|C|^2 \qquad x > x_R. \tag{15.15}$$

It is no longer the case that $J = 0$ everywhere. That is, the scattering state wave function in Eq. 15.14 includes a net positive flow of probability. We interpret this by saying that a particle approaching the interaction zone from the left may either be reflected or transmitted by the potential. The *transmission coefficient T* is the ratio of the transmitted flux to the incident flux:

$$T = \frac{k_R |C|^2}{k_L |A|^2}. \tag{15.16}$$

Similarly, the *reflection coefficient R* is the ratio of the reflected flux to the incident flux:

$$R = \frac{|B|^2}{|A|^2}. \tag{15.17}$$

Exercise 15.9 Why do k_R and k_L appear explicitly in the expression for T but not R? Give both the simple algebraic explanation and also a physical interpretation involving the meaning of k_R and k_L.

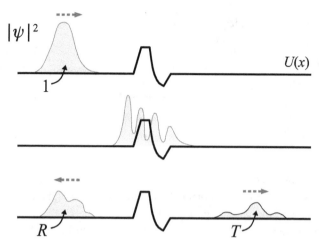

Fig. 15.5 A wave packet with well-defined energy scatters from an interaction zone. After a long time, the T and R coefficients give the probabilities for finding the particle to the right and left of the barrier.

Exercise 15.10 Show that $T + R = 1$. Is it possible for an incoming quantum particle with $E > U_L, U_R$ to be permanently "trapped" in the interaction zone?

Without a detailed knowledge of the potential $U(x)$ in the interaction zone, we cannot say much more about the transmission and reflection coefficients. Both T and R will be functions of the particle energy E. They can have some surprising properties. For example, we can have $T > 0$ even if $E < U_{max}$, the maximum value of the potential in the interaction zone. The particle may be transmitted even if it does not have sufficient energy to get "over the barrier." This phenomenon is called *quantum tunneling*.

Equally strange is the fact that we can have $R > 0$ even if $E > U_{max}$. This would not happen for a classical particle, which will certainly pass through the interaction zone if $E > U_{max}$. The classical situation can be summarized as follows:

- If $E < U_{max}$ then $R = 1$ and $T = 0$.
- If $E > U_{max}$ then $R = 0$ and $T = 1$.

For a quantum particle the situation is more complicated.

We can interpret T and R as the transmission and reflection *probabilities* for a particle of energy E. That is, suppose the initial state is described by a wave packet approaching the interaction zone from the left. This wave packet has an expectation value $\langle E \rangle$ for the energy, with an energy uncertainty ΔE that is very small.[2] Over time the wave packet strikes the interaction zone and evolves in a complicated way in the potential there. A long time later, the wave function has split into two parts moving away from the interaction zone to the left and right. The total probability for finding the particle to the left of the zone is about R (evaluated at the mean energy $\langle E \rangle$), and the total probability that the particle is found to the right of the zone is about T. This is illustrated in Fig. 15.5.

---

[2] Here "very small" means that the functions T and R do not vary significantly over the range of likely energies.

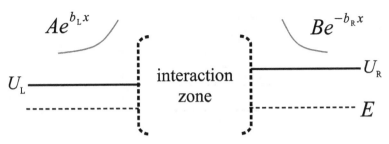

General form of a bound state.

Fig. 15.6

Bound states

Now we turn to the last case, in which E is less than both U_L and U_R. Since $E \geq U_{min}$, this can only happen if the interaction zone contains regions of lower potential. In this case, both the left-hand and right-hand regions have only exponential solutions. As previously discussed, we must exclude exponential solutions that increase without bound, since these cannot be approximated by normalized states. Thus, the wave function outside the interaction zone has the form

$$\psi(x) = \begin{cases} Ae^{b_L x} & x < x_L \\ Ce^{-b_R x} & x > x_R \end{cases}, \tag{15.18}$$

where $b_{L,R} = \sqrt{(2\mu/\hbar^2)(U_{L,R} - E)}$.

Exercise 15.11 Show that $J = 0$ for the bound state in Eq. 15.18.

Notice that $\psi \to 0$ as $x \to \pm\infty$. In fact, the bound state wave functions are normalizable and thus represent actual physical states.

Note that we are imposing a boundary condition on both sides of the interaction zone. This poses a bit of a puzzle. If we specify the value of A in Eq. 15.18, fixing both $\psi(x_0)$ and $\psi'(x_0)$ for some $x_0 < x_L$, then we can solve the differential equation in Eq. 15.4 and determine $\psi(x)$ everywhere. In general, this wave function will have the form $Ce^{-b_R x} + De^{b_R x}$ to the right of the interaction zone. How can we guarantee that the coefficient D winds up being zero, as required?

In general, we cannot guarantee this! When we solve Eq. 15.4 we usually get both $C \neq 0$ and $D \neq 0$. However, for certain discrete values of the energy E, it can happen that $D = 0$. Only these energies can be associated with physically acceptable wave functions. Thus, *the bound states of the potential have a discrete spectrum of energy eigenvalues.*

15.3 A potential step

We will now begin to apply our general results about reflecting, scattering, and bound states to some particular examples, chosen because they are especially easy to solve and because

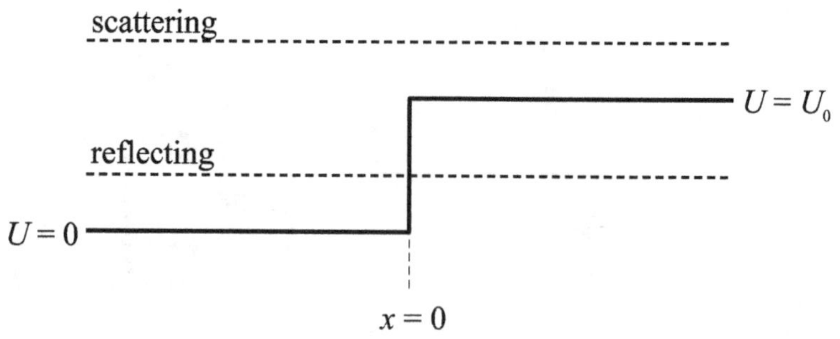

Fig. 15.7 A potential step, with typical energies for reflecting and scattering states.

they illustrate the types of behavior seen in more complex situations. The simplest of them is the potential step, for which $U(x)$ has the following form, illustrated in Fig. 15.7:

$$U(x) = \begin{cases} 0 & x < 0 \\ U_0 & x \geq 0 \end{cases}. \tag{15.19}$$

The interaction zone for the potential step consists of the single point $x = 0$, and the constants $U_L = 0$ and $U_R = U_0$. This discontinuous potential is of course an idealization of a situation in which $U(x)$ changes from 0 to U_0 within a small interval, where the particle experiences a strong force.

Solving the Schrödinger equation for this potential amounts to adapting our general solutions, "pasting" the left-hand and right-hand solutions together at $x = 0$. How do we do this? By remembering that both $\psi(x)$ and $\psi'(x)$ must be continuous at $x = 0$.

Consider first a reflecting state. We suppose that $U_0 > 0$ and that the energy $E < U_0$. Equation 15.12 gives the general form of the solution. Choosing the coefficient $A = 1$, we have

$$\psi(x) = \begin{cases} e^{ikx} + Be^{-ikx} & x < x_L \\ Ce^{-bx} & x > x_R \end{cases}, \tag{15.20}$$

where $k = \sqrt{(2\mu/\hbar^2)E}$ and $b = \sqrt{(2\mu/\hbar^2)(U_0 - E)}$. The continuity of $\psi(x)$ and $\psi'(x)$ will yield two conditions on the coefficients B and C:

$$\psi(x) \text{ continuous: } 1 + B = C,$$

$$\psi'(x) \text{ continuous: } ik(1 - B) = -bC. \tag{15.21}$$

From these we can easily determine the coefficients B and C:

$$B = \frac{ik + b}{ik - b},$$

$$C = \frac{2ik}{ik - b}. \tag{15.22}$$

Exercise 15.12 Show that $|B|^2 = 1$, and interpret this mathematical fact.

Exercise 15.13 Show that we can simplify Eq. 15.22 to obtain

$$B = \frac{i\sqrt{E} + \sqrt{U_0 - E}}{i\sqrt{E} - \sqrt{U_0 - E}},$$

$$C = \frac{2i\sqrt{E}}{i\sqrt{E} - \sqrt{U_0 - E}}. \tag{15.23}$$

As Eq. 15.23 indicates, the coefficients B and C are functions of the incident energy E. It is interesting to note the qualitative behavior of these. For $E \ll U_0$, we find that $|C|^2 \ll 1$. This means that there is little likelihood of finding the particle to the right of $x = 0$. On the other hand, suppose that E is very close to U_0, so that the incoming particle has almost enough energy to climb the step. Then C has a much larger value, and the particle is much more likely to be found in the classically forbidden region. Furthermore, the exponential coefficient b will be small, so that this likelihood extends very far beyond $x = 0$.

Exercise 15.14 Show that $\lim\limits_{E \to U_0^-} C = 2$.

Now let us assume that $E > U_0$, so that the particle is in a scattering state. Once again, we can use our previous work to write down the form of the wave function on either side of the step, then impose continuity of $\psi(x)$ and $\psi'(x)$ to solve for the coefficients. Letting $A = 1$ as before and simplifying our notation somewhat, Eq. 15.14 becomes

$$\psi(x) = \begin{cases} e^{ikx} + Be^{-ikx} & x < x_{\mathrm{L}} \\ Ce^{-ik'x} & x > x_{\mathrm{R}} \end{cases}, \tag{15.24}$$

where $k = \sqrt{(2\mu/\hbar^2)E}$ and $k' = \sqrt{(2\mu/\hbar^2)(E - U_0)}$. The continuity requirements are then

$$\psi(x) \text{ continuous: } 1 + B = C,$$

$$\psi'(x) \text{ continuous: } ik(1 - B) = ik'C, \tag{15.25}$$

which have solutions

$$B = \frac{k - k'}{k + k'},$$

$$C = \frac{2k}{k + k'}. \tag{15.26}$$

These are easily expressed in terms of E and U_0:

$$B = \frac{\sqrt{E} - \sqrt{E - U_0}}{\sqrt{E} + \sqrt{E - U_0}},$$

$$C = \frac{2\sqrt{E}}{\sqrt{E} + \sqrt{E - U_0}}. \tag{15.27}$$

These coefficients are both real.

What are the reflection and transmission coefficients for this potential? Since we have chosen $A = 1$, the reflection coefficient is

$$R = |B|^2 = \left(\frac{\sqrt{E} - \sqrt{E - U_0}}{\sqrt{E} + \sqrt{E - U_0}} \right)^2.$$
(15.28)

The transmission coefficient is just $T = 1 - R$.

How does R behave at different values of the energy? Here is a simple exercise that sheds some light on this.

Exercise 15.15

- Show that $\lim\limits_{E \to U_0^+} R = 1$.
- Evaluate R for $E = 2U_0$.
- Show that if $E \gg U_0$ then $R \approx 0$.

Suppose that U_0 is *negative* – that is, that the particle encounters a step "down" in potential energy. Then we still have $R > 0$. Classically speaking, this is nothing short of astonishing. We must imagine that a particle moving to the right experiences an impulsive force, also directed to the right – but nevertheless may *bounce back toward the left*. Indeed, the greater the "kick," the more likely such a paradoxical outcome is!

Exercise 15.16 Show that, if $-U_0 \gg E$, the reflection coefficient $R \approx 1$.

Part of the blame for this counter-intuitive result can, of course, be ascribed to the generally weird nature of quantum mechanics. But another part lies in our idealized problem. We do not expect an actual physical potential to be discontinuous. If our idealized $U(x)$ is replaced by one that has a smoother step down, then we find that the anomalous reflection probability is considerably reduced.

We close our discussion of the potential step function with the following observation:

Exercise 15.17 Prove that there are no bound states for a potential step.

15.4 Scattering from a square barrier

A square barrier is a potential of the form shown in Fig. 15.8:

$$U(x) = \begin{cases} 0 & x < 0 \\ U_0 & 0 \le x \le L. \\ 0 & x > L \end{cases}$$
(15.29)

The barrier is characterized by its "height" U_0 and its "width" L. The interaction zone here is the interval between 0 and L, and $U_L = U_R = 0$. Since the two outside potentials are equal, there are no reflecting states from the barrier. In this section, we will analyze the scattering states.

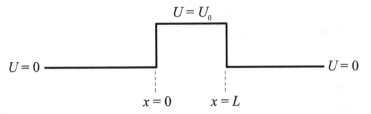

A square potential barrier with $U_0 > 0$.

To solve Eq. 15.4, we can paste together known solutions for the three regions of constant potential, using the fact that $\psi(x)$ and $\psi'(x)$ are continuous at both $x = 0$ and $x = L$. There are two different cases for the wave function in the interaction zone, depending on whether the energy E is greater or less than U_0. We will consider these two situations in turn.

First, suppose that $E < U_0$. The wave function for an energy eigenstate has the form

$$\psi(x) = \begin{cases} e^{ikx} + Be^{-ikx} & x < 0 \\ Ce^{bx} + De^{-bx} & 0 \le x \le L, \\ Fe^{ikx} & x > L \end{cases} \tag{15.30}$$

where $k = \sqrt{(2\mu/\hbar^2)E}$ and $b = \sqrt{(2\mu/\hbar^2)(U_0 - E)}$. The boundary conditions at $x = 0$ tell us that

$$\psi(x) \text{ continuous: } 1 + B = C + D,$$

$$\psi'(x) \text{ continuous: } ik(1 - B) = b(C - D), \tag{15.31}$$

while those at $x = L$ are

$$\psi(x) \text{ continuous: } Ce^{bL} + De^{-bL} = Fe^{ikL},$$

$$\psi'(x) \text{ continuous: } b\left(Ce^{bL} - De^{-bL}\right) = ikFe^{ikL}. \tag{15.32}$$

Equations 15.31 and 15.32 give us a system of four linear equations in the four unknown coefficients B, C, D, and F. We can solve this system and find the wave function everywhere.

What we are really interested in is the transmission coefficient $T = |F|^2$, so we want to find F. We can use Eq. 15.32 to solve for C and D in terms of F:

$$C = \frac{1}{2}e^{-bL}\left(1 + \frac{ik}{b}\right)e^{ikL}F,$$

$$D = \frac{1}{2}e^{bL}\left(1 - \frac{ik}{b}\right)e^{ikL}F. \tag{15.33}$$

We substitute these into Eq. 15.31, obtaining a pair of equations in B and F. Solving for F we obtain

$$F = \frac{4ikbe^{-ikL}}{(b + ik)^2 e^{-bL} - (b - ik)^2 e^{bL}}. \tag{15.34}$$

Now we can calculate T. It is actually a bit easier to write down the inverse of the transmission coefficient:

$$\frac{1}{T} = 1 + \frac{U_0^2}{4E(U_0 - E)} \sinh^2 bL, \qquad (15.35)$$

where we have expressed b and k in terms of E and U_0. Since $T > 0$ even though $E < U_0$, the particle may "tunnel" through the barrier with some probability.

The limiting behavior of Eq. 15.35 is worth noting. If the barrier is very thin, so that $bL \ll 1$, then $T \approx 1$ and the barrier is nearly "transparent." On the other hand, suppose that the barrier is very wide, with $bL \gg 1$. Then the hyperbolic sine function can be approximated by $\sinh bL \approx \frac{1}{2} e^{bL} \gg 1$. In this case,

$$T \approx \frac{16\,E(U_0 - E)}{U_0^2} e^{-2bL}. \qquad (15.36)$$

For wide barriers, the tunneling probability decreases exponentially with L.

Exercise 15.18 Fill in the algebraic details from Eq. 15.31 and 15.32 to the transmission coefficient in Eq. 15.35. Save plenty of paper for this one.

Exercise 15.19 An electron with energy 4 eV strikes a square barrier of height 6 eV. Show that a barrier of width 3 nm satisfies $bL \gg 1$. Find the approximate ratio of the transmission coefficients for two barriers of this height, one of width 4 nm and the other of width 3 nm.

Exercise 15.20 Find an expression for the reflection coefficient R for the square barrier when $E < U_0$.

Now we must consider the case where $E > U_0$. Then our solution must be of the form

$$\psi(x) = \begin{cases} e^{ikx} + Be^{-ikx} & x < 0 \\ Ce^{ik'x} + De^{-ik'x} & 0 \le x \le L \\ Fe^{ikx} & x > L \end{cases}, \qquad (15.37)$$

where $k = \sqrt{(2\mu/\hbar^2)E}$ and $k' = \sqrt{(2\mu/\hbar^2)(E - U_0)}$.

Exercise 15.21 Find the new boundary conditions at $x = 0$ and $x = L$, analogous to Eq. 15.31 and 15.32.

We can actually write down the solution for F by replacing the parameter b in Eq. 15.34 by ik'. This shortcut yields

$$F = \frac{4kk'e^{-ikL}}{(k' + k)^2 e^{-ik'L} - (k' - k)^2 e^{ik'L}}. \qquad (15.38)$$

From this, we can calculate the transmission coefficient T, which is most easily displayed as in Eq. 15.35:

$$\frac{1}{T} = 1 + \frac{U_0^2}{4E(E - U_0)} \sin^2 k'L. \qquad (15.39)$$

Once again we note that, for a thin barrier with $k'L \ll 1$, the transmission coefficient $T \approx 1$. For wider barriers, though, the situation has an interesting new feature. At certain discrete lengths, the $\sin^2 k'L$ factor is zero, and $T = 1$ exactly. That is, for certain combinations of E, U_0, and L, the barrier is perfectly transparent. This is called *resonant scattering*. The condition for resonant scattering is that $k'L = n\pi$, for $n = 1, 2, \ldots$ We can get a physical intuition for this by recalling that the wavelength within the barrier is $\lambda' = 2\pi/k'$. Then

$$L = \frac{n\pi}{k'} = \frac{n\lambda'}{2}. \tag{15.40}$$

For resonant scattering, the width of the barrier must be a half-integer number of full wavelengths:

$$n\pi = k'L = \left(\sqrt{\frac{2\mu}{\hbar^2}(E - U_0)} \right) L. \tag{15.41}$$

In terms of energy,

$$E - U_0 = \left(\frac{\pi^2 \hbar^2}{2\mu L^2} \right) n^2. \tag{15.42}$$

We recognize the right-hand expression as the energy $E_n^{(box)}$ of the nth stationary state of a particle in a box of length L (see Eq. 11.62). The condition for resonant scattering from the potential barrier is that $E - U_0 = E_n^{(box)}$ for some $n = 1, 2, \ldots$

Exercise 15.22 An electron strikes a potential barrier with a height of 1 eV and a width of 2 nm. For what energies does the electron experience resonant scattering?

Exercise 15.23 Does anything in our analysis of the square potential barrier change if $U_0 < 0$ – that is, we are scattering from a square potential *well*? Can there be resonant scattering from a square well?

Both tunneling and resonant scattering are illustrated in the graph of T versus E in Fig. 15.9.

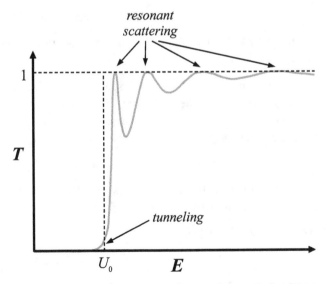

Fig. 15.9 Transmission coefficient T versus incident energy E for a particle scattering from a square barrier.

15.5 Bound states in a square well

We now turn our attention to the bound states in a square potential well. There are some slight shifts in our set-up and notation that will make things more convenient. The potential function is

$$U(x) = \begin{cases} U_0 & x < -a \\ 0 & -a \leq x \leq a \\ U_0 & x > a \end{cases} . \qquad (15.43)$$

That is, the potential well has width $L = 2a$ and is centered at the origin. Its depth is U_0. For a bound state, $E < U_0$.

Because the well is symmetric, the parity operator q commutes with the Hamiltonian H. This means that we can require the energy eigenstates to be parity eigenstates as well, i.e. we can consider only even and odd wave functions as solutions to Eq. 15.4.

Inside the well, the wave function has plane wave solutions e^{ikx} and e^{-ikx}. However, we wish to consider the even and odd combinations of these, which are the trigonometric forms $\cos kx$ and $\sin kx$. If we consider the even case, the wave function is

$$\psi(x) = \begin{cases} Ae^{bx} & x < -a \\ B \cos kx & -a \leq x \leq a \\ Ae^{-bx} & x > a \end{cases} , \qquad (15.44)$$

where $b = \sqrt{(2\mu/\hbar^2)(U_0 - E)}$ and $k = \sqrt{(2\mu/\hbar^2)E}$. The continuity of $\psi(x)$ and $\psi'(x)$ at $x = a$ yields the relations

$$\psi(x) \text{ continuous: } Ae^{-ba} = B \cos ka,$$

$$\psi'(x) \text{ continuous: } -bAe^{-ba} = -kB \sin ka. \qquad (15.45)$$

Exercise 15.24 What additional information do the boundary conditions at $x = -a$ provide?

The wave function $\psi(x)$ for a bound state is normalizable. For a given E (and thus k and b), this fact determines the coefficients A and B up to an overall phase. As we have seen, for a bound state only certain energy eigenvalues E will be possible; our task is to identify these. Dividing the $\psi'(x)$ boundary condition by the $\psi(x)$ condition, we obtain

$$b = k \tan ka. \qquad (15.46)$$

To understand the solutions to this transcendental equation, we introduce the variable $z = ka$. We can also write ba in terms of z:

Exercise 15.25 Show that

$$ba = \sqrt{z_0^2 - z^2}, \qquad (15.47)$$

where $z_0 = \left(\sqrt{(2\mu/\hbar^2)U_0} \right) a$.

Thus, Eq. 15.46 takes on a particularly simple form as a condition for z:

$$\sqrt{\left(\frac{z_0}{z}\right)^2 - 1} = \tan z. \tag{15.48}$$

What about the case of odd parity? Then

$$\psi(x) = \begin{cases} -Ae^{bx} & x < -a \\ B\sin kx & -a \le x \le a \\ Ae^{-bx} & x > a \end{cases} . \tag{15.49}$$

The boundary conditions at $x = a$ are

$$\psi(x) \text{ continuous: } Ae^{-ba} = B\sin ka,$$
$$\psi'(x) \text{ continuous: } -bAe^{-ba} = kB\cos ka, \tag{15.50}$$

the ratio of which yields

$$b = -k\cot ka. \tag{15.51}$$

Therefore, our combined equation for z is

$$\sqrt{\left(\frac{z_0}{z}\right)^2 - 1} = \begin{cases} \tan z & \text{(even)} \\ -\cot z & \text{(odd)} \end{cases} . \tag{15.52}$$

Solutions for z determine possible values for the energy $E = \frac{\hbar^2 z^2}{2\mu a^2}$.

We can identify solutions to this equation graphically, as shown in Fig. 15.10. In the diagram, curves for $\tan z$, $-\cot z$, and $\sqrt{\left(\frac{z_0}{z}\right)^2 - 1}$ are shown. The points of intersection of these curves will be values of z which satisfy Eq. 15.52. These solutions for z determine the allowed energies.

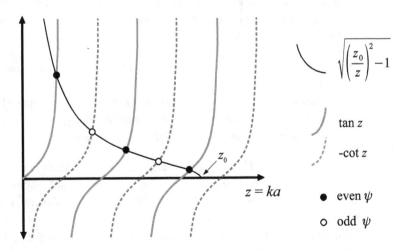

Graphical solution of Eq. 15.52 for even and odd parity wave functions.

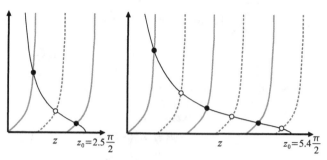

The number of bound state solutions depends on the value of z_0. For $z_0 = 2.5 \, (\pi/2)$, there are three solutions (two even and one odd). For $z_0 = 5.4 \, (\pi/2)$ there are six solutions (three even and three odd).

As we can see, there will only be a finite number of solutions to Eq. 15.52 for a given value of z_0. How many solutions will there be?

- For any value $z_0 > 0$, there is always at least one bound state.
- The graphs of $\tan z$ and $-\cot z$ cross the z-axis at the points $0, \pi/2, \pi$, etc. Therefore, if the value of z_0 satisfies

$$\frac{n\pi}{2} < z_0 < \frac{(n+1)\pi}{2}, \qquad (15.53)$$

for an integer n, then there are exactly $n+1$ bound states (see Fig. 15.11).

- The ground state (state of lowest energy) always has an even wave function. The wave functions for higher energies alternate between odd and even.

In fact, many of these points can be expressed very nicely by comparing the square well system to a free particle in a box with the same width $L = 2a$, having energies $E_n^{(\text{box})}$ for $n = 1, 2, \ldots$

Exercise 15.26 Show that if $E_n^{(\text{box})} < U_0 < E_{n+1}^{(\text{box})}$, then there are exactly $n+1$ bound states for the square well. (By convention, we take $E_0^{(\text{box})} = 0$, even though there is no $n = 0$ state of the particle in a box.)

Exercise 15.27 For the bound state energies E_n of the square well, show that $E_{n-1}^{(\text{box})} < E_n < E_n^{(\text{box})}$.

What are the wave functions for square-well bound states? Figure 15.12 shows them for a well with three bound states. Note that, since the energy eigenvalue equation (Eq. 15.4) is a real equation, we can find real solutions $\psi_n(x)$, as shown.

Suppose now that U_0 is extremely large, and we consider only those stationary states for which $E_n \ll U_0$. In other words, the energy "walls" around the system are much higher (in energy terms) than the energy of the particle. What sort of energies and bound states can we expect in this extreme situation?

Of course, there will be very many bound states, because Eq. 15.52 has very many solutions. The intersections of the curves in Fig. 15.10 occur near the vertical asymptotes

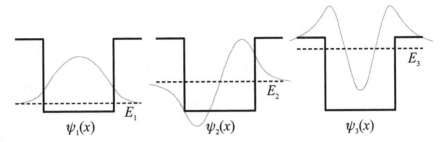

Fig. 15.12 Wave functions for the three bound states of a square well. As is often done, we have sketched the wave functions relative to their energy levels in the well.

of the $\tan z$ and $-\cot z$ functions; that is, near $z = n\pi/2$ for $n = 1, 2, \ldots$ This means that the energy levels with $E_n \ll U_0$ have approximate values

$$E_n \approx \frac{\pi^2 \hbar^2}{8\mu a^2} n^2 = E_n^{(\text{box})}, \tag{15.54}$$

the energies we found for the particle in a box of width $L = 2a$.

How about the wave functions? For a square well, each bound state wave function has exponential "tails" extending beyond the edges of the well. But if $E_n \ll U_0$, then these tails drop off extremely rapidly, relative to the well's width. This can be seen in the following exercise:

Exercise 15.28 If the energy $E_n \ll U_0$, show that $z_0 = ba \gg 1$. Thus, the exponential tails of the bound state will decrease almost to zero in a distance very much less than a.

In the limiting case $U_0 \to \infty$ (the *infinite* square well) the "tail" part of $\psi_n(x)$ is completely negligible. The wave function is effectively zero at the wall and everywhere beyond it.

In short, a particle in a very deep square potential well is a pretty good approximation to the particle in a box with perfectly impenetrable walls, provided we only consider low-lying energy levels. The very high potential barriers at $x = \pm a$ impose an approximate boundary condition that $|\psi(-a)| = |\psi(a)| \approx 0$. Though $\psi(x)$ is continuous at these points, the derivative $\psi'(x)$ changes so abruptly that it appears to be discontinuous. We end up with energy levels of the familiar form in Eq. 11.62.

Our analysis here sheds light on a question that we mentioned in Section 11.4, when we first introduced the particle in a box. There, we noted that the discontinuity in $\psi'(x)$ at the walls of the box might lead to a significant contribution to the total energy of the state, since the kinetic energy involves *second* derivatives of ψ. But we do not get such a contribution from the sharp "elbows" at the sides of the wave functions in a very deep finite potential well. This is because, while $\psi''(x)$ is very large there, $\psi(x)$ is extremely tiny, so the integral

$$\left\langle \frac{p^2}{2\mu} \right\rangle = -\frac{\hbar^2}{2\mu} \int_{-\infty}^{+\infty} \psi(x)\psi''(x)\,dx, \tag{15.55}$$

does not receive a significant contribution from this small region. We were justified in ignoring the "corners" in the wave functions for the idealized situation.

15.6 The variational method

Suppose we are looking for the bound-state energies for a given potential function $U(x)$. For any but the simplest potentials, we cannot solve the energy eigenvalue equation (Eq. 15.3) exactly. This means that we must resort to approximation techniques. In Chapter 17 we will develop a "perturbation" method that works well if $U(x)$ is close to a potential for which we can find the energy eigenvalues. In this section, though, we introduce a technique that can be used for estimating the ground state energy for almost any potential. This is the *variational method*.

The essential idea is simple. If E_0 is the energy of the ground state (a normalized bound state), then for any normalized state $|\psi\rangle$, it must be that $\langle E\rangle = \langle \psi | H | \psi\rangle \geq E_0$. This is very easy to see:

Exercise 15.29 Expand $|\psi\rangle$ in the energy eigenbasis for H and prove that $\langle E\rangle \geq E_0$.

Instead of finding the actual ground state, we could simply guess a "test state" $|\psi\rangle$ and calculate $\langle E\rangle$. We know that this expectation is an upper bound for E_0; but for a clever choice of the test state $|\psi\rangle$ we might hope that $\langle E\rangle$ and E_0 were reasonably close. But how do we choose a "clever" test state?

We do this by considering an entire continuous family of test states $|\psi_\alpha\rangle$, where α is a real parameter. If $E(\alpha) = \langle \psi_\alpha | H | \psi_\alpha\rangle$, then $E(\alpha) \geq E_0$ for all values of α. To find the best test state in our family, we simply find the α that minimizes $E(\alpha)$. That is, we solve

$$\frac{dE}{d\alpha} = \frac{d}{d\alpha} \langle \psi_\alpha | H | \psi_\alpha\rangle = 0. \qquad (15.56)$$

The resulting value of $E_{\min} = \min_\alpha E(\alpha)$ can be used as an (upper) estimate for the ground state energy E_0. This is the variational method.

The closeness of the estimate depends on how well we choose the family of test states $|\psi_\alpha\rangle$. On the one hand, we try to incorporate known properties of the ground state into the test states. If the potential $U(x)$ is a symmetric function, for instance, then we know that the ground state will have a symmetric wave function $\psi_0(x)$. We should therefore choose our test states to be states of even parity.

On the other hand, we wish to make our calculation of $E(\alpha)$ as simple as possible. Thus, we will choose test wave functions $\psi_\alpha(x)$ with relatively simple algebraic forms. Here we mention two popular choices of symmetric test wave functions.

Gaussian test states. In this case we choose our test states to have familiar Gaussian wave functions:

$$\psi_\alpha(x) = C_\alpha \exp\left(-\alpha^2 x^2\right), \qquad (15.57)$$

where the normalization coefficient $C_\alpha = (2\alpha^2/\pi)^{1/4}$.

Exercise 15.30 Verify that this choice of C_α yields a normalized wave function.

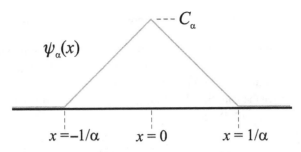

Fig. 15.13
Triangular test wave function $\psi_\alpha(x)$.

The parameter α measures how "narrow" the test wave function is. The energy expectation for this state is

$$E(\alpha) = \underbrace{\langle \psi_\alpha | \left(\frac{\mathrm{p}^2}{2\mu} \right) | \psi_\alpha \rangle}_{K(\alpha)} + \langle \psi_\alpha | U(\mathrm{x}) | \psi_\alpha \rangle . \qquad (15.58)$$

We can go ahead and calculate the kinetic energy term $K(\alpha)$, which will be the same for any problem using the Gaussian test states.

Exercise 15.31 For the Gaussian test states, show that

$$K(\alpha) = \frac{\hbar^2 \alpha^2}{2\mu} . \qquad (15.59)$$

Triangular test states. In this case we choose our test states to have "triangular" wave functions, as shown in Fig. 15.13. That is,

$$\psi_\alpha(x) = \left(\begin{array}{ll} C_\alpha(1 - \alpha |x|) & |x| \leq 1/\alpha \\ 0 & |x| > 1/\alpha \end{array} \right) . \qquad (15.60)$$

The constant C_α is determined by normalization.

Exercise 15.32 Show that we should choose

$$C_\alpha = \sqrt{\frac{3\alpha}{2}} . \qquad (15.61)$$

Again we will find it useful to calculate the expectation of the kinetic energy, which will be the same for any variational problem using these test states. This is somewhat complicated by the fact that the test wave functions are not smooth. The delta function can clarify our analysis.

Exercise 15.33 Show that it makes sense to write

$$\frac{d^2}{dx^2} |x| = 2\delta(x) . \qquad (15.62)$$

Exercise 15.34 For the triangular test states, show that

$$K(\alpha) = \frac{3\hbar^2 \alpha^2}{2\mu} . \qquad (15.63)$$

Explain how each of the three "corners" of $\psi_\alpha(x)$ contributes to this result.

For both types of test state, we chose $\psi_\alpha(x)$ to be a function of αx, where α has units of $(\text{length})^{-1}$. This is why the the two expressions for $K(\alpha)$ (Eq. 15.59 and 15.63) have the same form aside from numerical factors.

To see the variational method in action, let us apply it to the harmonic oscillator potential $U(x) = kx^2/2$. We have

$$\langle U \rangle = \langle \psi_\alpha | \, U(\mathrm{x}) \, | \psi_\alpha \rangle = \int_{-\infty}^{+\infty} |\psi_\alpha(x)|^2 \, \frac{kx^2}{2} \, dx. \tag{15.64}$$

For the Gaussian test state this becomes

$$\langle U \rangle = \frac{kC_\alpha^2}{2} \int_{-\infty}^{+\infty} x^2 \exp(-2\alpha^2 x^2) \, dx = \frac{k}{8\alpha^2}. \tag{15.65}$$

The energy expectation is thus

$$E(\alpha) = \frac{\hbar^2 \alpha^2}{2\mu} + \frac{k}{8\alpha^2}. \tag{15.66}$$

This is minimized when $E'(\alpha) = 0$, which happens at $\alpha = (k\mu/4\hbar^2)^{1/4}$. Thus, our estimate for the ground state energy is

$$E_0 \approx \min_\alpha E(\alpha) = \frac{1}{2} \hbar\omega, \tag{15.67}$$

where $\omega = \sqrt{k/\mu}$. This result is actually exact, since the ground state of the harmonic oscillator is in fact a Gaussian state. We got lucky!

Suppose instead we use the triangular test states. Then

$$\langle U \rangle = kC_\alpha^2 \int_0^{1/\alpha} x^2 (1 - \alpha x)^2 \, dx = \frac{k}{20\alpha^2}. \tag{15.68}$$

The energy expectation is

$$E(\alpha) = \frac{3\hbar^2 \alpha^2}{2\mu} + \frac{k}{20\alpha^2}. \tag{15.69}$$

At the minimum value, $E'(\alpha) = 0$, which occurs when $\alpha = (k\mu/30\hbar^2)^{1/4}$. Our estimate becomes

$$E_0 \approx \min_\alpha E(\alpha) = \sqrt{\frac{3}{10}} \, \hbar\omega = 0.548 \, \hbar\omega. \tag{15.70}$$

This is a remarkably good estimate, considering how unphysical the triangular test states are. Further applications of the variational method are found in the problems at the end of the chapter.

We can extend the variational method in several ways. It is not restricted to single particles moving in one dimension, and indeed it is often used in more complicated situations. We may choose a family of test states $|\psi(\alpha, \beta, \ldots)\rangle$ that depend on two or more parameters α, β, etc., minimizing the expectation function $E(\alpha, \beta, \ldots)$ over all these parameters.

Finally, it is sometimes possible to use the variational method to estimate, not just the ground state energy, but also the energy of one or more excited states. For example, if the potential $U(x)$ is a symmetric function, then we expect the ground state $|\psi_0\rangle$ to have even parity and the first excited state $|\psi_1\rangle$ to have odd parity. The excited state energy E_1 is the lowest energy among all odd parity states, which are all orthogonal to $|\psi_0\rangle$. If we consider only odd parity states in our family of test states, then $\langle E \rangle \geq E_1$ for every member of the family. The variational method using even parity states leads to an estimate of E_0, while using odd parity states leads to an estimate of E_1.

The main difficulty of the variational method is that we do not have a firm idea of how accurate its results are. As a practical matter, the variational method can yield a good estimate for the ground state energy even when the test states used are extremely crude. The variational method is therefore better at estimating energies than wave functions. If we keep this limitation in mind, the variational method is a very useful tool for understanding the quantum mechanics of particles in complicated potentials.

15.7 Parameters and scaling

A particle of mass μ moves in one dimension. Suppose that the Hamiltonian for the particle includes an adjustable parameter λ, which may appear in either the kinetic energy or the potential energy, or both. We write the Hamiltonian as H_λ. The energy eigenvalue E_λ and corresponding normalized eigenstate $|\psi_\lambda\rangle$ also depend on the parameter.[3] For any λ,

$$E_\lambda = \langle \psi_\lambda | H_\lambda | \psi_\lambda \rangle. \tag{15.71}$$

Let us assume that H_λ, E_λ, and $|\psi_\lambda\rangle$ are all smooth functions of λ. Then

$$\frac{dE_\lambda}{d\lambda} = \langle \psi_\lambda | \left(\frac{dH_\lambda}{d\lambda} \right) |\psi_\lambda\rangle + \left(\frac{d}{d\lambda} \langle \psi_\lambda | \right) H_\lambda |\psi_\lambda\rangle + \langle \psi_\lambda | H_\lambda \left(\frac{d}{d\lambda} |\psi_\lambda\rangle \right)$$

$$= \langle \psi_\lambda | \left(\frac{dH_\lambda}{d\lambda} \right) |\psi_\lambda\rangle + E_\lambda \left(\frac{d}{d\lambda} \langle \psi_\lambda |\psi_\lambda\rangle \right). \tag{15.72}$$

Since $\langle \psi_\lambda |\psi_\lambda\rangle = 1$ for all λ, we obtain

$$\frac{dE_\lambda}{d\lambda} = \langle \psi_\lambda | \left(\frac{dH_\lambda}{d\lambda} \right) |\psi_\lambda\rangle. \tag{15.73}$$

This is called the *Feynman–Hellmann* theorem.[4] It appears simple, but it has a remarkable number of applications.

---

[3] In the last section, $|\psi_\alpha\rangle$ was a test state for a fixed Hamiltonian H, and the resulting expectation $E(\alpha)$ was the basis for an estimate of the ground state energy. Now, $|\psi_\lambda\rangle$ is an exact eigenstate of H_λ with exact eigenvalue E_λ.

[4] And also, curiously enough, the *Hellmann–Feynman theorem*.

Consider, for instance, the quantum harmonic oscillator. We can write the Hamiltonian as

$$H = \frac{1}{2\mu}p^2 + \frac{1}{2}kx^2, \tag{15.74}$$

where μ is the particle mass and k is the elastic constant of the harmonic potential. The energy of the eigenstate $|n\rangle$ is

$$E_n^{(osc)} = \left(n + \frac{1}{2}\right)\hbar\sqrt{\frac{k}{\mu}}, \tag{15.75}$$

(see Eq. 15.2). We can treat μ and k as adjustable parameters and apply the Feynman–Hellmann theorem. For the parameter μ, we find that

$$\frac{dE_n^{(osc)}}{d\lambda} = \langle n| \left(\frac{dH}{d\mu}\right)|n\rangle,$$

$$-\frac{(n+1/2)\hbar}{\mu}\sqrt{\frac{k}{\mu}} = -\frac{1}{2\mu^2}\langle n| p^2 |n\rangle, \tag{15.76}$$

so that $\langle n| p^2 |n\rangle = (n+1/2)\hbar\sqrt{\mu k}$ for all n.

Exercise 15.35 Treat k as an adjustable parameter and show that

$$\langle n| x^2 |n\rangle = (n+1/2)\frac{\hbar}{\sqrt{\mu k}}, \tag{15.77}$$

for any stationary state $|n\rangle$ of a harmonic oscillator.

We previously used the properties of ladder operators to find these expectations (see Section 13.2, particularly Eq. 13.32). The Feynman–Hellmann theorem is a useful shortcut.

The Feynman–Hellmann theorem provides some insight into the scaling properties of energy eigenstates. Suppose that $|\psi\rangle$ is a solution to

$$H|\psi\rangle = \left(\frac{1}{2\mu}p^2 + U(x)\right)|\psi\rangle = E|\psi\rangle, \tag{Re 15.3}$$

for some energy eigenvalue E. Now modify the Hamiltonian by multiplying the potential energy and dividing the mass by the same scaling factor κ:

$$U(x) \rightarrow \kappa U(x), \qquad \mu \rightarrow \frac{\mu}{\kappa}. \tag{15.78}$$

If $\kappa > 1$, for example, we have transformed our system into one with a less massive particle moving in a deeper potential well. This has the overall effect of $H \rightarrow \kappa H$, and so the same $|\psi\rangle$ is an energy eigenstate with new energy κE.

Exercise 15.36 Apply the Feynman–Hellmann theorem to the parameter κ and show that it simply leads back to Eq. 15.3.

Now let us consider a less trivial example of scaling. Instead of changing the depth of our potential well, we will stretch the system spatially, so that $x \rightarrow \lambda x$. This will involve changes to the potential function and the mass, both involving the stretching parameter λ. To see how this ought to be done, consider the wave function $\psi(x) = \langle x | \psi \rangle$, which satisfies

$$-\frac{\hbar^2}{2\mu} \frac{d^2}{dx^2} \psi(x) + U(x)\psi(x) = E\psi(x). \qquad \text{(Re 11.17)}$$

If we simply replace x by λx in this equation, we get a rescaled version of the same:

$$-\frac{\hbar^2}{2\mu\lambda^2} \frac{d^2}{dx^2} \psi(\lambda x) + U(\lambda x)\psi(\lambda x) = E\psi(\lambda x). \qquad (15.79)$$

The function $\psi(\lambda x)$ can be renormalized, so that it becomes the wave function of an energy eigenstate state $|\psi_\lambda\rangle$ for a system with mass $\lambda^2 \mu$ moving in the potential $U(\lambda x)$. That is:

$$\mathsf{H}_\lambda |\psi_\lambda\rangle = \left(\frac{1}{2\lambda^2\mu} \mathsf{p}^2 + U(\lambda \mathsf{x}) \right) |\psi_\lambda\rangle = E |\psi_\lambda\rangle. \qquad (15.80)$$

Therefore, if $|\psi\rangle$ is an eigenstate of the original Hamiltonian with eigenvalue E, then $|\psi_\lambda\rangle$ is an eigenstate of the modified Hamiltonian *with the same eigenvalue*.

Exercise 15.37 Find the renormalization factor for $\psi(\lambda x)$.

The Feynman–Hellmann theorem now tells us that

$$\langle \psi_\lambda | \left(-\frac{1}{\lambda^3 \mu} \mathsf{p}^2 + \mathsf{x} U'(\lambda \mathsf{x}) \right) |\psi_\lambda\rangle = 0. \qquad (15.81)$$

If we set $\lambda = 1$, then for any bound state $|\psi\rangle$,

$$2 \langle \psi | \left(\frac{\mathsf{p}^2}{2\mu} \right) |\psi\rangle = \langle \psi | \mathsf{x} U'(\mathsf{x}) |\psi\rangle. \qquad (15.82)$$

This is called the *quantum virial theorem* for 1-D systems. It relates the expectation value of the kinetic energy K of the system to an expectation related to the potential energy. This is especially simple if the potential function is of the form $U(\mathsf{x}) = C\mathsf{x}^n$. Then the quantum virial theorem becomes

$$2 \langle K \rangle = n \langle U \rangle. \qquad (15.83)$$

Exercise 15.38 What does Eq. 15.83 tell us about the energy eigenstates of the harmonic oscillator?

Exercise 15.39 Suppose we know the bound states of a particle of mass μ moving in a potential $U(x)$. What can we say about the bound states of the same particle moving in the potential $\lambda^2 U(\lambda x)$? Apply your answer to the bound states of a finite square well.

Problems

Problem 15.1 A *delta function potential* has the form $U(x) = -K\,\delta(x)$ where $K > 0$ is a constant.

(a) Show that there is only one bound state for this potential.
(b) Find the energy eigenvalue and the wave function for this bound state.

There are several ways to approach this problem; one is to treat the delta function potential as a limiting case of a very narrow and deep square well potential.

Problem 15.2 As a follow-up to Problem 15.1, consider a situation in which the potential contains two delta functions:

$$U(x) = -K\,(\delta(x - a) + \delta(x + a)).\qquad(15.84)$$

How many orthogonal bound states does this potential have and what are their energies?

Problem 15.3 Calculate the reflection and transmission coefficients for scattering from the delta function potential in Problem 15.1.

Problem 15.4 Consider a stationary state for a potential that is zero outside the interaction zone. If $E > 0$, the wave function is

$$\psi(x) = \begin{cases} Ae^{ikx} + Be^{-ikx} & x < x_{\mathrm{L}} \\ Ce^{ikx} + De^{-ikx} & x > x_{\mathrm{R}} \end{cases},\qquad(15.85)$$

and something else within the interaction zone. We can write

$$\begin{pmatrix} B \\ C \end{pmatrix} = \mathbf{S} \begin{pmatrix} A \\ D \end{pmatrix},\qquad(15.86)$$

where $\mathbf{S}$ is a 2×2 complex matrix called the *scattering matrix* for the potential. The scattering matrix gives the "outgoing" amplitudes in terms of the "incoming" amplitudes. The matrix $\mathbf{S}$ depends on the potential function $U(x)$ and the particle energy E.

(a) Show that the scattering matrix $\mathbf{S}$ is unitary.
(b) Suppose the potential is an even function, so that $U(-x) = U(x)$. Show that $\mathbf{S}$ has the form

$$\mathbf{S} = \begin{pmatrix} \alpha & \beta \\ \beta & \alpha \end{pmatrix},\qquad(15.87)$$

for complex parameters α and β.

Problem 15.5 Consider a potential given by

$$U(x) = \begin{cases} \infty & x \le 0 \\ 0 & 0 < x < a. \\ U_0 & x > a \end{cases}\qquad(15.88)$$

This is a square well next to an impenetrable wall; physically, this might model a reflecting barrier that is slightly "sticky." Find the bound states for this potential, and give the conditions under which there are exactly n such states for $n = 0, 1, 2, \ldots$ (Hint: Compare this problem to the symmetric square well.)

Problem 15.6 Consider the following wave function:

$$\psi(x) = \begin{cases} A(1 + \cos kx) & -a < x < a \\ 0 & \text{elsewhere} \end{cases}, \tag{15.89}$$

where $ka = \pi$. Find a potential function and an energy value E so that this is a stationary state. Make a sketch of $U(x)$ and E.

The family of wave functions of this type, with a as an adjustable parameter, is useful in variational problems.

Problem 15.7 Suppose two potential functions $U_1(x)$ and $U_2(x)$ are both zero outside the same interaction zone, and that within the zone $U_1(x) \le U_2(x)$. Is it necessarily true that, at a given energy E, the transmission coefficients for the two potentials satisfy $T_1 \ge T_2$? If so, prove this result; if not, construct a counter-example.

Problem 15.8 Suppose $U(x)$ is an even (symmetric) function of x and has some bound states. Does the ground state (lowest energy bound state) of $U(x)$ always have an *even* parity wave function $\psi(x)$? Why or why not?

Problem 15.9 A particle of mass μ moves in a 1-D "triangular well" potential

$$U(x) = F|x|. \tag{15.90}$$

Use triangular test states to estimate the ground state energy of the particle as a function of C. Calculate a numerical result when the particle is an electron and $F = 20.0$ eV/nm.

Problem 15.10 As in the previous problem, consider a particle of mass μ moving in a potential given by Eq. 15.90.

(a) Use Gaussian test functions to estimate the ground state energy of the particle.
(b) We can use Gaussian functions to create a family of odd parity test states with wave functions

$$\phi_\alpha = C_\alpha x \exp\left(-\alpha^2 x^2\right). \tag{15.91}$$

Find the normalization constant C_α for this family and the kinetic energy function $K(\alpha)$.
(c) Use the odd parity test states to estimate the energy of the first excited state of the particle in this potential.

Problem 15.11 A quantum bouncer. A particle can move up and down along the z-axis. At $z = 0$ there is an impenetrable "floor," and above this floor the particle is subject to the potential $U(z) = \mu g z$, where μ is the particle's mass and g is the acceleration of gravity. Use test state wave functions of the form

$$\psi_\alpha(z) = A_\alpha z \exp(-\alpha z) \qquad (z > 0), \tag{15.92}$$

to estimate the minimum energy of a helium atom bouncing on a hard horizontal surface.

Problem 15.12 A particle is in a bound state of the finite square well potential given by Eq. 15.43. Show that

$$\frac{1}{\mu} \left\langle p^2 \right\rangle = a U_0 \, |\psi(a)|^2 . \tag{15.93}$$

Bound states in 3-D

16.1 Central potentials

In this chapter we will consider the stationary states of a particle of mass μ moving in three dimensions. (As we discussed in Section 14.2, this may in fact represent a relative position degree of freedom of a system of two or more particles, in which case μ is the reduced mass.) For a stationary state with energy E, the wave function $\psi(\vec{r})$ must satisfy

$$-\frac{\hbar^2}{2\mu}\nabla^2\psi + U(\vec{r})\psi = E\psi. \tag{Re 11.28}$$

We wish to find the possible eigenvalues E and the corresponding wave functions ψ for a given potential function $U(\vec{r})$. In fact, our focus will be even narrower, in two important respects. First, we will consider only *bound* stationary states, which are normalizable. Second, we will assume that the potential is a *central potential* $U(r)$ that depends only on the distance r from the coordinate origin. Even in such a restricted context, solving Eq. 11.28 is a matter of considerable physical interest!

Since the potential energy function depends only on r, it makes sense to do our analysis in spherical coordinates (r, θ, ϕ), as defined in Eq. 12.21 and 12.22.[1] In spherical coordinates, the Laplacian operator is

$$\nabla^2\psi = \frac{1}{r^2}\frac{\partial}{\partial r}\left(r^2\frac{\partial\psi}{\partial r}\right) + \frac{1}{r^2\sin\theta}\frac{\partial}{\partial\theta}\left(\sin\theta\frac{\partial\psi}{\partial\theta}\right) + \frac{1}{r^2\sin^2\theta}\frac{\partial^2\psi}{\partial\phi^2}. \tag{16.1}$$

Exercise 16.1 From the definition of the Laplacian in Cartesian coordinates (Eq. 11.27) and the partial derivatives in Eq. 12.23, derive Eq. 16.1.

We recognize part of this as the position representation of the orbital angular momentum operator L^2:

$$\mathsf{L}^2 : \psi \rightarrow -\hbar^2\left(\frac{1}{\sin\theta}\frac{\partial}{\partial\theta}\left(\sin\theta\frac{\partial\psi}{\partial\theta}\right) + \frac{1}{\sin^2\theta}\frac{\partial^2\psi}{\partial\phi^2}\right). \tag{Re 12.28}$$

We can draw the following conclusions:

---

[1] This is an excellent time to review the discussion of spherical coordinates and orbital angular momentum in Section 12.2.

- For a central potential $U(r)$, the Hamiltonian of the particle in the position representation depends only on the angular coordinates θ and ϕ via the orbital angular momentum operator L^2.
- Both L^2 and L_z, which depend only on the angular coordinates, commute with the Hamiltonian for the particle. Thus, we can find energy eigenstates that are also eigenstates of these angular momentum operators.

This simplifies our problem considerably. In Section 12.2 we established that the eigenstates of L^2 and L_z have wave functions of the form $R(r)\, Y_{lm}(\theta, \phi)$, where Y_{lm} is a spherical harmonic function. Given a function of this form, Eq. 11.28 becomes an equation for the radial function $R(r)$:

$$-\frac{\hbar^2}{2\mu} \left[\frac{1}{r^2} \frac{d}{dr} \left(r^2 \frac{dR}{dr} \right) - \frac{l(l+1)}{r^2} R \right] + U(r)R = ER. \tag{16.2}$$

Solving this ordinary differential equation will yield the stationary state wave functions and associated energy eigenvalues.

Exercise 16.2 What happened to the spherical harmonic $Y_{lm}(\theta, \phi)$ in Eq. 16.2?

What can we say about the functions $R(r)$ and the energies E that solve Eq. 16.2? Both may depend on the L^2 quantum number l, which appears in the equation. However, the L_z quantum number m does not appear in the equation, so we conclude that neither $R(r)$ nor E depends on m. For a given l, there may be many possible solutions to Eq. 16.2, which we can designate by a third radial quantum number n. The energy eigenstates $|nlm\rangle$ have wave functions

$$\langle r, \theta, \phi \,|nlm \rangle = \psi_{nlm}(r, \theta, \phi) = R_{nl}(r)\, Y_{lm}(\theta, \phi), \tag{16.3}$$

and energies E_{nl} (independent of m). The radial equation becomes

$$\frac{1}{r^2} \frac{d}{dr} \left(r^2 \frac{dR_{nl}}{dr} \right) + \frac{2\mu}{\hbar^2} \left(E_{nl} - U(r) - \frac{l(l+1)\hbar^2}{2\mu r^2} \right) R_{nl} = 0. \tag{16.4}$$

We can simplify this even further. First, we note that the term $\dfrac{l(l+1)\hbar^2}{2\mu r^2}$, though derived from the kinetic energy term in the Hamiltonian, formally appears like an l-dependent term in the potential energy. A potential of this form would provide a force pushing away from the origin, so that we call it the *centrifugal term*. We define the *effective potential* $U_{\text{eff}}(r)$ to be the sum of the real potential energy and the centrifugal term:

$$U_{\text{eff}}(r) = U(r) + \frac{l(l+1)\hbar^2}{2\mu r^2}, \tag{16.5}$$

(see Fig. 16.1). Rather than working with the function $R_{nl}(r)$ directly, it is sometimes easier to consider $u_{nl}(r) = rR_{nl}(r)$. This simplifies the derivatives somewhat.

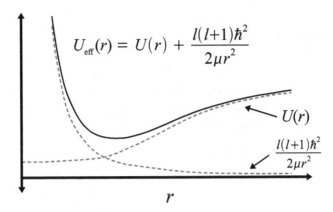

The effective potential is the sum of $U(r)$ and the centrifugal term, which depends on the angular momentum quantum number l. For $l > 0$ there is an infinite "centrifugal barrier" at $r = 0$.

Exercise 16.3 Show that $\dfrac{1}{r^2}\dfrac{d}{dr}\left(r^2\dfrac{dR_{nl}}{dr}\right) = \dfrac{1}{r}\dfrac{d^2u_{nl}}{dr^2}$.

We arrive at the equation

$$\frac{d^2u_{nl}}{dr^2} + \frac{2\mu}{\hbar^2}\left(E_{nl} - U_{\text{eff}}(r)\right)u_{nl} = 0. \tag{16.6}$$

This is exactly the same as the energy eigenvalue equation for a 1-D particle moving in the effective potential U_{eff}. (Compare it with Eq. 11.17 and 15.4.)

What boundary conditions does the function u_{nl} satisfy? The wave function is bounded at the origin, so we know that $u_{nl}(r) \to 0$ as $r \to 0$. As far as the function u_{nl} is concerned, there is an impenetrable "wall" at $r = 0$. The boundary condition as $r \to \infty$ requires slightly more thought. Since the bound state is normalized,

$$\iiint |\psi_{nlm}|^2\,dV = \iiint |R_{nl}|^2\,|Y_{lm}|^2\,r^2\sin\theta\,d\phi\,d\theta\,dr = 1, \tag{16.7}$$

where the integration includes all of 3-D space. The angular integrals are done by recognizing that the spherical harmonics have the normalization

$$\int_0^{2\pi} d\phi \int_0^{\pi} \sin\theta\,d\theta\,|Y_{lm}|^2 = 1. \tag{16.8}$$

Thus, our wave function normalization becomes

$$\int_0^{\infty} |R_{nl}|^2\,r^2 dr = \int_0^{\infty} |u_{nl}|^2\,dr = 1. \tag{16.9}$$

We conclude from this that $u_{nl}(r) \to 0$ as $r \to \infty$.

Because there are boundary conditions on u_{nl} in both directions, we expect the spectrum of possible energy eigenvalues E_{nl} to be discrete. See the discussion in Section 15.2.

We conclude our general discussion with a bit of nomenclature borrowed from atomic physics. The values $l = 0, 1, 2, 3, \ldots$ for the L^2 quantum number may be denoted by the letters S, P, D, F, etc. A state with $l = 0$ is called an S-state, one with $l = 1$ is called a P-state, and so on.

The S-states of a particle in a central potential are special in a couple of ways. First of all, since the spherical harmonic Y_{00} is a constant, these states have radially symmetric wave functions. Also, the effective potential $U_{\text{eff}}(r)$ for an S-state has no centrifugal term. Because such a term can only increase the energy of a state, it follows that the ground state of a particle in a central potential is always an S-state.

Exercise 16.4 Prove the last assertion by showing that, for any energy eigenstate having $l \neq 0$, there is an S-state with a lower expectation value $\langle E \rangle$ for its energy.

16.2 The isotropic oscillator

When exploring a new set of ideas, it is often useful to apply those ideas to a problem for which the answer is already known. For a particle moving in a central potential, we have a perfect example: the isotropic harmonic oscillator. In the isotropic oscillator, the potential energy function is

$$U(\vec{r}) = \frac{\mu\omega^2}{2}r^2 = \frac{\mu\omega^2}{2}\left(x^2 + y^2 + z^2\right). \tag{16.10}$$

This is at the same time both a central potential and a potential of the form $U_x(x) + U_y(y) + U_z(z)$. The three Cartesian coordinates are non-interacting degrees of freedom, each of which is a 1-D harmonic oscillator with classical frequency ω. We can construct independent ladder operators a_x, a_y, and a_z, with corresponding number operators

$$n_x = a_x^\dagger a_x, \qquad n_y = a_y^\dagger a_y, \qquad n_z = a_z^\dagger a_z. \tag{16.11}$$

The Hamiltonian for the system is

$$H = \hbar\omega\left(n_x + \frac{1}{2}\mathbf{1}\right) + \hbar\omega\left(n_y + \frac{1}{2}\mathbf{1}\right) + \hbar\omega\left(n_z + \frac{1}{2}\mathbf{1}\right). \tag{16.12}$$

The stationary states for this system are eigenstates of the total number operator $N = n_x + n_y + n_z$. This operator has eigenvalues $N = 0, 1, 2, \ldots$

There is some degeneracy in this system, of course. We can give a specific basis of stationary states by choosing the joint eigenstates $|n_x, n_y, n_z\rangle$ of the three Cartesian coordinate number operators. Then

$$H|n_x, n_y, n_z\rangle = \left(N + \frac{3}{2}\right)\hbar\omega|n_x, n_y, n_z\rangle, \tag{16.13}$$

where $N = n_x + n_y + n_z$. The degeneracies associated with the various energy levels – that is, the dimensions of the eigenspaces associated with each energy – can be worked out simply by enumerating the (n_x, n_y, n_z) combinations for each given N. For $N = 0, 1, 2, 3, \ldots$ the corresponding degeneracies are $1, 3, 6, 10, \ldots$

Exercise 16.5 Confirm these results and find the degeneracy for $N = 4$.

So much for the Cartesian analysis. The isotropic oscillator is also a central potential, so that we should be able to find energy eigenstates that are also eigenstates of the angular momentum operators L^2 and L_z. The eigenstates can be labeled $|nlm\rangle$, where n is a radial quantum number and l and m are angular momentum quantum numbers. Our previous analysis immediately yields the following facts:

- $L^2 |nlm\rangle = l(l+1)\hbar^2 |nlm\rangle$, and $L_z |nlm\rangle = m\hbar |nlm\rangle$.
- The energy of the state $|nlm\rangle$ is E_{nl}, which depends only on n and l. Since we already know from Eq. 16.13 that the energy is determined by the total quantum number N, we infer that this N is a function of n and l.
- The energy is not dependent on the L_z quantum number m. This leads to some degeneracy in the energy levels. There may also be additional degeneracy – that is, distinct values of n and l might yield the same E_{nl}.
- The state $|nlm\rangle$ will have a wave function of the form $R_{nl}(r)\, Y_{lm}(\theta, \phi)$, where the radial function R_{nl} is a solution to the radial equation, Eq. 16.4.

We could now proceed by solving this radial differential equation (Eq. 16.4) directly. In fact, we will follow exactly this strategy for the Coulomb potential in the next section. This time, however, it is more interesting to approach the problem algebraically.

The Cartesian coordinates and momenta can be written in terms of the corresponding ladder operators. Thus,

$$x = \sqrt{\frac{\hbar}{2\mu\omega}} \left(a_x + a_x^\dagger\right) \quad \text{and} \quad p_x = -i\sqrt{\frac{\hbar\mu\omega}{2}} \left(a_x - a_x^\dagger\right), \tag{16.14}$$

with similar expressions for the other two coordinates.

Exercise 16.6 Derive Eq. 16.14 from the analysis of the harmonic oscillator in Section 13.2.

The components of the angular momentum are $L_x = y p_z - z p_y$, etc. We can therefore write the angular momentum operators in terms of the Cartesian ladder operators.

Exercise 16.7 Use Eq. 16.14 (and the corresponding relations for the other coordinates) to show

$$L_x = i\hbar \left(a_y a_z^\dagger - a_y^\dagger a_z\right),$$

$$L_y = i\hbar \left(a_z a_x^\dagger - a_z^\dagger a_x\right),$$

$$L_z = i\hbar \left(a_x a_y^\dagger - a_x^\dagger a_y\right). \tag{16.15}$$

From these, we could work out a lengthy expression for L^2. But it is easier to postpone the algebra for a bit. We first establish two useful elementary facts about ladder operators.

Exercise 16.8 Suppose a and $a^\dagger$ are the ladder operators for an oscillator, and $n = a^\dagger a$ is the corresponding number operator. From the commutation relation $[a, a^\dagger] = 1$, show that (a) $[a^2, a^\dagger] = 2a$; and (b) $\left(a^\dagger\right)^2 a^2 = n^2 - n$.

Next, we define a *radial ladder operator* A by

$$A = a_x^2 + a_y^2 + a_z^2. \tag{16.16}$$

Why define A in this way? Intuitively, the Cartesian ladder operators appear in a "Pythagorean" way in this definition. More crucially, this operator commutes with all of the angular momentum components:

$$[L_x, A] = [L_y, A] = [L_z, A] = 0. \tag{16.17}$$

Exercise 16.9 Choose one of these commutators and confirm that it is zero.

It follows that $[L^2, A] = 0$ as well. The operator A has a definite effect on the energy – or, equivalently, on the total quantum number N. Suppose $|\psi\rangle$ is an eigenstate of N with eigenvalue N. Then $A|\psi\rangle$ and $A^\dagger|\psi\rangle$ are also N-eigenstates.

$$NA|\psi\rangle = (N-2)A|\psi\rangle \quad \text{and} \quad NA^\dagger|\psi\rangle = (N+2)A^\dagger|\psi\rangle. \tag{16.18}$$

We can see this in two distinct ways. The first would begin by calculating the commutator $[N, A] = -2A$, and then proceed from there. The second way to see this is simply to note that A involves *squares* of the Cartesian ladder operators. Thus, every term in A has the effect of changing the sum $N = n_x + n_y + n_z$ by exactly -2. The opposite is of course true for $A^\dagger$.

Now we can derive a compact expression for L^2. We leave the detailed algebra for Problem 16.4, and here simply state the final result:

$$L^2 = \hbar^2 \left(N(N+1) - A^\dagger A \right). \tag{16.19}$$

It is now straightforward to construct the states $|nlm\rangle$. For each choice of l and m, there is a "ladder" of states $|0lm\rangle$, $|1lm\rangle$, $|2lm\rangle$, and so on. Each state is an eigenstate of N (and thus the total energy) with eigenvalue N. Adjacent rungs of each ladder are connected by the A and $A^\dagger$ operators. The bottom "rung" of each ladder is defined by $A|0lm\rangle = 0$. What is the value of N for this state? We find that

$$L^2|0lm\rangle = \hbar^2 N(N+1)|0lm\rangle - \hbar^2 A^\dagger A|0lm\rangle,$$
$$l(l+1)\hbar^2|0lm\rangle = N(N+1)\hbar^2|0lm\rangle. \tag{16.20}$$

It follows that, for the bottom rung state, $N = l$. By applying $A^\dagger$ (and normalizing as necessary) we obtain $|1lm\rangle$, $|2lm\rangle$ and so on. Each time we apply $A^\dagger$, we increase N by 2. Thus $N = 2n + l$, and the energy is

$$E_{nl} = \left(N + \frac{3}{2}\right)\hbar\omega = \left(2n + l + \frac{3}{2}\right)\hbar\omega. \tag{16.21}$$

How does the overall pattern of energy levels, including degeneracy, compare with the pattern we found using Cartesian coordinates? Figure 16.2 shows the arrangement. There is a "ladder" of different values of n for each l, and each rung of this ladder corresponds to $2l + 1$ degenerate basis states differing in m. These are arranged according to the quantum number $N = 2n+l$. Thus, states on the same horizontal line (same N) have the same energy.

Energy levels $|nlm\rangle$ of the isotropic oscillator in 3-D, showing the degeneracy for each combination of n and l. At the right are the degeneracies for each $N = 2n + 1$.

The total degeneracy for each N-value is exactly as we found in the Cartesian analysis. (The eigenstates $|nlm\rangle$ are, however, generally superpositions of the various Cartesian $|n_x, n_y, n_z\rangle$ states. Some of the details are worked out in Problem 16.5.)

16.3 Hydrogen

In the hydrogen atom, a single electron of charge $-e$ moves around a nucleus (generally a proton), bound by the Coulomb attraction between them. Generalizing slightly, we consider a nucleus of arbitrary charge $+Ze$, where Z is the atomic number.[2] The potential energy function of the system is

$$U(r) = -\frac{kZe^2}{r},$$ (16.22)

where r is the distance between the nucleus and the electron (the magnitude of the relative position vector) and the Coulomb constant is $k = 8.99 \times 10^9$ N m^2/C^2. What are the stationary bound states and the corresponding energies for this system?

Note that $U(r) < 0$, with $U(r) \to 0$ as $r \to \infty$. Thus, the bound stationary states will be the energy eigenstates having $E < 0$. From our general analysis in Section 16.1, we know that we can make sure the eigenstates are at the same time eigenstates of L^2 and L_z. The wave functions are of the form $R(r)Y_{lm}(\theta, \phi)$, where $R(r)$ satisfies the radial equation

$$-\frac{\hbar^2}{2\mu} \left[\frac{1}{r^2} \frac{d}{dr} \left(r^2 \frac{dR}{dr} \right) - \frac{l(l+1)}{r^2} R \right] + U(r)R = ER,$$ (Re 16.2)

[2] Thus, our analysis will cover H atoms, He$^+$ ions, Li^{++} ions, etc.

where μ is the relative mass of the electron with respect to the nucleus, which is nearly equal to the electron mass. This leads to

$$\frac{d^2R}{dr^2} + \frac{2}{r}\frac{dR}{dr} + \frac{2\mu}{\hbar^2}\left(E + \frac{kZe^2}{r} - \frac{l(l+1)\hbar^2}{2\mu r^2}\right)R = 0. \tag{16.23}$$

Since the energy is negative, $E = -|E|$.

To work with this equation, we employ a trick we previously used in our discussion of the quantum circular billiard system in Section 11.5. We use combinations of the parameters in Eq. 16.23 to construct a dimensionless radial coordinate ρ:

$$\rho = \alpha r = \sqrt{\frac{8\mu|E|}{\hbar^2}}\, r. \tag{16.24}$$

Equation 16.23 can now be rewritten

$$\frac{d^2R}{d\rho^2} + \frac{2}{\rho}\frac{dR}{d\rho} - \frac{l(l+1)}{\rho^2}R + \left(\frac{\lambda}{\rho} - \frac{1}{4}\right)R = 0, \tag{16.25}$$

where the dimensionless parameter λ is

$$\lambda = \frac{kZe^2}{\hbar}\sqrt{\frac{\mu}{2|E|}}. \tag{16.26}$$

Exercise 16.10 Confirm that the coordinate ρ and the parameter λ are both dimensionless. Then verify Eq. 16.25.

Equation 16.25 has two "kinetic" terms involving derivatives of R and three "potential" terms. To learn more about its solutions, we will consider how they behave when the coordinate ρ is very small or very large.

First, suppose that $\rho \ll 1$. Then the centrifugal potential term will dominate (provided $l > 0$) and the other two will be negligible. If we suppose that the solution is given by a power series $R = \sum_n c_n \rho^n$, then the first non-zero term $c_s \rho^s$ also dominates. Then

$$\frac{d^2R}{d\rho^2} = s(s-1)c_s\rho^{s-2},$$

$$\frac{2}{\rho}\frac{dR}{d\rho} = 2sc_s\rho^{s-2},$$

$$\frac{l(l+1)}{\rho^2}R = l(l+1)c_s\rho^{s-2}, \tag{16.27}$$

and therefore

$$(s(s+1) - l(l+1))\, c_s\rho^{s-2} = 0. \tag{16.28}$$

Either $s = l$ or $s = -(l+1)$. The second possibility, however, would mean that R diverges as $\rho \to 0$, so we exclude it. We conclude that $R \propto \rho^l$ close to the origin.[3]

---

[3] This is a pretty general fact, and should hold whenever the potential $V(r)$ is "less singular" than $1/\rho^2$.

Next, consider what happens when $\rho \to \infty$. Then the only non-negligible terms from Eq. 16.25 yield

$$\frac{d^2 R}{d\rho^2} - \frac{1}{4} R = 0. \tag{16.29}$$

The independent solutions of this equations are $\exp(\rho/2)$ and $\exp(-\rho/2)$. The first solution cannot apply, because for a bound state we must have $R \to 0$ as $\rho \to \infty$.

All of this motivates us to write the general solution in the form

$$R = \rho^l f(\rho) \exp\left(-\frac{\rho}{2}\right), \tag{16.30}$$

where $f(\rho)$ is an as-yet-undetermined function. Equation 16.25 becomes an equation for f.

Exercise 16.11 Show that Eq. 16.25 becomes

$$\frac{d^2 f}{d\rho^2} + \left(\frac{2(l+1)}{\rho} - 1\right) \frac{df}{d\rho} + \left(\frac{\lambda - (l+1)}{\rho}\right) f = 0. \tag{16.31}$$

We solve this equation by letting $f(\rho) = \sum_{j=0}^{\infty} a_j \rho^j$. Equation 16.31 implies a recursion relation for the series coefficients a_j:

$$a_{j+1} = \frac{j + (l+1) - \lambda}{(j+1)(j+2(l+1))} a_j. \tag{16.32}$$

Exercise 16.12 Check this. Along the way, you will do some "reindexing" of the sums. For example, you have to rewrite the second derivative term

$$\sum_{j=0}^{\infty} j(j-1) a_j \rho^{j-2} \quad \text{as} \quad \sum_{j=0}^{\infty} (j+1)j \, a_{j+1} \rho^{j-1}. \tag{16.33}$$

Explain why this is okay.

As $k \to \infty$, both λ and l become negligible. Then $a_{j+1} \approx a_j/j$. But this is the same asymptotic recursion relation for $f(\rho) = \exp(\rho)$. This looks like a serious problem, since it would mean that $R \to \infty$ as $\rho \to \infty$, which is impossible for a normalized wave function.

The only way to avoid this unphysical disaster is to posit that the series for $f(\rho)$ *terminates* at some point – that is, to insist that $f(\rho)$ is a polynomial of finite degree. Then there exists an integer j such that $j + (l+1) - \lambda = 0$. This amounts to a restriction on the possible values of λ, namely that $\lambda = n$, for some integer $n \geq l+1$. We call n the *principal quantum number* for the system, and then rewrite Eq. 16.26 as

$$n^2 = \left(\frac{kZe^2}{\hbar}\right)^2 \frac{\mu}{2|E|}. \tag{16.34}$$

The bound state energy $E = -|E|$ depends only on the principal quantum number n. We can write

$$E_n = -\left(\frac{\mu k^2 Z^2 e^4}{2\hbar^2}\right)\frac{1}{n^2}. \tag{16.35}$$

The boundary condition that $R \to 0$ as $\rho \to \infty$ forces the energy levels of the atom to be quantized according to Eq. 16.35.

Exercise 16.13 For the actual hydrogen atom, $Z = 1$. Show that the constant factor in Eq. 16.35 has a value of 13.6 eV. Write down the energies of the lowest five levels of the atom, and sketch an energy-level diagram.

We can make things look a little simpler by defining a characteristic length, the *Bohr radius* a, to be

$$a = \frac{\hbar^2}{\mu k Z e^2}. \tag{16.36}$$

Then the dimensionless radial coordinate is $\rho = 2r/na$ and

$$E_n = -\frac{kZe^2}{2a}\frac{1}{n^2}. \tag{16.37}$$

Let us summarize our results. In addition to angular momentum quantum numbers l and m, the hydrogen wave function has a principal quantum number n. We can have $n = 1, 2, 3, \ldots$, and for a given value of n the angular momentum quantum number $l = 0, \ldots, n-1$. As usual, m ranges over $-l, \ldots, l$. The radial function R_{nl} depends on the quantum numbers n and l, but not m. Thus, we will have $R_{10}, R_{20}, R_{21}, R_{30}$, and so on. The energy eigenvalue of the state depends only on n, according to Eq. 16.35 (or 16.37). The general form of the radial function is

$$R_{nl} = A \left(\frac{r}{a}\right)^l \left(\text{polynomial in } \left(\frac{r}{a}\right)\right) \exp\left(-\frac{r}{na}\right), \tag{16.38}$$

where A is a normalization constant, a is the Bohr radius, and the polynomial function[4] has degree $n - (l+1)$.

We can obtain a good deal of qualitative information simply by inspecting Eq. 16.38. The asymptotic ($r \to \infty$) drop-off rate is determined by n, so that wave functions with higher n decrease more slowly with r. On the other hand, the behavior of the wave function near the origin is strongly influenced by l. For larger values of l, the wave function is shifted outward, away from the origin. This makes excellent sense, because a larger orbital angular momentum means a higher "centrifugal barrier" in the effective potential.

---

[4] These polynomials are well known in mathematical physics. For $l = 0$ they are the so-called Laguerre polynomials, and for $l > 0$ they are the associated Laguerre polynomials.

Here are the first few radial functions $R_{nl}(r)$:

$$R_{10}(r) = 2\,a^{-3/2}\,\exp(-r/a),$$

$$R_{20}(r) = \frac{1}{\sqrt{2}}\,a^{-3/2}\left(1 - \frac{1}{2}\left(\frac{r}{a}\right)\right)\exp(-r/2a),$$

$$R_{21}(r) = \frac{1}{\sqrt{24}}\,a^{-3/2}\left(\frac{r}{a}\right)\exp(-r/2a),$$

$$R_{30}(r) = \frac{2}{\sqrt{27}}\,a^{-3/2}\left(1 - \frac{2}{3}\left(\frac{r}{a}\right) + \frac{2}{27}\left(\frac{r}{a}\right)^2\right)\exp(-r/3a),$$

$$R_{31}(r) = \frac{8}{27\sqrt{6}}\,a^{-3/2}\left(\frac{r}{a}\right)\left(1 - \frac{1}{6}\left(\frac{r}{a}\right)\right)\exp(-r/3a),$$

$$R_{32}(r) = \frac{4}{81\sqrt{30}}\,a^{-3/2}\left(\frac{r}{a}\right)^2\exp(-r/3a), \tag{16.39}$$

where a is the Bohr radius from Eq. 16.36.

Exercise 16.14 Compare the radial functions in Eq. 16.39 to the general form in Eq. 16.38 and identify each of the parts.

Exercise 16.15 Why is there a factor of $a^{-3/2}$ in each $R_{nl}(r)$. Hint: What are the units of the wave function?

Exercise 16.16 Show that the radial function R_{20} is properly normalized.

Given the hydrogen stationary state wave functions $R_{nl}(r)Y_{lm}(\theta, \phi)$, we can compute various useful expectation values directly. However, many of these expectations can be found by other techniques, as we will now discuss.

16.4 Some expectations

Recall Eq. 16.6 for the radial function $u_{nl}(r) = rR_{nl}(r)$, which we repeat here in a slightly modified form:

$$-\frac{\hbar^2}{2\mu}\frac{d^2u_{nl}}{dr^2} + U_{\text{eff}}(r)u_{nl} = E_{nl}u_{nl}. \tag{16.40}$$

As we saw, this is exactly the energy eigenvalue equation for a 1-D particle of mass μ moving in the potential $U_{\text{eff}}(r)$, which for the hydrogen atom is

$$U_{\text{eff}}(r) = -\frac{kZe^2}{r} + \frac{l(l+1)\hbar^2}{2\mu r^2}, \tag{16.41}$$

(see Eq. 16.5). The energy eigenvalues E_{nl} are given by Eq. 16.35.

Let $|u_{nl}\rangle$ be the energy eigenstate of this equivalent 1-D problem. We can now apply the results of Section 15.7 to find certain expectation values for the hydrogen wave functions.

For instance, we can apply the Feynman–Hellmann theorem of Eq. 15.73 to the parameter Z and obtain

$$\langle u_{nl}| \left(-\frac{ke^2}{r} \right) |u_{nl}\rangle = -\left(\frac{\mu k^2 Z e^4}{\hbar^2} \right) \frac{1}{n^2}, \tag{16.42}$$

which in turn tells us that

$$\left\langle \frac{1}{r} \right\rangle = \left(\frac{\mu k Z e^2}{\hbar^2} \right) \frac{1}{n^2} = \frac{1}{n^2 a}. \tag{16.43}$$

Exercise 16.17 Apply the Feynman–Hellmann theorem to the mass parameter μ and obtain an expression for the average kinetic energy $\langle K \rangle$. (Remember, the physical kinetic energy includes one of the terms in the effective 1-D potential.)

We must apply the Feynman–Hellmann theorem with some care. For example, the angular momentum quantum number l appears as a parameter in Eq. 16.40. Applying the theorem to this parameter should yield an expectation value for $\langle 1/r^2 \rangle$. However, the energy E_{nl} is apparently independent of l, so we are tempted to conclude that $\langle 1/r^2 \rangle = 0$ – a plainly absurd result. What has gone wrong?

The answer is subtle. When we solved the radial equation for the hydrogen atom, we found there must be an integer j (the degree of a polynomial) such that $j + (l+1) = \lambda$, a dimensionless parameter related to the energy (defined in Eq. 16.26). The principal quantum number was the combination $n = j + (l+1)$. If we now vary the l parameter continuously, the integer j remains unchanged, but not the combination n. Thus

$$\frac{dE}{dl} = -\frac{kZe^2}{2a} \frac{d}{dl} \left(\frac{1}{n^2} \right) = \frac{kZe^2}{a} \frac{1}{n^3}. \tag{16.44}$$

Exercise 16.18 Verify this result. Then use it to show that

$$\left\langle \frac{1}{r^2} \right\rangle = \frac{2}{a^2} \frac{1}{(2l+1)n^3}, \tag{16.45}$$

for a stationary state of the hydrogen atom.

These results can be extended in various ways. For example, suppose we wish to find the expectation $\langle p^4 \rangle$ for the $|nlm\rangle$ state of the hydrogen atom. At first, this seems very challenging. However, we can write this as

$$p^4 = \left(p^2 \right)^2 = 4\mu^2 \left(H - U(r) \right)^2, \tag{16.46}$$

with the Hamiltonian H.

Exercise 16.19 Show that for the state $|nlm\rangle$,

$$\langle p^4 \rangle = 4\mu^2 \left(E_n^2 + 2kZe^2 E_n \left\langle \frac{1}{r} \right\rangle + k^2 Z^2 e^4 \left\langle \frac{1}{r^2} \right\rangle \right), \tag{16.47}$$

where E_n is given by Eq. 16.37. Do the algebra and write $\langle p^4 \rangle$ in terms of physical constants and the quantum numbers n and l.

Finally, we mention an extremely useful fact about the $|nlm\rangle$ states, called *Kramer's rule*, which relates the expectations $\langle r^k \rangle$ for various integers k. The derivation of the rule is somewhat involved, so we will merely state the rule here without giving its proof.

$$\frac{k+1}{n^2} \langle r^k \rangle - (2k+1)a \langle r^{k-1} \rangle + \frac{k}{4} \left((2l+1)^2 - k^2 \right) a^2 \langle r^{k-2} \rangle = 0. \qquad (16.48)$$

Exercise 16.20 Verify Kramer's rule for $k = 0$.

Problems

Problem 16.1 A free particle of mass μ moves in a *spherical box* of radius R. The wave function $\psi = 0$ for all $r \geq R$, but within the box its potential is $U = 0$. Find the ground state energy of this particle.

Problem 16.2 A particle moves in three dimensions subject to a delta function potential $U(\vec{r}) = -\epsilon\, \delta^3(\vec{r})$. This is a radially symmetric potential, so its ground state should be an S state. Find the ground state and its energy. Are there any other bound states?

Problem 16.3 In this problem we adapt the variational method of Section 15.6 to finding the lowest-energy solutions of the radial equation (Eq. 16.6), which we interpret as the energy eigenvalue equation for a particle moving in one dimension subject to the effective potential $U_{\text{eff}}(r)$. Since the radial function $R(r) = u(r)/r \propto r^l$ as $r \to 0$, we choose the test function

$$u_\alpha(r) = A_\alpha r^{l+1} e^{-\alpha r}, \qquad (16.49)$$

for $r \geq 0$. We will consider only the cases $l = 0, 1, 2$.

(a) For each l, find A_α so that $u(r)$ is properly normalized.
(b) Calculate the expectation values for the kinetic and centrifugal terms in the equivalent 1-D Hamiltonian.
(c) Suppose the actual potential $U(r) = kr^2/2$, the isotropic oscillator potential. Use the variational method to estimate the minimum energies for S, P, and D states. Compare your results to the exact oscillator energies.
(d) Do the same for the hydrogen atom potential. (The $l = 0$ result is actually correct. Why?)

Problem 16.4 For the isotropic oscillator, show that the angular momentum operator L^2 is given by

$$L^2 = \hbar^2 \left(N(N+1) - A^\dagger A \right), \qquad (\text{Re } 16.19)$$

where N and A are as defined in Section 16.2.

Problem 16.5 The isotropic oscillator has radial energy eigenstates $|nlm\rangle$ which can be written in terms of Cartesian energy eigenstates $|n_x, n_y, n_z\rangle$.

(a) Write the three radial states $|01m\rangle$ as superpositions of the Cartesian states $|1, 0, 0\rangle$, $|0, 1, 0\rangle$, and $|0, 0, 1\rangle$.

(b) Write the radial states $|10m\rangle$ and $|02m\rangle$ as superpositions of Cartesian basis states.

Problem 16.6 The size of the atomic nucleus is of the order of 10^{-4} times the Bohr radius a, which is why we can regard it as a point charge in our analysis. Nuclear interactions only take place over very short ranges.

Some unstable nuclei decay by *electron capture*, in which an orbital electron is absorbed by the nucleus, transforming a proton into a neutron. (A neutrino is emitted in this process.) The captured electron essentially always comes from a 1S orbital state. Why?

Problem 16.7 Tritium (^{3}H) is an unstable isotope of hydrogen, which can undergo beta decay and become helium-3 (^{3}He). The decay happens so fast that the wave function of the orbiting electron is unchanged. However, an energy eigenstate for ^{3}H ($Z = 1$) is not one for ^{3}He ($Z = 2$). Suppose the electron is in its ground state in the tritium atom immediately before the decay. Calculate the probability that it will be found in the ground state of the helium atom immediately afterwards.

Problem 16.8 Use Kramer's rule to find $\langle r \rangle$ for the $|nlm\rangle$ state of hydrogen. Check the result for the $|210\rangle$ state by doing the appropriate integral.

17 Perturbation theory

17.1 Shifting the energy levels

In physics, it often happens that the problem we wish to solve is close to – but not quite the same as – a problem we actually *can* solve. For example, for motions close to a stable equilibrium point, a classical 1-D particle can usually be regarded as a harmonic oscillator, because the potential function $U(x)$ is approximately a parabola near its minimum. This fact accounts for the very great importance of the harmonic oscillator in mechanics. The same is true for a quantum particle.[1] We can solve the classical and quantum harmonic oscillator problems very well by now. Yet most real potential functions are only *approximately* parabolic around an equilibrium point. How can we make our analysis more exact?

The general strategy for this type of quantum problem is known as *perturbation theory*. We imagine that the Hamiltonian is composed of two parts, an "unperturbed" part H_0 and a *perturbation* H_P, both of which are independent of time. (We are therefore doing *time-independent perturbation theory*.) That is,

$$H = H_0 + H_P. \tag{17.1}$$

For the harmonic oscillator example, H_0 is the exact harmonic oscillator Hamiltonian, and H_P is the small difference between the real potential U and the local parabolic approximation. We assume that we know the stationary states $|n\rangle$ and the corresponding energy levels ε_n for the unperturbed Hamiltonian H_0; for now, we shall suppose that these levels are non-degenerate. Our job in time-independent perturbation theory is to approximate the energy levels E_n and stationary states $|\psi_n\rangle$ for the full Hamiltonian H. In fact, what we will do is develop a sequence of better and better approximations to the exact answer, a sort of "power series" in the perturbation. If the perturbation is small, then we may be able to make do with only the first couple of corrections to the unperturbed ε_n and $|n\rangle$.

To see how this works, introduce a parameter λ and let

$$H_\lambda = H_0 + \lambda H_P. \tag{17.2}$$

---

[1] True, with the following reservation. Unlike the classical oscillator, the quantum harmonic oscillator has a natural length scale $\sqrt{\hbar/\mu\omega}$. Even the ground state is spread out over a region of this size. The harmonic oscillator approximation will only make sense in the quantum case if the potential is nearly parabolic over an interval wider than this.

For the unperturbed situation, $\lambda = 0$, and for the situation of interest, $\lambda = 1$. The role of the parameter λ is to smoothly interpolate between these.[2] The Hamiltonian, stationary states and energy eigenvalues all become functions of λ and satisfy

$$(\mathsf{H}_0 + \lambda \mathsf{H}_\mathsf{P}) \, |\psi_n(\lambda)\rangle = E_n(\lambda) \, |\psi_n(\lambda)\rangle \,, \tag{17.3}$$

for all λ. We can express the states $|\psi_n(\lambda)\rangle$ in terms of the unperturbed $|n\rangle$ basis:

$$|\psi_n(\lambda)\rangle = |n\rangle + \sum_{k \neq n} c_{kn}(\lambda) \, |k\rangle \,, \tag{17.4}$$

where $c_{kn}(\lambda) = \langle k \, | \psi_n(\lambda) \rangle$. For algebraic convenience, we have chosen not to normalize the vector $|\psi_n(\lambda)\rangle$, but it nevertheless is a solution to Eq. 17.3. Now write $E_n(\lambda)$ and the coefficients $c_{kn}(\lambda)$ as power series in λ:

$$c_{kn}(\lambda) = \lambda \, c_{kn}^{(1)} + \lambda^2 c_{kn}^{(2)} + \dots \,,$$

$$E_n(\lambda) = E_n^{(0)} + \lambda E_n^{(1)} + \lambda^2 E_n^{(2)} + \dots \tag{17.5}$$

For $\lambda = 1$, this reduces to $c_{kn} = c_{kn}^{(1)} + c_{kn}^{(2)} + \dots$ and $E_n = E_n^{(0)} + E_n^{(1)} + E_n^{(2)} + \dots$

Exercise 17.1 Why did we not include a λ^0 term in the power series for $c_{kn}(\lambda)$?

Equation 17.3 becomes

$$(\mathsf{H}_0 + \lambda \mathsf{H}_\mathsf{P}) \left(|n\rangle + \lambda \, c_{kn}^{(1)} + \dots \right) = \left(E_n^{(0)} + \lambda E_n^{(1)} + \dots \right) \left(|n\rangle + \lambda \, c_{kn}^{(1)} + \dots \right). \tag{17.6}$$

Each side of Eq. 17.6 is a function of λ expressed as a power series. For two power series to be equal, the terms involving corresponding powers of λ must be equal. The $\mathcal{O}\left(\lambda^0\right)$ terms in Eq. 17.6 yield the relation

$$\mathsf{H}_0 \, |n\rangle = E_n^{(0)} \, |n\rangle \,. \tag{17.7}$$

This holds provided that we identify the zeroth-order energy term $E_n^{(0)} = \varepsilon_n$, the unperturbed energy.

The first-order ($\mathcal{O}\left(\lambda^1\right)$) terms give us

$$\mathsf{H}_0 \sum_{k \neq n} c_{kn}^{(1)} \, |k\rangle + \mathsf{H}_\mathsf{P} \, |n\rangle = E_n^{(1)} \, |n\rangle + E_n^{(0)} \sum_{k \neq n} c_{kn}^{(1)} \, |k\rangle \,. \tag{17.8}$$

Recalling that $E_n^{(0)} = \varepsilon_n$ and $\mathsf{H}_0 \, |k\rangle = \varepsilon_k \, |k\rangle$, we arrive at the following:

$$\mathsf{H}_\mathsf{P} \, |n\rangle = E_n^{(1)} \, |n\rangle + \sum_{k \neq n} (\varepsilon_n - \varepsilon_k) \, c_{kn}^{(1)} \, |k\rangle \,. \tag{17.9}$$

Equation 17.9 tells us a great deal. For instance, suppose we act on both sides of this equation with the dual vector $\langle n |$. Since $\langle n | k \rangle = 0$ for $k \neq n$, we find

$$E_n^{(1)} = \langle n | \, \mathsf{H}_\mathsf{P} \, |n\rangle \,. \tag{17.10}$$

---

[2] The λ parameter also serves as a "marker" for the perturbation H_P. Wherever a factor of λ appears in an expression involving H_λ, the operator H_P also appears. Counting "powers of λ" amounts to counting powers of the perturbation H_P.

The first-order energy correction $E_n^{(1)}$ is just the expectation of H_P for the unperturbed eigenstate. This is a very useful result – one that may, in fact, seem somewhat familiar:

Exercise 17.2 Discuss the connection between Eq. 17.10 and the Feynman–Hellmann theorem of Section 15.7.

What about the first-order correction to the state vector?

Exercise 17.3 From Eq. 17.9 show that

$$c_{kn}^{(1)} = \frac{\langle k| H_P |n\rangle}{\varepsilon_n - \varepsilon_k},\tag{17.11}$$

for $k \neq n$.

We notice, first of all, that this depends on our assumption that the unperturbed energy spectrum is non-degenerate, i.e. that $\varepsilon_n \neq \varepsilon_k$ for $k \neq n$. Second, it is worth noting a fact about the normalization of this state.

Exercise 17.4 Show that the vector $|n\rangle + \lambda \sum_{k \neq n} c_{kn}^{(1)} |k\rangle$ is in fact normalized, if we ignore terms that are $\mathcal{O}\left(\lambda^2\right)$ and higher. What constant factor would normalize this vector exactly?

Equations 17.10 and 17.11 describe the first-order effects of the perturbation H_P on the energy and stationary state. We can write

$$E_n \approx \varepsilon_n + \langle n| H_P |n\rangle,$$

$$|\psi_n\rangle \approx |n\rangle + \sum_{k \neq n} \frac{\langle k| H_P |n\rangle}{\varepsilon_n - \varepsilon_k} |k\rangle,\tag{17.12}$$

to first order in the perturbation H_P.

Exercise 17.5 This is the world's easiest perturbation theory problem. Let $H_P = \eta \mathbf{1}$. Show that the first-order energy correction is $E_n^{(1)} = \eta$, while the first-order state correction is exactly zero. Explain why this result makes intuitive sense. What do you expect for higher-order terms?

Can we go further? We can, and sometimes we must. It often happens that, because of symmetry, the first-order energy corrections $E_n^{(1)}$ vanish. The first non-zero change in the energy levels will be the second-order correction $E_n^{(2)}$. This arises from the λ^2 terms in Eq. 17.6:

$$\sum_{k \neq n} c_{kn}^{(2)} E_k^{(0)} |k\rangle + \sum_{k \neq n} c_{kn}^{(1)} H_P |k\rangle$$

$$= E_n^{(0)} \sum_{k \neq n} c_{kn}^{(2)} |k\rangle + E_n^{(1)} \sum_{k \neq n} c_{kn}^{(1)} |k\rangle + E_n^{(2)} |n\rangle.\tag{17.13}$$

Exercise 17.6 Apply $\langle n|$ to Eq. 17.13 and show that

$$E_n^{(2)} = \sum_{k \neq n} \frac{\left|\langle k| H_P |n\rangle\right|^2}{\varepsilon_n - \varepsilon_k}.\tag{17.14}$$

We can continue the perturbation expansion by identifying higher-order terms in Eq. 17.6. The next-order state and energy terms are worked out in Problem 17.1.

Equation 17.14, which can sometimes be the lowest-order non-zero shift in the energy levels, has some interesting features. For example:

Exercise 17.7 Show that $E_0^{(2)}$ is never positive for the ground state $|\psi_0\rangle$. That is, this correction never increases the ground state energy.

Because of the denominator $\varepsilon_n - \varepsilon_k$ in Eq. 17.14, the contribution to $E_n^{(2)}$ will be larger for nearby energy levels. Furthermore, these energy levels tend to "repel" one another. Consider $|m\rangle$ and $|n\rangle$ such that $\langle m| H_P |n\rangle \neq 0$, and suppose that $\varepsilon_m < \varepsilon_n$. Then

- $E_m^{(2)}$ contains a negative term $\left| \langle m| H_P |m\rangle \right|^2 /(\varepsilon_m - \varepsilon_n)$, while
- $E_n^{(2)}$ contains a positive term $\left| \langle n| H_P |n\rangle \right|^2 /(\varepsilon_n - \varepsilon_m)$.

These terms tend to move the m and n levels apart (though of course other terms may counteract this effect).

17.2 Dealing with degeneracy

In our analysis so far, we have relied on the assumption that the spectrum of the unperturbed Hamiltonian H_0 is non-degenerate. The trouble that degeneracy causes can be appreciated through a very simple example. Suppose that our quantum system is a qubit, and that the unperturbed Hamiltonian is zero: $H_0 = 0$. Every qubit state is an eigenstate of this Hamiltonian with energy $\varepsilon = 0$. We choose the standard basis states $|0\rangle$ and $|1\rangle$ to serve as our eigenbasis.

Now the system is perturbed by $H_P = \eta X$, where X is the Pauli operator. Because $\langle 0| X |0\rangle = \langle 1| X |1\rangle = 0$, it appears that there are no first-order changes in the energies. But this is not right! The total Hamiltonian is ηX, which has eigenvalues $\pm\eta$. These are non-zero to first-order in the size of the perturbation. Where did we go wrong?

The problem becomes obvious when we try to calculate the first-order change in the states. For instance, the quantity

$$c_{10}^{(1)} = \frac{\langle 1| \eta X |0\rangle}{\varepsilon_0 - \varepsilon_1} = \frac{\eta}{\varepsilon_0 - \varepsilon_1}, \qquad (17.15)$$

is infinite. Our first-order equation (Eq. 17.9) does not make sense.

The unperturbed system was degenerate, so we had some freedom of choice in specifying the unperturbed eigenbasis. We picked the standard $\{ |0\rangle , |1\rangle \}$ basis. But this was a poor choice! To have any hope of avoiding infinities, we need to choose our unperturbed basis states so that $\langle k| H_P |n\rangle = 0$ whenever $\varepsilon_n = \varepsilon_k$. Then the offending terms will not appear in Eq. 17.9.

How can we do this? The general answer is fairly involved; here we will describe how things work out in two common situations. First, it could be that the unperturbed

Hamiltonian and the perturbation actually commute with each other:

$$[H_0, H_P] = 0. \tag{17.16}$$

Then our course of action is clear. We choose our H_0 eigenstates so that they are also H_P eigenstates. In our qubit example with the ηX perturbation, this means we choose to use the $|\pm\rangle$ basis. The degeneracy of the $|\pm\rangle$ states is "lifted" by the perturbation, so that

$$E_{\pm}^{(1)} = \langle \pm | \eta X | \pm \rangle = \pm \eta, \tag{17.17}$$

as expected. The degeneracy of the unperturbed situation is due to a symmetry – a kind of rotational symmetry in this case – that is "broken" by the perturbation. We choose our unperturbed basis states with this broken symmetry in mind.

Let us consider a more general situation, but suppose that the spectrum of H_0 has a two-fold degeneracy. That is, there is a two-dimensional H_0 eigenspace spanned by states $|a\rangle$ and $|b\rangle$ for which $\varepsilon_a = \varepsilon_b = \varepsilon$. Consider the matrix

$$\mathbf{W} = \begin{pmatrix} W_{aa} & W_{ab} \\ W_{ba} & W_{bb} \end{pmatrix} = \begin{pmatrix} \langle a| H_P |a\rangle & \langle a| H_P |b\rangle \\ \langle b| H_P |a\rangle & \langle b| H_P |b\rangle \end{pmatrix}. \tag{17.18}$$

(This is a submatrix of a matrix representation of H_P.) If the $|a\rangle$ and $|b\rangle$ basis states are the "right" ones, then $\mathbf{W}$ is diagonal, and the diagonal elements are the first-order energy corrections $W_{aa} = E_a^{(1)}$ and $W_{bb} = E_b^{(1)}$.

But what if the basis we chose happens to be a "wrong" one? It is still true that the first-order energy corrections are the *eigenvalues* of the matrix $\mathbf{W}$. We can find these by solving the characteristic equation:

$$0 = \det \begin{pmatrix} W_{aa} - E^{(1)} & W_{ab} \\ W_{ba} & W_{bb} - E^{(1)} \end{pmatrix}$$

$$= (W_{aa} - E^{(1)})(W_{bb} - E^{(1)}) - W_{ab} W_{ba}. \tag{17.19}$$

The quadratic formula gives us the two first-order energy corrections:

$$E_{\pm}^{(1)} = \frac{1}{2} \left(W_{aa} + W_{bb} \pm \sqrt{(W_{aa} + W_{bb})^2 + 4\,|W_{ab}|^2} \right). \tag{17.20}$$

Problems 17.6 and 17.7 explore these ideas in a simple situation with two-fold degeneracy.

17.3 Perturbing the dynamics

Once again, we suppose our unperturbed Hamiltonian H_0 is independent of time and has eigenstates $|n\rangle$ with energies ε_n. Now, however, the perturbation H_P is *not* assumed to be time-independent. Since the total Hamiltonian may now be varying with time, it may not have stationary states and fixed energy levels. In this situation, we will seek to find out how the perturbation affects the *time evolution* of quantum states of the system. The perturbation may induce *transitions* between the unperturbed eigenstates $|n\rangle$ – states that would be entirely stationary in the absence of $H_p(t)$. Determining which transitions are possible and calculating their likelihood is the domain of *time-dependent perturbation theory*.

Time evolution is easy to work out in the unperturbed situation. We can write

$$|\psi(t)\rangle = \sum_k a_k \, e^{-i\varepsilon_k t/\hbar} \, |k\rangle , \qquad (17.21)$$

where the coefficients a_k do not depend on time. All of the time-dependence of the state lies in the evolving phase factors $e^{-i\varepsilon_k t/\hbar}$.

Exercise 17.8 Show that $|\psi(t)\rangle$ in Eq. 17.21 satisfies the Schrödinger equation

$$i\hbar \frac{d}{dt} |\psi(t)\rangle = H_0 |\psi(t)\rangle . \qquad (17.22)$$

Also show that $a_k = \langle k \, | \, \psi(0) \rangle$.

To restate a point we have made before, a knowledge of the system's stationary states $|n\rangle$ and energy levels ε_n will suffice to describe the system's time evolution.

Now we include the effect of the perturbation $H_p(t)$, which might vary with time. As before, we introduce the interpolation parameter λ, so that we are in fact considering the one-parameter family of Hamiltonians of the form $H_\lambda = H_0 + \lambda H_p(t)$. For any λ, we can still describe the time evolution of a state $|\psi(t)\rangle$ using the $|k\rangle$ basis states, as in Eq. 17.21. However, the coefficients a_k are now functions of time:

$$|\psi(t)\rangle = \sum_k a_k(t) \, e^{-i\varepsilon_k t/\hbar} \, |k\rangle . \qquad (17.23)$$

To figure out the time evolution of the state, we need to know how the coefficients $a_k(t)$ evolve under the action of the perturbed Hamiltonian. This is determined by the Schrödinger equation:

$$i\hbar \frac{d}{dt} |\psi(t)\rangle = (H_0 + \lambda H_p(t)) \, |\psi(t)\rangle . \qquad (17.24)$$

Substituting Eq. 17.23 into this equation, we obtain

$$\sum_k e^{-i\varepsilon_k t/\hbar} \left(i\hbar \frac{da_k}{dt} + \varepsilon_k a_k \right) |k\rangle = \sum_k e^{-i\varepsilon_k t/\hbar} (\varepsilon_k a_k + \lambda a_k H_p) \, |k\rangle . \qquad (17.25)$$

Now we apply $e^{i\varepsilon_n t/\hbar} \langle n|$ to arrive at a governing differential equation for $a_n(t)$:

$$i\hbar \frac{da_n}{dt} = \lambda \sum_k e^{i(\varepsilon_n - \varepsilon_k)t/\hbar} a_k(t) \, \langle n| \, H_p(t) \, |k\rangle . \qquad (17.26)$$

The coefficient $a_n(t)$ depends on the parameter λ. We write it as a power series:

$$a_n(t) = a_n^{(0)}(t) + \lambda \, a_n^{(1)}(t) + \lambda^2 a_n^{(2)}(t) + \dots \qquad (17.27)$$

As in the time-independent theory, $\lambda = 0$ represents the unperturbed situation and $\lambda = 1$ is the actual situation of interest. If we substitute this into Eq. 17.26 and insist on equality of terms with like powers of λ, we arrive at a remarkably neat sequence of equations:

$$ i\hbar \frac{da_n^{(0)}}{dt} = 0, \tag{17.28} $$

$$ i\hbar \frac{da_n^{(1)}}{dt} = \sum_k e^{i(\varepsilon_n - \varepsilon_k)t/\hbar} a_k^{(0)}(t) \, \langle n| \, H_P(t) \, |k \rangle, \tag{17.29} $$

$$ i\hbar \frac{da_n^{(2)}}{dt} = \sum_k e^{i(\varepsilon_n - \varepsilon_k)t/\hbar} a_k^{(1)}(t) \, \langle n| \, H_P(t) \, |k \rangle, \tag{17.30} $$

and so on. Each higher-order term $a_k^{(m+1)}(t)$ is obtained by solving a differential equation involving all of the previous $a_n^{(m)}(t)$s together with the perturbation H_P. In many situations, it might suffice to include only the first couple of iterations of this process – the "first-order" solution $a_k(t) \approx a_n^{(0)}(t) + a_n^{(1)}(t)$.

Let us focus on a specific type of problem. At $t = 0$, our quantum system is in the initial state $|i\rangle$, one of the unperturbed stationary states. Therefore, $a_i(0) = 1$ and $a_k(0) = 0$ for $k \neq i$. We are interested in finding the amplitude $a_n(t)$ that the system will be found in another basis state $|n\rangle$ at a later time $t > 0$. The resulting transition probability $P(i \rightarrow n) = |a_n(t)|^2$.

We do have the freedom to assign the initial conditions $a_k(0)$ as we wish among the terms of various orders: $a_k^{(m)}(0)$. If there were no perturbation ($\lambda = 0$), then only the $a_k^{(0)}$ terms would be present; and even in the perturbed case, it makes sense to include as much as possible in the lowest-order terms. Thus, we adopt the following initial conditions:

- $a_i^{(0)}(0) = 1$ and $a_k^{(0)}(0) = 0$ for $k \neq i$. (In short, $a_k^{(0)}(0) = \delta_{ki}$.)
- $a_k^{(m)}(0) = 0$ for all orders $m \geq 1$.

Equation 17.28 tells us that the zeroth-order coefficients $a_k^{(0)}$ will all be constant over time. Equation 17.29, applied to $a_n^{(1)}$, tells us that

$$ i\hbar \frac{da_n^{(1)}}{dt} = e^{i(\varepsilon_n - \varepsilon_i)t/\hbar} \, \langle n| \, H_P(t) \, |i \rangle. \tag{17.31} $$

Exercise 17.9 In Eq. 17.29 there is a sum over basis states $|k\rangle$, but no such sum appears in Eq. 17.31. Where did it go?

We can write down the solution to Eq. 17.31:

$$ a_n^{(1)}(t) = \frac{1}{i\hbar} \int_0^t e^{i(\varepsilon_n - \varepsilon_i)t'/\hbar} \, \langle n| \, H_P(t') \, |i \rangle \, dt'. \tag{17.32} $$

Since we are considering $n \neq i$ (so that $a_n^{(0)} = 0$), this is the first-order transition amplitude from $|i\rangle$ to $|n\rangle$.

A kicked oscillator. Consider a harmonic oscillator consisting of a particle of mass μ moving in a harmonic potential with elastic constant k and frequency $\omega = \sqrt{k/\mu}$. Initially, the oscillator is in its ground state $|0\rangle$. At $t = 0$, the particle is subjected to an impulsive "kick" in the positive direction. We can describe this kick by the perturbation

$$H_p(t) = -\kappa \delta(t) x. \tag{17.33}$$

Exercise 17.10 Explain why Eq. 17.33 represents a "kick" of the type described. Show that κ is the total impulse provided by the kick.

If we recall that the position operator $x = \sqrt{\hbar/2\mu\omega}(a^\dagger + a)$, it is easy to evaluate Eq. 17.32:

Exercise 17.11 Show that for the kicked oscillator, all of the amplitudes $a_k^{(1)}$ are zero except

$$a_1^{(1)} = \frac{i\kappa}{\sqrt{2\hbar\mu\omega}}. \tag{17.34}$$

What is the first-order probability that the quantum oscillator is excited by the kick?

Now suppose the perturbation H_p is constant over time, at least for $t > 0$. The first-order transition amplitude from $|i\rangle$ to $|n\rangle$ is

$$
\begin{aligned}
a_n^{(1)}(t) &= \frac{1}{i\hbar} \langle n| H_p |i\rangle \int_0^t e^{i(\varepsilon_n - \varepsilon_i)t'/\hbar} dt' \\
&= \frac{2}{i\hbar} \langle n| H_p |i\rangle \, e^{i\omega_{ni}t/2} \left(\frac{\sin(\omega_{ni}t)}{\omega_{ni}} \right),
\end{aligned}
\tag{17.35}
$$

where $\omega_{ni} = (\varepsilon_n - \varepsilon_i)/\hbar$. The transition probability is $\left| a_n^{(1)}(t) \right|^2$. We will instead look more closely at the *mean transition rate* $w(i \to n)$, which is the transition probability divided by the time t:

$$w(i \to n) = \frac{4}{\hbar^2} \left| \langle n| H_p |i\rangle \right|^2 \left(\frac{\sin^2(\omega_{ni}t)}{\omega_{ni}^2 t} \right). \tag{17.36}$$

This expression leads us to several valuable insights.

First, the first-order transition rate is zero if the matrix element $\langle n| H_p |i\rangle = 0$. Such processes are known as *forbidden transitions*. They can still occur, of course, due to higher-order terms in the perturbation series. But if H_p is small, the higher-order terms will generally be smaller than first-order ones. Forbidden transitions will be substantially less likely than unforbidden ones.

Next, consider the function $\sin^2(\omega t/2)/\omega^2 t$, where ω stands for ω_{ni}. A graph of this function of ω is found in Fig. 17.1. The function is small unless $|\omega t| \lesssim 1$. In our problem, this means that the transition rate $w(i \to n)$ is suppressed if $|\varepsilon_n - \varepsilon_i| t \gg \hbar$. What does this mean? Unless the transition is very rapid (so that t is small), this function enforces an approximate conservation of the unperturbed energy: the transition rate is negligible unless $\varepsilon_n \approx \varepsilon_i$.

Exercise 17.12 How is this connected to the time–energy uncertainty relation discussed in Section 5.3?

As t increases, this function gets both narrower and higher. The area under the curve, however, remains constant.

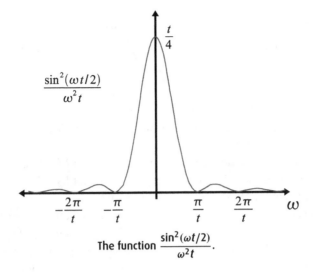

Fig. 17.1

The function $\dfrac{\sin^2(\omega t/2)}{\omega^2 t}$.

Exercise 17.13 Show that

$$\int_{-\infty}^{+\infty} \frac{\sin^2(\omega t/2)}{\omega^2 t}\, d\omega = \frac{\pi}{2}. \tag{17.37}$$

(This is a difficult integral; the easiest way to evaluate it is to consider the function $f(x) = (\sin x)/x$, find its Fourier transform, and apply Parseval's Theorem. See Appendix B.)

For a large value of t, we can regard this as an approximation of the delta function:

$$\frac{\sin^2(\omega t/2)}{\omega^2 t} \longrightarrow \frac{\pi}{2}\,\delta(\omega) = \frac{\pi\hbar}{2}\,\delta(\varepsilon_n - \varepsilon_i). \tag{17.38}$$

Thus, as $t \to \infty$,

$$w(i \to n) = \frac{2\pi}{\hbar}\,\left|\langle n|\, \mathsf{H_p}\, |i\rangle\right|^2\, \delta(\varepsilon_n - \varepsilon_i). \tag{17.39}$$

The delta function reflects the fact that energy is conserved in possible transitions.[3]

It often happens that the quantum system of interest has a great many states that are degenerate or nearly degenerate in energy. We describe the spectrum of such a system by a smooth *density of states* function $\rho(\varepsilon)$. The number dN of distinct energy eigenstates in the small energy range $d\varepsilon$ is $dN = \rho(\varepsilon)\, d\varepsilon$. Discrete sums over energy levels can be regarded as integrals:

$$\sum_n (\cdots) \longrightarrow \int (\cdots)\, \rho(\varepsilon)\, d\varepsilon. \tag{17.40}$$

We are usually interested in the total transition rate W from $|i\rangle$ to any final state $|n\rangle$ in the allowed narrow range of energies. This will be $W = \displaystyle\int w\, \rho(\varepsilon)\, d\varepsilon$, where ε is the energy of

---

[3] Since the total Hamiltonian $\mathsf{H_0} + \mathsf{H_p}$ is time-independent, the actual total energy is exactly conserved. The energy we mean here is that due to the unperturbed Hamiltonian $\mathsf{H_0}$.

the final state. If we let $\overline{\left|\langle n|\,\mathsf{H}_{\mathrm{P}}\,|i\rangle\right|^2}$ represent the *average* squared magnitude of the matrix element for final states in the energy range, we obtain a total transition rate

$$W = \frac{2\pi}{\hbar}\,\overline{\left|\langle n|\,\mathsf{H}_{\mathrm{P}}\,|i\rangle\right|^2}\,\rho(\varepsilon_i). \qquad (17.41)$$

Equation 17.41 is a simple result with a remarkably wide range of applications. Enrico Fermi used it to great effect in nuclear and particle physics, so that it has become known as *Fermi's golden rule*.

We can also derive a useful variant of Eq. 17.41. Suppose that the states of our system are characterized both by their energy ε and by other observable attributes, which we collectively denote by α. A free particle in space, for example, has stationary states characterized by the particle's energy and its direction of motion. We can give a more refined description of its spectrum by letting the density of states depend on both properties: $\rho(\varepsilon, \alpha)$. The number of states dN in a small range of these variables is $dN = \rho(\varepsilon, \alpha)\,d\varepsilon\,d\alpha$. We are interested in the total transition rate to states $|n\rangle$ whose α_n values lie in a narrow range $d\alpha$. This is

$$dW = \frac{2\pi}{\hbar}\,\overline{\left|\langle n|\,\mathsf{H}_{\mathrm{P}}\,|i\rangle\right|^2}\,\rho(\varepsilon_i, \alpha_n)\,d\alpha. \qquad (17.42)$$

(We have written dW as an infinitesimal to reflect the fact that only final states within $d\alpha$ are to be counted.) This refined version of Fermi's golden rule allows us to give a more detailed description of the transitions of the system.

Equations 17.41 and 17.42 indicate that it will be useful to work out the density of states ρ for various quantum systems. For example, a 1-D harmonic oscillator with frequency ω has energy levels that are uniformly separated by $\hbar\omega$. An energy range $d\varepsilon$ (tiny on the macroscopic scale, but large compared to the level spacing $\hbar\omega$) contains $dN = d\varepsilon/\hbar\omega$ distinct levels. Thus, $\rho(\varepsilon)$ has the constant value $1/\hbar\omega$.

Exercise 17.14 Consider a free particle of mass μ moving in a 1-D box of length L. Show that the density of states is

$$\rho(\varepsilon) = \frac{L}{\pi\hbar}\sqrt{\frac{\mu}{2\varepsilon}}. \qquad (17.43)$$

Now let us consider a free particle of mass μ moving in 3-D. To simplify our calculation somewhat, we will make the rather peculiar assumption that the 3-D space is periodic in each Cartesian direction. That is, each (x, y, z) coordinate is in effect a circle of circumference L, as described in Section 11.3. The wave function $\psi(x, y, z)$ for the particle must therefore be periodic in each coordinate. The total volume of this periodic space is $V = L^3$. At a later stage in our analysis, we will remove the periodic condition by allowing $V \to \infty$.

The three Cartesian directions are independent degrees of freedom. The quantum system therefore has three integer quantum numbers n_x, n_y, and n_z, which we can regard as the components of a quantum number vector $\vec{n} = n_x\hat{x} + n_y\hat{y} + n_z\hat{z}$. The stationary states are labeled by a discrete wave vector $\vec{k}$, where

$$\vec{k} = \frac{2\pi}{L}\vec{n}. \qquad (17.44)$$

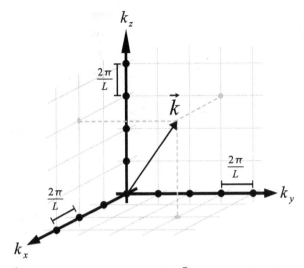

Fig. 17.2 Allowed values of $\vec{k}$ lie on a discrete cubical lattice in $\vec{k}$-space. The vector shown may be designated $\vec{k}_{233}$.

The possible values of $\vec{k}$ are arranged as the points of a 3-D cubical lattice with spacing $2\pi/L$ (see Fig. 17.2). The density of states in $\vec{k}$-space is therefore $V/(2\pi)^3$ per unit $\vec{k}$-volume. For a given wave vector $\vec{k}$, the stationary state wave function is $\psi(\vec{r}) = V^{-3/2}\exp(i\vec{k}\cdot\vec{r})$, and its energy is $\varepsilon = \hbar^2 k^2/2\mu$.

We describe $\vec{k}$-space by a radial coordinate k together with angular coordinates θ and ϕ that we collectively designate Ω. The $\vec{k}$-volume element in these coordinates is $k^2\,dk\,d\Omega$. (Here $d\Omega$ represents an element of "solid angle," a tiny range of directions given by $d\Omega = \sin\theta\,d\theta\,d\phi$.) The number of states dN in this $\vec{k}$-volume is

$$dN = \frac{V}{(2\pi)^2}\,k^2\,dk\,d\Omega = \frac{\mu V}{(2\pi)^3\hbar^3}\sqrt{2\mu\varepsilon}\,d\varepsilon\,d\Omega, \qquad (17.45)$$

from which we recognize the density of states

$$\rho(\varepsilon, \Omega) = \frac{\mu V}{(2\pi)^3\hbar^3}\sqrt{2\mu\varepsilon}. \qquad (17.46)$$

Exercise 17.15 Write k and dk in terms of ε and $d\varepsilon$ to fill in the gaps in the derivation of Eq. 17.46.

17.4 Cross-sections

In a common type of experiment, a stream of particles is aimed at a target, producing certain detectable events. The events may be anything: particle absorption, an induced reaction in the target, or the scattering of the incoming particles in some direction. The mean rate at which a particular type of event occurs is proportional to the intensity of the incoming

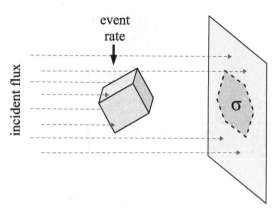

event
rate

incident flux

σ

Fig. 17.3 Cross-section for impact on an obstacle.

stream – that is, to the beam's *flux*, the number of particles in the stream per unit area per unit time. Then

$$\left(\begin{array}{c} \text{event} \\ \text{rate} \end{array} \right) = \sigma \times \left(\begin{array}{c} \text{incident} \\ \text{flux} \end{array} \right), \tag{17.47}$$

where the quantity σ is called the *cross-section* for this type of event. The cross-section has units of (length)^2, or area.

The term "cross-section" comes from an elementary classical example. Suppose the stream of particles encounters a solid obstacle, and we count how many particles impact the obstacle's surface. The cross-section σ for impact is precisely the geometric cross-sectional area of the obstacle, the size of its projection onto a plane perpendicular to the stream. The cross-section σ of the obstacle is, in effect, the size of its shadow (see Fig. 17.3).

More generally, the cross-section is a probabilistic quantity that expresses the likelihood that an event will take place. Instead of a whole stream of particles, we can consider a single incident quantum particle. The event that is counted is a transition from the incident state to another distinguishable state, and the event rate is just the transition probability rate W discussed in the previous section. The incident particle flux is the probability flux $\vec{J}$ defined in Eq. 11.29. The quantum cross-section for the event is defined by the relation $W = \sigma J$.

Consider the problem of particle scattering from a target described by a fixed potential $U(\vec{r})$. We treat the scattering potential as a perturbation to the free particle Hamiltonian, and consider only the first-order contribution to the scattering probability. This is called the *Born approximation*. The initial state of the particle is the momentum eigenstate $|\vec{p}\rangle$. As in the previous section, we imagine that the 3-D space is periodic in each of its Cartesian directions, with a total volume V. Then the normalized wave function associated with the state $|\vec{p}\rangle$ is

$$\psi(\vec{r}) = \langle \vec{r} | \vec{p} \rangle = \frac{1}{\sqrt{V}} \, e^{i\vec{p}\cdot\vec{r}/\hbar}. \tag{17.48}$$

The energy of this state is $\varepsilon = p^2/2\mu$. The probability flux is

$$\vec{J} = \frac{\vec{p}}{\mu V} = \frac{1}{V}\sqrt{\frac{2\varepsilon}{\mu}}\,\hat{k}, \tag{17.49}$$

where $\hat{k}$ is a unit vector in the direction of $\vec{p}$, the initial direction of motion of the particle.

Exercise 17.16 Derive Eq. 17.49. (This follows from the result of Exercise 11.5, so if you have not already worked that exercise, do so now.)

We wish to calculate the cross-section for the particle to scatter into another momentum eigenstate $|\vec{p}\,'\rangle$. Since our previous analysis tells us that the final energy of the particle equals its initial energy ($\varepsilon' = \varepsilon$), the final and incident momenta $|\vec{p}\,'\rangle$ and $|\vec{p}\rangle$ can differ only in direction. So, to be precise, we will calculate the total cross-section for scattering into any direction within a small solid angle $d\Omega$ about a given direction of interest, which we label Ω'.

The "refined" version of Fermi's golden rule (Eq. 17.42) tells us that the total rate for this scattering process is

$$dW = \frac{2\pi}{\hbar}\,\overline{\left|\langle \vec{p}\,'|\,U(\vec{r})\,|\vec{p}\rangle\right|^2}\,\rho(\varepsilon,\Omega')\,d\Omega. \tag{17.50}$$

All of the final momenta $\vec{p}\,'$ in the average $\overline{(\cdots)}$ are nearly the same, so the matrix elements $\langle\vec{p}\,'|\,U(\vec{r})\,|\vec{p}\rangle$ are essentially equal in the average. We can calculate this matrix element in the position representation:

$$\langle \vec{p}\,'|\,U(\vec{r})\,|\vec{p}\rangle = \frac{1}{V}\iiint e^{i\vec{q}\cdot\vec{r}}U(\vec{r})\,d^3r, \tag{17.51}$$

where $\vec{q} = \vec{p} - \vec{p}\,'$ is called the *momentum transfer*.

Exercise 17.17 Is $\vec{q}$ the momentum transfer from the target to the scattered particle, or the other way around?

Exercise 17.18 Express Eq. 17.51 in terms of the 3-D Fourier transform of the potential function, which we can write $\widetilde{U}(\vec{k})$.

The cross-section for the scattering event is $d\sigma$, which is written as an infinitesimal because it is proportional to the range of scattering directions $d\Omega$. The cross-section is related to the rate by $dW = J\,d\sigma$. Combining all this with the density of states from Eq. 17.46, we obtain

$$\frac{1}{V}\sqrt{\frac{2\varepsilon}{\mu}}\,d\sigma = \frac{2\pi}{\hbar}\frac{1}{V^2}\left|\iiint e^{i\vec{q}\cdot\vec{r}}U(\vec{r})\,d^3r\right|^2\frac{\mu V}{(2\pi)^3\hbar^3}\sqrt{2\mu\varepsilon}\,d\Omega. \tag{17.52}$$

The same net factor of $1/V$ occurs in each side, so that our final relation does not depend on the volume of our periodic space, and we can take the limit $V \to \infty$. The factor of $\sqrt{\varepsilon}$ is also the same on each side, and cancels. Our goal is to find the *differential cross-section*, the cross-section per unit solid angle:

$$\frac{d\sigma}{d\Omega} = \frac{\mu^2}{4\pi^2\hbar^4} \left| \iiint e^{i\vec{q}\cdot\vec{r}} U(\vec{r})\, d^3r \right|^2. \tag{17.53}$$

Note that the differential cross-section depends on the scattering direction only through its dependence on the momentum transfer $\vec{q}$.

Equation 17.53 is an extremely useful conclusion, as we can see from a couple of simple examples.

Scattering by slow particles. Suppose the potential $U(\vec{r})$ is zero outside a region of radius R about the origin, and that the incident momentum is low enough that $pR \ll \hbar$. Then the exponential factor in Eq. 17.53 will be essentially the same for any scattering direction, and so

$$\frac{d\sigma}{d\Omega} = \frac{\mu^2}{4\pi^2\hbar^4} \left| \iiint U(\vec{r})\, d^3r \right|^2. \tag{17.54}$$

This is independent of the scattering angle, so the scattering process is isotropic. The scattering from a small potential well ($U(\vec{r}) < 0$) is the same as from a potential "bump" ($U(\vec{r}) > 0$). (This expression for the differential cross-section should hold good for *small angle* scattering even for fast particles, since it depends only on the momentum transfer $\vec{q}$ satisfying $qR \ll \hbar$.)

For slow particles, the total cross-section for scattering in all directions (except the incident one) will be

$$\sigma = \int \left(\frac{d\sigma}{d\Omega}\right) d\Omega = \frac{\mu^2}{\pi\hbar^4} \left| \iiint U(\vec{r})\, d^3r \right|^2. \tag{17.55}$$

Exercise 17.19 Find the differential and total scattering cross-sections for a delta-function potential $\epsilon\delta^3(\vec{r})$ in the Born approximation.

Scattering from a crystal. For a single "dimple" in the potential at the origin, the scattering cross-section is

$$\left(\frac{d\sigma}{d\Omega}\right)_1 = \frac{\mu^2}{4\pi^2\hbar^4} |u|^2, \tag{17.56}$$

where $u = \iiint e^{i\vec{q}\cdot\vec{r}/\hbar} U\, d^3r$. This potential "dimple" might represent a single atom. Now suppose the particle scatters from a collection of N such atoms located at various points in space: $\vec{r}_1, \vec{r}_2, \ldots, \vec{r}_N$. The amplitude u will be replaced by a sum of N terms, each of them corresponding to a shifted version of the original potential. That is,

$$\left(\frac{d\sigma}{d\Omega}\right)_N = \frac{\mu^2}{4\pi^2\hbar^4} |K(\vec{q})\, u|^2 = |K(\vec{q})|^2 \left(\frac{d\sigma}{d\Omega}\right)_1, \tag{17.57}$$

where

$$K(\vec{q}) = \sum_n e^{i\vec{q}\cdot\vec{r}_n/\hbar}. \tag{17.58}$$

The cross-section is multiplied by the factor $|K|^2$, which we can write

$$|K(\vec{q})|^2 = N + \sum_{\substack{m,n \\ m \neq n}} e^{i\vec{q}\cdot(\vec{r}_m - \vec{r}_n)/\hbar}. \tag{17.59}$$

This may be as small as zero if the phase factors in Eq. 17.58 exactly cancel out. If the atoms are randomly distributed in space, then the complex phases in Eq. 17.59 will typically add up to about zero, and the total cross-section will be roughly N times larger than for a single atom – just as if the scattering from each atom contributed equally (and incoherently) to the total probability.

The greatest possible scattering enhancement factor is $|K|^2 = N^2$, in which case the scattering probability is N^2 times greater than any single atom would produce. This is due to constructive interference in the scattering amplitudes from the individual atoms. Equation 17.59 tells us that this will happen if and only if $\vec{q} \cdot (\vec{r}_m - \vec{r}_n)/\hbar$ is a multiple of 2π for every pair of atoms m and n. What does this mean? If we choose the z-axis to be parallel to $\vec{q}$, then the condition means that $z_m - z_n$ is always an integral multiple of $2\pi\hbar/q$. In other words, the atoms all lie in a set of parallel planes – planes of constant z – that are perpendicular to $\vec{q}$ and separated by multiples of $2\pi\hbar/q$.

Suppose the atoms are somehow arranged in a set of parallel planes separated by d. We do an experiment in which we send in particles at an angle θ to these planes, and we look for scattering at the same "reflected" angle. The momentum transfer is perpendicular to the planes and has a magnitude $q = 2p \sin\theta$. Our analysis tells us that we will have a very strong likelihood of scattering in this direction provided that $d = 2\pi n\hbar/q$ for some integer $n = 1, 2, \ldots$ That is,

$$n\frac{2\pi\hbar}{p} = n\lambda = 2d \sin\theta. \tag{17.60}$$

Here $\lambda = 2\pi\hbar/p$, the de Broglie wavelength of the incident particles. Equation 17.60 is known as *Bragg's Law* and governs the scattering of X-rays and electrons from crystals. The regular arrangement of atoms in the crystal means that the atoms lie in various sets of parallel planes, with different orientations and values of d (see Fig. 17.4). The intensity

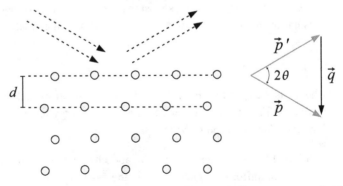

Fig. 17.4 Scattering from the regular arrangement of atoms in a crystal. The scattering probability will be greatest if the separation d of the atomic planes satisfies $d = 2\pi n\hbar/|\vec{q}|$.

of the scattered beam will be greatest if the condition in Eq. 17.60 holds for some set of atomic planes. In this way, the diffraction of X-rays or electrons allows us to measure d for various planes and map out the atomic structure of the crystal.

Exercise 17.20 Suppose 75% of the N atoms lie in planes perpendicular to the momentum transfer $\vec{q}$ and separated by $d = 2\pi\hbar/q$, but the other 25% of the atoms lie in other planes exactly halfway between these. Find the scattering enhancement factor $|K|^2$. What if the proportions were 50% and 50%?

Problems

Problem 17.1 Consider the perturbed Hamiltonian $H_0 + H_P$, where the energy spectrum of H_0 is non-degenerate. Derive expressions for the second-order state coefficients $c_{kn}^{(2)}$ and the third-order energy correction $E_n^{(3)}$.

Problem 17.2 Consider a 1-D harmonic oscillator with $U(x) = \frac{1}{2}kx^2$. We perturb this system by adding a term bx to the potential.

(a) Show that the perturbed problem is still a harmonic oscillator, with shifts in the equilibrium position and minimum energy. What are these shifts?
(b) Show that $E_n^{(1)} = 0$ for the harmonic oscillator $|n\rangle$ states.
(c) Find the first-order correction to the state $|n\rangle$. Use this to calculate $\langle x \rangle$ for the perturbed ground state.
(d) Find $E_n^{(2)}$.
(e) Explain how the results of parts (c) and (d) make sense in light of your results in part (a).

Problem 17.3 Calculate the first-order shift in the energies of a harmonic oscillator due to the addition of an anharmonic term Cx^4 to the potential.

Problem 17.4 A free particle moves in 1-D between impenetrable walls at $x = 0$ and $x = L$. The particle is moving so fast that we cannot entirely neglect the effects of relativity. This has the approximate effect of adding a term to the kinetic energy:

$$H_P = -\frac{p^4}{8\mu^3 c^2}, \tag{17.61}$$

where c is the speed of light. (Consult a textbook on special relativity to find the origin of this term.) Find the first-order changes to the energy levels and the stationary state wave functions due to this perturbation.

Problem 17.5 A particle of mass μ and charge q is bound in a central potential $U(r)$. An external uniform magnetic field in the z-direction is applied to the system. The orbital angular momentum of the particle leads to a magnetic moment that couples to the field. (This is roughly sketched in Section 2.2 and Problem 2.4.) The perturbation is $H_P = -(qB_z/2\mu)L_z$.

Describe in detail how this perturbation lifts the degeneracy of the energy levels of the system. Calculate the magnitude of the "level splitting" for an electron in a 0.5 T field.

In atomic systems, the resulting change in the observed atomic spectrum is called the "normal" Zeeman effect. Because of electron spin, however, what is actually observed in real atoms is usually the more complicated "anomalous" Zeeman effect.

Problem 17.6 A particle moves on a circle of circumference L, as discussed in Section 11.3. The particle is subject to a perturbing Hamiltonian $H_p = \beta p$, where p is the momentum.[4] Find the first-order corrections to the energy levels and stationary states of the particle on a circle.

Problem 17.7 A particle moves on a circle of circumference L, as discussed in Section 11.3. In addition to the kinetic energy, the particle is subject to a weak potential $U(x) = \eta \cos(2\pi x/L)$. Explain why the term "tilted circle" is a good description of this situation. Then find the energy levels of this system to first order in η.

Problem 17.8 Apply the impulsive perturbation of Eq. 17.33 to an oscillator that is initially in an excited state $|n\rangle$. To first order, what is the relative likelihood (ratio of probabilities) that the oscillator's energy is increased or decreased by the kick?

Problem 17.9 Consider a quantum system subject to an oscillating perturbation:

$$H_p = A e^{-i\omega t} + A^\dagger e^{i\omega t}, \qquad (17.62)$$

where A need not be Hermitian. Show that for large t this perturbation leads to first-order transitions that may either absorb or emit an energy $\hbar\omega$, and that the first-order rates for these are

$$w_{abs}(i \to n) = \frac{2\pi}{\hbar} \left| \langle n| A |i\rangle \right|^2 \delta(\varepsilon_n - \varepsilon_i - \hbar\omega),$$

$$w_{em}(i \to n) = \frac{2\pi}{\hbar} \left| \langle n| A^\dagger |i\rangle \right|^2 \delta(\varepsilon_n - \varepsilon_i + \hbar\omega). \qquad (17.63)$$

This is sometimes used as a *semiclassical* approach to analyzing the interaction of an atom with the electromagnetic field, in which the field is treated as an external classical perturbation to the (quantum) atom. A passing electromagnetic wave subjects the atom to an oscillating potential gradient, so that the atom may either absorb or emit an energy $\hbar\omega$.

Problem 17.10 A particle scatters from a Gaussian potential well

$$U(\vec{r}) = -U_0 \exp\left(-\frac{r^2}{\lambda^2}\right). \qquad (17.64)$$

The potential is radially symmetric; U_0 measures its depth and λ its width. Using the Born approximation, find the differential cross-section $d\sigma/d\Omega$ as a function of the incident energy ε and the angle θ between the initial and final momenta of the particle.

---

[4] Such a perturbation could arise from an external magnetic field.

Quantum information processing

18.1 Quantum circuits

If the state of a quantum system is a kind of information, then the dynamics of that system is a kind of information processing. This is the basic idea behind *quantum computing*, which seeks to exploit the physics of quantum systems to do useful information processing. In this chapter we will acquaint ourselves with a few of the ideas of quantum computing, using the idealized *quantum circuit* model. Then we will turn our attention to an actual quantum process, *nuclear magnetic resonance*, that can be understood as a realization of quantum information processing.

In a quantum circuit, we have a set of n qubit systems whose dynamical evolution is completely under our control. We represent the evolution as a sequence of unitary operators, each of which acts on one or more of the qubits. Graphically, we represent the qubits as a set of horizontal lines, and the various stages of their evolution as a series of boxes or other symbols showing the structure of the sequence of operations. In such a diagram, time runs from left to right[1] (see Fig. 18.1.). The n qubits form a kind of "computer," and its overall evolution amounts to a "computation." The key idea is that very complicated unitary operations on the n qubits can be built up step-by-step from many simple operations on one, two or a few qubits.

The individual steps of the evolution – the elementary operations on one, two or a small number of qubits – are usually called *quantum gates*, by analogy with the basic elements of digital circuit design. The simplest gates are unitary operators U that affect single qubits. Among these are the identity $\mathbf{1}$ and the Pauli operators X, Y, and Z.

Exercise 18.1 The ordinary (classical) logic gate NOT inverts the value of a bit. Show that X is the Pauli operator most closely analogous to this. How does it differ from Y?

The Z gate multiplies the $|1\rangle$ basis state by -1, changing its phase relative to the $|0\rangle$ basis state. We can generalize this by defining the *phase gate* $\Phi(\alpha)$ with a matrix representation

$$\Phi(\alpha) = \begin{pmatrix} 1 & 0 \\ 0 & e^{i\alpha} \end{pmatrix}. \tag{18.1}$$

Then $Z = \Phi(\pi)$.

---

[1] This can lead to some confusion in translating between diagrams and algebraic expressions. Recall that the operator product ABC should be interpreted from right to left: first perform C, then B, then A. The diagram for such a sequence of operators would look like $-\boxed{C}\boxed{B}\boxed{A}-$.

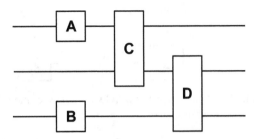

Fig. 18.1 A quantum circuit for three qubits. First, A is applied to the first qubit and B to the second qubit; then C is applied to the first and second qubits; and finally D is applied to the second and third qubits.

Another particularly useful elementary gate is the so-called *Hadamard gate* H, which has the matrix representation

$$\mathbf{H} = \frac{1}{\sqrt{2}} \begin{pmatrix} 1 & 1 \\ 1 & -1 \end{pmatrix}. \tag{18.2}$$

Like the Pauli gates, H is its own inverse: $H^2 = \mathbf{1}$.

Exercise 18.2 Here are three easy exercises about the Hadamard gate H.

(a) What interferometer element from Section 2.1 corresponds to H? How about the phase gate $\Phi(\alpha)$?
(b) Show how to construct an X gate using two H gates and one Z gate. Express your answer as a circuit diagram like Fig. 18.1 (with only one line).
(c) If an H gate is applied to each of n qubits initially in the state $|0^n\rangle = |0\rangle^{\otimes n}$, show that the result is a uniform superposition of all 2^n basis states of the qubits.

Now let us consider two-qubit gates. Among the simplest and most useful of these are the *controlled gates*. In controlled gates, one qubit serves as the *control* qubit and one as the *target* qubit. Roughly speaking, the gate applies a unitary operator to the target provided that the control qubit is in the state $|1\rangle$. However, no measurement is involved; rather, we have a unitary operator of the form

$$\text{controlled-U} = |0\rangle\langle0| \otimes \mathbf{1} + |1\rangle\langle1| \otimes \mathsf{U}. \tag{18.3}$$

Exercise 18.3 Show that if U is unitary, then so is controlled-U.

In a quantum circuit diagram, the controlled-U gate is represented as shown in Fig. 18.2. The controlled-X gate, also known as the *controlled-NOT* or CNOT gate, is particularly simple and useful in designing quantum circuits. It is also shown in Fig. 18.2.

Exercise 18.4 The symbol $\oplus$ represents modulo-2 addition (XOR) of two bits. Show that the action of the CNOT gate on basis states $|a, b\rangle$ is described by CNOT $|a, b\rangle = |a, b \oplus a\rangle$. Which of these qubits acts as the control and which the target?

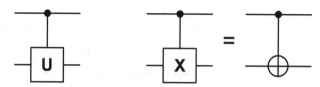

Fig. 18.2 The controlled-U gate, together with two symbols for the CNOT gate. The control qubits are above and the target qubits below.

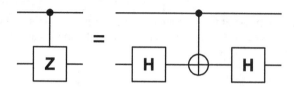

Fig. 18.3   Making a controlled-Z gate from a CNOT and two H gates.

With respect to the standard product basis ($|00\rangle$, $|01\rangle$, etc.) the CNOT gate has a matrix representation

$$\mathbf{CNOT} = \begin{pmatrix} 1 & 0 & 0 & 0 \\ 0 & 1 & 0 & 0 \\ 0 & 0 & 0 & 1 \\ 0 & 0 & 1 & 0 \end{pmatrix}. \tag{18.4}$$

At first, it appears that there is a fundamental asymmetry between the control and target qubits in the CNOT gate. The control qubit appears unchanged by the interaction, but its state affects the target qubit. We might express this by saying that the CNOT gate has "one-way information flow" from the control to the target. This idea, however, is quite wrong in the quantum context, as the following exercise shows:

Exercise 18.5 Suppose that each qubit in a CNOT gate is acted upon by H both before and after the gate. Show that the result is also a CNOT gate, but with the target and control qubits reversed.

Exercise 18.6 Consider the controlled-Z gate, a close cousin of the CNOT. Is there really any distinction between target and control for this gate?

Exercise 18.7 Verify that Fig. 18.3 shows how to make a controlled-Z gate from a CNOT gate and two H gates. How would you make a CNOT from a controlled-Z gate and some H gates?

Not every two-qubit gate is a controlled gate, but they are useful as building blocks. For instance:

Exercise 18.8 The SWAP gate just exchanges the states of two qubits, so that $|a, b\rangle \rightarrow |b, a\rangle$. Show how to make a SWAP gate from three CNOTs.

How does one approach such problems? It is always possible to analyze the action of a quantum circuit by multiplying matrix representations of the gates involved. With the

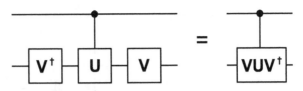

Fig. 18.4
An identity for controlled gates.

assistance of a computer, various designs for the circuit can be tried out relatively quickly. Alternately, we can simply work out the action of a circuit on all possible standard basis inputs. (This may be especially easy when many of the gates are controlled operations.) These approaches are handy for small numbers of qubits; but the size of the matrices and the number of basis states both grow exponentially with the number of qubits in the system. Even worse, the number of possible *combinations* of elementary gates grows even more rapidly, so mere trial-and-error is unlikely to be a successful strategy.

It is better, whenever possible, to use the algebraic properties of the gate operators. Efficient quantum circuit designers keep these properties firmly in mind. For instance, the CNOT gate is its own inverse, so that two identical CNOTs in a row will "cancel out." In Exercise 18.8 our task is to construct a SWAP gate from three CNOTs. There are eight possible ways to put together three CNOT gates for two qubits, depending on how the control and target qubits are chosen for each gate. Six of these ways involve successive identical CNOT gates, so each one reduces to a single CNOT. The remaining two ways are both solutions to the exercise, as can be quickly verified.

Another common technique in quantum circuit design is to place an operation U between another operation V and its inverse. Both Exercises 18.5 and 18.7 use this idea, as does the following exercise, which is a generalization of Exercise 18.7.

Exercise 18.9 Prove the equality shown in Fig. 18.4. This provides a method for constructing new controlled gates from old ones.

We can use certain quantum gates to construct other, more complicated unitary operations. What relatively simple gates belong in our basic "toolbox" for building quantum circuits? One answer is contained in a pair of remarkable theorems about quantum circuits, which we will now state without proof:

- Every unitary operator on n qubits can be constructed from one- and two-qubit gates.
- Every two-qubit quantum gate can be constructed from one-qubit gates and CNOT gates.

The first statement tells us that we need only to be able to manipulate our qubits singly and in pairs to achieve any sort of unitary evolution we wish. The second fact makes things even simpler: we can make do with just a single type of two-qubit gate, the CNOT. Note that CNOT gates and single-qubit unitary operators suffice to construct any sort of quantum evolution. They form a *universal* set of gates.

Exercise 18.10 Think of another two-qubit gate that could replace CNOT in this universal set.

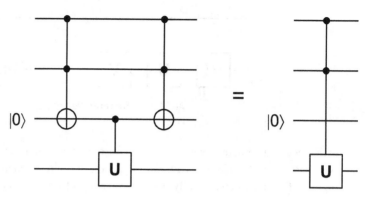

Fig. 18.5   **Using Toffoli gates to create a double-controlled gate.**

Exercise 18.11 Prove that SWAP gates and single qubit unitary operators do *not* form a universal set. That is, show that some unitary operators cannot be constructed from those components alone.

This is very different from the classical situation. The classical analogs to unitary gates are the *reversible* classical logic gates. These have the same number of input and output bits, and there is a one-to-one correspondence between the input and output bit states.[2] One- and two-bit reversible classical gates are not enough to build any reversible operation on n bits. For classical gates, any universal set must contain at least one gate with three or more bits.

Among the many useful and interesting quantum gates on three or more qubits are the *multiple-controlled gates*. The simplest of these is the *quantum Toffoli gate*, which is a "double-controlled NOT" gate: all basis input states are unchanged except for

$$|110\rangle \rightarrow |111\rangle \quad \text{and} \quad |111\rangle \rightarrow |110\rangle. \tag{18.5}$$

Roughly speaking, the X operator acts on the target qubit only if both control qubits are in the state $|1\rangle$. The construction of the Toffoli gate from simpler gates is the subject of Problem 18.4.

Once we have the Toffoli gate, it is not difficult to construct any double-controlled gate. The easiest way illustrates a useful technique in quantum circuit design. We allow the use of one or more *work qubits*, which are required to start out and end up in the state $|0\rangle$, but which can participate in the circuit. The work qubits are "scratch space" for the computation; the requirement that they experience no net change in state guarantees that the net evolution of the remaining qubits will be unitary. (See the discussion in Section 9.2.) The construction shown in Fig. 18.5 is easily checked.

Exercise 18.12 First, check the circuit diagram in Fig. 18.5 to make sure the work qubit is returned to its original state $|0\rangle$. Then build a "triple-controlled NOT" gate out of Toffoli gates, perhaps using one or more work qubits.

---

[2] Both our X gate and CNOT are coherent "quantized" versions of classical reversible gates. The Z, Y, $\Phi(\alpha)$, H, and controlled-Z gates, since they involve superpositions and relative phases, are not.

18.2 Computers and algorithms

We have seen how to create unitary operators for n qubits from a sequence of much simpler elementary operations, a process that is conveniently described by the quantum circuit model. In effect, a quantum circuit expresses the *program* or *algorithm* that is to be executed by an n-qubit quantum computer. Such computers function coherently, transforming superpositions of input states to superpositions of output states. During the computation, the different components of the computer may become highly entangled. For this to take place, the computer must remain informationally isolated from its environment – a physical requirement that usually becomes harder to meet in systems with many qubits. This is one of the main obstacles to the practical realization of large-scale quantum computation.

We can describe the quantum computer's program by a quantum circuit. To give a more complete picture of the computer, we need to discuss two further issues, the input and the output:

- We will assume that all of the n qubits in the computer are initially in the basis state $|0\rangle$. If any other input state is desired, its preparation will be the first part of the program.
- At the end of the unitary evolution, we suppose that individual qubits are measured in the standard $\{|0\rangle, |1\rangle\}$ basis. If some other basic measurement is desired, then the last part of the program will be a unitary rotation of that basis into the standard basis.

Both of these deserve some further comment. We can use one- and two-qubit gates to transform the initial $|0^n\rangle$ state into a more general state of n qubits. For instance, the two-qubit circuit shown in Fig. 18.6 creates the maximally entangled Bell state $|\Phi_+\rangle$ from the standard input state $|00\rangle$.

Exercise 18.13 Design circuits to create each of the other three Bell states $|\Phi_-\rangle$, $|\Psi_+\rangle$, and $|\Psi_-\rangle$, starting with the standard input $|00\rangle$.

Exercise 18.14 Design a two-qubit circuit to prepare the state $\cos\theta\,|00\rangle + \sin\theta\,|11\rangle$ from the standard input, using only single-qubit gates and CNOTs.

Exercise 18.15 Design a three-qubit circuit to prepare the Greenberger–Horne–Zeilinger state $|GHZ\rangle$ discussed in Section 6.7.

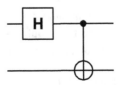

Fig. 18.6   Quantum circuit that creates the Bell state $|\Phi_+\rangle$ from a standard input.

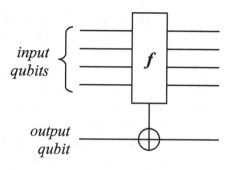

Fig. 18.7   **Unitary evaluation of the function f on a set of input qubits.**

Problem 18.5 describes a situation in which an entangled output measurement is transformed into a standard measurement. It is also useful to reflect on the following exercise:

Exercise 18.16 How could we realize a generalized positive-operator measurement in our quantum computer, if we can make only standard basis measurements? (You might wish to spend a few moments reviewing Section 9.5.)

Now let us take a somewhat higher-level view. Suppose f is a mathematical function that takes a string a of k bits as input, calculating a single bit $f(a)$ as output. Any function calculated by a classical digital computer is made up of functions of this type. A quantum computer executing a program to calculate such a function undergoes this dynamical evolution on $k + 1$ of its qubits:

$$\left|0^k, 0\right\rangle \longrightarrow |a, 0\rangle \longrightarrow |a, f(a)\rangle. \tag{18.6}$$

The first stage prepares the input string a as a basis state of the first k qubits, and the second stage evaluates the function f and writes the result in the last qubit. (The value $f(a)$ can be read by a standard measurement on the output state.) The function evaluation is a coherent "program" on a quantum computer, represented by a unitary operator on $k + 1$ qubits: $|a, f(a)\rangle = \mathsf{U}_f |a, 0\rangle$.

But suppose the initial state of the output qubit is $|1\rangle$ rather than $|0\rangle$? It cannot be that $\mathsf{U}_f |a, 1\rangle = |a, f(a)\rangle$, because this would mean that two orthogonal initial states ($|a, 0\rangle$ and $|a, 1\rangle$) map to the same final state. Instead, we will assume that the calculation of f corresponds to the following evolution:

$$|a, b\rangle \longrightarrow \mathsf{U}_f |a, b\rangle = |a, b \oplus f(a)\rangle, \tag{18.7}$$

where $b \oplus f(a)$ is the sum modulo 2 of the bits b and $f(a)$. This amounts to an "f-controlled NOT" gate, in which the target qubit is negated depending on a (possibly complicated) function f of the k control qubits. A graphical representation of this is shown in Fig. 18.7.

Exercise 18.17 Explain why Eq. 18.7 is unitary for any function f, and why it reduces to Eq. 18.6 when the output qubit is initially $|0\rangle$.

We can use the quantum evaluation of f as a subroutine in larger quantum programs. This is particularly useful if we are designing a program to determine some property of the function f itself. For instance, we may wish to know whether f is constant, or whether $f(0^k) = 0$, etc. To answer such a question, we may need to evaluate f one or more times. In a quantum program, the f-controlled NOT gate from Eq. 18.7 may appear one or more times.

If we actually know more details about the function f – if we can "get inside" the f subroutine – then we might be able to answer our question without ever evaluating f at all. But let us suppose that f is complicated enough that this short-cut is impractical. Then the subroutine that calculates f acts like a "black box" that may be used to evaluate f but cannot be further analyzed. In the language of computer science, our subroutine is regarded as an *oracle* for the function f. Our problem of determining some property of f is thus an *oracle problem*.

In an oracle problem, the key question is how many times we must evaluate the function f (execute the f subroutine) to determine the property in question. For instance, suppose we want to know whether the value of f is constant (either 0 or 1). We might learn that the answer is "no" with as few as two evaluations, if $f(a) \neq f(a')$. But to answer the question conclusively – particularly if the answer is "yes" – we may have to evaluate the function 2^k times. If the number k of input bits is large, then this could take a long time. It may even be impractical to answer the question at all with the computational resources available.

What does this have to do with quantum mechanics? If we are evaluating f using a unitary f-controlled NOT gate on a quantum computer, then the program can act, not merely on the basis state $|a\rangle$ representing a single input string a, but on a coherent superposition of many basis states. This new way of "interrogating" the function f may allow a new and more efficient way of solving an oracle problem.

The simplest example of this is an abstract problem introduced by David Deutsch and Richard Jozsa. In the Deutsch–Jozsa problem, the function f is guaranteed to be either *constant* or *balanced*. For a constant function, $f(a)$ is always 0 or always 1 for all k-bit strings a. For a balanced function, $f(a)$ is 0 or 1 for equally many (2^{k-1}) values of a. How many times must we evaluate f to determine which of these possibilities is true?

A classical computer, which is forced to evaluate f on its inputs one by one, might have to evaluate up to $2^{k-1} + 1$ (one more than half) of the possible inputs to be sure of the answer. After all, if we have evaluated only 2^{k-1} inputs and have always obtained the result 0, the remaining 2^{k-1} possible inputs might either be 0 (for a constant function) or 1 (for a balanced function). The difficulty of solving this oracle problem with certainty on a classical computer is thus exponential in k.

Deutsch and Jozsa showed that, on the contrary, a quantum computer can solve this oracle problem with only *one* coherent evaluation of f.

Before we examine the Deutsch–Jozsa quantum algorithm in detail, we should work out two useful results.

Exercise 18.18 The f-controlled NOT gate of Eq. 18.7 (and Fig. 18.7) can be transformed into an "f-controlled Z" gate, as shown in Fig. 18.8. Show that this gate has the effect

$$|a, 0\rangle \to |a, 0\rangle \qquad \text{and} \qquad |a, 1\rangle \to (-1)^{f(a)} |a, 1\rangle . \qquad (18.8)$$

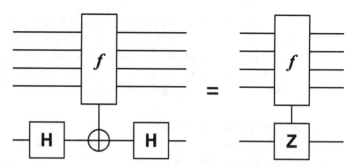

Fig. 18.8 **How to make an f-controlled Z gate.**

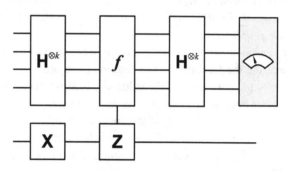

Fig. 18.9 **The Deutsch–Jozsa algorithm for the function f. The final measurement is in the standard basis.**

Exercise 18.19 Suppose k qubits start out in the basis state $|a\rangle$, where a is a k-bit string. We apply the Hadamard gate H to each of the qubits. Show that

$$H^{\otimes k} |a\rangle = \frac{1}{2^{k/2}} \sum_c (-1)^{a \cdot c} |c\rangle, \tag{18.9}$$

where $a \cdot c = a_1 c_1 + \ldots + a_k c_k$, the number of places in which the two bit strings are both 1.

Now we are ready to begin our analysis of the Deutsch–Jozsa algorithm, which is laid out in Fig. 18.9. The $k + 1$ qubits start out in the state $\left|0^k, 0\right\rangle$.

- The input state is prepared by applying a Hadamard gate to each of the k input qubits, and by using X to convert the output qubit to $|1\rangle$:

$$\left|0^k, 0\right\rangle \longrightarrow \frac{1}{2^{k/2}} \sum_a |a, 1\rangle. \tag{18.10}$$

- The function f is evaluated once using the f-controlled Z gate. Note first that this evaluates f on a superposition of all possible inputs, and second that we arrange for f to affect the *phases* of the terms in the superposition. Thus:

$$\frac{1}{2^{k/2}} \sum_a |a, 1\rangle \longrightarrow \frac{1}{2^{k/2}} \sum_a (-1)^{f(a)} |a, 1\rangle. \tag{18.11}$$

- We once more perform the Hadamard gate on all of the k input qubits. Equation 18.9 tells us how this works. Our final state is

$$\frac{1}{2^k} \sum_a (-1)^{f(a)} \left(\sum_c (-1)^{a \cdot c} |c, 1\rangle \right) = \sum_c \left(\frac{1}{2^k} \sum_a (-1)^{a \cdot c + f(a)} \right) |c, 1\rangle . \quad (18.12)$$

- Finally, we measure the k input qubits in the standard basis and determine an experimental value of the bit string c.

What is the probability that the result of our final measurement of c is 0^k? Equation 18.12 tells us that this is

$$P\left(0^k\right) = \left| \frac{1}{2^k} \sum_a (-1)^{f(a)} \right|^2 . \quad (18.13)$$

If the function f is constant, the 2^k terms in the probability amplitude are all of the same sign, and $P(0^k) = 1$. On the other hand, if the function is balanced, then the terms in the probability amplitude exactly cancel out, so that $P(0^k) = 0$. The measurement of c either yields $00 \ldots 0$ or something else, and this determines with certainty whether f is constant or balanced.

There are a number of remarkable features of the Deutsch–Jozsa algorithm. Its general structure amounts to an "interference experiment" among all of the different possible computations of f. (This is why we had f affect the phases of these terms in the superposition.) Somewhat curiously, the final answer to our oracle problem lies not in the function output qubit (which remains in the state $|1\rangle$) but in the final state of the k input qubits.

The Deutsch–Jozsa problem is a highly artificial example, of course, but it illustrates the essential point of quantum computing. A quantum computer can sometimes determine a *global* property of a function far more efficiently than any classical computer could do. This idea is the basis for several remarkable results in quantum computer science. The most famous of these is the quantum factoring algorithm discovered by Peter Shor in 1994. A quantum computer running Shor's algorithm can find the factors of a large integer in exponentially fewer computational steps than any known classical algorithm. The theoretical power of quantum computing has inspired a great deal of effort to realize it in the laboratory. For more details concerning the theory and practise of quantum computing, consult a more specialized text.[3]

18.3 Nuclear spins

In our discussion of quantum computers, we made a series of assumptions regarding our system of n qubits:

---

[3] Our favorite remains *Quantum Computation and Quantum Information* by Michael A. Nielsen and Isaac L. Chuang (Cambridge: Cambridge University Press, 2000). Several other, more recent texts are also excellent.

- The system is completely informationally isolated from its surroundings during the computation.
- We have complete control over the system's dynamics, including the ability to impose one- and two-qubit evolution operators at will. When a qubit is not being manipulated in this way, it "stays put," neither evolving on its own nor interacting with its neighbors.
- We are able to prepare a pure $|0^n\rangle$ state at the beginning of the computation.
- We can make an ideal measurement of the standard basis on one or more qubits at the end of the computation.

Finding a real physical system for which these assumptions hold, even approximately, is no easy task. This is why no quantum computer with many qubits has yet (at the time of this writing) been demonstrated in the laboratory.

For the remainder of this chapter, we will discuss systems of nuclear spins that can be controlled and measured via *nuclear magnetic resonance* (NMR). The NMR technique provides detailed control over the quantum dynamics of one or more nuclei, making it an excellent laboratory for exploring the ideas of quantum computing. It is tremendously useful in its own right as a tool for studying the structure of molecules, and it has applications in medical imaging and other scanning technologies. We, on the other hand, will principally regard NMR as a natural instance of quantum information processing, which we can to some extent control. For us, NMR is quantum computing "in the wild."

We introduced nuclear spins as long ago as Section 2.3, where we considered a single spin-1/2 nucleus in an external magnetic field. Here we summarize that discussion, bringing the terminology and the notation up to date. Throughout our exploration of NMR, we will adopt "natural units" in which $\hbar = 1$. In such units, energy and frequency are the same thing.[4]

The nucleus has a magnetic moment $\vec{\mu} = \gamma \vec{S}$, where γ is a constant that depends on the type of nucleus and $\vec{S}$ is the nuclear spin. The magnetic moment couples to an external magnetic field $\vec{B}$ that is parallel to the positive z-axis. The nuclear spin then evolves subject to the Hamiltonian

$$H_0 = -\vec{\mu} \cdot \vec{B}_0 = -\frac{\Omega}{2} Z, \tag{18.14}$$

where $\Omega = \gamma B$ is the Larmor frequency. For a proton in a 10.00 T magnetic field, Ω has a value of about 2.675×10^9 s^{-1}, corresponding to a circular frequency of 425.7 MHz. The Hamiltonian H_0 causes the spin state to precess about the z-axis at the frequency Ω. This precession may be described by the evolution operator

$$U_0(t) = \exp(-iH_0 t)$$

$$= \exp\left(i\frac{\Omega t}{2} Z\right) = \cos\left(\frac{\Omega t}{2}\right) \mathbf{1} + i\sin\left(\frac{\Omega t}{2}\right) Z. \tag{18.15}$$

---

[4] This is a simplification that both a theorist and an experimentalist can love. Our mathematical expressions will be clearer and more elegant, pleasing the theorist. The experimentalist recognizes that in NMR systems one is always most concerned with the control and measurement of the frequencies of radio-frequency signals, rather than energies.

Exercise 18.20 Verify Eq. 18.15, either by working out the power series for $\exp(i\alpha Z)$ or by using matrix representations.

The Hamiltonian of an actual nuclear spin, however, is not completely determined by the external field. The local molecular environment and interactions with other nuclei contribute to the total Hamiltonian, which we will write $H_0 + H$. The spin state $|\psi(t)\rangle$ is governed by the Schrödinger equation

$$i\frac{d}{dt}|\psi(t)\rangle = (H_0 + H)|\psi(t)\rangle. \tag{18.16}$$

For the external magnetic fields used in NMR experiments, the Hamiltonian H can be regarded as a small perturbation of H_0. That is, the spin precession at the Larmor frequency Ω is much more rapid than state changes produced by H. In this case, it is convenient to adopt the *rotating frame picture*, which compensates for the simple precession of the spin state due to H_0. We define the rotating frame state

$$\left|\hat{\psi}(t)\right\rangle = U_0(t)^\dagger|\psi(t)\rangle. \tag{18.17}$$

Exercise 18.21 If the perturbation $H = 0$, show that the rotating frame state $\left|\hat{\psi}(t)\right\rangle$ will be constant over time.

The perturbation H will produce changes in $\left|\hat{\psi}(t)\right\rangle$. A rotating frame version of the Schrödinger picture governs such changes:

$$i\frac{d}{dt}\left|\hat{\psi}(t)\right\rangle = -H_0\left|\hat{\psi}(t)\right\rangle + U_0(t)^\dagger(H_0 + H)U_0(t)\left|\hat{\psi}(t)\right\rangle. \tag{18.18}$$

Exercise 18.22 First show that $i\frac{d}{dt}U_0(t) = H_0 U_0(t)$. Then verify Eq. 18.18.

We now define the *rotating frame Hamiltonian* $\hat{H}(t) = U_0(t)^\dagger H U_0(t)$, so that our evolution equation for $\left|\hat{\psi}(t)\right\rangle$ takes the familiar form

$$i\frac{d}{dt}\left|\hat{\psi}(t)\right\rangle = \hat{H}(t)\left|\hat{\psi}(t)\right\rangle. \tag{18.19}$$

The rotating frame Hamiltonian $\hat{H}$ is a function of time; but notice that the original H might also be time-dependent.

How is the rotating frame Hamiltonian $\hat{H}(t)$ different from the original H? Let us write H in terms of the Pauli operators, which form a basis for the operators on the qubit Hilbert space $\mathcal{Q}$,

$$H = \varepsilon_0 \mathbf{1} + \varepsilon_x X + \varepsilon_y Y + \varepsilon_z Z. \tag{18.20}$$

The Pauli operators are affected by the rotation operators as follows:

$$U_0^\dagger \mathbf{1} U_0 = \mathbf{1}, \qquad U_0^\dagger X U_0 = \cos\Omega t\, X + \sin\Omega t\, Y,$$

$$U_0^\dagger Z U_0 = Z, \qquad U_0^\dagger Y U_0 = \cos\Omega t\, Y - \sin\Omega t\, X. \tag{18.21}$$

Exercise 18.23 This is a fine opportunity to review the algebra of the Pauli operators. First, confirm as many of the following identities as you need in order to be confident in applying them:

$$X^2 = 1, \quad YZ = -ZY = iX, \quad YXY = ZXZ = -X,$$

$$Y^2 = 1, \quad ZX = -XZ = iY, \quad ZYZ = XYX = -Y,$$

$$Z^2 = 1, \quad XY = -YX = iZ, \quad XZX = YZY = -Z. \tag{18.22}$$

Use these results and Eq. 18.15 to verify Eq. 18.21.

The rotating frame Hamiltonian is therefore

$$\hat{H}(t) = \varepsilon_0 \mathbf{1} + (\varepsilon_x \cos \Omega t - \varepsilon_y \sin \Omega t)X$$

$$+ (\varepsilon_x \sin \Omega t + \varepsilon_y \cos \Omega t)Y + \varepsilon_z Z. \tag{18.23}$$

If the original H is constant or varies only slowly, the X and Y terms in Eq. 18.23 oscillate extremely rapidly about zero. On average, they make almost no contribution to the time evolution of the system in the rotating frame. (Problem 18.7 makes this point in more detail.) For practical purposes, therefore, we can ignore these terms. We can also ignore the $\varepsilon_0 \mathbf{1}$ term, which displaces all system energies by ε_0 but has no other dynamical significance. The effective Hamiltonian in the rotating frame reduces to

$$\hat{H}(t) = \varepsilon_z Z \quad \text{(effective)}. \tag{18.24}$$

The main effect of the local molecular environment on the nuclear spin is to create an additional precession in the rotating frame. Two nuclei of the same type located in different parts of a molecule may experience different perturbing Hamiltonians and therefore precess at slightly different net frequencies. This is the *chemical shift* of the nuclear spins, a source of information about the detailed structure of molecules.

Consider a molecule containing two spin-1/2 nuclei, which may or may not be of the same type. Each nuclear spin would freely rotate in the external field at its own Larmor frequency Ω_1 or Ω_2. The rotating frame is defined by the evolution operator

$$U_0^{(12)}(t) = \exp\left(\frac{i\Omega_1 t}{2} Z_1\right) \exp\left(\frac{i\Omega_2 t}{2} Z_2\right). \tag{18.25}$$

A general perturbation H can be expanded in an operator basis that includes $\mathbf{1}^{(1)} \otimes \mathbf{1}^{(2)}$, $\mathbf{1}^{(1)} \otimes X^{(2)}$, $X^{(1)} \otimes Y^{(2)}$, and so on. In the rotating frame, most of these terms are rapidly oscillating and do not contribute to the average rotating frame dynamics. Only four of the terms do not average to zero. The $\mathbf{1}^{(1)} \otimes \mathbf{1}^{(2)}$ term yields a constant displacement of the energy and may be ignored. The remaining effective Hamiltonian is

$$\hat{H}^{(12)} = \varepsilon_1 Z^{(1)} + \varepsilon_2 Z^{(2)} + \varepsilon_{12} Z^{(1)} \otimes Z^{(2)}. \tag{18.26}$$

The first two terms are chemical shifts (by ε_1 and ε_2) of the two nuclei. The remaining term represents an interaction between the two nuclear spins. As it happens, there is just such an interaction between nearby nuclei in real molecules. This *J-coupling* is mediated by shared electrons in chemical bonds between atoms. Measurement of the effects of J-coupling can help determine how various atoms in a molecule are bonded togther – another important way that NMR is used to probe molecular structure.[5]

It is clear from this discussion that we ought not to interpret the "rotating frame" as a spinning coordinate system in 3-D space. Rather, it represents a rotating view of the Hilbert space describing the quantum system. For a pair of spins, there is no reason why the "frames" in $\mathcal{H}^{(1)}$ and $\mathcal{H}^{(2)}$ cannot be rotating at entirely different frequencies.

What if the perturbing Hamiltonian H is itself a rapidly oscillating function of time? Then the rotating frame Hamiltonian in Eq. 18.23 will average to zero, except in one important case. Consider

$$H = A(t) \left(\cos(\Omega't - \phi)X - \sin(\Omega't - \phi)Y \right). \tag{18.27}$$

This is the Hamiltonian of a nuclear magnetic moment subjected to a pulse of radio waves of frequency Ω' that is circularly polarized in the xy-plane. The amplitude of the wave is described by the envelope function $A(t)$ and its phase is determined by ϕ. Equation 18.23 becomes

$$\hat{H} = A(t) \Big(\cos\left((\Omega - \Omega')t + \phi\right) X + \sin\left((\Omega - \Omega')t + \phi\right) Y \Big). \tag{18.28}$$

Exercise 18.24 Recall the formulas for $\cos(\alpha - \beta)$ and $\sin(\alpha - \beta)$ and use them to derive Eq. 18.28.

If Ω' is far from the Larmor frequency Ω, then these terms will oscillate rapidly and their contribution will be negligible. But suppose instead that the frequencies are exactly matched: $\Omega' = \Omega$. Then the rotating frame Hamiltonian is

$$\hat{H} = A(t)(\cos\phi\, X + \sin\phi\, Y). \tag{18.29}$$

This does not oscillate rapidly and it can have a very large effect on the dynamics of the rotating frame state $\left| \hat{\psi}(t) \right\rangle$.

Exercise 18.25 Suppose we subject the spin to a radio pulse with the frequency Ω but an opposite circular polarization. In this case,

$$H = A(t) \left(\cos(\Omega t - \phi)X + \sin(\Omega t - \phi)Y \right). \tag{18.30}$$

What is the effect of such a perturbation in the rotating frame?

This is precisely the phenomenon of *nuclear magnetic resonance*. By matching the frequency of a radio pulse to that of a precessing nuclear spin, we can "steer" the state of the spin in the rotating frame. The phase ϕ of the pulse determines whether this term is proportional to X, Y, or a combination; the amplitude function $A(t)$ describes how $\hat{H}$ can be turned on and off. By subjecting the spin to resonant radio pulses with precisely controlled ϕ and $A(t)$, we can produce various unitary transformations.

In general, consider a Hamiltonian of the form $f(t)K$, where K is a fixed operator. What unitary operator U corresponds to the action of this Hamiltonian from time 0 to τ? We already know from Eq. 5.38 how to answer this question for a *constant* Hamiltonian. We

---

[5] With more spins, we might expect terms of the form Z, Z $\otimes$ Z, Z $\otimes$ Z $\otimes$ Z, and so on. In practice, however, only the one- and two-spin terms are significant.

therefore imagine dividing the time interval into n smaller intervals of length Δt_i, during which $f(t)$ essentially has the constant value $f(t_i)$. Then

$$U \approx \exp\left(-i(f(t_n)K)\,\Delta t_n\right) \cdots \exp\left(-i(f(t_1)K)\,\Delta t_1\right). \qquad (18.31)$$

The operators in these exponentials commute with each other, since they are all multiples of K. We can therefore combine them as follows:

$$U \approx \exp\left(-i\left(\sum_i f(t_i)\Delta t_i\right)K\right). \qquad (18.32)$$

Taking $n \to \infty$ and $\Delta t_i \to 0$, we obtain the exact expression

$$U = \exp\left(-i\left(\int_0^\tau f(t)dt\right)K\right). \qquad (18.33)$$

For a precessing spin, let $\hat{R}_x(\beta)$ and $\hat{R}_y(\beta)$ represent the unitary operators (in the rotating frame) produced by resonant pulses with phases $\phi = 0$ and $\phi = \pi/2$, respectively, such that[6]

$$\int_{-\infty}^{+\infty} A(t)\,dt = \frac{\beta}{2}. \qquad (18.34)$$

Exercise 18.26 Show that

$$\hat{R}_x(\beta) = e^{-i(\beta/2)X} = \cos(\beta/2)\mathbf{1} - i\sin(\beta/2)X,$$
$$\hat{R}_y(\beta) = e^{-i(\beta/2)Y} = \cos(\beta/2)\mathbf{1} - i\sin(\beta/2)Y. \qquad (18.35)$$

(You may wish to recall Exercise 18.20.)

Exercise 18.27 (a) First show that $\hat{R}_x(\pi) = -iX$ and $\hat{R}_y(\pi) = -iY$. (b) Now compute matrix representations for $\hat{R}_x(\pi/2)$ and $\hat{R}_y(\pi/2)$. (c) Find a product of these operators that yields a multiple of the Pauli Z operator. (d) Do the same for the qubit Hadamard gate operator of Eq. 18.2. (e) Describe your answers in parts (c) and (d) as sequences of resonant radio pulses applied to a precessing nuclear spin.

An overall phase factor in a unitary operator is not important, since it only affects the irrelevant overall phase of the state vector $|\psi\rangle$. Such a phase will make no difference in the density operator $|\psi\rangle\langle\psi|$, or in any statistical prediction. Thus, $\hat{R}_x(\pi) = -iX$ is physically equivalent to the operator X, etc. We can see this by considering how the operations affect the Bloch vector describing a state. Both $-iX$ and X yield the same rotation of the Bloch sphere – in fact, a rotation by π about the X axis.

We can use radio-frequency pulses to produce any rotation about the X or Y axes. By composing such rotations, we can create any rotation of the Bloch sphere. Therefore, up to an irrelevant overall phase factor, we can produce any single-qubit unitary operation on a nuclear spin by a sequence of resonant radio pulses.

---

[6] The factor of 2 is chosen so that this definition agrees with the rotation operators for a spin-1/2 system. See, for example, Eq. 5.56.

18.4 NMR in the real world

Nuclear magnetic resonance provides an unparalleled degree of control over the dynamics of a quantum system. However, to understand actual NMR experiments – including those that use nuclear spins for quantum information processing – we must first describe some of the real-world complications of such systems.

The magnetic field. The external magnetic field $\vec{B}$ is generated by a large set of coils, often superconducting. The sample is held in a container that is as free of irregularities and magnetic impurities as practicable. However, the field may not quite be uniform over the sample volume – a fact which, if uncorrected, would produce an undesirable "spread" in the precessional frequencies of the sample spins.[7]

To correct for this, fine adjustments to the field can be made with several much smaller coils, called *shims*. Also, in many NMR systems the sample can be rapidly rotated within the apparatus, which helps to "average out" the effects of an inhomogeneous field.

The initial state. NMR experiments are always performed on huge ensembles of molecules rather than individual systems. Even a small sample can easily contain 10^{18} molecules of a given type. The spins in the sample are initially in thermal equilibrium at the ambient temperature T, described by a canonical density operator. Consider for instance an isolated proton. In the absence of an external field $\vec{B}$, its two energy levels are degenerate, so the canonical state is uniform: $\rho = \dfrac{1}{2}\mathbf{1}$. But even an extremely large field will not change the situation much.

Exercise 18.28 We can write the canonical density operator for a spin in a magnetic field as $\rho = \lambda_0 \, |0\rangle\langle 0| + \lambda_1 \, |1\rangle\langle 1|$, where the ratio of populations is

$$\frac{\lambda_0}{\lambda_1} = \frac{e^{-E_0/k_{\mathrm{B}}T}}{e^{-E_1/k_{\mathrm{B}}T}} = \exp\left(\frac{\hbar\Omega}{k_{\mathrm{B}}T}\right), \tag{18.36}$$

where Ω is the Larmor frequency. Show that, for a 10.00 T field at room temperature ($T = 300$ K), this ratio differs from 1 by less than 10^{-5}. Which of the two states has the larger population? (We have restored the factor of $\hbar$ in Eq. 18.36 to make it easy to do this calculation in conventional units.)

The concept of a pseudopure state was introduced back in Problem 8.9. This is a mixed state that differs only very slightly from a uniform state on a d-dimensional Hilbert space. The mixed state

$$\pi = \left(\frac{1-\eta}{d}\right)\mathbf{1} + \eta \, |\psi\rangle\langle\psi|, \tag{Re 8.84}$$

is the pseudopure state for the pure state $|\psi\rangle\langle\psi|$, with purity η. Although $|\psi\rangle\langle\psi|$ and π are very different, the pseudopure state π can for many purposes serve as a "stand-in" for $|\psi\rangle\langle\psi|$.

---

[7] On the other hand, variations of the magnetic field can be desirable. If a spatially extended sample is placed in a non-uniform field, then the precession frequency of a nucleus will depend on its location within the sample. This fact is the basis for an important technological application of NMR, magnetic resonance imaging.

Exercise 18.29 If $|\psi\rangle\langle\psi|$ is a qubit state represented by the Bloch vector $\vec{a}$, show that the associated pseudopure state π is represented by the Bloch vector $\eta\vec{a}$.

In Problem 8.9 you were asked to show that the dynamical evolution of the pseudopure π tracks that of the associated pure state $|\psi\rangle\langle\psi|$, and that for any traceless observable A, the pseudopure state has $\langle A \rangle = \text{Tr}\,\pi\text{A} = \eta\,\langle\psi|\,\text{A}\,|\psi\rangle$.

Exercise 18.30 If you have not already worked Problem 8.9, do so now.

Pure states cannot arise in room-temperature NMR, but pseudopure states are either naturally available or can be constructed experimentally. (See Problem 18.8 for more details.)

The rotating frame. Our theoretical analysis of the dynamics of nuclear spins relies on a "rotating frame" that precesses at the Larmor frequency Ω. The radio pulses used to control the spins, for instance, are defined relative to this frame. In a real NMR experiment, this frame is established by an external high-precision oscillator. For instance, the difference between resonant pulses that produce $\hat{R}_x(\beta)$ and $\hat{R}_y(\beta)$ spin rotations is simply a matter of timing – of the relative phase of the radio signals to the fixed external oscillator. Any small discrepancy between the oscillator frequency and the actual spin precession frequency will appear as an extra chemical shift for the spins in the experiment.

Measurement. The last stage of a real NMR experiment is called *free induction decay* (FID), in which the spins are allowed to precess in the external field, and the resulting radio-frequency signals are detected. These signals, like those used to control the spins, can be treated in a completely classical way. Our ensemble of N spins has a net magnetization vector

$$\vec{M}(t) = N\gamma\left\langle\vec{S}\right\rangle, \tag{18.37}$$

where $\vec{S} = (\hbar/2)(\langle X \rangle\,\hat{x} + \langle Y \rangle\,\hat{y} + \langle Z \rangle\,\hat{z})$. As the spins precess, this magnetization vector also precesses, which produces a small but detectable radio signal at the precession frequency. This radio signal provides certain types of information about the ensemble of spins:

- Since the spins precess about the z-axis, the $\langle Z \rangle$ term makes no contribution to the FID signal.
- Types of spin with different precession frequencies each contribute to the FID signal. Thus, it is the Fourier transform of this signal that is most often studied. Note that FID signals from different types of nuclei, or similar nuclei with different chemical shifts, can readily be distinguished in frequency space. (Other effects, such as J-coupling between spins, can also be observed in the details of the Fourier transform of the FID signal.)
- By examining the amplitude of the FID signal and its phase relative to the fixed external oscillator, we can determine both $\langle X \rangle$ and $\langle Y \rangle$ for the spin ensemble in the rotating frame.

Decoherence and decay. The evolution of the nuclear spin state is not perfectly unitary. In the standard basis, the populations of the density operator approach the equilibrium populations with a time constant T_1. This is called *longitudinal* relaxation.

The coherences of the operator in this basis decay toward zero with a time constant T_2. This is called *transverse* relaxation. Problem 18.9 discusses these two processes in more detail.

Both T_1 and T_2 depend on the details of the system, including the temperature and the strength of the external field. Typically, T_2 is much shorter than T_1, and so decoherence is the first important non-unitary effect. As long as the timescale τ of the experiment satisfies $\tau \ll T_2$, then we can regard the system's evolution as effectively unitary. In real experiments, T_2 can be several hundred milliseconds, which is long enough to apply quite complicated sequences of radio pulses.

Atoms and molecules. Real experiments are done with real molecules, which in turn are made out of real atoms. Here are some of the relevant atoms for some useful NMR experiments:

Hydrogen. There are two naturally occurring isotopes, ^{1}H and ^{2}H. The ^{1}H nucleus is a proton with spin-1/2. The deuterium (^{2}H) nucleus (also known as D) contains both a proton and a neutron and has spin-1. The natural abundance of deuterium is only around 0.015%. However, it is a very important tool in liquid-state NMR. If a molecule of interest is dissolved in ordinary water, any NMR signal from its protons would be completely overwhelmed by the signal of the protons in the water. This problem can be avoided by using D_2O rather than H_2O as a solvent.

Carbon. Most carbon nuclei are ^{12}C, which has spin-0 and is thus "invisible" to NMR. Around 1% of carbon nuclei are ^{13}C, which has spin-1/2. We can make the carbon nuclei in a molecule more visible in NMR experiments by preparing the test sample with ^{13}C instead of ^{12}C. (The Larmor frequency for ^{13}C in a 10.00 T magnetic field is 6.726×10^8 s^{-1}, corresponding to a circular frequency of 107.0 MHz, almost exactly one-fourth that of a proton.)

Oxygen. Almost all oxygen nuclei are ^{16}O, which has spin-0. The isotope ^{17}O has spin-5/2, but its natural abundance is only about 0.04%. Oxygen can usually be ignored in an NMR experiment.

Chlorine. The two naturally occurring isotopes of chlorine are ^{35}Cl and ^{37}Cl, each of which has spin-3/2. More to the point, these two types of nuclei have significant electric quadrupole moments, which strongly couple them to the electric fields within a molecule. In a liquid sample, the molecules are rapidly tumbling, producing rapid rotations in chlorine nuclei. On average, these nuclei make no contribution to the nuclear spin Hamiltonian and thus can be effectively ignored in liquid-state NMR experiments.

From these atoms, we can build up molecules. One of the simplest molecules for an NMR experiment is chloroform ($CHCl_3$), consisting of a carbon atom with one hydrogen and three chlorine atoms (see Fig. 18.10). If the carbon atom is the more abundant ^{12}C, then the proton is a nearly-isolated qubit. If we make the molecule using ^{13}C, then it is a simple two-qubit system. Because the Larmor frequencies for ^{1}H and ^{13}C are very different, these qubits can be controlled and measured independently. The J-coupling of the hydrogen and carbon nuclei in the effective Hamiltonian (Eq. 18.26) for chloroform is about $\varepsilon_{12} = 1350$ s^{-1}.

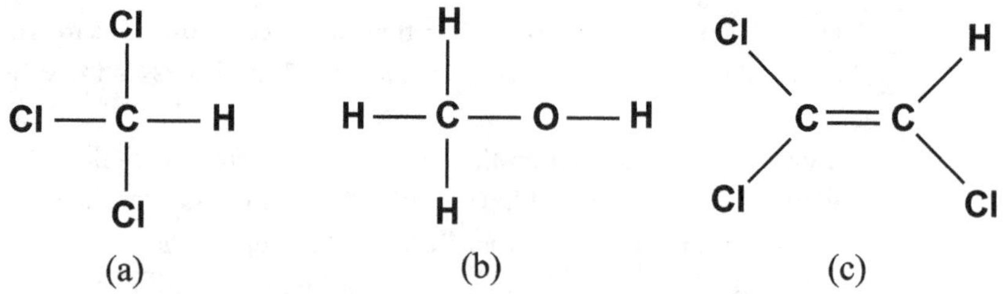

Fig. 18.10   Molecular structures of (a) chloroform, (b) methanol, and (c) trichloroethylene.

To do experiments on more qubits, we need more complex molecules. Consider methanol (CH_3OH), also shown in Fig. 18.10. Taking the most common isotopes for each atom (^1H, ^{12}C, and ^{16}O) there are potentially four qubits in the methanol molecule, one for each hydrogen atom. However, three of these – the ones bonded to the carbon atom – have exactly the same chemical environment and are *magnetically equivalent*. They cannot be individually "addressed" by resonant radio pulses or by measurements.[8]

Trichloroethylene (C_2HCl_3) is a more promising candidate. If we make this molecule using ^{13}C atoms, then it will have three possible qubits. The two carbon atoms have different chemical environments and therefore are separated in frequency by a small chemical shift. If we are careful, we can address these qubits individually.

18.5 Pulse sequences

Using resonant radio-frequency pulses, we can produce any desired single-qubit operation on a nuclear spin (up to an overall phase). For at least some molecules, the various qubits can be individually addressed by tuning the frequency of these pulses. To implement quantum information processing in an NMR system, we need to find a way to make quantum gates like CNOT, in which two qubits interact in a specified way. We do this by exploiting the J-coupling between nuclear spins. The molecules are subjected to a *pulse sequence* consisting of rapid pulses that produce nearly instantaneous changes in the states of various spins, interspersed with much longer time delays during which the spins are allowed to interact. The particular pattern of pulses and delays amounts to a kind of "program" for the particular operation.

The obvious problem is that *all* of the chemical potentials and J-couplings of the spins act at the same time. We cannot simply "turn off" the inconvenient parts of the system Hamiltonian $\hat{H}$ during the delays. How can we possibly design a pulse sequence to implement a particular quantum gate like CNOT? Although we cannot "turn off" parts of the

---

[8] The NMR characteristics of the three-hydrogen methyl group are obviously of considerable interest when NMR is used to explore molecular structure and dynamics. Our point here is simply that the three-fold symmetry complicates matters in using NMR to realize quantum information processing.

Hamiltonian, there is a common technique of NMR that does something remarkably similar. This is the technique of *refocusing*.

This idea is easiest to understand in the case of a single nuclear spin. This has an internal effective Hamiltonian

$$\hat{H} = \varepsilon Z. \tag{18.38}$$

We would like to find a simple way to compensate for the action of this chemical shift. If the Hamiltonian acts for a time t, the resulting unitary operator is

$$V(t) = \exp(-i\varepsilon t Z). \tag{18.39}$$

Suppose, however, we apply pulses before and after this time interval, each producing an impulsive unitary operation X (up to an overall phase). Because $XZX = -Z$, we have

$$XV(t)X = \exp(i\varepsilon t Z), \tag{18.40}$$

and so $XV(t)XV(t) = \mathbf{1}$. The process we envision has four stages. (1) The spin evolves for a time interval t. (2) The spin is rapidly "flipped" via X. (3) The spin evolves for a second equal time interval. (4) Finally, the spin is flipped again using X. The net result is that the second half of the evolution exactly reverses the first half, regardless of the value of the chemical shift ε or the time t.

A dramatic illustration of this is the *spin echo* experiment. After an initial $\hat{R}_x(\pi/2)$ pulse, the spins in a sample are rotated to a direction in the xy-plane. They should experience free induction decay, and the magnetization of the sample should decay with a time constant T_2. However, if the magnetic field is not exactly uniform, then there will be different Larmor frequencies for different parts of the sample. This will cause the spin precessions to desynchronize, and the overall magnetization will appear to decay much more rapidly than T_2. The inhomogeneity of the magnetic field appears to lead to an irreversible relaxation of the system.

However, suppose we apply an X pulse at a time t. If $t \ll T_2$, then by the time $2t$ all of the spins will once again be in their original direction. The magnetization will reappear after its apparent decay. This spin revival, involving perhaps 10^{18} individual nuclei, can be quite dramatic.

Exercise 18.31 Explain how a spin echo experiment can distinguish between the effects of free induction decay and desynchronization due to field inhomogeneities. What is the fundamental difference in these two processes?

Now consider a molecule with two nuclear spins. In the rotating frame, the effective internal Hamiltonian of the system is

$$\hat{H}^{(12)} = \varepsilon_1 Z^{(1)} + \varepsilon_2 Z^{(2)} + \varepsilon_{12} Z^{(1)} \otimes Z^{(2)}. \tag{Re 18.26}$$

We wish to exploit the J-coupling term $\varepsilon_{12} Z^{(1)} \otimes Z^{(2)}$ without having to worry about the two chemical potential terms. We can do this by applying a refocusing operation on both spins. The J-coupling term is affected by both refocusings, and thus undergoes no net cancellation.

Exercise 18.32 Let $V(t) = \exp(-i\hat{H}^{(12)}t)$. Show that

$$(X^{(1)} \otimes X^{(2)})V(t)(X^{(1)} \otimes X^{(2)})V(t) = \exp(-2i\varepsilon_{12}Z^{(1)} \otimes Z^{(2)}t). \qquad (18.41)$$

This is exactly as if only the J-coupling acts over the time $2t$.

Exercise 18.33 Describe a refocusing procedure that cancels out both the $Z^{(2)}$ and $Z^{(1)} \otimes Z^{(2)}$ terms from the evolution due to $\hat{H}^{(12)}$, but leaves the effect of the $Z^{(1)}$ chemical shift term.

Also see Problem 18.10 for a more complex situation.

Given a non-zero J-coupling coefficient ε_{12}, we can use refocusing to create the net time-evolution operator

$$U_{12} = \exp\left(-i\theta Z^{(1)} \otimes Z^{(2)}\right)$$
$$= \cos\theta \, \mathbf{1} - i\sin\theta \, Z^{(1)} \otimes Z^{(2)}, \qquad (18.42)$$

for any desired $\theta \geq 0$.

Exercise 18.34 Verify Eq. 18.42. Show that the standard product basis vectors evolve according to

$$U_{12}|00\rangle = e^{-i\theta}|00\rangle, \quad U_{12}|01\rangle = e^{i\theta}|01\rangle,$$
$$U_{12}|10\rangle = e^{i\theta}|10\rangle, \quad U_{12}|11\rangle = e^{-i\theta}|11\rangle. \qquad (18.43)$$

Since we can use resonant pulses to produce any single-qubit unitary (up to overall phase), we can produce $U_1 = \exp(i\theta Z^{(1)})$ and $U_2 = \exp(i\theta Z^{(2)})$. The product $U_1 U_2 U_{12}$ acts on standard basis elements by

$$|00\rangle \rightarrow e^{i\theta}|00\rangle, \quad |01\rangle \rightarrow e^{i\theta}|01\rangle,$$
$$|10\rangle \rightarrow e^{i\theta}|10\rangle, \quad |11\rangle \rightarrow e^{-3i\theta}|11\rangle . c \qquad (18.44)$$

Choosing $\theta = \pi/4$, this becomes (up to an irrelevant overall phase of $e^{i\pi/4}$),

$$|00\rangle \rightarrow |00\rangle, \quad |01\rangle \rightarrow |01\rangle,$$
$$|10\rangle \rightarrow |10\rangle, \quad |11\rangle \rightarrow -|11\rangle . \qquad (18.45)$$

We recognize the operation in Eq. 18.43 as the controlled-Z gate from Section 18.1. Therefore, by a pulse sequence including time delays, using refocusing to isolate the J-coupling term, we can produce a unitary transformation that is, up to phase, equivalent to the controlled-Z. From this we can easily construct a CNOT gate.

Exercise 18.35 Explain how the solution to Exercise 18.7 also tells us how to construct CNOT from controlled-Z.

Therefore, any quantum circuit on a set of nuclear spins can in principle be implemented via resonant pulses and J-couplings in an NMR system. Our practical limitations will be (1) the choice of a molecule with suitable couplings whose spins may be individually addressed by frequency, and (2) the requirement that all of the operations must be accomplished in a time much less than the decoherence time T_2 for the spins. Luckily, this may be as long as several seconds, giving us time for scores or hundreds of individual gates.

Exercise 18.36 Estimate the time required to perform a CNOT gate on the two nuclear spins in chloroform. Compare this to T_2 for the ^{13}C nucleus, which is about 0.3 s. (T_2 is much longer for the ^1H nucleus in chloroform.)

A more serious limitation comes from the use of pseudopure rather than pure states. We make pseudopure states from thermal states as suggested in Problem 18.8. The resulting purity, and thus the NMR signals, decreases exponentially in the number of qubits. Thus, the NMR approach to quantum computing we have outlined has an effective limit of about 10–12 qubits.

Nuclear magnetic resonance is a highly developed experimental technique, with an extremely rich range of applications. For a deeper look at NMR, consult a specialized text.[9]

Problems

Problem 18.1 We have stated that any n-qubit unitary operation U can be constructed from single-qubit operations and CNOT gates acting between pairs of qubits. But suppose that, for technical reasons, we can only use CNOT gates between *adjacent* qubits. Moreover, the CNOT between qubits k and $k + 1$ must always have k as the control and $k + 1$ as the target. (All single-qubit operations are still allowed.)

(a) Even under these restrictions, are we still able to accomplish any n-qubit unitary U? Prove your answer.
(b) For $n = 3$, construct a quantum circuit obeying these restrictions that performs a CNOT operation with qubit 3 and the control and qubit 1 as the target.

Problem 18.2 Design a controlled-$\Phi(\alpha)$ gate on a pair of qubits using only CNOTs and single-qubit gates. This is a useful generalization of the controlled-Z.

Problem 18.3 This problem tells how to construct multiply-controlled qubit gates.

(a) Prove the identity in Fig. 18.11.
(b) Use this same idea to show how to construct a triple-controlled unitary gate for four qubits, starting from a version controlled by only one qubit.

Problem 18.4 Use the identity from Problem 18.3 (see Fig. 18.11) to construct the quantum Toffoli gate from one- and two-qubit gates. Why does this procedure not work for the *classical* Toffoli gate?

Problem 18.5 When describing dense coding and quantum teleportation in Section 7.4, we made use of the Bell measurement, a basic measurement using the entangled Bell basis states. Design a section of a quantum computer that performs this measurement on a pair of input qubits. That is, the outputs of the final standard measurements on the two qubits

---

[9] We recommend *Spin Dynamics: Basics of Nuclear Magnetic Resonance*, 2nd edn, by Malcolm H. Levitt (Chichester: John H. Wiley and Sons, 2008) as a particularly readable and comprehensive introduction.

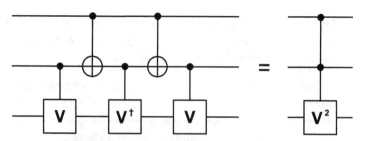

Fig. 18.11   How to make a doubly-controlled quantum gate. See Problem 18.3.

(00, 01, 10, and 11) identify which Bell state is present in the input. Use only single-qubit gates and CNOTs in your design. The symbol for a final standard-basis measurement on a qubit is

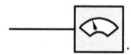

.

Problem 18.6 Here are two extensions of the discussion of the Deutsch–Jozsa problem.

(a) If we do not insist on certainty, a *probabilistic* quantum computer can solve the Deutsch–Jozsa oracle problem pretty efficiently. Suppose a function f is given and we evaluate m inputs chosen at random. We conclude that the function is constant if all of the $f(a)$ values are equal, and balanced otherwise. What is the probability that we would incorrectly identify a constant function as balanced? A balanced function as constant? Discuss the relation between this and the "fair coin/fraudulent coin" example in Appendix A.

(b) Suppose the function f is either *nearly* constant (with only a small number r of exceptions) or *nearly* balanced (with 0 and 1 outputs equal to within a small number r). Calculate the probability that the Deutsch–Jozsa quantum algorithm yields an incorrect answer.

Problem 18.7 An oscillating Hamiltonian. Suppose that the Hamiltonian of a quantum system is of the form $H_c \cos \Omega t$. Show that the resulting time evolution is described by the operator

$$U(t_1, t_2) = \exp\left(-iH_c \int_{t_1}^{t_2} \cos \Omega t\, dt\right), \tag{18.43}$$

where we have set $\hbar = 1$. You should use the fact that the Hamiltonians at various times, though different, at least commute with each other.

Suppose that the oscillation frequency Ω satisfies $\Omega \gg |\varepsilon|$ for all eigenvalues ε of H_c. Show that $U(t_1, t_2) \approx 1$ for all times. In short, a rapidly oscillating Hamiltonian is approximately equivalent to a zero Hamiltonian.

Problem 18.8

(a) Using only CNOT gates, devise a quantum circuit for two qubits that leaves the state $|00\rangle$ unaffected but takes $|01\rangle \rightarrow |11\rangle \rightarrow |10\rangle \rightarrow |01\rangle$.

(b) In a system of two nuclear spins, the canonical state is not a pseudopure state. Assuming that we can make any unitary operation we need by using pulse sequences and delays, describe a general method by which the behavior of an ensemble of these systems, averaged over several different experiments, is the same as that of a pseudopure state for $|00\rangle\langle00|$.

(c) Calculate the effective purity of your "synthetic pseudopure" state in part (b) for a pair of protons at room temperature in a 10.00 T magnetic field. Compare this to the purity of the canonical state of a single proton under these conditions.

(d) Repeat (b) and (c) for a system of three protons.

Problem 18.9 This problem considers a simplified model of the longitudinal and transverse relaxation times for a spin in an NMR experiment. Suppose the spin evolves according to the Lindblad equation with zero Hamiltonian H and three Lindblad operators $\lambda |0\rangle\langle1|$, $\lambda |1\rangle\langle0|$, and μZ. The first two operators represent longitudinal relaxation and the third represents decoherence.

(a) Show that the equilibrium state is $\frac{1}{2}\mathbf{1}$.

(b) Show that the coherences of the density operator in the standard basis decay exponentially toward zero with time constant T_2. Find T_2 in terms of λ and μ.

(b) Show that the populations of the density operator decay exponentially toward 1/2 with time constant T_1. Find T_1 in terms of λ and μ.

(c) Prove that $T_2 \leq 2T_1$.

Problem 18.10 Suppose a molecule contains three nuclear spins. All three spins have chemical shifts, and each pair has a J-coupling. Devise a refocusing technique to eliminate the effects of all terms in the effective Hamiltonian except for the $Z^{(1)} \otimes Z^{(2)}$ term. (Assume that the three spins can be individually addressed by frequency. The easiest solution involves four equal time delays.)

Problem 18.11 Fill in the details of the argument in Section 18.5, and devise a pulse sequence that implements the CNOT gate on two nuclear spins with a J-coupling ε_{12}. For simplicity, you may ignore the chemical shifts in the effective Hamiltonian. On the other hand, you should write all of the single-spin operations as products of $\hat{R}_x(\alpha)$ and $\hat{R}_y(\alpha)$ operators.

Problem 18.12 Write a brief essay discussing how the characteristics of NMR systems match up to the list of "quantum computing assumptions" at the start of Section 18.3.

19 Classical and quantum entropy

19.1 Classical entropy

In this chapter we will generalize classical and quantum concepts of entropy and take a deeper look at their relation to information. We begin with the classical entropy H. This was introduced in Chapter 1 as a practical measure of the amount of information in a message. If we consider a source that generates M equally likely messages, we define the entropy H to be

$$H = \log M. \tag{Re 1.2}$$

This definition has a number of intuitively appealing properties. For example, if two messages are independent, the entropy of the joint message is the sum of the individual entropies. This entropy is also related to the thermodynamical entropy of classical statistical mechanics.

We wish to generalize Eq. 1.2 to situations in which the various messages are not equally likely. Consider, for instance, an English text, which we will regard as a sequence of independent letters and spaces. If we ignore punctuation and capitalization, there are 27 distinct symbols. If the letters were equally likely the entropy would be $H = \log 27 = 4.75$ bits per letter. But in English, some letters (space, e, t, a) are far more probable than others (x, q, z). How can this fact be taken into account?

A limited version of this question was addressed in Problem 1.1, using an argument that we restate in more general form here. Our "messages" are taken to be possible values of an abstract random variable X, so that the value x has probability $p(x)$. We will further suppose that all of the probabilities are rational numbers, which we can write in terms of a common denominator: $p(x) = n_x/N$. These probabilities might arise in the following way:

- There is some larger message Z with N equally likely values.
- The values of Z are partitioned into subsets labeled by the values of X. The subset designated x has n_x equally likely outcomes.

The total probability of the x subset is $p(x)$. To specify the value of Z, we must specify two pieces of data: (1) the value x of X, and (2) which of the n_x outcomes in that subset actually occurs. We know that the overall entropy $H(Z) = \log N$, and that for each particular x (occurring with probability $p(x)$) the entropy of the subset is $\log n_x$. It is reasonable to suppose that

$$H(Z) = \log N = H(X) + \sum_x p(x) \log n_x. \tag{19.1}$$

Solving for $H(X)$ we find

$$H(X) = -\sum_x p(x) \log p(x). \tag{19.2}$$

Exercise 19.1 Verify the result of Eq. 19.2.

Since any irrational probabilities can always be approximated by rational ones, we postulate that Eq. 19.2 holds for all discrete probability distributions over X.

The general definition of entropy in Eq. 19.2 is due to Shannon and forms one of the bases of the classical theory of information.[1] The definition has a number of excellent properties. For instance, if X has M possible values and all of them are equally likely, then $H(X) = \log M$, in agreement with Eq. 1.2. This is in fact the maximum possible value for the entropy:

Exercise 19.2 If X has M possible values, show that $H(X)$ is largest when all of these values are equally likely. (Hint: Write H as a function of the probabilities, not forgetting that the probabilities sum to one and thus $p(x_M) = 1 - p(x_1) - \ldots - p(x_{M-1})$. Maximize this function.)

Exercise 19.3 What if the probability distribution over X contains some x for which $p(x) = 0$? Then the corresponding term in Eq. 19.2 appears to be undefined. To resolve this, show that

$$\lim_{p \to 0} p \log p = 0, \tag{19.3}$$

justifying the convention that sets $0 \log 0 = 0$ in the sum. This means that impossible values for X make no contribution to $H(X)$.

Exercise 19.4 Show that $H(X) \geq 0$, and that $H(X) = 0$ if and only if one value x has probability 1. What is the significance of this fact if we are to regard $H(X)$ as a measure of the "information content of X"?

Exercise 19.5 If we roll two six-sided dice and add the numbers on their upper faces, the sum could be any value between 2 and 12. Find the probability for each sum value, and calculate the entropy of the distribution. How does this compare to $\log 6$ (the entropy for one die) and $\log 36$ (the entropy of the full two-dice configuration, not just the sum)?

See also Problem 19.2, in which you are to find the letter entropy for English.

Although we have defined the entropy $H(X)$ with respect to a random variable X, we should emphasize that it is really a function of the underlying probability distribution of X. This makes sense if we regard $H(X)$ as a measure of information content. The value of X can be encoded in the value of another variable Y; in this new representation, the message x is now designated by the encoded version y, but $p(y) = p(x)$. Since X and Y are merely different representations of the same information, it makes sense that $H(X) = H(Y)$.

---

[1] Shannon did his work in the 1940s, but the analog of Eq. 19.2 in classical statistical mechanics had been discovered more than half a century earlier by J. Willard Gibbs. This is in fact why Shannon called his H by the name "entropy."

Joint and conditional entropies

Consider a pair of random variables X and Y with a joint distribution $p(x,y)$. This has a joint entropy defined in the usual way:

$$H(X,Y) = -\sum_{x,y} p(x,y) \log p(x,y). \tag{19.4}$$

From the joint distribution, we can derive various other important distributions. As noted in Appendix A, we sum the joint probabilities to obtain marginal distributions for X and Y:

$$p(x) = \sum_y p(x,y), \qquad p(y) = \sum_x p(x,y). \tag{Re A.5}$$

From these we may calculate the entropies $H(X)$ and $H(Y)$. For a given Y-value y with $p(y) > 0$, we can also define the conditional probability distribution over X

$$p(x|y) = \frac{p(x,y)}{p(y)}, \tag{Re A.6}$$

from which we can calculate an entropy

$$H(X|y) = -\sum_x p(x|y) \log p(x|y). \tag{19.5}$$

This depends on the particular Y-value y that is given. The *conditional entropy* $H(X|Y)$ is defined to be the average over all such values, weighted by the probability with which they appear:

$$H(X|Y) = \sum_y p(y) H(X|y) = -\sum_{x,y} p(x,y) \log p(x|y). \tag{19.6}$$

From this, it is easy to arrive at the useful identities

$$H(X|Y) = H(X,Y) - H(Y),$$
$$H(X,Y) = H(Y) + H(X|Y). \tag{19.7}$$

Equation 19.7 has a satisfying interpretation. The information content of X and Y together equals that of Y alone plus that of X given Y.

Exercise 19.6 Show also that, for random variables X, Y, and Z,

$$H(X,Y|Z) = H(Y|Z) + H(X|Y,Z). \tag{19.8}$$

This is sometimes called the *chain rule* for conditional entropies.

Since $H(X|Y) \geq 0$ (as the average of non-negative terms), Eq. 19.7 implies that $H(X,Y) \geq H(Y)$. This property is sometimes called the *monotonicity* of classical entropy. If we include more variables, the entropy – our measure of the information in the joint outcome – can never actually get smaller.

One of the fundamental properties of the entropy function is called *Gibbs's inequality*. For two distributions $p(x)$ and $q(x)$ for the same random variable X,

$$-\sum_x p(x) \log p(x) \leq -\sum_x p(x) \log q(x), \tag{19.9}$$

where equality holds if and only if the two distributions are the same: $p(x) = q(x)$. The proof of this is not difficult, and is left for Problem 19.1. A great many facts about the entropy function can be derived from Gibbs's inequality.

Exercise 19.7 Suppose that X has M possible values. Use Gibbs's inequality with $q(x) = 1/M$ to give an alternate proof that X is maximized (at $\log M$) for the uniform distribution.

This motivates the definition of the *relative entropy* $D(p||q)$ between two distributions:

$$D(p||q) = \sum_x p(x) \log \frac{p(x)}{q(x)}. \tag{19.10}$$

Gibbs's inequality tells us that $D(p||q) \geq 0$, with equality if and only if the two distributions are the same. (If there exists an x for which $q(x) = 0$ but $p(x) > 0$, then we formally write $D(p||q) = \infty$.) The relative entropy is a kind of non-symmetric measure of the "distance" between two probability distributions.

The *mutual information* of random variables X and Y, written $H(X : Y)$, is defined to be

$$H(X : Y) = H(X) + H(Y) - H(X, Y) = \sum_{x,y} p(x, y) \log \frac{p(x, y)}{p(x)p(y)}. \tag{19.11}$$

Exercise 19.8 Prove the second equality in Eq. 19.11.

Exercise 19.9 Use Gibbs's inequality with $q(x, y) = p(x) p(y)$ to show that $H(X : Y) \geq 0$ for any joint distribution over X and Y. Further prove that $H(X : Y) = 0$ if and only if X and Y are independent.

Exercise 19.9 establishes that the mutual information is useful as a measure of the *correlation* of X and Y. From this it also follows that $H(X) \geq H(X, Y) - H(Y) = H(X|Y)$. In other words, determining the value of the variable Y can never (on average) increase the entropy of the variable X.

Exercise 19.10 We previously pointed out that $H(X, Y) \geq H(Y)$. Under what condition will equality hold? Try to describe this condition plainly as a relationship between the variables X and Y.

Convexity and concavity

A real function f is said to be *convex* if, for any two input values x and y and any parameter λ such that $0 \leq \lambda \leq 1$,

$$f(\lambda x + (1 - \lambda)y) \leq \lambda f(x) + (1 - \lambda)f(y). \tag{19.12}$$

We can recognize a convex function by its graph. A straight chord line joining any two points on the graph of f must lie above that graph.

Exercise 19.11 Draw a picture to illustrate this idea, and show the relation between the graphical characterization and the definition in Eq. 19.12.

A sufficient condition for f to be convex is for $f''(x) \geq 0$ for all x.

It is straightforward to generalize Eq. 19.12. Suppose $x_1, \ldots, x_n$ are inputs of the function f, and the numbers $\lambda_1, \ldots, \lambda_n$ are non-negative numbers that add up to 1. The convexity of f means that

$$f\left(\sum_k \lambda_k x_k\right) \leq \sum_k \lambda_k f(x_k). \tag{19.13}$$

The numbers x_k might be the values of a random variable X, with the λ_ks being the corresponding probabilities. Then the convexity of f means that

$$f(\langle X\rangle) \leq \langle f(X)\rangle. \tag{19.14}$$

This useful fact is called *Jensen's inequality*.

Exercise 19.12 Use Jensen's inequality to prove the familiar fact that $\langle X^2\rangle \geq \langle X\rangle^2$.

A function $g(x)$ is said to be *concave* if $-g(x)$ is convex. To put it another way, a function is concave provided that the direction of the inequality is reversed in Eq. 19.12 (and thus also in Eq. 19.13 and 19.14).

Exercise 19.13 Prove that the sum of two or more convex (or concave) functions is also convex (or concave).

Consider an abstract random variable X with probability distribution $p(x)$. We could consider the numbers $p(x)$ also to be the values of a numerical random variable, which we denote P_X. This new variable has a mean

$$\langle P_X\rangle = \sum_x p(x)^2. \tag{19.15}$$

The function $f(y) = -\log(y)$ is a convex function for all $y > 0$. Jensen's inequality therefore yields an inequality for the entropy:

$$H(X) = -\sum_x p(x) \log p(x)$$

$$= \langle -\log P_X\rangle$$

$$\geq -\log\langle P_X\rangle = -\log\left(\sum_x p(x)^2\right). \tag{19.16}$$

Exercise 19.14 Suppose that X and X' are independent, identically-distributed random variables. Use Eq. 19.16 to show that the probability the two agree is at least

$$\Pr(X = X') \geq 2^{-H(X)}. \tag{19.17}$$

The function $g(y) = -y \log y$ is concave for $0 \leq y \leq 1$. Since the entropy function is the sum of such terms, it too is a concave function of the probability distribution. That is, suppose $p_1(x)$ and $p_2(x)$ are two probability distributions over the same set, and let $0 \leq \lambda \leq 1$. Then we can make an average probability distribution

$$p(x) = \lambda p_1(x) + (1 - \lambda) p_2(x). \tag{19.18}$$

The concavity of the entropy function means that the entropies of these distributions satisfy

$$H = -\sum_x p(x)\log p(x) \geq \lambda H_1 + (1-\lambda)H_2, \qquad (19.19)$$

where H_i is the entropy of the distribution $p_i(x)$. The entropy of the average distribution is always at least as great as the average of the two entropies.

This concavity property of entropy is closely related to a fact we have already seen.

Exercise 19.15 Suppose that X and Y are two random variables with a joint distribution. For each y, we have the conditional distribution $p(x|y)$ over x. The ys themselves occur with probabilities $p(y)$, leading to an average distribution $p(x)$. Use the concavity of the entropy function to obtain the result that $H(X) \geq H(X|Y)$.

19.2 Classical data compression

The value of the numerical measure of entropy lies in its relation to particular information tasks. As we first saw in Section 1.1, the entropy H for uniform distributions was the minimum number of bits required to represent a message faithfully. (To approach the goal of using H bits per message, we must encode entire blocks of independent messages together. The main point remains.) Does such a relation still hold when the underlying probability distributions are not uniform?

The brief answer is *yes*, but to understand this answer we need to examine the problem in more detail. We shall begin with a binary example. Suppose we have a binary message source that produces 0s and 1s with very different probabilities: $p(0) = 0.9$ and $p(1) = 0.1$. The entropy for this source is $H = 0.469$ bits. This seems to suggest that we might faithfully code n messages from the source using around $n/2$ binary digits, if n is large enough.

Yet this appears to be impossible. There are 2^n possible strings of n independent messages from the source, and we cannot represent them all with fewer than n bits.

On the other hand, do we really need to represent *every* possible n-message string in our code? Some of these strings are not very likely to occur. Consider the string $111\cdots 11$, which has the probability $(0.1)^n$. If n is reasonably large, this may be quite negligible. We might be prepared to tolerate *errors* in our code, provided that the total probability of error can be made small enough.[2]

As we will see, for large n it is possible to identify a *typical set* T of message strings. Our code will be designed to represent strings in T faithfully, but will fail for other strings. If there are N_T strings in T, our code requires about $\log N_T$ bits (and thus $(\log N_T)/n$ bits per message). The probability that a randomly generated string is in T is P_T. Thus, our efforts will be successful if we can choose N_T such that $(\log N_T)/n \to H$ bits per message, while nevertheless $P_T \to 1$, as n becomes large.

---

[2] We do not need to give a particular universal definition for a "small enough" probability of error. Instead, we require that, once we establish an upper error limit ϵ – tiny, but non-zero – we can make sure the total probability of error is less than ϵ.

Consider strings of ten outputs of our source. We will choose as our typical set T the set of strings with "about" 90% zeroes – to be specific, the number of zeroes is between seven and ten (inclusive). Recall that

$$\text{Number of } n\text{-strings with } k \text{ zeroes} = \binom{n}{k} = \frac{n!}{k!(n-k)!},$$

$$\text{Probability of each such string} = p(0)^k p(1)^{n-k}. \tag{19.20}$$

The following sequence of exercises shows what happens:

Exercise 19.16 There are $2^{10} = 1024$ ten-bit message strings that the source might produce. Show that the typical set T described above has only $N_T = 176$ members, but has a total probability P_T in excess of 0.98. How many bits per message do we need to represent the strings in T?

Exercise 19.17 If you have a computer system suitable for doing the calculation, consider strings of $n = 100$ messages. For T, include all strings with between 82 and 98 zeroes, inclusive. (Note that the most likely single message, $0000 \cdots 00$, is not in T.) Show that $P_T > 0.99$, but that the messages in T can be encoded using only about 0.65 bits per message.

Exercise 19.18 Consider strings of $n = 1000$ messages. Let T consist of all such strings with between 870 and 930 zeroes. Show that P_T is even closer to 1 than in the previous exercise, but we only require about 0.55 bits per message to represent the strings in T faithfully.

The typical set T is *exponentially smaller* than the set of all n-message strings. Nevertheless, it is nearly certain that the particular string actually generated by the source lies within T. This combination of "small size" and "high probability" allows us to use many fewer than n bits to represent the message, while maintaining an exceedingly small likelihood of error.

To make our argument more general and rigorous, we need a systematic way to choose the set T, so that in the limit $n \to \infty$ it has the requisite properties. From Appendix A recall the following basic theorem:

Weak Law of Large Numbers. Let $X_1, X_2, \ldots, X_n$ be independent, identically distributed numerical random variables, each with finite mean μ and variance σ^2. Given any $\delta, \epsilon > 0$, for sufficiently large n,

$$\Pr\left(\left| \left(\frac{1}{n} \sum_k x_k \right) - \mu \right| > \epsilon \right) < \delta. \tag{Re A.22}$$

For large n, the numerical average of the x_is will almost certainly be very close to the theoretical mean μ.

Now we once again use the probabilities to construct a numerical random variable. Given the random variable X, we can define a numerical random variable that takes on values $-\log p(x)$. This new random variable is sometimes called the *surprise* of X, denoted Σ_X, because its value is large when $p(x)$ is small (and thus x would be quite surprising). Note that we use the probability distribution $p(x)$ to fix both the possible values of Σ_X and their probabilities. The mean of Σ_X is exactly $H(X)$ as defined in Eq. 19.2; and if we further assume that the underlying set of X is finite, Σ_X must also have finite variance.

Now consider a sequence of n independent instances of X. It is convenient to denote such a sequence by $\vec{X} = (X_1, \ldots, X_n)$. The probability $p(\vec{x}) = p(x_1) \cdots p(x_n)$, and so

$$\Sigma_{\vec{X}} = -\log p(\vec{x}) = -\log p(x_1) - \ldots - \log p(x_n) = \Sigma_{X_1} + \ldots + \Sigma_{X_n}. \quad (19.21)$$

The surprise is thus an "additive" random variable when we create sequences of independent Xs. We can now apply the Weak Law of Large Numbers to the surprise. Given $\delta, \epsilon > 0$, then for large enough n

$$P\left(\left| -\frac{\log p(\vec{x})}{n} - H(X) \right| > \epsilon \right) < \delta. \quad (19.22)$$

For such large n, we define the typical set T to be the set of sequences for which

$$n(H(X) - \epsilon) \leq -\log p(\vec{x}) \leq n(H(X) + \epsilon). \quad (19.23)$$

Equation 19.22 establishes that a sequence $\vec{x}$ is in this set T with probability $P_T > 1 - \delta$, so T is a "high probability" set. But how big is T? Since no individual $\vec{x}$ in T has a probability less than $2^{-n(H(X)+\epsilon)}$, the total number of sequences must satisfy

$$N_T \leq 2^{n(H(X)+\epsilon)}. \quad (19.24)$$

Exercise 19.19 Verify the derivation of this result from Eq. 19.22. Why is the inequality here "$\leq$" rather than "$<$"?

Given any $\delta, \epsilon > 0$, for sufficiently large n we can find a typical set T of n-strings of X such that $P_T > 1 - \delta$, and $\log N_T \leq n(H(X) + \epsilon)$.

Intuitively, we say that the typical set T of n-sequences has a total probability that is about equal to 1; that there are about $2^{nH(X)}$ sequences in T; and that all of these sequences have about the same probability $p(\vec{x}) \approx 2^{-nH(X)}$. These simple informal statements conceal a good deal of mathematical machinery within the word "about"! Nevertheless, they are useful as heuristic principles.

Since T has total probability near 1, we will be satisfied to devise a code that faithfully represents sequences in T but fails for all others. Since T has about $2^{nH(X)}$ elements, we can represent those elements uniquely by sequences of about $nH(X)$ bits. More exactly, we can construct a code for sequences in T that uses between $n(H(X)+\epsilon)$ bits and $n(H(X)+\epsilon)+1$ bits. The number of bits per message is thus no greater than $H(X) + \epsilon + 1/n$. By suitable choice of ϵ and n, we can make this as close as we like to $H(X)$ itself.

We can therefore devise a code for long strings of X messages that has a probability of error less than δ, but which only requires about $H(X)$ bits per message. The entropy $H(X)$ quantifies the physical resources (bits) sufficient to faithfully represent X. This fact is called the (classical) *data compression theorem*.

The data compression theorem is a fundamental principle of classical information theory. In effect, it gives an operational meaning to the entropy $H(X)$ in terms of a definite communication task. But the particular scheme used to prove the data compression theorem is not necessarily a practical one. As Exercises 19.16, 19.17, and 19.18 suggest, we might need to encode very long sequences to compress our data to about $H(X)$ bits per message. Furthermore, even though we know that a good code exists, it might be hard to describe it in detail or to apply it efficiently. Real codes for data compression are generally based on a different approach. Problem 19.3 suggests one such alternative.

19.3 Quantum entropy

Suppose a quantum system is in a mixed state described by the density operator ρ. The mixed state may arise either because we are uncertain about the system's preparation, or because the system is entangled with other systems. When the ρ is a uniform density operator on a subspace of dimension d in the Hilbert space, we have defined the quantum entropy of the system to be

$$S = \log d. \tag{Re 8.56}$$

This quantum entropy is analogous to the classical entropy H for a uniform distribution. Just as we generalized H to more general distributions, we will now generalize S to more general density operators.

First, we should point out that the quantum entropy S has some very different properties from classical entropy. For instance, if X and Y are two random variables, then $H(X, Y) \geq H(Y)$. The classical entropy of the whole is at least as great as the classical entropy of the part. But consider an ebit, a pair of qubits A and B in a maximally entangled state. For each qubit, the subsystem state is a uniform density operator over the whole Hilbert space Q, and so $S(A) = S(B) = 1$. But the joint system AB is in a pure state, so that $S(AB) = 0$. Because of entanglement, the quantum entropy of the whole can be less than the quantum entropy of the part.[3]

The general definition of entropy in the quantum case was given by John von Neumann. For a system Q with density operator ρ, the entropy is

$$S(A) = -\operatorname{Tr} \rho \log \rho. \tag{19.25}$$

The logarithm $\log \rho$ of the operator ρ can be defined using a power series, as was done for the operator exponential in Eq. 5.39. The resulting expression is independent of basis but is easiest to compute in a basis that diagonalizes ρ. If the eigenvalues of ρ are ρ_k, then

$$S(Q) = -\sum_k \rho_k \log \rho_k. \tag{19.26}$$

The quantum (von Neumann) entropy is just the classical (Shannon) entropy of the eigenvalues of ρ.

---

[3] Note that we associate the entropy $S(A)$ with the system A, though it is a function only of the density operator $\rho^{(A)}$. In the same way we associate the classical entropy $H(X)$ with the random variable X, though it depends only on the set of probabilities.

Exercise 19.20 The quantum system Q is equally likely to be prepared in the states $|a\rangle$ and $|b\rangle$, so that the density operator for the mixture is

$$\rho = \frac{1}{2}\left(|a\rangle\langle a| + |b\rangle\langle b|\right). \tag{19.27}$$

However, $|a\rangle$ and $|b\rangle$ are not necessarily orthogonal. Suppose that their inner product $\langle a|b\rangle$ happens to be real. Show that the quantum entropy is

$$S(Q) = -\rho_+ \log \rho_+ - \rho_- \log \rho_-, \tag{19.28}$$

where $\rho_\pm = \frac{1}{2}(1 \pm \langle a|b\rangle)$. If the two states are equally likely, why is the quantum entropy not necessarily equal to 1? What is the entropy if the two state vectors are nearly parallel in the Hilbert space?

Suppose we apply a basic measurement to Q with basis elements $|a_n\rangle$. The probability of the outcome a_n is

$$p(a_k) = \sum_k |\langle a_k|k\rangle|^2 \rho_k, \tag{19.29}$$

which we can regard as a probability distribution over a random variable A. Consider the matrix whose components are $M_{nk} = |\langle a_k|k\rangle|^2 \geq 0$. Normalization gives us that

$$\sum_n M_{nk} = 1 \qquad \text{and} \qquad \sum_k M_{nk} = 1. \tag{19.30}$$

In other words, both the rows and columns of M_{nk} can be regarded as probability distributions. A matrix with this property is called *doubly stochastic*.

The classical entropy $H(A)$ is

$$H(A) = -\sum_n \left(\sum_k M_{nk}\rho_k\right) \log \left(\sum_l M_{nl}\rho_l\right). \tag{19.31}$$

Each term in the n-sum has the form $f(\langle R\rangle)$, where R is a random variable with values ρ_k and probability M_{nk} (for a given n), and where $f(x) = x \log x$, a convex function. Jensen's inequality yields

$$H(A) \geq -\sum_n \left(\sum_k M_{nk}\rho_k \log \rho_k\right). \tag{19.32}$$

Since $\sum_n M_{nk} = 1$, a reordering of the sums yields

$$H(A) \geq S(Q). \tag{19.33}$$

For example, suppose Q is a qubit with $\rho^{(Q)} = |0\rangle\langle 0|$. Then $S(Q) = 0$. A measurement in any basis except the standard basis $\{|0\rangle, |1\rangle\}$ will yield $H(A) > 0$.

Exercise 19.21 Explain why Eq. 19.33 can be restated thus: *The Shannon entropy of the populations of ρ can be no less than the von Neumann entropy of ρ.*

Klein's inequality

The same mathematical trick involving a doubly-stochastic matrix allows us to prove an analog of Gibbs's inequality, which in the quantum case is known as *Klein's inequality*. Suppose ρ and σ are two density operators for Q such that

$$\rho = \sum_k \rho_k \, |r_k\rangle \langle r_k| \quad \text{and} \quad \sigma = \sum_n \sigma_n \, |s_n\rangle \langle s_n|. \tag{19.34}$$

Now consider the quantity

$$-\operatorname{Tr} \rho \log \sigma = -\sum_k \rho_k \langle r_k| \log \sigma \, |r_k\rangle$$

$$= -\sum_k \rho_k \langle r_k| \left(\sum_n \log \sigma_n \, |s_n\rangle \langle s_n| \right) |r_k\rangle$$

$$= -\sum_k \rho_k \left(\sum_n M_{nk} \log \sigma_n \right), \tag{19.35}$$

where $M_{nk} = |\langle s_n | r_k \rangle|^2$, a doubly stochastic matrix. Applying Jensen's inequality to the logarithm gives

$$-\operatorname{Tr} \rho \log \sigma \geq -\sum_k \rho_k \log \left(\sum_n M_{nk} \sigma_n \right)$$

$$= -\sum_k \rho_k \log \tau_k, \tag{19.36}$$

where $\tau_k = \sum_n M_{nk} \sigma_n$. By Gibbs's inequality,

$$-\sum_k \rho_k \log \tau_k \geq -\sum_k \rho_k \log \rho_k, \tag{19.37}$$

and therefore

$$-\operatorname{Tr} \rho \log \sigma \geq -\operatorname{Tr} \rho \log \rho. \tag{19.38}$$

This is Klein's inequality.

Exercise 19.22 The logarithm is *strictly* concave, which means that $\langle \log X \rangle \leq \log \langle X \rangle$ with equality if and only if one value x of X has $p(x) = 1$. From this (and from the equality condition for Gibbs's inequality) show that the two sides of Eq. 19.38 are equal if and only if $\rho = \sigma$.

Equation 19.38 motivates us to define the *quantum relative entropy*

$$D(\rho||\sigma) = \operatorname{Tr} \rho \log \rho - \operatorname{Tr} \rho \log \sigma. \tag{19.39}$$

Klein's inequality then tells us that $D(\rho||\sigma) \geq 0$, with equality if and only if $\rho = \sigma$.

Exercise 19.23 Under what circumstances is $D(\rho||\sigma) = \infty$?

Like the classical relative entropy between distributions, the quantum relative entropy can be considered as an asymmetrical "distance" measure between density operators.

Properties of quantum entropy

Some of the quantities we defined for the classical entropy have straightforward analogs in the quantum domain. For example, a composite system AB has both a joint entropy $S(AB)$ and subsystem entropies $S(A)$ and $S(B)$. The *quantum mutual information* is

$$S(A{:}B) = S(A) + S(B) - S(AB). \tag{19.40}$$

Exercise 19.24 Show that $\log(E^{(A)} \otimes F^{(B)}) = (\log E^{(A)}) \otimes \mathbf{1}^{(B)} + \mathbf{1}^{(A)} \otimes (\log F^{(B)})$ for positive operators $E^{(A)}$ and $F^{(B)}$.

Exercise 19.25 Use the result of the previous exercise to show that

$$S(A{:}B) = D\left(\rho^{(AB)} || \rho^{(A)} \otimes \rho^{(B)}\right) \geq 0. \tag{19.41}$$

The quantum mutual information $S(A{:}B)$ is a measure of "how far" the state $\rho^{(AB)}$ is from being a product of subsystem states – a measure, in short, of the degree of correlation of systems A and B. It follows that

$$S(A,B) \leq S(A) + S(B), \tag{19.42}$$

a property known as the *subadditivity* of the entropy.

What about a *conditional* quantum entropy $S(A|B)$? We might of course simply define

$$S(A|B) = S(AB) - S(B). \tag{19.43}$$

But this quantity is not always positive, so it cannot be an "average" of other entropies (as we had in the definition of $H(X|Y)$ in Eq. 19.6). The point is that we have no obvious way to define a *conditional state* $\rho^{(A|B)}$, a state of system A in which system B somehow acts as a "given."[4] So the mathematical definition in Eq. 19.43 – though acceptable from a purely mathematical point of view – is potentially more misleading than enlightening.

If A and B are in a pure state $|\Psi^{(AB)}\rangle$, then the density operators $\rho^{(A)}$ and $\rho^{(B)}$ have exactly the same non-zero eigenvalues (see Exercise 8.14). Thus,

$$S(A) = S(B) \qquad (\rho^{(AB)} \text{ pure}). \tag{19.44}$$

This entropy is a measure of the entanglement of A and B. The general fact in Eq. 19.44 is useful for deriving even more properties of the entropy. For example, suppose the composite system AB is in the mixed state $\rho^{(AB)}$, and we introduce a third system C that purifies AB. Subadditivity (Eq. 19.42) for system AC means that $S(A) + S(C) \geq S(AC)$. Since ABC is in a pure state, complementary subsystems have the same entropy. That is, $S(A) = S(BC)$, $S(AC) = S(B)$ and so on. Thus,

$$S(A) + S(AB) \geq S(B), \tag{19.45}$$

---

[4] We might try to define $S(A|B)$ by considering the conditional states of A due to a particular measurement of B. This is, in effect, the entropy of A conditioned on the *classical* variable that represents the B-measurement outcome. This depends on the choice of measurement. It is not Eq. 19.43.

which implies that $S(AB) \geq S(A) - S(B)$. The corresponding fact with A and B reversed yields

$$|S(A) - S(B)| \leq S(AB), \tag{19.46}$$

a result known as the *triangle inequality* for entropy.

Three deep facts

We will introduce three more general properties of quantum entropy. These properties are related to one another – we will derive the second and third from the first – but their ultimate proof involves mathematical techniques beyond the scope of this book. On the other hand, they are exceedingly useful for understanding the information properties of open quantum systems, so we will state them here and use them as needed.

The first and most basic property is known as the *monotonicity of relative entropy*. Given two states $\rho^{(AB)}$ and $\sigma^{(AB)}$ of a composite system AB,

$$D\left(\rho^{(AB)}||\sigma^{(AB)}\right) \geq D\left(\rho^{(A)}||\sigma^{(A)}\right). \tag{19.47}$$

That is, the relative entropy "distance" between two quantum states cannot be increased by disregarding one of the subsystems.

This has a direct corollary for open quantum systems. Suppose ρ and σ are two possible initial states of a system Q. This system interacts with an environment E (initially in the state $|0\rangle\langle 0|$), producing a generalized evolution map $\mathcal{E}$ for Q-states: $\mathcal{E}(\rho) = \mathrm{Tr}_{(E)} U(\rho \otimes |0\rangle\langle 0|) U^\dagger$. Unitary evolution does not change the relative entropy of two states. Therefore

$$D(\rho||\sigma) = D\left(\rho \otimes |0\rangle\langle 0| \,\middle|\middle|\, \sigma \otimes |0\rangle\langle 0|\right)$$
$$= D\left(U(\rho \otimes |0\rangle\langle 0|)U^\dagger \,\middle|\middle|\, U(\sigma \otimes |0\rangle\langle 0|)U^\dagger\right)$$
$$D(\rho||\sigma) \geq D(\mathcal{E}(\rho)||\mathcal{E}(\sigma)). \tag{19.48}$$

Therefore, relative entropy of two states cannot be increased by open system evolution. Such evolution is *contractive* with respect to relative entropy.

From Eq. 19.47 we discover our second fundamental property of entropy, called the *monotonicity of quantum mutual information*. In Eq. 19.41, the quantum mutual information is given in terms of the relative entropy. It follows that, for any state of a composite quantum system ABC,

$$S(A:BC) \geq S(A:B). \tag{19.49}$$

Exercise 19.26 Fill in the steps to derive the monotonicity of quantum mutual information from the monotonicity of relative entropy.

This too has a corollary for open systems. Suppose the composite system RQ starts out in a state with a quantum mutual information $S(R:Q)$. Appending an environment system E in the state $|0\rangle\langle 0|$, we have $S(R:QE) = S(R:Q)$. If the QE subsystem evolves in a unitary way, this will not affect the mutual information, a fact that we may write $S(R:QE') = S(R:QE)$. Finally, taking the partial trace over the environment E, we find that the open system evolution of Q cannot increase its quantum mutual information with R:

$$S(R:Q) \geq S(R:Q').$$ (19.50)

We obtain our third fundamental property of the entropy by rewriting Eq. 19.49 in terms of subsystem entropies:

$$S(A) + S(BC) - S(ABC) \geq S(A) + S(B) - S(AB).$$ (19.51)

Rearranging the terms, we arrive at the *strong subadditivity* of the quantum entropy:

$$S(ABC) + S(B) \leq S(AB) + S(BC).$$ (19.52)

This is the quantum counterpart of a classical inequality involving three random variables. The proof of the classical inequality is not difficult (see Problem 19.4), but it relies on the use of conditional entropies that have no quantum analog. Strong subadditivity nonetheless holds in the quantum context as well.

19.4 Quantum data compression

The strong formal parallel between the classical and quantum entropy functions will allow us to adapt much of the discussion of classical data compression (Section 19.2) to the quantum realm. We begin by exploring the quantum notion of "typicality."

Suppose we have a composite quantum system $\vec{Q} = Q_1 Q_2 \ldots Q_n$ consisting of n independent copies of Q. The overall system is in a product state, and each subsystem is described by an identical density operator $\rho^{(Q)}$,

$$\rho^{\vec{Q}} = \rho^{(Q_1)} \otimes \cdots \otimes \rho^{(Q_n)}.$$ (19.53)

This is the quantum analog of a sequence of n independent, identically distributed random variables. To make this analogy even more obvious, we write

$$\rho^{(Q)} = \sum_x p(x) |x\rangle \langle x|,$$ (19.54)

where the kets $|x\rangle$ are eigenstates of $\rho^{(Q)}$ with eigenvalue $p(x)$. Then

$$\rho^{\vec{Q}} = \sum_{x_1 \ldots x_n} p(x_1 \ldots x_n) |x_1 \ldots x_n\rangle \langle x_1 \ldots x_n|,$$ (19.55)

where $p(x_1 \ldots x_n) = p(x_1) \cdots p(x_n)$. The eigenvalues of $\rho^{\vec{Q}}$ act exactly like the probabilities of sequences of random variables – sequences, in effect, of $\rho^{(Q)}$ eigenstates. We can now apply our previous results about typical sequences to the eigenvalues and eigenstates of $\rho^{(Q)}$. The facts proved there do not change their mathematical form, but they take on new meaning.

Given $\epsilon, \delta > 0$, for sufficiently large n we have the following:

- There exists a typical set T of sequences $x_1 \ldots x_n$. This is associated with a *typical subspace* $\mathcal{T}$ of the tensor product space $\mathcal{H}^{\otimes n}$ spanned by the eigenstates $|x_1 \ldots x_n\rangle$. The projection Π onto $\mathcal{T}$ can be written

$$\Pi = \sum_{x_1 \ldots x_n \in T} |x_1 \ldots x_n\rangle\langle x_1 \ldots x_n|. \tag{19.56}$$

- The probability $\Pr(x_1 \ldots x_n \in T) > 1 - \delta$. This means that

$$\operatorname{Tr} \rho^{(Q)} \Pi > 1 - \delta. \tag{19.57}$$

- The number of sequences in the typical set T satisfies $N_T < 2^{n(H(X)+\epsilon)}$, where H is the Shannon entropy of the distribution $p(x)$ (Eq. 19.24). In the quantum case, this means that the dimension of the typical subspace satisfies

$$N_T = \dim \mathcal{T} < 2^{n(S+\epsilon)}, \tag{19.58}$$

where $S = S(Q)$, the quantum entropy of Q. Just as the typical set T is generally an exponentially small subset of the set of all sequences, so $\mathcal{T}$ is an exponentially small subspace of $\mathcal{H}^{\otimes n}$.

- The sequence probabilities $p(x_1 \ldots x_n)$ lie in a range around $2^{-nH(X)}$ (see Eq. 19.22). In the quantum case, the eigenstates $|x_1 \ldots x_n\rangle$ that span $\mathcal{T}$ have eigenvalues satisfying

$$n(S - \epsilon) \leq -\log p(\vec{x}) \leq n(S + \epsilon). \tag{19.59}$$

(In fact, this is the criterion by which we construct $\mathcal{T}$ in the first place.)

A simple heuristic picture emerges. For large enough n, the state $\rho^{\otimes n}$ for n independent systems has associated with it a typical subspace $\mathcal{T}$. This subspace contains almost all of the "weight" of $\rho^{\otimes n}$ and has a dimension of about 2^{nS}. In this subspace, the eigenvalues of $\rho^{\otimes n}$ are all about 2^{-nS}. In many respects, and with appropriate mathematical caveats given above, the density operator $\rho^{\otimes n}$ behaves as a uniform density operator on $\mathcal{T}$.

Now let us consider a problem of quantum information transfer – or rather, a problem of quantum information storage and retrieval. This is, in the language of Section 7.5, a Type II process. We start with a composite system RQ in an entangled pure state $|\Psi^{(RQ)}\rangle$. The quantum information of Q is transferred to another system C for storage; during retrieval, the information is transferred back to Q. Schematically,

$$RQ \xrightarrow{\text{storage}} RC \xrightarrow{\text{retrieval}} RQ, \tag{19.60}$$

where R takes no part in the process, but merely serves to "anchor" the entanglement. If the systems Q and C were of the same kind, then the storage and retrieval could be accomplished by applying the unitary operator U that exchanges Q and C states:

$$\mathsf{U} |\psi^{(Q)}, \phi^{(C)}\rangle = |\phi^{(Q)}, \psi^{(C)}\rangle. \tag{19.61}$$

(This is like the qubit SWAP gate of Exercise 18.8.) Then, if C starts out in the state $|\alpha^{(C)}\rangle$, the storage process is

$$|\Psi^{(RQ)}\rangle \otimes |\alpha^{(C)}\rangle \longrightarrow |\Psi^{(RC)}\rangle \otimes |\alpha^{(Q)}\rangle. \tag{19.62}$$

In the retrieval process, the system Q must start out in a fixed state $|\beta^{(Q)}\rangle$ – in other words, we retrieve the information into a "fresh copy" of Q. The retrieval process is

$$\left|\Psi^{(RC)}\right\rangle \otimes \left|\beta^{(Q)}\right\rangle \longrightarrow \left|\Psi^{(RQ)}\right\rangle \otimes \left|\beta^{(C)}\right\rangle, \qquad (19.63)$$

and the RQ state is exactly restored. The entanglement fidelity of this process is $F_e = 1$.

But suppose that C is a smaller system than Q, so that $\dim \mathcal{H}^{(C)} < \dim \mathcal{H}^{(Q)}$. Then it will not generally be possible to store and retrieve the entangled state perfectly. Instead, we will describe a procedure for doing so *approximately*, with F_e near 1.

We choose a subspace $\mathcal{C}$ of $\mathcal{H}^{(Q)}$ such that $\dim \mathcal{C} \leq \dim \mathcal{H}^{(C)}$. We call $\mathcal{C}$ the *coding subspace*. Let Π_c be the projection onto the coding subspace $\mathcal{C}$ and $\Pi_n = \mathbf{1} - \Pi_c$ be the projection onto the orthogonal *non-coding subspace* $\mathcal{C}^\perp$. The initial RQ state can be written

$$\left|\Psi^{(RQ)}\right\rangle = \left|\Psi_c^{(RQ)}\right\rangle + \left|\Psi_n^{(RQ)}\right\rangle, \qquad (19.64)$$

where $\left|\Psi_c^{(RQ)}\right\rangle = \Pi_c \left|\Psi^{(RQ)}\right\rangle$ and $\left|\Psi_n^{(RQ)}\right\rangle = \Pi_n \left|\Psi^{(RQ)}\right\rangle$.

The first step of our procedure is to make an ideal measurement on Q defined by the projections Π_c and Π_n. This measurement is assumed to be minimally disturbing, so that the projection postulate holds. The "coding" result is obtained with probability

$$P_c = \left\langle \Psi_c^{(RQ)} \middle| \Psi_c^{(RQ)} \right\rangle = \operatorname{Tr} \rho^{(Q)} \Pi_c, \qquad (19.65)$$

where $\rho^{(Q)} = \operatorname{Tr}_{(R)} |\Psi^{(RQ)}\rangle \langle \Psi^{(RQ)}|$. If this result occurs, nothing more is done. But if the "non-coding" result occurs, then the Q-state is reset to a standard state $|0^{(Q)}\rangle$ in $\mathcal{C}$.

Exercise 19.27 Show that the procedure described changes the RQ-state according to

$$\left|\Psi^{(RQ)}\right\rangle\!\left\langle\Psi^{(RQ)}\right| \to \hat{\rho}^{(RQ)} = \left|\Psi_c^{(RQ)}\right\rangle\!\left\langle\Psi_c^{(RQ)}\right| + \left|0^{(Q)}\right\rangle\!\left\langle 0^{(Q)}\right| \otimes \sigma^{(R)}, \qquad (19.66)$$

where $\sigma^{(R)} = \operatorname{Tr}_{(R)} \left|\Psi_n^{(RQ)}\right\rangle\!\left\langle\Psi_n^{(RQ)}\right|$.

The new RQ-state $\hat{\rho}^{(RQ)}$ has the property that $\operatorname{Tr} \hat{\rho}^{(Q)} \Pi_c = 1$. That is, the new Q-state "lies entirely within" the coding subspace $\mathcal{C}$. This state *can* be perfectly stored and retrieved by an operation that exchanges the states within the coding subspace $\mathcal{C}$ with states in a subspace of $\mathcal{H}^{(C)}$. The following exercise gives a definite prescription:

Exercise 19.28 Let $\{\,|k^{(Q)}\rangle\}$ be a basis for $\mathcal{C}$ that includes the vector $|0^{(Q)}\rangle$. Let $\{\,|k^{(C)}\rangle\}$ be a corresponding orthonormal set in $\mathcal{H}^{(C)}$ (which we can find because $\dim \mathcal{H}^{(C)} \geq \dim \mathcal{C}$). Define a unitary operator U so that

$$U\left|k^{(Q)}, l^{(C)}\right\rangle = \left|l^{(Q)}, k^{(C)}\right\rangle. \qquad (19.67)$$

Show, first, that such a U exists. Then show that this operation will allow perfect storage and retrieval of the state $\hat{\rho}^{(RQ)}$, using C as a substitute for Q.

The overall process can be summarized this way:

$$\left|\Psi^{(RQ)}\right\rangle\!\left\langle\Psi^{(RQ)}\right| \xrightarrow{\text{measurement}} \hat{\rho}^{(RQ)} \xrightarrow{\text{storage}} \hat{\rho}^{(RC)} \xrightarrow{\text{retrieval}} \hat{\rho}^{(RQ)}. \qquad (19.68)$$

The entanglement fidelity is

$$F_e = \left\langle \Psi^{(RQ)} \middle| \hat{\rho}^{(RQ)} \middle| \Psi^{(RQ)} \right\rangle$$
$$= \left| \left\langle \Psi^{(RQ)} \middle| \Psi_c^{(RQ)} \right\rangle \right|^2$$
$$= P_c^2, \tag{19.69}$$

where P_c is the probability that the first measurement found the state within the coding subspace. If $P_c > 1 - \eta$, then $F_e = P_c^2 > 1 - 2\eta$.

We have now established the following important result:

General compression lemma. Given an entangled state $|\Psi^{(RQ)}\rangle$, suppose we can find a subspace $\mathcal{C}$ of $\mathcal{H}^{(Q)}$ with projection operator Π_c so that

$$\mathrm{Tr}\, \rho^{(Q)} \Pi_c > 1 - \eta, \tag{19.70}$$

for some $\eta > 0$. Also suppose we have a system C available with dim $\mathcal{H}^{(C)} \geq \dim \mathcal{C}$. Then we can store and retrieve the entanglement of Q with R (Type II quantum information) by means of C with an entanglement fidelity

$$F_e > 1 - 2\eta. \tag{19.71}$$

Once we have reached the general compression lemma, we do not have far to go to prove a theorem about quantum data compression. Instead of a single pair of systems RQ, we imagine a collection of n such systems $(RQ)^{\otimes n}$, each of which is in the same entangled state $|\Psi^{(RQ)}\rangle$. The entropy of each Q system is $S = S(Q)$. The state of all n Qs together is $\rho^{\vec{Q}} = (\rho^{(Q)})^{\otimes n}$. Let $\epsilon, \delta > 0$. For sufficiently large n, we can find a typical subspace $\mathcal{T}$ of $(\mathcal{H}^{(Q)})^{\otimes n}$ with projection Π such that

$$\mathrm{Tr}\, \rho^{\vec{Q}} \Pi > 1 - \delta/2 \quad \text{and} \quad \dim \mathcal{T} < 2^{n(S+\epsilon)}. \tag{19.72}$$

We will use $\mathcal{T}$ as our coding subspace.

For our coding system C we will choose a collection of N qubits, where $n(S + \epsilon) \leq N < n(S + \epsilon) + 1$. Then we have that dim $\mathcal{H}^{(C)} = 2^N \geq \dim \mathcal{T}$, as required. The general compression lemma tells us that we can store and retrieve the entanglement of $Q^{\otimes n}$ with $R^{\otimes n}$ using N qubits with an entanglement fidelity $F_e > 1 - \delta$. This uses

$$\frac{N}{n} < \frac{n(S + \epsilon) + 1}{n} = S(Q) + \epsilon + \frac{1}{n}, \tag{19.73}$$

qubits per RQ pair. We can make ϵ and $1/n$ as small as we like. Thus, $S(Q)$ gives the physical resources (qubits) sufficient to store the (Type II) quantum information in Q with entanglement fidelity F_e approaching 1. This fact is called the *quantum data compression theorem*.

Just as classical data compression gave the Shannon entropy $H(X)$ an operational significance, so quantum data compression does the same for the von Neumann entropy $S(Q)$. It is just the first of many ways in which the entropy plays a decisive role in the theory of quantum information. Before exploring these further developments, though, we will see how the quantum entropy and its properties apply to the thermodynamics of quantum systems.

Exercise 19.29 Discuss how the quantum data compression theorem applies to Type I (unknown state) quantum information in Q.

19.5 Entropy and thermodynamics

In Sections 8.6, 8.7, and 9.3, we used the density operator description to analyze the thermodynamics of quantum systems. For uniform density operators (as in the micro-canonical state of Eq. 8.57), the thermodynamic entropy was

$$S_\theta = k_B \ln 2 \; S, \tag{19.74}$$

(see Eq. 8.59). Now that we have a more general definition of S, we can discuss the thermodynamic entropy of a broader class of states. We write

$$S_\theta = -k_B \ln 2 \; \mathrm{Tr}\, \rho \log \rho = -k_B \mathrm{Tr}\, \rho \ln \rho. \tag{19.75}$$

Denote by ω the canonical state of the system at temperature T, given in Eq. 8.70. This has entropy

$$
\begin{aligned}
S_\theta \text{ (canonical)} &= -k_B \mathrm{Tr} \left(\frac{e^{-H/k_B T}}{\mathcal{Z}} \right) \ln \left(\frac{e^{-H/k_B T}}{\mathcal{Z}} \right) \\
&= \frac{1}{T} (\mathrm{Tr}\, \omega H) + k_B \ln \mathcal{Z} \, (\mathrm{Tr}\, \omega) \\
&= \frac{E_\theta}{T} + k_B \ln \mathcal{Z},
\end{aligned}
\tag{19.76}
$$

where $E_\theta = \langle E \rangle$ for the canonical state.

Now consider another state ρ with the same expected energy as ω, so that $\mathrm{Tr}\, \rho H = \mathrm{Tr}\, \omega H = E_\theta$. According to Klein's inequality (Eq. 19.38) its entropy satisfies

$$
\begin{aligned}
S_\theta = -k_B \mathrm{Tr}\, \rho \ln \rho &\leq -k_B \mathrm{Tr}\, \rho \ln \left(\frac{e^{-H/k_B T}}{\mathcal{Z}} \right) \\
&= \frac{E_\theta}{T} + k_B \ln \mathcal{Z},
\end{aligned}
\tag{19.77}
$$

with equality if and only if $\rho = \omega$. The canonical state ω is therefore the density operator having maximum entropy for a given thermodynamic energy E_θ.

The canonical state ω is the equilibrium state of a system with Hamiltonian H in contact with a reservoir at temperature T. In Section 9.4 we defined the free energy of ω to be $F_\theta = -k_B T \ln \mathcal{Z}$. Equation 19.76 simplifies this to

$$F_\theta = E_\theta - TS_\theta. \tag{19.78}$$

This relation permits us to extend the definition of F_θ to an arbitrary state ρ. The free energy is a function of three things: the quantum state ρ, the system Hamiltonian H, and the reservoir temperature T. Equation 19.78 becomes

$$F_\theta(\rho, \mathsf{H}, T) = \operatorname{Tr}\rho\mathsf{H} + k_\mathrm{B}T \operatorname{Tr}\rho \ln \rho. \tag{19.79}$$

This leads in turn to a remarkable connection between free energy and relative entropy. Given H and T, consider the relative entropy of a state ρ with respect to the canonical state ω

$$D(\rho\|\omega) = \operatorname{Tr}\rho\log\rho - \operatorname{Tr}\rho\log\omega$$

$$= -S - \operatorname{Tr}\rho\log\left(\frac{e^{-\mathsf{H}/k_\mathrm{B}T}}{\mathcal{Z}}\right). \tag{19.80}$$

Multiplying by $k_\mathrm{B}\ln 2$ shifts to thermodynamic quantities:

$$k_\mathrm{B}\ln 2\, D(\rho\|\omega) = -S_\theta + \frac{1}{T}\operatorname{Tr}\rho\mathsf{H} + k_\mathrm{B}\ln\mathcal{Z}. \tag{19.81}$$

From this we can see that

$$k_\mathrm{B}T\ln 2\, D(\rho\|\omega) = F_\theta(\rho, \mathsf{H}, T) - F_\theta(\omega, \mathsf{H}, T). \tag{19.82}$$

Exercise 19.30 Explain how the three terms on the right-hand side of Eq. 19.81 give rise to just two terms on the right-hand side of Eq. 19.82.

Since the relative entropy is never negative, $F_\theta(\rho, \mathsf{H}, T) \geq F_\theta(\omega, \mathsf{H}, T)$, with equality only in the case where $\rho = \omega$. The equilibrium (canonical) state is therefore the state of minimum free energy for a given H and T.

By what process do we create a non-equilibrium state ρ? Suppose for instance that we have some particles moving in an enclosure. In the equilibrium state these particles are equally likely to be found anywhere within the volume. How do we arrange a situation in which all of the particles are certain to be in the right-hand half of the available volume? The simplest procedure is to introduce a moveable wall in the enclosure and push all the molecules to the right side. Then the wall can be suddenly removed, leaving the particles (momentarily) occupying only half of the original space.

Let us treat this more abstractly. Initially, we have a Hamiltonian H and a canonical state ω at temperature T. The Hamiltonian is changed to H', so that ρ is the canonical state for the modified Hamiltonian and the same temperature. We can imagine that this is done isothermally – that is, in a gradual way while maintaining equilibrium with a reservoir system. Finally, the Hamiltonian is changed back to H, so rapidly that the system state does not have time to evolve. We can represent this two-stage process as follows:

$$\omega, \mathsf{H} \xrightarrow{\text{isothermal}} \rho, \mathsf{H}' \xrightarrow{\text{sudden}} \rho, \mathsf{H}. \tag{19.83}$$

During each stage the system does work on its surroundings. Section 9.4 tells us how to compute this work for either stage. For the first (isothermal) process, Eq. 9.44 tells us that the work done by the system is

$$W_1 = F_\theta(\omega, \mathsf{H}, T) - F_\theta(\rho, \mathsf{H}', T). \tag{19.84}$$

The second (sudden) stage is even easier. Equation 9.40 tells us that $W_2 = \text{Tr}\,\rho H' - \text{Tr}\,\rho H$. During this stage the entropy of the state remains constant, so according to Eq. 19.79 the work W_2 must also be the difference of two free energies:

$$W_2 = F_\theta(\rho, H', T) - F_\theta(\rho, H, T). \tag{19.85}$$

The total work $W = W_1 + W_2$ done by the system is

$$W = F_\theta(\omega, H, T) - F_\theta(\rho, H, T) = -\Delta F_\theta. \tag{19.86}$$

This can never be positive. Net external work $\mathbb{W} = -W$ must be done *on* the system to shift it from the equilibrium state ω to a non-equilibrium ρ.

Note that the process in Eq. 19.83 is a *reversible* one. Imagine that we somehow begin with the state ρ and the Hamiltonian H. We now reverse the order of our operations: first a sudden change of the Hamiltonian to H', then a slow, isothermal return to H. In this reversed process, the state of the system will also exactly reverse its history, ending up at ω. The work done by the system will be exactly $-W$.

We can generalize this to any sort of change of state.

Exercise 19.31 Initially, a system's Hamiltonian is H_a and its state is ρ_a; at the end of a process these are H_b, ρ_b, respectively.

(a) Show that this change can be accomplished by a sequence of sudden changes in the Hamiltonian and isothermal processes with a reservoir at temperature T.
(b) In the process from part (a), show that the work done by the system is

$$W = F_\theta(\rho_a, H, T) - F_\theta(\rho_b, H, T) = -\Delta F_\theta. \tag{19.87}$$

What if we adopted a different procedure for changing the state from ρ_a to ρ_b? How is the work W for this alternative process related to the free energy change? This question requires a more careful analysis of the entropy changes of a quantum system.

When a density operator changes over time, how does its thermodynamic entropy S_θ change? To answer this, we first need to clear up a technical issue about time derivatives of operators.

Exercise 19.32 Suppose that the operator A is time-dependent with rate of change $\dot{A}$. The operators A and $\dot{A}$ do not necessarily commute.

(a) Explain why, in general, $\dfrac{d}{dt}A^n \neq nA^{n-1}\dot{A}$.
(b) Now show that, nevertheless, $\dfrac{d}{dt}\text{Tr}\left(A^n\right) = \text{Tr}\left(\dfrac{d}{dt}A^n\right) = n\,\text{Tr}\left(A^{n-1}\dot{A}\right)$.
(c) If the function f is given by a power series, show that

$$\frac{d}{dt}\text{Tr}\left(f(A)\right) = \text{Tr}\left(\frac{d}{dt}f(A)\right) = \text{Tr}\left(f'(A)\,\dot{A}\right). \tag{19.88}$$

The final part of this exercise is the tool we need. For a time-dependent state ρ the entropy changes at a rate

$$\dot{S}_\theta = -k_{\mathrm{B}}\mathrm{Tr}\left((\ln\rho - 1)\dot{\rho}\right) = -k_{\mathrm{B}}\mathrm{Tr}\,\dot{\rho}\ln\rho, \qquad (19.89)$$

since $\mathrm{Tr}\,\dot{\rho} = 0$.

Exercise 19.33 Suppose ρ evolves according to Eq. 8.51. Show that $\dot{S}_\theta = 0$.

Consider a system that is interacting with a reservoir at temperature T, and suppose that the evolution of its state ρ can be described by the Lindblad equation.[5] We will write this equation in somewhat simplified form:

$$\dot{\rho} = \frac{1}{i\hbar}[\mathsf{H}, \rho] + \mathcal{L}(\rho), \qquad (19.90)$$

where the superoperator $\mathcal{L}$ includes the contributions of the Lindblad operators. Both H and $\mathcal{L}$ may depend on time; but imagine that both are "frozen" at some particular stage and consider the resulting evolution of the system. We now make an important assumption about the system's evolution, which we may designate the *equilibrium assumption*. We first assume that the canonical state ω, which represents a long-term equilibrium state of the system interacting with the reservoir, is a *fixed point* of the evolution:

$$\dot{\omega} = \frac{1}{i\hbar}[\mathsf{H}, \omega] + \mathcal{L}(\omega) = 0. \qquad (19.91)$$

Exercise 19.34 Show that this in fact means that $\mathcal{L}(\omega) = 0$.

We further assume that any non-equilibrium initial state will approach the canonical one over time. Thus, ω is the *only* fixed point of Eq. 19.90.

Consider another state ρ and how it evolves according to the "frozen" evolution. Equation 19.48 says that the relative entropy $D\left(\rho\|\omega\right)$ cannot increase with time:

$$\frac{d}{dt}D\left(\rho\|\omega\right) = \mathrm{Tr}\,\dot{\rho}\log\rho - \mathrm{Tr}\,\dot{\rho}\log\omega \le 0. \qquad (19.92)$$

The rate of change of the system's thermodynamic entropy is therefore

$$\dot{S}_\theta \ge -k_{\mathrm{B}}\mathrm{Tr}\,\dot{\rho}\ln\left(\frac{e^{-\mathsf{H}/k_{\mathrm{B}}T}}{\mathcal{Z}}\right) = \frac{1}{T}\mathrm{Tr}\,\dot{\rho}\mathsf{H}. \qquad (19.93)$$

From our discussion in Section 9.4, particularly the analysis leading up to Eq. 9.38, we recognize $\mathrm{Tr}\,\dot{\rho}\mathsf{H}$ as P_Q, the rate at which heat flows into the system. Thus, in a small interval of time dt, the thermodynamic entropy of a system in contact with a reservoir at temperature T changes by

$$dS_\theta \ge \frac{\text{\dj}Q}{T}, \qquad (19.94)$$

where $\text{\dj}Q$ is the heat transfer to the system. Equality holds if and only if the state $\rho = \omega$ — that is, if the system is in equilibrium with the reservoir. If we now consider a full time-dependent process (maintaining at every stage the equilibrium assumption about the time

---

[5] The Lindblad equation supposes that at any given moment we can disregard any pre-existing correlations between the system and the reservoir. This is consistent with the general supposition that the reservoir's properties are the same no matter how the system changes.

evolution), we obtain a relation between the change in entropy and the total heat Q added to the system:

$$\Delta S_\theta \geq \frac{Q}{T},$$ (19.95)

with equality for isothermal processes, i.e. those in which the system remains in the canonical state throughout.

Equation 19.95 is exactly the result needed to answer our question about general state transformations. The state evolves by $\rho_a \to \rho_b$ while the system Hamiltonian is changed according to $H_a \to H_b$. The work done by the system in the process is

$$W = -\Delta E_\theta + Q \leq -(\Delta E_\theta - T \Delta S_\theta) = -\Delta F_\theta.$$ (19.96)

The type of procedure from Exercise 19.31, consisting of sudden Hamiltonian changes (having zero heat transfer) and gradual isothermal processes, therefore maximizes W – or equivalently, *minimizes* the external work $\mathbb{W}$ done on the system to effect the change. Other processes in which the system spends time away from equilibrium will require more external work to accomplish.

19.6 Bits and work

Suppose Alice wishes to send a classical message to Bob by way of a quantum system Q. Alice's message is a classical random variable A. The message a arises with probability $p(a)$, and the entropy $H(A)$ measures the information content of the message. For the message a, Alice prepares Q in the *signal state* ρ_a and then conveys the system to Bob. If Q undergoes any noise or distortion, then we include these effects in the definition of ρ_a; these are the density operators of the *received* signals. The *average signal state* is

$$\rho = \sum_a p(a)\rho_a.$$ (19.97)

Bob decodes the message by making a measurement on Q, the result of which is another random variable B. The particular value b of B is associated with a positive operator E_b, one element of a generalized measurement (see Section 9.5). The conditional probabilities are given by the quantum rule

$$p(b|a) = \text{Tr}\,\rho_a E_b.$$ (19.98)

From this we can find the joint distribution $p(a,b) = p(a)p(b|a)$. How much information is transmitted from Alice to Bob?

Although the conditional probabilities in Eq. 19.98 arise from quantum mechanics, this is essentially a problem of classical information theory. The classical measure of the information conveyed from Alice to Bob is the mutual information

$$H(A:B) = H(A) - H(A|B).$$ (19.99)

This is the average by which the entropy of A is reduced by the result of Bob's measurement. If $H(A : B) = H(A)$, then $H(A|B) = 0$ and Bob is able to determine Alice's message with certainty from his measurement result.[6] The value of $H(A : B)$ will depend on Bob's choice of *decoding observable* given by the operators E_b. For a given set of signal states ρ_a and probabilities $p(a)$, how well can Bob do?

To answer this question, we introduce a way of representing the measurement process as the generalized evolution of an open quantum system. To the Q we add a *measurement apparatus* M, initially in a standard state $|0\rangle$. The different measurement results b correspond to distinguishable (hence orthogonal) states $|b\rangle$ of the apparatus. These states are the "measurement record" in M. For a given input Q-state ρ, the measurement process should have the effect

$$\rho \otimes |0\rangle\langle 0| \longrightarrow \hat{\sigma}_b \otimes |b\rangle\langle b| \qquad \text{with probability } p(b) = \text{Tr}\,\rho E_b. \qquad (19.100)$$

(The final Q-states $\hat{\sigma}_b$ may be anything; we do not require the measurement process to obey the projection rule.) Thus, we want to design an operation for which

$$\mathcal{E}(\rho \otimes |0\rangle\langle 0|) = \sum_b \sigma_b \otimes |b\rangle\langle b|, \qquad (19.101)$$

where $\sigma_b = (\text{Tr}\,\rho E_b)\,\hat{\sigma}_b$, normalized by the probabilities.

Here is the simplest superoperator that does the trick. For each b define the Kraus operator

$$A_b = E_b^{1/2} \otimes |b\rangle\langle 0|. \qquad (19.102)$$

(The operator $E_b^{1/2}$ is the unique positive operator whose square is E_b.) Then

$$\sum_b A_b^\dagger A_b = \sum_b E_b \otimes |0\rangle\langle 0| = \mathbf{1} \otimes |0\rangle\langle 0|. \qquad (19.103)$$

If we append the additional Kraus operator $\mathbf{1} \otimes (\mathbf{1} - |0\rangle\langle 0|)$, then the resulting set is properly normalized. The action of the superoperator described by the $\{A_b, A_0\}$ set is

$$\mathcal{E}(\rho \otimes |0\rangle\langle 0|) = \sum_b \left(E_b^{1/2} \rho\, E_b^{1/2} \right) \otimes |b\rangle\langle b|. \qquad (19.104)$$

Since $\text{Tr}\,E_b^{1/2} \rho\, E_b^{1/2} = \text{Tr}\,\rho E_b$, the proposed superoperator $\mathcal{E}$ meets our specifications.

As Problem 9.1 shows, any normalized set of Kraus operators can be realized as unitary evolution on a larger system including an environment E. We can either think of $\mathcal{E}$ as arising from a probabilistic measurement process or from unitary evolution involving the

---

[6] The mutual information also plays a central role in the *noisy coding theorems* of classical information theory. Given a noisy channel from Alice to Bob with a mutual information $H(A : B)$ between input and output, we can use this channel repeatedly to send about $H(A : B)$ bits per use with a probability of error $P_E \to 0$.

environment.[7] Therefore, the results we have derived for open system dynamics can be applied to the measurement.

Having made Bob's measurement more "tangible," let us do the same for Alice's preparation. We suppose that Alice possesses a system R that contains a record of which message she wishes to send. The initial joint state of RQ is thus a mixture of states of the form $|a\rangle\langle a| \otimes \rho_a$, where the $|a\rangle$ are orthogonal record states of R:

$$\rho^{(RQ)} = \sum_a p(a) |a\rangle\langle a| \otimes \rho_a. \qquad (19.105)$$

Let $S(Q)$ and $S(Q|a)$ be the quantum entropies of the average signal state ρ and the individual signal state ρ_a, respectively.

Exercise 19.35 Derive the following for the state in Eq. 19.105:

$$S(R) = H(A),$$
$$S(RQ) = H(A) + \sum_a p(a)S(Q|a),$$
$$S(R{:}Q) = S(Q) - \sum_a p(a)S(Q|a). \qquad (19.106)$$

From the last result, also prove that the quantum entropy is a concave function of density operators.

Now let us put both ends together. Initially, the joint system RQM is in the mixed state $\rho^{(RQ)} \otimes |0^{(M)}\rangle\langle 0^{(M)}|$. Bob's measurement then occurs, described by the superoperator $\mathcal{E}$ acting on the QM subsystem. This produces the state

$$\mathcal{E}^{(QM)}(\rho^{(RQM)}) = \sum_{a,b} p(a) |a\rangle\langle a| \otimes \left(E_b^{1/2} \rho_a E_b^{1/2}\right) \otimes |b\rangle\langle b|. \qquad (19.107)$$

Now we discard the system Q (taking the partial trace) and consider only RM, the joint system of the preparation record and the measurement record. Its state is

$$\rho^{(RM)'} = \sum_{a,b} p(a,b) |a\rangle\langle a| \otimes |b\rangle\langle b|. \qquad (19.108)$$

Although this is a quantum state, it is a mixture of product basis states. The final system entropies are just the Shannon entropies for the input and output variables:

$$S(R') = H(A), \qquad S(M') = H(B), \qquad S(RM') = H(A,B). \qquad (19.109)$$

---

[7] Indeed, according to the *many-worlds interpretation* of quantum mechanics, first proposed in 1957 by Hugh Everett III, *every* measurement process should be regarded as unitary evolution on a larger system including measuring devices, the environment, and the observers themselves. This is a fascinating concept with far-reaching implications for the nature of physical reality. However, one need not adopt all of its philosophical ramifications to make use of the two equivalent ways of thinking about the same superoperator $\mathcal{E}$!

Thus, $H(A : B) = S(\text{R:M}')$. The mutual information of the input A and output B is just the (quantum) mutual information of the physical systems that record these variables.[8]

This is the fact we need. Since M is initially in a product state with RQ, $S(\text{R:Q}) = S(\text{R:QM})$. Equation 19.50 tells us that the quantum mutual information cannot be increased by open system dynamics on one subsystem; thus, after the QM measurement process we must have $S(\text{R:QM}) \geq S(\text{R:QM}')$. Similarly, the monotonicity of quantum mutual information (Eq. 19.49) means that $S(\text{R:QM}') \geq S(\text{R:M}')$. From this chain of inequalities we find that, for any choice of decoding observable that Bob might make,

$$H(A : B) \leq S(\text{Q}) - \sum_a p(a)S(\text{Q}|a). \qquad (19.110)$$

This remarkable inequality was proved in 1973 by Aleksandr Holevo and is known as *Holevo's theorem*.

We should stress that Holevo's theorem is an *inequality*. It often happens that, for a given collection of Alice's signal states, $H(A : B)$ does not approach the bound for *any* choice of Bob's measurement. Nevertheless, Eq. 19.110 provides a useful fundamental limit on the ability of quantum systems to convey classical information.

For instance, suppose that Alice's signal states are all pure states $|\phi_a\rangle\langle\phi_a|$, so that $S(\text{Q}|a) = 0$. Then Holevo's theorem tells us that the information conveyed is never greater than $S(\text{Q})$. If the signal states are orthogonal to each other, $S(\text{Q}) = H(A)$; also, a suitable choice of measurement basis (namely, the $|\phi_a\rangle$ states themselves) will allow Bob to read the message perfectly. Thus, $H(A : B) = H(A) = S(\text{Q})$. On the other hand, if the states are not very distinguishable, the density operator will have an entropy $S(\text{Q})$ much less than $H(A)$. In this case, we know that $H(A : B) < H(A)$, and so Bob will not be able to read the message reliably.

Exercise 19.36 Comment qualitatively on the connection between Holevo's theorem and the basic decoding theorem of Section 4.1. What about the extension of this result to mixed states, found in Section 8.3?

Exercise 19.37 Suppose Alice chooses signal states of a qubit according to the BB84 scheme described in Section 4.4. Calculate $H(A)$, $S(\text{Q})$, and the mutual information $H(A:B)$ that Bob can achieve by measuring either Z or X. You should find that $H(A:B) < S(\text{Q}) < H(A)$.

Holevo's theorem has an interesting connection to relative entropy. The right-hand side of Eq. 19.110 is sometimes designated χ. It is not hard to show that

$$\chi = S(\text{Q}) - \sum_a p(a)S(\text{Q}|a) = \sum_a p(a)D\left(\rho_a||\rho\right). \qquad (19.111)$$

Exercise 19.38 Verify this equality.

----

[8] We use the "prime" symbol ($'$) to indicate that the systems involved have undergone some dynamical evolution. The prime applies to all of the systems in the expression $S(\cdots ')$, since the entropies are evaluated for all systems after the evolution has taken place. (For system R, in this particular instance, $S(\text{R}') = S(\text{R})$.)

The Holevo bound χ is the average relative entropy "distance" of the signal state from the average signal state. Now consider another state σ of Q and examine the "χ-like" quantity

$$\sum_a p(a)D\left(\rho_a||\sigma\right) = \sum_a p(a)\mathrm{Tr}\,\rho_a \log \rho_a - \mathrm{Tr}\,\rho \log \sigma. \qquad (19.112)$$

Adding and subtracting $\mathrm{Tr}\,\rho \log \rho$, we obtain *Donald's identity*:

$$\sum_a p(a)D\left(\rho_a||\sigma\right) = \chi + D\left(\rho||\sigma\right). \qquad (19.113)$$

Alice must prepare the system Q in one of several possible signal states. What thermodynamic resources are required for her to do so? We assume that she does all of her operations in the presence of a thermal reservoir at temperature T. Furthermore, Q has an internal Hamiltonian H. Alice can induce short-term departures from this Hamiltonian as she prepares the system, but afterwards it must revert to H.

What is the initial state of Q? If this state were anything other than the canonical equilibrium state ω, then Alice would be able to use Q itself as a source of useful work. This would complicate our accounting. It is more reasonable to suppose that Alice is presented with Q in the state ω, and must invest work to change this to another state.

Recall our discussion in Section 19.5. To create the signal state ρ_a from ω, the work input is

$$\mathbb{W}_a \geq F_\theta(\rho, \mathrm{H}, T) - F_\theta(\omega, \mathrm{H}, T), \qquad (19.114)$$

where equality can be achieved via a sequence of sudden and gradual (isothermal) modifications of H. But Eq. 19.82 tells us that this is related to the quantum relative entropy:

$$\mathbb{W}_a \geq (k_{\mathrm{B}}T \ln 2)\, D\left(\rho_a||\omega\right). \qquad (19.115)$$

The average work required will be

$$\bar{\mathbb{W}} = \sum_a p(a)\mathbb{W}_a \geq k_{\mathrm{B}}T \ln 2 \sum_a p(a)D\left(\rho_a||\omega\right). \qquad (19.116)$$

We apply Donald's identity to this, noting that $D\left(\rho||\omega\right) \geq 0$:

$$\bar{\mathbb{W}} \geq (k_{\mathrm{B}}T \ln 2)\, \chi. \qquad (19.117)$$

Holevo's theorem then allows us to conclude that

$$\bar{\mathbb{W}} \geq (k_{\mathrm{B}}T \ln 2)\, H(A : B). \qquad (19.118)$$

For every bit of information that Alice is able to convey to Bob (given by the mutual information $H(A : B)$), Alice must expend at least $k_{\mathrm{B}}T \ln 2$ of work on average. The minimum work cost of communication in the presence of a thermal reservoir is $k_{\mathrm{B}}T \ln 2$ per bit. (At a typical room temperature of 300 K, this is about 0.04 eV per bit. Actual communication systems generally expend much more energy than this!)

Exercise 19.39 Examine the conditions for equality in the various stages of our argument, and devise a scheme whereby Alice is able to convey information to Bob with exactly $k_B T \ln 2$ per bit.

Exercise 19.40 If Alice uses signal states ρ_a that are not very far from the equilibrium state ω, then she can reduce the work cost of creating the signals to almost nothing. Why is this not a good strategy for communicating to Bob?

Exercise 19.41 Suppose the initial state of Q is not an equilibrium state. As part of her work cost for communication, Alice should include the work that she *could* have obtained from Q if she had not used it to send messages to Bob. (Economists call this idea *opportunity cost.*) Show that the minimum work cost is still at least $k_B T \ln 2$ per bit.

Problems

Problem 19.1 Suppose $p(x)$ and $q(x)$ are two distributions for the same random variable X.

(a) Show that $\log z \leq (z - 1)/\ln 2$ for all $z > 0$.
(b) Use part (a) to prove Gibbs's inequality:

$$\sum_x p(x) \log \frac{p(x)}{q(x)} \geq 0. \qquad \text{(Re 19.9)}$$

(c) Show that equality holds if and only if $p(x) = q(x)$ for all values of x – that is, the two distributions are identical.

Be sure to comment on how your arguments apply when one or the other probability happens to be zero.

Problem 19.2 Look up the letter frequencies for the 26 English letters and the space, and use these frequencies to calculate the single-letter entropy of English. How much less is this than $\log 27 = 4.75$ bits per symbol?

Problem 19.3 Variable-length codes. Let us suppose that we are representing the value of a random variable X using binary codewords. We give ourselves the freedom to choose codewords of different lengths for different xs. The codeword C_x for x has a length L_x.

We will code the string $(x_1, \ldots, x_n)$ in the binary sequence $C_{x_1} \cdots C_{x_n}$. For this long binary sequence to be readable, however, we must be able to figure out where the "breaks" are. For instance, if $C_1 = 1$ and $C_2 = 110$, then when we encounter the symbol 1 in the binary sequence we will find it difficult to tell whether it is C_1 or the first part of C_2.

We can avoid this problem if we insist on using a *prefix-free code*, one in which no codeword C_x is an initial segment of another codeword $C_{x'}$. Then the codewords are always "recognizable" as the binary sequence is read. But this puts a restriction on the possible lengths of the codewords.

(a) Given a prefix-free code for a finite X, show that codeword lengths must satisfy the *Kraft inequality*:

$$\sum_x 2^{-L_x} \leq 1. \tag{19.119}$$

(b) Suppose we have a set of proposed lengths L_x (all positive integers) that satisfy the Kraft inequality. Show that it is possible to find binary codewords C_x with lengths L_x forming a prefix-free code.

(c) For any prefix-free code, show that the mean length satisfies

$$\sum_x p(x)L_x \geq H(X). \tag{19.120}$$

Under what conditions is equality possible?

(d) For any distribution $p(x)$, show that it is possible to find a prefix-free binary code for which the mean length satisfies

$$\sum_x p(x)L_x \leq H(X) + 1. \tag{19.121}$$

Thus, though some messages might use more bits, the *average* number of bits per message is not too much greater than $H(X)$.

Problem 19.4 Given random variables X, Y, and Z, show that

$$H(X,Y,Z) + H(Y) \leq H(X,Y) + H(Y,Z). \tag{19.122}$$

Hint: For each value y of Y, consider $H(X:Z|y)$.

Problem 19.5 Calculate the quantum entropy S of the qubit state described by a Bloch vector $\vec{a}$. Your answer should only depend on the magnitude $|\vec{a}|$. Why?

Problem 19.6 Adapt the calculations in Exercises 19.16, 19.17, and 19.18 into a series of detailed worked examples of quantum data compression. What entangled state is being transmitted? What are the resources used and the entanglement fidelity achieved in each case?

Problem 19.7 Cooling an oscillator. A ladder system with Hamiltonian $H = \varepsilon a^\dagger a$ is in contact with a reservoir system at temperature T, and is initially in its equilibrium canonical state $\omega = Z^{-1} \exp(-H/k_B T)$. We wish to change the system's state to ρ, which would be the equilibrium state at temperature $T/2$.

(a) Calculate the change in free energy associated with this process. Derive approximate expressions for ΔF_θ that apply for $\varepsilon \ll k_B T$ and $\varepsilon \gg k_B T$.

(b) To achieve this process, we can control the parameter ε in H. (Roughly speaking, this is equivalent to adjusting the elastic constant k in a harmonic oscillator potential.) Describe a control procedure $\varepsilon(t)$ that accomplishes the desired state change with the minimum necessary external work on the system.

(c) How is the work in part (b) related to the change ΔE_θ in the energy of the system?

Problem 19.8 Suppose we add a constant to the Hamiltonian of a system: $H \rightarrow H + \varepsilon \mathbf{1}$. Explain how the following things change: the canonical state ω at a given temperature, the partition function $\mathcal{Z}$ for this state, its entropy S_θ, and its free energy F_θ.

Problem 19.9 Suppose Alice can make various signal states ρ_a, each of which has a "cost function" C_a. (This cost function is not necessarily given by any system observable on Q.) The average cost over all signals is $\bar{C}$. One of the possible signals ρ_0 is special and has zero cost: $C_0 = 0$. Define the *cost-effectiveness* of Alice's code to be the ratio $\chi/\bar{C}$. Show that this is bounded above by

$$\frac{\chi}{\bar{C}} \leq \max_a \left(\frac{D\left(\rho_a \| \rho_0\right)}{C_a} \right). \tag{19.123}$$

How does this result relate to the thermodynamic cost of communication in Eq. 19.118? *Optional:* Find probabilities $p(a)$ that approach this bound on $\chi/\bar{C}$.

Error correction

20.1 Undoing errors

Suppose the random variable X represents a message that Alice wishes to send to Bob. What Bob actually receives from the communication system is Y. It is not necessarily the case that X and Y are equal. Bob can, however, perform further *data processing* on Y to obtain $\hat{X}$, his estimate of Alice's message. The probability of error is

$$P_E = \Pr(\hat{X} \neq X) = \sum_{\substack{x, \hat{x} \\ x \neq \hat{x}}} p(x, \hat{x}). \tag{20.1}$$

Under what circumstances can Bob reconstruct Alice's message reliably, that is, with $P_E = 0$?

This is the problem of (classical) *error correction*. Noise and distortion may introduce differences between X and Y, and we wish to know when these effects can be overcome by a subsequent operation on Y.

We can envision a corresponding quantum communication problem. Alice begins with the system Q, which is in a pure entangled state $|\Psi^{(RQ)}\rangle$ with another system R. During the communication process, system Q evolves as an open system according to some evolution superoperator $\mathcal{E}$, so that the new joint state of RQ is $\rho^{(RQ)'}$. Under what circumstances can Bob, who has access only to Q, use a second evolution $\mathcal{D}$ to restore RQ to the original state? If $\rho^{(RQ)''}$ is the final state, when can Bob make the entanglement fidelity

$$F_e = \langle \Psi^{(RQ)} | \, \rho^{(RQ)''} \, | \Psi^{(RQ)} \rangle = 1? \tag{20.2}$$

That is, when is *quantum error correction* possible?

The problem of error correction is one of the basic concerns of information theory, and we will only be able to scratch the surface of the subject. In both the classical and the quantum case, we will find necessary and sufficient conditions for perfect error correction, and we will derive additional results about classical and quantum data compression that amount to converses of the theorems of Chapter 19. We will also take a brief look at particular methods – *error-correcting codes* – that can be used to achieve reliable communication even in the presence of noise.

Error correction can be recast as a more general question: When can dynamical evolution be *inverted* by subsequent dynamics? As we will see, the answer to this question in the quantum case will shed new light on the exchange of information between quantum systems and their surroundings.

20.2 Classical communication and error correction

A *Markov process* or *Markov chain* is a sequential probabilistic process in which each stage takes as input only the output of the previous stage. This seems like a trivial definition, but it actually entails a number of important consequences. Suppose, for example, that the random variables X, Y, and Z form a Markov chain, a fact we represent thus: $X \to Y \to Z$. The value of Y depends on X through some conditional probability distribution $p(y|x)$. The value of Z then depends only on the value of Y, but not at all on the X value, except through its influence on Y. That is, $p(z|x,y) = p(z|y)$, and so

$$p(x,y,z) = p(x)\,p(y|x)\,p(z|y). \tag{20.3}$$

(The conditional distribution $p(z|y)$, of course, might be quite different from $p(y|x)$, though for simplicity we have used the same letter p to denote them.) One sometimes says that this sequence of variables has "no memory." Later stages depend only on the present random variable, not on any earlier ones.

Exercise 20.1 For a Markov chain, show that X and Z are independent of each other, given Y:

$$p(x,z|y) = p(x|y)p(z|y). \tag{20.4}$$

Markov chains are very commonly used to model natural stochastic processes such as Brownian motion. Our interest stems from their application to information processes. The progression from Alice's input message X to Bob's received signal Y to Bob's estimate $\hat{X}$ is a Markov chain. This is because Bob can only use Y as a basis to obtain $\hat{X}$.

In this communication context, the input distribution $p(x)$ for X values represents the statistical properties of Alice's message. The conditional probabilities $p(y|x)$ characterize the *communication channel* from Alice to Bob, while the $p(\hat{x}|y)$ probabilities describe Bob's *data processing*. Our basic question is this: given an input distribution $p(x)$ and the channel described by $p(y|x)$, how well can Bob reconstruct Alice's message?

We begin by noting a general property of Markov chains. The classical version of strong subadditivity (Problem 19.4) implies that, for any three random variables X, Y, and Z,

$$H(X,Y,Z) + H(Z) \leq H(X,Z) + H(Y,Z). \tag{20.5}$$

From this it follows that $H(X|Z) \geq H(X|Y,Z)$. If $X \to Y \to Z$ is a Markov chain, then Eq. 20.4 also tells us that $H(X,Z|Y) = H(X|Y) + H(Z|Y)$. Thus

$$H(X,Y,Z) + H(Y) = H(X,Y) + H(Y,Z), \tag{20.6}$$

and so $H(X|Y) = H(X|Y,Z)$. We conclude that

$$H(X|Y) \leq H(X|Z). \tag{20.7}$$

In terms of the mutual information, we can write this as

$$H(X) \geq H(X:Y) \geq H(X:Z). \tag{20.8}$$

Equation 20.8 is called the *data processing inequality*. It expresses an "irreversibility" in the Markov chain: once mutual information with X has been lost at some stage of the process, it cannot be restored at any later stage.

Exercise 20.2 By similar methods, prove the *data pipelining inequality* for the Markov chain $X \to Y \to Z$:

$$H(Z:Y) \geq H(Z:X). \tag{20.9}$$

For the communication process $X \to Y \to \hat{X}$, the data processing inequality means that $H(X:\hat{X}) \leq H(X:Y)$. Bob's estimate $\hat{X}$ can be no more strongly correlated with X (in an information sense) than Y, despite his data processing. To put the same idea in different terms, if $H(X|Y) > 0$ then $H(X|\hat{X}) > 0$ as well, so that Bob cannot reliably deduce the value of X from the estimate $\hat{X}$. Thus, the error probability $P_E = 0$ only if $H(X|\hat{X}) = H(X|Y) = 0$. As we will now show, there is also a more general connection between P_E and the conditional entropy.

We introduce a new random variable, the "error" variable E, with values e (error, when $x \neq \hat{x}$) and s (success, when $x = \hat{x}$). These have overall probabilities P_E and $1 - P_E$, respectively. Since the value of E is completely determined by the values of X and $\hat{X}$,

$$\begin{aligned}
H(X|\hat{X}) &= H(X, E|\hat{X}) \\
&= H(X|\hat{X}, E) + H(E|\hat{X}) \\
&\leq H(X|\hat{X}, E) + H(E).
\end{aligned} \tag{20.10}$$

The entropy $H(E) = h(P_E)$, where $h(p) = -p\log p - (1-p)\log(1-p)$ is the *binary entropy function* for the probability p. The conditional entropy

$$H(X|\hat{X}, E) = P_E H(X|\hat{X}, e), \tag{20.11}$$

because $H(X|\hat{X}, s) = 0$. Suppose that the random variable X has N possible values. Then, given that an error has been made, it must be that $H(X|\hat{X}, e) \leq \log(N-1)$. From all of this we can write down *Fano's inequality*, which states that

$$H(X|\hat{X}) \leq h(P_E) + P_E \log(N-1). \tag{20.12}$$

For a given $H(X|\hat{X}) > 0$, Eq. 20.12 gives an implicit lower bound on P_E. A somewhat weaker (but often useful) version of this is

$$H(X|\hat{X}) \leq 1 + P_E \log N, \tag{20.13}$$

see also Problem 20.1.

Exercise 20.3 We could replace $H(X|\hat{X})$ with $H(X|Y)$ in both Eq. 20.12 and 20.13. Why?

Let us sum up the conclusions we can now draw about the $X \to Y \to \hat{X}$ communication process:

- If $H(X|Y) > 0$, then $H(X|\hat{X})$ must also be non-zero by Eq. 20.7; so by Fano's inequality (Eq. 20.12) we know that $P_E > 0$. Thus, we can have $P_E = 0$ only if $H(X|Y) = 0$ as well.

- If $H(X|Y) = 0$, we cannot conclude that $P_E = 0$, since Bob might introduce errors during data processing. However, we do know that, for every possible value y of Y, $H(X|y) = 0$. This means there must be a unique possible value of x for the given y. Bob can therefore design a data processing function that maps y to this x value, and achieve $P_E = 0$.

The relation between entropy and error probability amounts to a limitation on Bob's ability to do error correction on his received signal. Perfect ($P_E = 0$) error correction is possible if and only if $H(X|Y) = 0$, or equivalently $H(X : Y) = H(X)$.

We can use these results to derive a counterpart to the classical data compression theorem of Section 19.2. The aim in that discussion was to store the values of a sequence of independent, identically distributed variables X_i ("messages") using bits. We proved that, if $H(X) + \epsilon$ bits are available per message, then we can store n messages in $n(H(X) + \epsilon)$ bits with $P_E \to 0$ as $n \to \infty$.

In the Markov chain $X \to Y \to \hat{X}$, X represents the initial message, Y represents the state of the storage bits, and $\hat{X}$ represents the retrieved message. The data compression theorem applies if we have more than $H(X)$ bits per message. But suppose we have only $K < H(X)$ bits available to store the message. Then $H(Y) \leq K$, and

$$H(X|Y) = H(X, Y) - H(Y) \geq H(X) - K. \qquad (20.14)$$

Equation 20.13 (with Exercise 20.3) leads to

$$H(X) - K \leq 1 + P_E \log N, \qquad (20.15)$$

where N is the number of possible X values.

Now replace X with a whole sequence of n independent messages, and Y with the state of nK available bits. There are N^n possible X-sequence values, and

$$n(H(X) - K) \leq 1 + nP_E \log N. \qquad (20.16)$$

Rearranging terms we find that

$$P_E \geq \frac{H(X) - K}{\log N} - \frac{1}{n \log N}. \qquad (20.17)$$

The term $(H(X) - K)/\log N > 0$ and is independent of n; the other term tends toward zero as $n \to \infty$. Therefore, as $n \to \infty$, we cannot have that $P_E \to 0$. Using fewer than $H(X)$ bits per message, we cannot compress the data so that the probability of error approaches zero for large n. This fact is known as the *weak converse* of the data compression theorem.[1]

Exercise 20.4 In what sense is this a "converse" to the data compression theorem?

---

[1] The *strong converse* of the theorem shows that $P_E \to 1$ as $n \to \infty$. Hence, the weak converse states that errors cannot be made asymptotically impossible, and the strong converse states that errors are asymptotically certain. The stronger statement requires more mathematical machinery to prove, and we will omit it in our discussion.

20.3 Quantum communication and error correction

A *quantum Markov process* or *quantum Markov chain* is a process during which a system Q undergoes two stages of open system evolution, which we can represent by Q $\rightarrow$ Q' $\rightarrow$ Q''. Given an input state ρ of Q, the subsequent states are given by superoperators:

$$\rho \longrightarrow \rho' = \mathcal{E}(\rho) \longrightarrow \rho'' = \mathcal{D}(\rho'). \tag{20.18}$$

The quantum Markov property means that the environment seen by Q in the second stage ($\mathcal{D}$) is not initially correlated with the system, notwithstanding the interaction between system and environment in the first stage ($\mathcal{E}$). Effectively, the two stages involve independent environments. In a quantum Markov process, it is the environment that has no memory.

Exercise 20.5 Review the discussion and derivation of the Lindblad equation in Section 9.3. What role does the Markov hypothesis play in Lindblad-type dynamics?

In the quantum communication context, we introduce a "bystander" system R and suppose that RQ is initially in a pure state $|\Psi^{(RQ)}\rangle$. The quantum Markov process is RQ $\rightarrow$ RQ' $\rightarrow$ RQ'' (though R's own dynamical evolution is trivial). Given the superoperator $\mathcal{E}$ describing the first stage of this process, how well can the original state be recovered by a second, independent evolution $\mathcal{D}$?

Our first step is to find a *quantum data processing inequality* that allows us to diagnose irreversible losses of information in the process. In fact, we already have the basis for such an inequality. In Eq. 19.50 we saw that the quantum mutual information between two systems could not be increased by open system evolution. Thus,

$$S(\text{R:Q}) \geq S(\text{R:Q'}) \geq S(\text{R:Q''}). \tag{20.19}$$

This is analogous to Eq. 20.8 for the classical mutual information. As we saw in Section 19.6, the quantum mutual information $S(\text{R:Q})$ can be related to the amount of classical information conveyed in a quantum communication channel. However, it does not seem to be a very good measure of the amount of *quantum* information present in the initial entanglement of Q with R. As we saw in Section 19.4, in a pure state the amount of entanglement is most reasonably measured by the entropy $S(\text{Q})$. But the quantum mutual information of such a state is

$$S(\text{R:Q}) = S(\text{R}) + S(\text{Q}) - S(\text{RQ}) = 2S(\text{Q}), \tag{20.20}$$

twice as large as we wish.

One way to express the problem is to say that the mutual information measures total *correlation* rather than *entanglement*. To suggest a better measure, note that each mutual information in Eq. 20.19 involves the same term $S(\text{R})$, which remains unchanged by Q's evolution. If we subtract this term from each expression, we would arrive at something with the same meaning. Define the *coherent information* of Q to be

$$I(\text{Q}\,;\text{R}) = S(\text{R:Q}) - S(\text{R}) = S(\text{Q}) - S(\text{RQ}). \tag{20.21}$$

This is a quantity whose classical analog $(H(Y) - H(X, Y) = -H(X|Y))$ can never be positive, but which may be positive for entangled quantum systems. The initial pure entangled state $|\Psi^{(RQ)}\rangle$ has $I(Q ; R) = S(Q)$, as we would like. These facts motivate us to adopt coherent information $I(Q ; R)$ as an "information measure" of the entanglement of Q with R.

Exercise 20.6 Suppose R and Q are qubits. What are the quantum mutual information and the coherent information for (a) the product state $|\phi, \psi\rangle$; (b) the correlated mixed state $\frac{1}{2}(|00\rangle\langle00| + |11\rangle\langle11|)$; and (c) the Bell state $|\Phi_+\rangle$?

Equation 20.19 now implies that

$$S(Q) \geq I(Q' ; R) \geq I(Q'' ; R), \tag{20.22}$$

(where we have placed the primes on Q to remind ourselves that the bystander system R "stands aloof" from the evolution). This will be our *quantum data processing inequality*.

Exercise 20.7 QR begins in a pure entangled state, and then Q interacts with the environment E (initially in pure state $|0^{(E)}\rangle$). Show that after this interaction

$$I(Q' ; R) = S(RE') - S(E'). \tag{20.23}$$

After the first stage $\mathcal{E}$ of evolution, we say that perfect quantum error correction is possible if the exact original state of RQ can be restored by the second stage $\mathcal{D}$. This can happen only if no coherent information is lost during $\mathcal{E}$: $I(Q' ; R) = S(Q)$. (This is a necessary condition for perfect error correction; however, we do *not* yet know that it is sufficient. See below.)

We can prove a stronger connection between the coherent information and the entanglement fidelity F_e, analogous to Fano's inequality. Suppose we specify a particular value for the entanglement fidelity after the action of $\mathcal{D}$:

$$F_e = \langle \Psi^{(RQ)}| \rho^{(RQ)''} |\Psi^{(RQ)}\rangle. \tag{20.24}$$

What is the largest possible value for $S(RQ'')$? The maximum entropy will be achieved when F_e is the largest eigenvalue of $\rho^{(RQ)''}$, and all the other eigenvalues are equal. How many eigenvalues are there? If $\dim \mathcal{H}^{(Q)} = d$, then we can take $\dim \mathcal{H}^{(R)} = d$ as well, since the original entangled state $|\Psi^{(RQ)}\rangle$ occupies at most a d-dimensional subspace of $\mathcal{H}^{(R)}$. Thus, $\dim \mathcal{H}^{(RQ)} = d^2$ and so

$$S(RQ'') \leq h(F_e) + (1 - F_e)\log(d^2 - 1), \tag{20.25}$$

where $h(p)$ is the binary entropy function. The joint entropy $S(RQ'')$ is related to the loss of coherent information. Noting that $S(Q) = S(R) = S(R'')$,

$$S(Q) - I(Q'' ; R) = S(R'') - S(Q'') + S(RQ''). \tag{20.26}$$

By the triangle inequality (Eq. 19.46), $S(R'') - S(Q'') \leq S(RQ'')$. Thus,

$$S(Q) - I(Q'' ; R) \leq 2S(RQ''). \qquad (20.27)$$

This yields the *quantum Fano's inequality* (together with a weaker, but often useful bound):

$$\frac{1}{2}\left(S(Q) - I(Q'' ; R)\right) \leq h(F_e) + (1 - F_e)\log(d^2 - 1) \qquad (20.28)$$

$$\leq 1 + 2(1 - F_e)\log d. \qquad (20.29)$$

If $I(Q' ; R) < S(Q)$ after the evolution $\mathcal{E}$, then by the quantum data processing inequality (Eq. 20.22) it must be that $S(Q) - I(Q'' ; R) > 0$. The quantum Fano's inequality then gives a non-trivial upper bound on F_e.

We used Fano's inequality in Section 20.2 to prove a weak converse to the classical data compression theorem. We will do the same with the quantum version. In data compression, the system Q is replaced with a composite system QC, where C is a collection of K qubits. Initially, C is in a fixed "zero" state $|0^{(C)}\rangle$.

- In the first stage of the process (described by $\mathcal{E}$), the initial entanglement of Q with R is transferred to C, and the state of Q is reset to some fixed $|0^{(Q)}\rangle$. Thus, $I(QC' ; R) = I(C' ; R)$.
- In the second stage of the process (described by $\mathcal{D}$), the entanglement is transferred back to Q, and C is reset to the state $|0^{(C)}\rangle$. Thus, $I(QC'' ; R) = I(Q'' ; R)$.

Therefore,

$$S(R) = S(Q) \geq I(C' ; R) \geq I(Q'' ; R). \qquad (20.30)$$

The overall loss of coherent information is

$$S(Q) - I(Q'' ; R) \geq S(Q) - I(C' ; R) = S(Q) - S(C') + S(RC''). \qquad (20.31)$$

By the triangle inequality, $S(RC'') \geq S(R) - S(C')$. As before, $S(R) = S(Q)$; and since C is composed of K qubits, $S(C') \leq K$. Thus,

$$S(Q) - I(Q'' ; R) \geq 2\left(S(Q) - K\right). \qquad (20.32)$$

Using Eq. 20.29, we obtain

$$S(Q) - K \leq 1 + 2(1 - F_e)\log d. \qquad (20.33)$$

Suppose we are compressing the entanglement (Type II quantum information) of n copies of Q with n copies of R into nK qubits. Then the entropy is $S(Q^n) = nS(Q)$ and the dimension of the Hilbert space is d^n. We obtain

$$1 - F_e \geq \frac{S(Q) - K}{2\log d} - \frac{1}{2n\log d}. \qquad (20.34)$$

As $n \to \infty$, the second term on the right tends toward zero. Therefore, if $K < S(Q)$, we cannot have that $F_e \to 1$ as $n \to \infty$. We cannot achieve asymptotically perfect quantum data compression with fewer than $S(Q)$ qubits per system. This is the weak converse of the quantum data compression theorem.

We have seen that if $I(Q'\,;R) < S(Q)$ after the evolution $\mathcal{E}$, no further operation $\mathcal{D}$ can accomplish perfect ($F_e = 1$) quantum error correction. Now suppose $I(Q';R) = S(Q)$, so that there has been no loss of coherent information. In the corresponding classical situation ($H(X:Y) = H(X)$) we could always accomplish perfect error correction. Is this possible in the quantum context?

Introduce explicitly the environment E, initially in the pure state $|0^{(E)}\rangle$, whose interaction with Q produces the evolution described by $\mathcal{E}$. If there has been no loss of coherent information at this stage, then

$$S(Q) + S(RQ') - S(Q') = S(R') + S(E') - S(RE') = 0. \tag{20.35}$$

Thus, RE must be a product state after the first stage of the dynamical evolution. We can write

$$\rho^{(RE)'} = \rho^{(R)} \otimes \sigma^{(E)'}, \tag{20.36}$$

where $\rho^{(R)}$ is unchanged from the initial state.

Write the initial state of RQ using the Schmidt decomposition:

$$\left|\Psi^{(RQ)}\right\rangle = \sum_k \sqrt{\lambda_k}\, \left|k^{(R)}, k^{(Q)}\right\rangle, \tag{20.37}$$

where λ_ks are the eigenvalues of $\rho^{(R)}$, and the $\{\,|k^{(R)}\rangle\}$ and $\{\,|k^{(Q)}\rangle\}$ are orthonormal sets in $\mathcal{H}^{(R)}$ and $\mathcal{H}^{(Q)}$, respectively. After the interaction of Q and E, the overall state is still a pure state. We can write this using a Schmidt decomposition between Q and RE:

$$\left|\Psi^{(RQE)'}\right\rangle = \sum_{k,a} \sqrt{\lambda_k \mu_a}\, \left|\phi_{ka}^{(Q)}\right\rangle \otimes \left|k^{(R)}, a^{(E)}\right\rangle. \tag{20.38}$$

The μ_as are the eigenvalues of $\sigma^{(E)'}$, and we have used the fact that the RE state is a product state. We also note that the $\{\,|\phi_{ka}^{(Q)}\rangle\}$ states form an orthonormal set.

Exercise 20.8 Show that the state of RQ at this stage is

$$\rho^{(RQ)'} = \sum_a \mu_a \left|\Psi_a^{(RQ)}\right\rangle\!\left\langle\Psi_a^{(RQ)}\right|, \tag{20.39}$$

where the states $\left|\Psi_a^{(RQ)}\right\rangle$ are

$$\left|\Psi_a^{(RQ)}\right\rangle = \sum_k \sqrt{\lambda_k}\, \left|k^{(R)}, \phi_{ka}^{(Q)}\right\rangle. \tag{20.40}$$

Each of the $\left|\Psi_a^{(RQ)}\right\rangle$ states in Exercise 20.8 already bears a strong resemblance to the original state $|\Psi^{(RQ)}\rangle$. We therefore define the second-stage evolution $\mathcal{D}$ by a set of Kraus operators for Q:

$$A_a = \sum_k \left|k^{(Q)}\right\rangle\!\left\langle\phi_{ka}^{(Q)}\right|. \tag{20.41}$$

This is properly normalized, since

$$\sum_a A_a^\dagger A_a = \sum_{k,l,a} \left|\phi_{ka}^{(Q)}\right\rangle \left\langle k^{(Q)}\,\middle|\,l^{(Q)}\right\rangle \left|\phi_{la}^{(Q)}\right\rangle = \sum_{k,a} \left|\phi_{ka}^{(Q)}\right\rangle\!\left\langle\phi_{ka}^{(Q)}\right|, \tag{20.42}$$

which is the identity 1 on the subspace spanned by the $\left|\phi_{ka}^{(Q)}\right\rangle$ vectors. (We can include an additional A_0 if necessary to cover the orthogonal subspace.) These operators act according to

$$A_a \left|\Psi_b^{(RQ)}\right\rangle = \delta_{ab} \left|\Psi^{(RQ)}\right\rangle, \tag{20.43}$$

and therefore

$$\mathcal{D}\left(\rho^{(RQ)\prime}\right) = \sum_a A_a \rho^{(RQ)\prime} A_a^\dagger = \left|\Psi^{(RQ)}\right\rangle\!\left\langle\Psi^{(RQ)}\right|, \tag{20.44}$$

so that the entanglement fidelity $F_e = 1$. We have achieved perfect quantum error correction.

To sum up, we have found that perfect quantum error correction is possible if and only if there is no loss of coherent information in the communication process. In the case of data compression, using fewer than $S(Q)$ qubits per system guarantees such a loss, which in turn guarantees that the entanglement fidelity F_e of the process cannot approach 1 as $n \to \infty$. The situation is strongly reminiscent of the classical theory, except that the coherent information $I(Q ; R)$ – which measures entanglement – plays the role that mutual information $H(X : Y)$ plays in the classical context.

20.4 Error-correcting codes

To give a sense of how error correction actually works, we will take a brief look at the subject of *error-correcting codes*. These are ways of representing information so that it is more resistant to degradation.

Let us begin with a classical example. Suppose that a binary digit is subject to a process that "flips" its value with probability p. That is,

$$\begin{aligned} p(0|0) &= 1 - p, & p(0|1) &= p, \\ p(1|0) &= p, & p(1|1) &= 1 - p. \end{aligned} \tag{20.45}$$

We will assume that $0 \leq p \leq \frac{1}{2}$. Note that $p = \frac{1}{2}$ completely randomizes the value of the binary digit.[2] If we encode a bit of information into this variable and then read it after the process is complete, the probability of making an error is $P_E = p$.

The simplest idea is to encode our information into several binary digits, which can be used to check each other. For instance, we can use three binary digits, with "codewords" 0_C and 1_C:

$$0_C = 000, \qquad 1_C = 111. \tag{20.46}$$

We encode one bit of information into the values of three binary digits using this code. This enables us to correct for certain possible errors. If a single bit is flipped, then the correct

---

[2] A situation in which $p > \frac{1}{2}$ can be converted to one in which $p < \frac{1}{2}$ by always inverting the output.

value can still be determined by a "majority vote" of the three bits. This error correction procedure is

$$000, 001, 010, 100 \rightarrow 0_c, \qquad 011, 101, 110, 111 \rightarrow 1_c. \qquad (20.47)$$

If we know that an error occurs on only one of the three bits, then this error correction procedure is foolproof. A somewhat more realistic situation is one in which each binary digit is independently subject to error with probability p. Then the net probability of error is just the likelihood that two or three errors occur:

$$P_E = 3p^2(1 - p) + p^3 = 3p^2 - 2p^3 \leq 3p^2. \qquad (20.48)$$

For any p between 0 and $\frac{1}{2}$, the probability of error $P_E < p$. Thus, our three-bit code improves the reliability of the information transfer.

For instance, suppose $p = 0.1$. Using the three-bit code, we can achieve $P_E = 3(0.1)^2 - 2(0.1)^3 = 0.028$, almost a fourfold improvement in error likelihood.

Exercise 20.9 Compute P_E for the three-bit code when $p = 0.25, p = 0.05$, and $p = 0.001$. By what factor does the three-bit code reduce P_E in each case?

Also see Problem 20.3.

The three-bit code is the simplest example, but there are many other codes with various error-correction capabilities. For instance, the five-bit code

$$0_c = 00000, \qquad 1_c = 11111, \qquad (20.49)$$

using the "majority vote" error-correcting procedure, can correct any *two* single-bit errors.

Exercise 20.10 Compute P_E for the five-bit code when $p = 0.1$ (assuming independent errors on each bit).

To code two bits, we could use the three-bit code (Eq. 20.46) twice, representing the bits using six binary digits. This code would allow us to correct single-bit errors in separate codewords (though not two errors in the same codeword). If we need to correct only one single-bit error overall, we can make do with fewer binary digits, as the following five-bit code illustrates:

$$\begin{aligned} 00_c &= 00000, & 01_c &= 00111, \\ 10_c &= 11001, & 11_c &= 11110. \end{aligned} \qquad (20.50)$$

Exercise 20.11 Show that we can correct any single-bit error with this code. What role does the fifth bit play?

A useful tool for analyzing binary codes is the *Hamming distance* between two binary sequences. If A and B are binary sequences of length n, then

$$D(A, B) = \text{\# of places in which } A \text{ and } B \text{ differ.} \qquad (20.51)$$

Thus, $D(011, 101) = 2$. Clearly $D(A, A) = 0$ and $D(A, B) = D(B, A)$. Furthermore:

Exercise 20.12 Show that, for any sequences A, B, and C, the Hamming distance satisfies the metric triangle inequality:

$$D(A, B) + D(B, C) \geq D(A, C). \tag{20.52}$$

If we start out with a codeword A, and n single bit errors occur, we end up with a string C such that $D(A, C) = 1$.

The *minimum distance* $D_{\min}$ of a binary code is the minimum value of $D(A, B)$ for distinct codewords A and B. The value of $D_{\min}$ determines how many single bit errors can be corrected:

- If $D_{\min} = 1$, then there are two codewords A and B that differ only in one bit. Each could arise from a single bit error on the other, so such errors could not be recognized and corrected.
- If $D_{\min} = 2$, the nearest codewords A and B differ in two bits. A single bit error can always be *detected*; but it may not be possible to *correct* such an error. Single bit errors on A and B could give rise to the same output string C, for which $D(A, C) = D(B, C) = 1$. It is therefore impossible to reliably determine which of the two codewords gave rise to C.
- If $D_{\min} \geq 3$, then a single bit error in codeword A gives rise to a string C such that $D(A, C) = 1$ but $D(B, C) \geq 2$ for any other codeword B. Given C, we can determine the unique A that could produce it by a single error. We can therefore correct any such error.

Exercise 20.13 Show that a binary code can correct up to n single bit errors if and only if $D_{\min} \geq 2n + 1$.

Exercise 20.14 Determine $D_{\min}$ for the binary codes in Eq. 20.46, 20.49, and 20.50. Apply the criterion in Exercise 20.13 to determine how many single bit errors can be corrected in each case.

Our errors have been symmetric (the processes $0 \to 1$ and $1 \to 0$ are equally likely) and independent. We can defend against more complicated types of error with more complicated codes. The general lesson is this: By using more bits, we can reduce the probability of error and communicate more reliably in the presence of noise.

Now we turn to *quantum error-correcting codes*. In the quantum problem we want to find ways to preserve the entanglement fidelity of the quantum information, even if the evolution of the system is not unitary.

Once again, we will consider a particular instructive example. Suppose that a qubit Q is subject to *partial decoherence*, a process represented by two Kraus operators $\sqrt{1-p}\,\mathbf{1}$ and $\sqrt{p}\,\mathsf{Z}$, for $0 \leq p \leq \frac{1}{2}$. That is,

$$\mathcal{E}(\rho) = (1-p)\rho + p\,\mathsf{Z}\rho\mathsf{Z}. \tag{20.53}$$

We can understand this qubit operation as a unitary relative phase flip in the standard basis (Z) that occurs with probability p.

For $p = 0$, the process introduces no decoherence. If $p = \frac{1}{2}$, the decoherence is complete, and any superposition input state of Q evolves into a simple mixture of $|0\rangle\langle 0|$ and $|1\rangle\langle 1|$. But even in this case, the states $|0\rangle\langle 0|$ and $|1\rangle\langle 1|$ are themselves completely unaffected by $\mathcal{E}$. If these states are used to represent a classical bit of classical data, the action of $\mathcal{E}$ introduces no error into that data. Thus, a perfectly noiseless channel of classical information may be very noisy as a channel of quantum information.

On the other hand, suppose we use the states $|+\rangle\langle +|$ and $|-\rangle\langle -|$ to represent the classical bit. Since

$$Z|+\rangle\langle +|Z = |-\rangle\langle -| \quad \text{and} \quad Z|-\rangle\langle -|Z = |+\rangle\langle +|, \tag{20.54}$$

the action of $\mathcal{E}$ is to produce a classical "bit flip" error with probability p, as in Eq. 20.45. Decoherence in the $\{|0\rangle, |1\rangle\}$ basis is the same thing as a random exchange of the basis states of the conjugate $\{|+\rangle, |-\rangle\}$ basis.

We introduce the system R (also a qubit) and suppose that RQ is initially an ebit in the Bell state $|\Phi_+\rangle = \frac{1}{\sqrt{2}}(|00\rangle + |11\rangle)$. The initial entropy $S(Q) = 1$. After the action of $\mathcal{E}$ on Q the joint state is

$$\rho^{(RQ)'} = \frac{1}{2}(|00\rangle\langle 00| + |11\rangle\langle 11|) + \left(\frac{1-2p}{2}\right)(|00\rangle\langle 11| + |11\rangle\langle 00|). \tag{20.55}$$

Exercise 20.15 Show that $F_e = 1 - p$ for this process. Also show that the entropy of $\rho^{(RQ)'}$ is $S(RQ') = h(p)$. (This is not difficult if you first note that the eigenstates of $\rho^{(RQ)'}$ are $|\Phi_+\rangle$ and $|\Phi_-\rangle = Z^{(Q)}|\Phi_+\rangle$.) What is the coherent information $I(Q' ; R)$?

Can we find a way to represent the entanglement between R and Q, perhaps using several qubits, that is more resistant to the decoherence of the $\mathcal{E}$ process? The fact that decoherence is related to bit flip suggests that we might adapt the three-bit classical code of Eq. 20.46. Suppose R is entangled with three independent qubits, a composite system we designate RQ^3. For the three Q qubits, consider the codeword basis states

$$|0_c\rangle = |+++\rangle \qquad |1_c\rangle = |---\rangle. \tag{20.56}$$

These codewords span a two-dimensional subspace of $\mathcal{Q} \otimes \mathcal{Q} \otimes \mathcal{Q}$, associated with the projection operator $\Pi_c = |0_c\rangle\langle 0_c| + |1_c\rangle\langle 1_c|$. The initial entangled state of RQ^3 is

$$|\Psi\rangle = \frac{1}{\sqrt{2}}\left(|0^{(R)}, 0_c\rangle + |1^{(R)}, 1_c\rangle\right). \tag{20.57}$$

Note that R and Q^3 initially form an ebit, as before, but the entanglement is spread among the three Q qubits. Each of the Q qubits independently undergoes the partial decoherence process $\mathcal{E}$. The overall operation $\mathcal{E} \otimes \mathcal{E} \otimes \mathcal{E}$ can be regarded as a probabilistic application of various possible unitary operators:

- The identity $\mathbf{1}$, which applies with probability $(1 - p)^3$.
- The operators $Z^{(1)}$, $Z^{(2)}$, and $Z^{(3)}$, each of which applies with probability $p(1 - p)^2$.
- Operators $Z^{(1)}Z^{(2)}$, etc., each of which applies with probability $p^2(1 - p)$; and the operator $Z^{(1)}Z^{(2)}Z^{(3)}$ which applies with probability p^3.

Exercise 20.16 Write down an operator–sum representation for $\mathcal{E} \otimes \mathcal{E} \otimes \mathcal{E}$ and explain how the terms correspond to the descriptions given above.

We design our quantum error correction procedure to correct any single-qubit error in the process – that is, to correct for $\mathbf{1}$, $Z^{(1)}$, $Z^{(2)}$, or $Z^{(3)}$. We can do this for the following reason:

Exercise 20.17 Show that Π_C, $Z^{(1)} \Pi_C Z^{(1)}$, $Z^{(2)} \Pi_C Z^{(2)}$, and $Z^{(3)} \Pi_C Z^{(3)}$ are projections onto orthogonal subspaces of $\mathcal{Q} \otimes \mathcal{Q} \otimes \mathcal{Q}$.

Thus, we could make an incomplete measurement using these projections and determine which single-qubit error operator has acted; then we can correct for its effect. We can summarize this by giving Kraus operators for the error correction procedure $\mathcal{D}$:

$$D_0 = \Pi_C, \qquad D_2 = \Pi_C Z^{(2)},$$
$$D_1 = \Pi_C Z^{(1)}, \qquad D_3 = \Pi_C Z^{(3)}. \tag{20.58}$$

Exercise 20.18 First, show that the set of Kraus operators $\{D_k\}$ is properly normalized. Then show that

$$D_1 = |+++\rangle\langle-++| + |---\rangle\langle+--|, \tag{20.59}$$

and write down similar expressions for D_2 and D_3.

Now suppose we start with initial RQ^3 state $|\Psi\rangle$ from Eq. 20.57. The trio of Q qubits is subject to the process $\mathcal{E} \otimes \mathcal{E} \otimes \mathcal{E}$ and then to the error correction procedure described by D. What is the overall entanglement fidelity F_e?

This is not hard to work out if we adopt the "probabilistic unitary" picture of $\mathcal{E} \otimes \mathcal{E} \otimes \mathcal{E}$. If there is no error ($\mathbf{1}$) or a single qubit error ($Z^{(1)}$, $Z^{(2)}$, or $Z^{(3)}$), then the procedure $\mathcal{D}$ exactly restores the original state $|\Psi\rangle$. If there are two or three qubit errors ($Z^{(1)}Z^{(2)}$, etc.) then the error correction procedure misidentifies this error and instead restores the state

$$|\Psi'\rangle = \frac{1}{\sqrt{2}}\left(|0^{(R)}, 1_C\rangle + |1^{(R)}, 0_C\rangle\right), \tag{20.60}$$

which is orthogonal to $|\Psi\rangle$. The final state of RQ^3 is

$$\rho' = F_e |\Psi\rangle\langle\Psi| + (1 - F_e)|\Psi'\rangle\langle\Psi'|, \tag{20.61}$$

where the entanglement fidelity $F_e = \langle\Psi| \rho' |\Psi\rangle$.

Exercise 20.19 Show that $F_e = 1 - 3p^2 + 2p^3$. Make a graph of F_e versus p for (a) the original situation where R is entangled with a single qubit Q, and (b) the error-correction situation where R is entangled with a trio of independent qubits.

The example of quantum error correction we have analyzed has a formal resemblance to classical error correction. But the two concepts are quite different. In classical error correction, we defend against error using *redundancy* – essentially, by making many copies of the same information. This is not possible for quantum information! Instead, the input state $|\Psi\rangle$ of Eq. 20.57 stores quantum information (entanglement with R) in highly entangled states of many qubits.

We have devised a particular code that helps to defend against a single type of quantum error. Other codes with other characteristics are possible. A quantum error-correcting code using n qubits involves a codeword subspace of $Q^{\otimes n}$. The entangled state of R with the n qubits only involves states in the codeword subspace. Correctable errors map the codeword subspace to an orthogonal subspace, an effect that can be identified and corrected by a subsequent operation.

Exercise 20.20 Devise a code using five Q qubits that can correct up to two qubit "phase flip" errors. (Base this on the five-bit code in Eq. 20.49.)

Exercise 20.21 Suggest a code to defend against the error process

$$\mathcal{E}'(\rho) = (1-p)\rho + p\,\mathsf{X}\rho\mathsf{X}. \tag{20.62}$$

See also Problem 20.5.

20.5 Information and isolation

In this section we will explore the close connection between quantum error correction and the idea of an informationally isolated system. This will shed additional light on the no-cloning theorem, the security of quantum cryptography, and the isolation theorem of Chapter 9. To begin with, though, we must generalize the idea of a conditional state to mixed states of a composite quantum system.

Let AB be a composite quantum system and let $\rho^{(AB)}$ be some joint mixed state. We can always choose a basis $\{\,|n\rangle\,\}$ of A-states and write

$$\rho^{(AB)} = \sum_{m,n} |m\rangle\langle n| \otimes \sigma_{mn}, \tag{20.63}$$

for some B-operators σ_{mn}. These operators are given by the partial inner products

$$\sigma_{mn} = \langle m|\,\rho^{(AB)}\,|n\rangle. \tag{20.64}$$

Under what circumstances is the joint state $\rho^{(AB)}$ a simple product state $\rho^{(A)} \otimes \rho^{(B)}$? If it is a product state, then for all m, n

$$\sigma_{mn} = \langle m|\,\rho^{(A)}\,|n\rangle\,\rho^{(B)} = \rho_{mn}\rho^{(B)}. \tag{20.65}$$

All of the σ_{mn} operators are multiples of the same operator $\rho^{(B)}$. Conversely, if this is true for all m, n, then

$$\rho^{(AB)} = \sum_{m,n} |m\rangle\langle n| \otimes \left(\rho_{mn}\rho^{(B)}\right) = \left(\sum_{m,n} \rho_{mn}\,|m\rangle\langle n|\right) \otimes \rho^{(B)}, \tag{20.66}$$

a product state.

Exercise 20.22 Show that $\sigma_{mn} = \sigma_{nm}^{\dagger}$ for all m, n. What does this tell us about σ_{nn}?

Fix a particular $\rho^{(AB)}$, which may or may not be a product state. Suppose a generalized measurement is performed on A, and that the result α is associated with the positive operator $E_\alpha^{(A)}$. The probability of this result is $p(\alpha) = \mathrm{Tr}_{(A)}\rho^{(A)}E_\alpha^{(A)}$, where $\rho^{(A)} = \mathrm{Tr}_{(B)}\rho^{(AB)}$. We define the *conditional state* of B given α to be

$$\rho_\alpha^{(B)} = \frac{1}{p(\alpha)}\, \mathrm{Tr}_{(A)}\rho^{(AB)}E_\alpha^{(A)}. \tag{20.67}$$

How do we know that this is a suitable definition? Consider separate measurements performed on A and B. The joint measurement is described by positive operators of the form $E_\alpha^{(A)} \otimes F_\beta^{(B)}$. The conditional probability $p(\beta|\alpha)$ is

$$\begin{aligned} p(\beta|\alpha) &= \frac{1}{p(\alpha)} p(\alpha,\beta) = \frac{1}{p(\alpha)}\, \mathrm{Tr}\, \rho^{(AB)}\left(E_\alpha^{(A)} \otimes F_\beta^{(B)}\right) \\ &= \frac{1}{p(\alpha)} \mathrm{Tr}_{(B)}\left(\mathrm{Tr}_{(A)}\rho^{(AB)}E_\alpha^{(A)}\right)F_\beta^{(B)} \\ &= \mathrm{Tr}_{(B)}\rho_\alpha^{(B)}F_\beta^{(B)}. \end{aligned} \tag{20.68}$$

Thus, the density operator $\rho_\alpha^{(B)}$ as defined in Eq. 20.67 correctly predicts the conditional probability of any measurement result on B.

What do the conditional states tell us about whether $\rho^{(AB)}$ is a product state? One connection is very easy:

Exercise 20.23 If $\rho^{(AB)} = \rho^{(A)} \otimes \rho^{(B)}$, show that $\rho_\alpha^{(B)} = \rho^{(B)}$ for any $E_\alpha^{(A)}$.

Now suppose that every A measurement result (associated with $E_\alpha^{(A)}$) leads to the same conditional state $\rho^{(B)}$. Express the joint state according to Eq. 20.63. Then:

Exercise 20.24 Let $E_\alpha^{(A)} = |n\rangle\langle n|$ for some A-basis ket $|n\rangle$ and show that

$$\sigma_{nn} = p(\alpha)\rho^{(B)}. \tag{20.69}$$

Exercise 20.25 Let $E_\alpha^{(A)} = \frac{1}{2}(|n\rangle + |m\rangle)(\langle n| + \langle m|)$; show that for some x_{mn},

$$\sigma_{mn} + \sigma_{nm} = x_{mn}\,\rho^{(B)}. \tag{20.70}$$

Exercise 20.26 Let $E_\alpha^{(A)} = \frac{1}{2}(|n\rangle + i\,|m\rangle)(\langle n| - i\langle m|)$; show that for some y_{mn},

$$\sigma_{mn} - \sigma_{nm} = y_{mn}\,\rho^{(B)}. \tag{20.71}$$

From these exercises, we can see that $\sigma_{mn} = \rho_{mn}\rho^{(B)}$ for all m,n. Therefore, the joint state $\rho^{(AB)}$ is a product state if and only if the conditional state of B equals $\rho^{(B)}$ for any A measurement result α.

Exercise 20.27 Compare this discussion of conditional states with the one in Section 6.4. What is the same? What is different?

In Section 20.3, we discussed a situation in which the quantum system RQ is initially in an entangled state $|\Psi^{(RQ)}\rangle$, and then Q interacts with an environment system E initially in the state $|0^{(E)}\rangle$. We were interested in how well the Type II quantum information (entanglement) in Q is preserved in this process. We showed that the perfect quantum error correction was possible if and only if no coherent information was lost: $I(Q';R) = S(Q)$. This happens if and only if the system R and the environment E are in a product state $\rho^{(RE)'} = \rho^{(R)} \otimes \sigma^{(E)'}$ (Eq. 20.36). We have now shown that this is equivalent to saying that, if we make a measurement on R after the time evolution, the conditional state of E is independent of the result.

We are considering two successive operations on the initial state $|\Psi^{(RQ)}, 0^{(E)}\rangle$: (1) Q and E interact, and (2) a measurement is made on R. Because R does not participate in the interaction of Q and E, the order of these operations does not matter. Suppose then we first make a measurement on R, whose result α specifies a conditional initial state $|\psi_\alpha^{(Q)}\rangle$ of Q; then Q and E interact. We have shown that the final state $\sigma_\alpha^{(E)'}$ of the environment is independent of α. Conversely, if this state is independent of α for all choices of R-measurement, and thus for all choices of possible conditional state $|\psi_\alpha^{(Q)}\rangle$ of Q, then perfect quantum error correction is possible. We can undo the effect of $\mathcal{E}$ (due to the QE interaction) by a second operation $\mathcal{D}$. The map $\mathcal{E}$ is *invertible* on this set of possible Q states.

What conditional Q states are possible given an initial RQ state $|\Psi^{(RQ)}\rangle$? Let the *support* supp G of a Hermitian operator G be the subspace spanned by non-zero eigenvectors of G – that is, the subspace orthogonal to the kernel ker G.

Exercise 20.28 Given $|\Psi^{(RQ)}\rangle$, show that any state of Q can arise as a conditional state from an R-measurement, provided that state lies in supp $\rho^{(Q)} = $ supp $\mathrm{Tr}_{(R)} |\Psi^{(RQ)}\rangle\langle\Psi^{(RQ)}|$.

Perfect quantum error correction for $|\Psi^{(RQ)}\rangle$ means invertibility of $\mathcal{E}$ on supp $\rho^{(Q)}$, and this in turn means that Q is informationally isolated for input states in this subspace.

Exercise 20.29 Carefully explain the meaning of the last sentence.

In fact, we have now established a large set of interconnected facts regarding an evolution superoperator $\mathcal{E}$ given by an interaction between Q and its environment E:

- $\mathcal{E}$ is invertible on a subspace of $\mathcal{H}^{(Q)}$ if and only if Q is informationally isolated on that subspace.
- The isolation theorem of Section 9.2 states that the evolution of Q is unitary if and only if Q is informationally isolated on all of $\mathcal{H}^{(Q)}$. Thus, unitary evolution maps are the only ones that are invertible on the whole Hilbert space of Q.
- Q is *not* informationally isolated if and only if the final E-state does depend, however slightly, on the initial Q-state. This happens whenever there is a loss of coherent information in the process.
- In a quantum cryptographic set-up, we can imagine that the eavesdropper has access only to E – that is, the "environment" also includes any devices used by the eavesdropper to monitor the communication. If Q is informationally isolated, the eavesdropper is completely "shut out" and the communication channel is perfectly private. This happens if and only if quantum error correction is possible.

The final point – the link between cryptographic security and quantum error correction – gives an insight into how that security may be rigorously established. If Alice and Bob can do

quantum error correction on their systems, then they can guarantee that the eavesdropper
Eve is excluded. (A more sophisticated result connects *approximate* error correction to
approximate cryptographic security.)

Quantum error correction is therefore more than a potentially useful technique for quantum communication systems. It provides a vital link in a web of fundamental connections
between information, unitarity, and cryptographic security. Quantum information is pre-
served if and only if no information escapes into the environment. If this is true for all
possible initial states, then the evolution of the system is unitary (and thus maintains superpositions). In any case, no physical record is created in the outside world to indicate the state
of the system. To repeat the slogan of Section 1.2: *Quantum mechanics is what happens
when nobody is looking.*

Problems

Problem 20.1 Equation 20.12 gives a lower bound on P_E that is not very easy to use.
Equation 20.13 is not useful for small values of $H(X|\hat{X})$. In this problem, we will derive
another lower bound on P_E.

(a) For $0 \le x \le 1$, show that $-x \ln x \le \dfrac{1}{2}$ and $-x \ln x \le 1 - x$.
(b) Use the inequalities in part (a) to show that $h(p) \le 2 \ln 2 \sqrt{p}$, where $h(p)$ is the binary
    entropy function.
(c) From Eq. 20.12 prove that

$$P_E \ge \left(\frac{H(X|\hat{X})}{\log 4N} \right)^2. \tag{20.72}$$

(d) Suppose X is a four-bit message ($N = 4$) and $H(X|\hat{X}) = 0.5$ bit. Use Eq. 20.12, 20.13,
    and 20.72 to compute lower bounds for the probability of error P_E. Which gives the
    most stringent bound? How difficult are these bounds to compute?

Problem 20.2 Suppose Alice's message X has N possible values that are equally likely, so
that $H(X) = \log N$. She represents these as pure states of a quantum system with a Hilbert
space dimension d. Bob makes a measurement on the system, obtaining the result Y, from
which he constructs his estimate $\hat{X}$.

(a) Use Holevo's theorem (Eq. 19.110) to show that $H(X : Y) \le \log d$.
(b) Use Fano's inequality (Eq. 20.12) to derive a lower bound on P_E when $N = 4$ and
    $d = 2$.
(c) How does your result compare with the basic decoding theorem of Section 4.1?

Problem 20.3 Let X be a binary input with 0 and 1 equally likely, so that $H(X) = 1$ bit.

(a) Suppose the bit is subject to the bit-flip process described by Eq. 20.45 with $p = 0.1$.
    Let Y be the output of this process. Calculate $H(X : Y)$. By how much does this fall
    short of $H(X)$?

(b) Now suppose the bit is encoded into three bits according to Eq. 20.46, and that each of these bits is independently subject to Eq. 20.45. The output Y is the final value of the three output bits. Calculate $H(X : Y)$ for $p = 0.1$.

(c) In each case, compare the actual probability of error with the bound on P_E obtained from Fano's inequality.

Consider the three-bit code from Eq. 20.46. Let X be the input value (0 and 1 being equally likely) and Y be the output

Problem 20.4 Our system Q consists of two qubits: R is initially in the Bell state $|\Phi_+\rangle$ with one qubit, while the other is in the state $|0\rangle$. This system undergoes an evolution by the superoperator $\mathcal{E}$. Calculate the coherent information $I(Q';R)$ if

(a) the evolution $\mathcal{E}$ swaps the states of the two Q qubits with probability p;
(b) the evolution $\mathcal{E}$ applies the $X \otimes X$ operator to Q with probability p;
(c) the evolution $\mathcal{E}$ performs a CNOT operation on Q (with the first system as control) with probability p; and
(d) the evolution $\mathcal{E}$ performs a controlled-Z operation on Q with probability p.

In each case, assume that $\mathcal{E}$ otherwise does not affect the state of Q. In which cases is perfect quantum error correction possible?

Problem 20.5 In 1995, Peter Shor introduced the idea of quantum error correction. He suggested a nine-qubit code with the following codewords:

$$|0_c\rangle = \frac{1}{2\sqrt{2}}(|000\rangle + |111\rangle) \otimes (|000\rangle + |111\rangle) \otimes (|000\rangle + |111\rangle),$$

$$|1_c\rangle = \frac{1}{2\sqrt{2}}(|000\rangle - |111\rangle) \otimes (|000\rangle - |111\rangle) \otimes (|000\rangle - |111\rangle).$$

(a) Show that we can use the Shor code to correct any single-qubit unitary error involving the error operators Z, X, or XZ.
(b) Show that the Shor code can also correct any single-qubit non-unitary error in which the qubit state is reset to $|0\rangle$.

For each of these, we need to show that the various possible errors map the codeword subspace to orthogonal subspaces, which can be identified by an incomplete measurement and corrected by some operation.[3]

---

[3] In fact, the Shor code can correct any single-qubit error. (You may regard the proof of this statement as an optional part (c) for this problem.)

Appendix A **Probability**

A.1 Random variables

Probability is an essential idea in both information theory and quantum mechanics. It is a highly developed mathematical and philosophical subject of its own, worthy of serious study. In this brief appendix, however, we can only sketch a few elementary concepts, tools, and theorems that we use elsewhere in the text.

In discussing the properties of a collection of sets, it is often useful to suppose that they are all subsets of an overall "universe" set $\mathcal{U}$. The universe serves as a frame within which unions, intersections, complements, and other set operations can be described. In much the same way, the ideas of probability exist with a frame called a *probability space* Σ. For simplicity, we will consider only the *discrete* case. Then Σ consists of an underlying set of points and an assignment of probabilities. The set is called a *sample space* and its elements are *events*. The probability function, or *probability distribution*, assigns to each event e a real number $P(e)$ between 0 and 1, such that the sum of all the probabilities is 1.

The probability $P(e)$ is a measure of the likelihood that event e occurs. An impossible event has $P(e) = 0$ and a certain event would have $P(e) = 1$; in other cases, $P(e)$ has some intermediate value.

The probability space itself contains all possible events that may occur, identified in complete detail, which makes it too elaborate for actual use. Suppose we flip a coin. The probability space contains events like, "The coin came up heads, the weather was sunny, etc." and also "The coin came up heads, the weather was rainy, etc." We need a way of disregarding the weather and other extraneous factors, identifying each of these events as a "heads" outcome for our experiment. An *abstract random variable* X is a way of labeling the events in the sample space by various possible *values* x. For each value x, we have a whole collection of events in the sample space that are assigned that value: $V_x = \{e : X(e) = x\}$. We can now define the probability of a value x:

$$p(x) = \sum_{e \in V_x} P(e). \tag{A.1}$$

These probabilities satisfy $0 \leq p(x) \leq 1$ and $\sum_x p(x) = 1$. In our coin-flip example, either $X(e) = $ "heads" or $X(e) = $ "tails" for every event e, and we can define the probabilities $p(\text{heads})$ and $p(\text{tails})$ for the two values according to Eq. A.1.

Strictly speaking, the random variable X is a labeling function for the sample space. But informally we can regard it as a set of values equipped with its own probability distribution.

Thus we say that $x \in X$ for a value x. We have probabilities, not only for the values in X, but also for whole subsets of X. The probability of $A \subseteq X$ is

$$p(A) = \sum_{x \in A} p(x). \tag{A.2}$$

Exercise A.1 If $A \subseteq B \subseteq X$, prove that $p(A) \leq p(B)$. Under what circumstances does equality hold?

Exercise A.2 Suppose A and A^c are complementary subsets of X. Show that $p(A^c) = 1 - p(A)$.

We can also describe the probability that a particular property holds for X. Given a property $\mathbf{Q}(x)$ (a statement about x values that might be true or false) there is a subset Q of X containing exactly those x values for which $\mathbf{Q}(x)$ holds. Then $\Pr(\mathbf{Q}(x)$ is true$) = p(Q)$.

When the values of X are real numbers – representing the numerical value of some measurement, perhaps – then we say that X is a *numerical random variable*. We will have much more to say about these in the next section.

A *joint random variable* (X, Y) has values that are pairs (x, y) of values in X and Y, which are random variables in the same underlying probability space. To take a simple example, the day's weather may be characterized by temperature X and precipitation Y. The joint probability $p(x, y)$ is

$$p(x, y) = \sum_{x \in V_{x,y}} P(e), \tag{A.3}$$

where $V_{x,y} = \{e : X(e) = x \text{ and } Y(e) = y\}$. (Note that a joint distribution can be defined only if X and Y are random variables in the same probability space.)

Once we have the joint distribution, we can simply accept it as a distribution over the pairs of values (x, y). Consider, for example, the following distribution:

$p(x, y)$	warm	cold
dry	0.4	0.2
rainy	0.1	0.3.

$$\tag{A.4}$$

From the joint distribution we can calculate the distributions for the individual variables X and Y. These *marginal distributions* are given by

$$p(x) = \sum_{y} p(x, y), \qquad p(y) = \sum_{x} p(x, y). \tag{A.5}$$

In our weather example, the probability $p(\text{warm}) = 0.4 + 0.1 = 0.5$.

Exercise A.3 What is the probability that the day will be rainy?

Exercise A.4 Show that the probability $p(x)$ calculated from the joint distribution via Eq. A.5 must be equal to that calculated directly from the underlying probability space according to Eq. A.1.

From joint distributions we can also derive *conditional probabilities*. Suppose we select some value y for Y with $p(y) \neq 0$. Then we define the conditional probability for x given y to be

$$p(x|y) = \frac{p(x,y)}{p(y)}. \tag{A.6}$$

For each possible y this yields a probability distribution over X, representing the likelihood of various values of x provided that we know y has occurred. Two random variables are said to be *independent* if knowledge of one does not affect the probabilities assigned to the other – that is, if $p(x|y) = p(x)$ for every possible y. In this case,

$$p(x,y) = p(x|y)\,p(y) = p(x)\,p(y), \tag{A.7}$$

and the joint distribution over (X, Y) is exactly the product of the marginal distributions over X and Y.

Exercise A.5 From Eq. A.4, what is $p(\text{cold}|\text{rainy})$? Are the temperature and precipitation independent random variables?

We often consider a sequence of independent, identically distributed random variables: $\vec{X} = (X_1, \ldots, X_n)$. (As always, these must be regarded as random variables in the same "universe," the same underlying probability space.) A particular outcome of this joint variable is a sequence $\vec{x} = (x_1, \ldots, x_n)$. The probability of this sequence is

$$p(\vec{x}) = p(x_1) \cdots p(x_n), \tag{A.8}$$

where each marginal distribution is a duplicate of the same $p(x)$.

Exercise A.6 Suppose the weather on successive days is a sequence of independent, identically distributed random variables with the distribution given by Eq. A.4. What is the probability that it will be warm and dry for an entire week?

Probability theory provides rational methods for making inferences based on the data. Suppose the variable Y represents different theoretical hypotheses, so that $p(y)$ is an estimate of the likelihood of hypothesis y. (This is sometimes called the *a priori* probability for y.) The variable X represents the result of some experiment. Each hypothesis y produces a conditional predicted distribution $p(x|y)$. The joint distribution is $p(x,y) = p(x|y)\,p(y)$.

Now we perform the experiment and observe a particular result x. How does this observation change our judgment of the likelihood of the various hypotheses? In other words, what is the *a posteriori* probability $p(y|x)$? This turns out to be

$$p(y|x) = \frac{p(x,y)}{p(x)} = \frac{p(x|y)\,p(y)}{\sum_{y'} p(x|y')p(y')}. \tag{A.9}$$

This result, known as *Bayes's theorem*, tells us how to adjust our initial distribution over Y based on X data.

To see Bayes's theorem in action, imagine that we have a bag containing many coins. Most of these are ordinary, fair coins that are equally likely to come up "heads" or "tails" when flipped. However, 10% of the coins are of the fraudulent, two-headed variety for which $p(\text{heads}) = 1$. We draw a coin at random from the bag. In the absence of any further data, we assign the *a priori* probability $p(\text{fraud}) = 0.1$ to the hypothesis that the coin has two heads. The conditional probabilities are

$$p(\text{heads}|\text{fair}) = 0.5, \quad p(\text{heads}|\text{fraud}) = 1.0,$$
$$p(\text{tails}|\text{fair}) = 0.5, \quad p(\text{tails}|\text{fraud}) = 0.0. \tag{A.10}$$

Now suppose the coin is flipped and comes up "heads." This bit of data changes our assessment of the likelihood that the coin is fraudulent:

$$p(\text{fraud}|\text{heads}) = \frac{p(\text{heads}|\text{fraud})p(\text{fraud})}{p(\text{heads}|\text{fraud})p(\text{fraud}) + p(\text{heads}|\text{fair})p(\text{fair})}$$

$$= \frac{1.0 \times 0.1}{(1.0 \times 0.1) + (0.5 \times 0.9)} = 0.182. \tag{A.11}$$

Given the result "heads" of a single coin flip, the likelihood that the coin has two heads nearly doubles.

Exercise A.7 What adjustment to the fair/fraud probabilities should we make if the coin comes up "tails"?

Exercise A.8 Suppose a coin is drawn from the bag and flipped three times, coming up heads each time. What is $p(\text{fraud}|\text{heads, heads, heads})$? How many times must the coin come up "heads" before the likelihood of a fraudulent coin exceeds 99%?

A.2 Numerical random variables

A numerical random variable X labels events by values that are real numbers x. For such a variable, we define the *mean* or *expected value* or *expectation value* of X, denoted $\langle X \rangle$, to be

$$\langle X \rangle = \sum_x x p(x), \tag{A.12}$$

which is the sum of the various possible values of X, weighted by the probabilities. The phrase "expected value," of course, is slightly misleading, since the number $\langle X \rangle$ may not be one of the possible numerical values of X. Intuitively, $\langle X \rangle$ is what we expect to obtain "on average, in the long run" if we repeat our experiment many times. We will make this intuition rigorous below.

Suppose we have a pair of numerical random variables X and Y in the same underlying probability space. We can create new numerical variables from the sum $X + Y$, product XY, etc., which also have means:

Exercise A.9 Show that $\langle X + Y \rangle = \langle X \rangle + \langle Y \rangle$ for any pair X and Y. Show that $\langle XY \rangle \neq \langle X \rangle \langle Y \rangle$ in general, but that equality does hold for independent variables.

How far is a variable X from its mean $\mu = \langle X \rangle$? We could imagine situations in which the values of X cluster tightly around μ, and other situations in which the X values are very far from μ. We would like a measure of how "spread out" the values of X are likely to be. The difference $X - \mu$ is itself a random variable, for which we may calculate the mean:

$$\langle X - \mu \rangle = \langle X \rangle - \mu = 0. \tag{A.13}$$

This ends up zero because $\langle X - \mu \rangle$ can be either positive or negative depending on the particular x, and such terms cancel out in the sum. We should instead consider the squared difference $(X - \mu)^2$, which is never negative. The mean of this variable is called the *variance* of X, denoted $(\Delta X)^2$ or $\mathrm{var}(X)$:

$$\mathrm{var}(X) = \left\langle (X - \mu)^2 \right\rangle = \left\langle X^2 \right\rangle - 2\mu \langle X \rangle + \mu^2 = \left\langle X^2 \right\rangle - \langle X \rangle^2. \tag{A.14}$$

The variance is a measure of how "spread out" the possible values of X are. A closely related measure is the *standard deviation*, which is $\Delta X = \sqrt{\mathrm{var}(X)}$.

Exercise A.10 Show that $\left\langle X^2 \right\rangle = \langle X \rangle^2$ if and only if the numerical value of X is certain.

Exercise A.11 A six-sided die is rolled. Let F be the number (1–6) shown on its upper face. Calculate $\langle F \rangle$, $\mathrm{var}(F)$, and ΔF.

Suppose we have a sequence of independent, identically distributed numerical random variables $(X_1, \ldots, X_n)$. Each variable has the same mean $\langle X \rangle$ and variance $\mathrm{var}(X) = \left\langle X^2 \right\rangle - \langle X \rangle^2$. Consider a new variable $\bar{X}$, the statistical average of the X_is:

$$\bar{X} = \frac{X_1 + \ldots + X_n}{n}. \tag{A.15}$$

A single experiment to find $\bar{X}$ would include n independent experiments to measure X_i, after which the average in Eq. A.15 is calculated. So what are the mean and variance of this new random variable $\bar{X}$? It is easy to see that the mean is $\langle \bar{X} \rangle = \langle X \rangle$. The variance requires more work. First, we calculate $\left\langle \bar{X}^2 \right\rangle$:

$$\left\langle \bar{X}^2 \right\rangle = \frac{1}{n^2} \left\langle (X_1 + \ldots + X_n)^2 \right\rangle$$

$$= \frac{1}{n^2} \sum_i \left\langle X_i^2 \right\rangle + \frac{1}{n^2} \sum_{\substack{i,j \\ i \neq j}} \langle X_i X_j \rangle. \tag{A.16}$$

The variables are identical, so that each term $\left\langle X_i^2 \right\rangle = \left\langle X^2 \right\rangle$. They are also independent, so $\langle X_i X_j \rangle = \langle X \rangle^2$ for $i \neq j$; there are $n^2 - n$ such terms. The variance of $\bar{X}$ is

$$\mathrm{var}(\bar{X}) = \frac{1}{n} \left\langle X^2 \right\rangle + \left(\frac{n^2 - n}{n^2} \right) \langle X \rangle^2 - \langle X \rangle^2 = \frac{\mathrm{var}(X)}{n}. \tag{A.17}$$

The average $\bar{X}$ of many independent "samples" of X has the same mean as X but a smaller variance. The observed values of the average $\bar{X}$ must be clustered more closely about the mean value $\langle X \rangle$.

Let us probe this idea more carefully. From now on let X be a numerical random variable with mean $\mu = \langle X \rangle$ and variance $\sigma^2 = \text{var}(X)$. For the moment, also suppose that $X \geq 0$; that is, no value x of X is negative. Then for any $\eta > 0$,

$$\Pr(x > \eta) = \sum_{x > \eta} p(x)$$

$$< \sum_{x > \eta} \frac{x}{\eta} p(x)$$

$$\leq \frac{1}{\eta} \sum_x x p(x) = \frac{\mu}{\eta}. \tag{A.18}$$

This is called *Markov's inequality*. For non-negative X, the likelihood that x is very much larger than its mean – i.e. that $x > \eta \gg \mu$ – has to be small.

Using Markov's inequality we can establish a more general result. We now no longer assume that $X \geq 0$; however, the random variable $(X - \mu)^2$ is still non-negative, with a mean of σ^2. We can apply Markov's inequality to this variable. For any $\eta > 0$,

$$\Pr\left((x - \mu)^2 > \eta^2\right) < \frac{\sigma^2}{\eta^2}. \tag{A.19}$$

(For later clarity we have used η^2 instead of η.) Rewriting the condition in the probability, we obtain

$$\Pr\left(|x - \mu| > \eta\right) < \frac{\sigma^2}{\eta^2}, \tag{A.20}$$

a fact known as *Chebyshev's inequality*. If $\eta \gg \sigma$ (the standard deviation), Eq. A.20 tells us that it is unlikely for X to be more than η removed from its mean μ.

The final step is to apply Chebyshev's inequality to $\bar{X}$, the average value of n independent Xs. We are especially interested in the case when η is small, so we shall rename this parameter ϵ. The variance of $\bar{X}$ is σ^2/n. Thus, for any $\epsilon > 0$,

$$\Pr\left(|\bar{x} - \mu| > \epsilon\right) \leq \frac{\sigma^2}{n\epsilon^2}. \tag{A.21}$$

Given ϵ and any $\delta > 0$, we can choose $n > \sigma^2/\delta\epsilon^2$, making the right-hand side itself no larger than δ. We arrive at a very important result, which we state formally:

Weak Law of Large Numbers. Let $X_1, X_2, \ldots, X_n$ be independent, identically distributed numerical random variables, each with finite mean μ and variance σ^2. Given any $\delta, \epsilon > 0$, for sufficiently large n,

$$\Pr\left(\left|\left(\frac{1}{n}\sum_k x_k\right) - \mu\right| > \epsilon\right) < \delta. \tag{A.22}$$

For large n, the likelihood that the average value $\bar{x}$ is farther than ϵ from $\langle X \rangle$ is less than δ.

The Weak Law of Large Numbers is one of the most significant theorems in probability theory. Simply put, it states that a statistical average (taken over independent samples) almost certainly approximates the probabilistic mean, provided that the sample size n is large enough. This is exactly what we meant at the beginning of the section when we claimed that $\langle X \rangle$ is what we expect to obtain "on average, in the long run" if we repeat our experiment many times.

Exercise A.12 Apply the Weak Law of Large Numbers to the statistical estimation of probability. If we count the number of times a value x occurs in a large number of independent experiments, the frequency of occurrence is most likely close to the probability $p(x)$. Phrase this in a rigorous way and prove it. What numerical random variable are we considering?

The facts of probability presented in this appendix are enough to start with. A few more results are described in the main chapters, such as continuous distributions in Section 10.1 and the classical entropy function in Sections 19.1 and 19.2. For more sophisticated developments, including analysis of important special distributions, consult a textbook on probability theory.

Problems

Problem A.1 Consider a simple binary experiment such that $p(0) = q$ and $p(1) = p = 1 - q$. Some n independent trials of this experiment are performed, and we are interested in the random variable K that counts the number of 1s that occur. K has $n + 1$ possible values $0, 1, \ldots, n$.

(a) Show that the probability distribution for K is given by

$$p(k) = \binom{n}{k} p^k q^{n-k}, \tag{A.23}$$

where $\binom{n}{k} = \dfrac{n!}{k!(n-k)!}$ is the binomial coefficient, the number of ways that k items can be selected from among n items. Equation A.23 is called the *binomial distribution*.

(b) Find the mean and the variance of the binomial distribution. (You can do this by computing a difficult summation, but there is a much easier way!)

(c) *Stirling's formula* provides an approximation of $n!$ for large n:

$$n! \approx \sqrt{2\pi n}\, n^n\, e^{-n}. \tag{A.24}$$

Use this to find an approximate formula for the binomial coefficient. Once you have done this, estimate the value of k that maximizes the probability $p(k)$ for large n. (The idea is to treat k as a continuous variable and apply the usual methods of calculus.) Does your result agree with your intuition?

Appendix B **Fourier facts**

B.1 The delta function

> ... [E]ven the most precise sciences normally work with more or less ill-understood approximations toward which the scientist must maintain an appropriate skepticism .... The physicist rightly dreads precise argument, since an argument which is only convincing if precise loses all its force if the assumptions upon which it is based are slightly changed, while an argument which is convincing though imprecise may well be stable under small perturbations of its underlying axioms. (Jack Schwartz, *The Pernicious Influence of Mathematics on Science*)

In this appendix, we review some techniques of mathematical physics. We present the mathematics in "physics style" – that is, with apparent disregard for the mathematical niceties. We will use "functions" whose properties cannot be matched by any actual function. We will exchange the order of limit operations by commuting integrals, derivatives, and infinite sums, all without any apparent consideration of the deep analytical issues involved. If the math police gave out tickets for reckless deriving, we would probably get one.

Why risk it? Often, the "reckless" derivation is a useful shorthand for a more sophisticated (and rigorous) chain of mathematical reasoning. An ironclad proof of a result may have to deal with many technical issues that, though necessary to close all of the logical loopholes, act to obscure the central ideas. The less formal approach is therefore both briefer and more revealing. Finally, we should remember (as the above quotation reminds us) that the mathematical objects used in physics are already drastic idealizations of real physical entities. Mathematics is capable of a level of rigor that is simply inappropriate for such rough approximations to reality.

Does this mean that mathematical rigor should be banished from physics? Not at all. Any chain of logical reasoning, even an informal one, must avoid the pitfalls of fallacy and self-contradiction. Ultimately, the only defence against these is careful logic. Generations of mathematicians have labored to make rigorous the customary tools of physics, from calculus to the path integrals of quantum field theory. We as physicists are confident that our tools are safe and effective in daily use, but the ultimate source of that confidence is their vast body of meticulous, rigorous mathematical reasoning.

There is no better illustration of the "physics style" than the *delta function* $\delta(x)$, introduced into mathematical physics by P. A. M. Dirac. This function has two key properties:

- $\delta(x) = 0$ for all $x \neq 0$.
- $\displaystyle\int_{-\infty}^{+\infty} \delta(x)\,dx = 1$.

These are clearly problematic as rigorous statements! Intuitively, we think of $\delta(x)$ as a function whose graph is a thin "spike" at the origin. Even though the spike is extremely narrow, it is also very high, so that the area under the curve is 1.

The delta function $\delta(x)$ is actually a shorthand for a limit involving a family of spike functions that become narrower and higher. Expressions involving $\delta(x)$ stand for much more complex limiting expressions. However, it is more convenient to pretend that $\delta(x)$ is an actual function that may be manipulated in the usual way. This works surprisingly well.

Exercise B.1 Show that $\delta(-x)$ has exactly the same two defining properties as $\delta(x)$. We can thus regard $\delta(x)$ as an "even" function: $\delta(x) = \delta(-x)$. (This is our shorthand for the fact that we can consider only *symmetric* spike functions in our limit.)

The most useful property of the delta function is the *sampling property*. Suppose we have a continuous function $f(x)$ defined on an interval C containing a. We consider the integral

$$I = \int_C f(x)\delta(x-a)\,dx. \tag{B.1}$$

The integrand is zero (since $\delta(x-a)$ is zero) for all values of x except $x = a$. We can thus replace $f(x)$ by $f(a)$ inside the integral, since this has no effect on the value of the integrand at any point). Furthermore, we can replace the integral over C by an integral from $-\infty$ to ∞, since the delta function is zero outside of C. Then

$$I = \int_{-\infty}^{+\infty} f(a)\delta(x-a)\,dx = f(a)\int_{-\infty}^{+\infty} \delta(x-a)\,dx = f(a). \tag{B.2}$$

The sampling property of the delta function is thus

$$f(a) = \int_C f(x)\delta(x-a)\,dx. \tag{B.3}$$

That is, the factor $\delta(x-a)$ in the integrand causes the integral to "pick out" the value $f(a)$ from the function on the interval C.

A second useful property involves the derivative $\delta'(x)$. If $f(x)$ is differentiable on the interval C containing a, then

$$f'(a) = -\int_C f(x)\delta'(x-a)\,dx. \tag{B.4}$$

We call this the *derivative property* of the delta function.

Exercise B.2 Obtain the derivative property via integration by parts.

The delta function $\delta(x)$ can be seen as a continuous version of the Kronecker delta δ_{mn}. The delta function "collapses" integral expressions in much the same way that δ_{mn} "collapses" discrete sums.

B.2 Fourier series

Suppose that the function f is periodic with period L. That is, $f(x + L) = f(x)$ for all x. Among such functions are the basic "sinusoidal" functions

$$u_n(x) = e^{ik_n x}, \tag{B.5}$$

where $k_n = \dfrac{2\pi n}{L}$ and n is an integer, which may be positive or negative. These have an orthogonality property over the interval from $-L/2$ to $+L/2$:

$$\int_{-L/2}^{+L/2} u_m^*(x)\, u_n(x)\, dx = \begin{cases} L & m = n \\ 0 & m \neq n \end{cases}. \tag{B.6}$$

Exercise B.3 Derive Eq. B.6 from the definition of $u_n(x)$.

Our essential idea, first advanced by Joseph Fourier, is that *any* reasonable periodic function $f(x)$ can be written as a sum of sinusoidal components:

$$f(x) = \sum_n c_n e^{ik_n x}, \tag{B.7}$$

where $n = \dots, -1, 0, +1, \dots$ This is called the *Fourier series* for $f(x)$. The coefficients are given by

$$c_n = \frac{1}{L} \int_{-L/2}^{L/2} f(x) e^{-ik_n x}\, dx. \tag{B.8}$$

Exercise B.4 Under the assumption that $f(x)$ is given by the Fourier series in Eq. B.7, use the orthogonality property in Eq. B.6 to derive the Fourier coefficients in Eq. B.8.

The Fourier coefficients provide an alternative way of describing the periodic function $f(x)$. Various properties of the function are related to properties of the coefficients:

- If $f(x)$ is real, then $c_{-n} = c_n^*$.
- If $f(x)$ is even, then $c_{-n} = c_n$.
- If $f(x)$ is odd, then $c_{-n} = -c_n$, and as a consequence $c_0 = 0$.

B.3 Fourier transforms

In the Fourier series for functions that are periodic with period L, the adjacent values of k_n are separated by

$$\Delta k = \frac{2\pi}{L}. \tag{B.9}$$

We define $\tilde{f}(k_n)$ to be related to the Fourier coefficient c_n by

$$\tilde{f}(k_n) = \frac{L}{\sqrt{2\pi}} c_n. \tag{B.10}$$

Then

$$f(x) = \sum_n \frac{\sqrt{2\pi}}{L} \tilde{f}(k_n) e^{ik_n x} = \frac{1}{\sqrt{2\pi}} \sum_n \tilde{f}(k_n) e^{ik_n x} \Delta k, \tag{B.11}$$

and

$$\tilde{f}(k_n) = \frac{1}{\sqrt{2\pi}} \int_{-L/2}^{L/2} f(x) e^{-ik_n x} \, dx. \tag{B.12}$$

If we let the period $L \to \infty$ then the discrete variable k_n becomes a continuous k. In this limit, we have

$$f(x) = \frac{1}{\sqrt{2\pi}} \int_{-\infty}^{+\infty} \tilde{f}(k) e^{ikx} \, dk, \tag{B.13}$$

$$\tilde{f}(k) = \frac{1}{\sqrt{2\pi}} \int_{-\infty}^{+\infty} f(x) e^{-ikx} \, dx. \tag{B.14}$$

The function $\tilde{f}(k)$ is called the *Fourier transform* of $f(x)$. Equations B.13 and B.14 give the relationship of $f(x)$ and $\tilde{f}(k)$.

The Fourier transform of the delta function is particularly simple:

$$\tilde{\delta}(k) = \frac{1}{\sqrt{2\pi}} \int_{-\infty}^{+\infty} \delta(x) e^{-ikx} \, dx = \frac{1}{\sqrt{2\pi}}. \tag{B.15}$$

This means that we can use Eq. B.13 to give a highly useful formula for the delta function:

$$\delta(x) = \frac{1}{2\pi} \int_{-\infty}^{+\infty} e^{ikx} \, dk. \tag{B.16}$$

We can use the delta function to prove the uniqueness of the Fourier transform. Suppose $g(k)$ is a function such that $f(x) = \frac{1}{\sqrt{2\pi}} \int_{-\infty}^{+\infty} g(k) e^{ikx} \, dx$. Then

$$\tilde{f}(k) = \frac{1}{\sqrt{2\pi}} \int_{-\infty}^{+\infty} f(x) e^{-ikx} \, dx$$

$$= \frac{1}{2\pi} \int_{-\infty}^{+\infty} \left(\int_{-\infty}^{+\infty} g(k') e^{ik'x} \, dk' \right) e^{-ikx} \, dk. \tag{B.17}$$

Exchanging the order of integration and combining terms, we obtain

$$\tilde{f}(k) = \frac{1}{2\pi} \int_{-\infty}^{+\infty} g(k') \left(\int_{-\infty}^{+\infty} e^{i(k'-k)x} \, dx \right) dk$$

$$= \int_{-\infty}^{+\infty} g(k') \delta(k' - k) \, dk$$

$$\tilde{f}(k) = g(k). \tag{B.18}$$

Any function $g(k)$ that yields $f(x)$ via the Fourier integral in Eq. B.13 must in fact be the Fourier transform $\tilde{f}(k)$.

Exercise B.5 All of the factors of 2π in this section are determined by the 2π in Eq. B.10. If we had omitted this factor in the definition of $\widetilde{f}(k_n)$, how would our expressions change?

It is handy to have a couple of other examples of the Fourier transform in our "toolkit." Consider a function that is a rectangular pulse centered on the origin:

$$h(x) = \begin{cases} 0 & |x| > 1 \\ 1 & |x| \le 1 \end{cases}. \tag{B.19}$$

The Fourier transform can be computed pretty easily

$$\widetilde{h}(k) = \int_{-1}^{1} e^{-ikx}\, dx$$

$$= \frac{1}{-ik}\left(e^{-ik} - e^{ik}\right)$$

$$\widetilde{h}(k) = 2\,\frac{\sin k}{k}. \tag{B.20}$$

The function $\dfrac{\sin k}{k}$ is sometimes denoted $\operatorname{sinc}(k)$.

Exercise B.6 Suppose we consider a rectangular pulse of width $2a$ instead of 2, for some $a > 0$. What is the resulting Fourier transform?

Also consider a function of the form

$$f(x) = e^{-b|x|}. \tag{B.21}$$

This has a pointed "peak" of unit height at $x = 0$, and drops off exponentially in both directions. Its Fourier transform is

$$\widetilde{f}(k) = \sqrt{\frac{2}{\pi}}\,\frac{b}{b^2 + k^2}. \tag{B.22}$$

Note that a narrower $f(x)$ (given by a greater value of the exponential parameter b) leads to a wider $\widetilde{f}(k)$.

Exercise B.7 Derive $\widetilde{f}(k)$ in Eq. B.22.

The Fourier transform defined by Eq. B.13 and B.14 has a number of elementary properties, which can be used to greatly extend the range of functions whose transforms we can compute:

- *Linearity.* If $f(x) = af_1(x) + bf_2(x)$, then $\widetilde{f}(k) = a\widetilde{f_1}(k) + b\widetilde{f_2}(k)$.
- *Shift property.* The Fourier transform of $f(x - x_0)$ is $e^{-ikx_0}\widetilde{f}(k)$.
- *Modulation property.* The Fourier transform of $e^{-ik_0x}f(x)$ is $\widetilde{f}(k - k_0)$.
- *Dilation property.* If $\alpha > 0$ is a real parameter, the Fourier transform of $f(\alpha x)$ is $\dfrac{1}{\alpha}\widetilde{f}\left(\dfrac{k}{\alpha}\right)$.

Linearity follows from the linearity of the integral in Eq. B.14. The other properties are easy to prove by appropriate changes of variable.

Exercise B.8 Prove the modulation property of the Fourier transform.

Now suppose that $f(x)$ has Fourier transform $\widetilde{f}(k)$. What is the Fourier transform of the derivative $f'(x)$?

$$
\begin{aligned}
f'(x) &= \frac{d}{dx} \int_{-\infty}^{+\infty} \widetilde{f}(k) e^{ikx} \, dk \\
&= \int_{-\infty}^{+\infty} \widetilde{f}(k) \left(\frac{\partial}{\partial x} e^{ikx} \right) dk \\
&= \int_{-\infty}^{+\infty} ik \widetilde{f}(k) e^{ikx} \, dk.
\end{aligned}
\tag{B.23}
$$

From the uniqueness of Fourier transforms, we see from this that the Fourier transform of the derivative $f'(x)$ is just $ik\widetilde{f}(k)$. We can call this the *derivative property* of the Fourier transform.

Exercise B.9 What is the Fourier transform of $f''(x)$? How about the function $\delta'(x - a)$?

A very deep fact about Fourier transform is known as *Parseval's theorem*. Suppose we have two functions $f(x)$ and $g(x)$, and consider the integral

$$
I = \int_{-\infty}^{+\infty} f^*(x) g(x) \, dx.
\tag{B.24}
$$

How can we write this using Fourier transforms? We can figure this out using the properties of the Fourier transform, exchanging the order of integration on the way:

$$
\begin{aligned}
I &= \frac{1}{2\pi} \int_{-\infty}^{+\infty} \left(\int_{-\infty}^{+\infty} \widetilde{f}(k) e^{ikx} dk \right)^* \left(\int_{-\infty}^{+\infty} \widetilde{g}(k') e^{ik'x} \, dk' \right) dx \\
&= \int_{-\infty}^{+\infty} dk \int_{-\infty}^{+\infty} dk' \widetilde{f}^*(k) \widetilde{g}(k') \left(\frac{1}{2\pi} \int_{-\infty}^{+\infty} e^{i(k'-k)x} \, dx \right) \\
&= \int_{-\infty}^{+\infty} dk \int_{-\infty}^{+\infty} dk' \widetilde{f}^*(k) \widetilde{g}(k') \delta(k' - k).
\end{aligned}
\tag{B.25}
$$

From this we find that

$$
\int_{-\infty}^{+\infty} f^*(x) g(x) \, dx = \int_{-\infty}^{+\infty} \widetilde{f}^*(k) \widetilde{g}(k) \, dk.
\tag{B.26}
$$

This is Parseval's theorem. As a simple consequence, it follows that, for all $f(x)$,

$$
\int_{-\infty}^{+\infty} |f(x)|^2 \, dx = \int_{-\infty}^{+\infty} |\widetilde{f}(k)|^2 \, dk.
\tag{B.27}
$$

Parseval's theorem has implications for the position and momentum wave functions introduced in Chapter 10. See, for example, Eq. 10.45.

B.4 Application to PDEs

The Fourier transform is so amazingly useful that we cannot hope to do more than suggest a few of its applications. One such application is to the solution of differential equations.

Suppose a substance is diffusing in one dimension through a uniform medium. Then the concentration of the substance is represented by a function $c(x, t)$ depending both on location x and time t. The concentration function obeys the 1-D *diffusion equation*, which is

$$\frac{\partial c}{\partial t} = D \frac{\partial^2 c}{\partial x^2}, \tag{B.28}$$

where D is a constant called the *diffusion constant*. How can we find a general solution to this partial differential equation?

The answer lies in re-expressing $c(x, t)$ by its Fourier transform[1] $\tilde{c}(k, t)$. Taking the Fourier transform of both sides of Eq. B.28 gives us

$$\frac{\partial}{\partial t} \tilde{c}(k, t) = D(-ik)^2 \tilde{c}(k, t) = -k^2 D \tilde{c}(k, t). \tag{B.29}$$

This is now a differential equation only in t, and can be solved analytically for each value of k:

$$\tilde{c}(k, t) = \tilde{c}(k, 0) \, e^{-k^2 D t}. \tag{B.30}$$

Each part of the Fourier transform $\tilde{c}(k, t)$ decays independently in an exponential way. The rate of decay is determined by D and k^2. Components with higher spatial frequencies k will decay away more rapidly, which is why the $c(x, t)$ function "smooths out" over time. The main point here is that, given any initial concentration function $c(x, 0)$, we can find $\tilde{c}(k, 0)$ and then work out the concentration – both $\tilde{c}(k, t)$ and $c(x, t)$ – for any later time t.

Exercise B.10 Suppose the initial concentration of the diffusing material is a Gaussian:

$$c(x, 0) = c_0 e^{-a_0 x^2}. \tag{B.31}$$

Show that the concentration at a later time is also a Gaussian distribution. Write an explicit solution for $c(x, t)$. (You will need the results for Gaussian functions from the next appendix.)

---

[1] Notice that only the x variable is undergoing a Fourier transform to k, and the time variable t is left alone.

Appendix C **Gaussian functions**

A *Gaussian* function has the general form

$$g(x) = e^{-ax^2}, \tag{C.1}$$

for $a > 0$. This is the famous "bell curve" function, which is important in probability theory.

The Gaussian has a finite integral:

$$\int_{-\infty}^{+\infty} e^{-ax^2} dx = \sqrt{\frac{\pi}{a}}. \tag{C.2}$$

The fact that we can evaluate this integral is slightly surprising, since there is no elementary antiderivative of e^{-ax^2}. The definite integral is done using a famous and very clever trick. We actually compute the square of the integral:

$$\left(\int_{-\infty}^{+\infty} e^{-ax^2} dx \right)^2 = \left(\int_{-\infty}^{+\infty} e^{-ax^2} dx \right) \left(\int_{-\infty}^{+\infty} e^{-ay^2} dy \right)$$

$$= \int_{-\infty}^{+\infty} dx \int_{-\infty}^{+\infty} dy \, e^{-a(x^2+y^2)}$$

$$= \int_{\mathbb{R}^2} e^{-a(x^2+y^2)} \, dA. \tag{C.3}$$

This is the integral over the whole plane of the function $e^{-a(x^2+y^2)}$. But we can do this integral in any coordinate system. If we choose polar coordinates (r, θ), then the integrand is e^{-ar^2}, which is radially symmetric. For the area element dA we choose thin circular rings, which have an area $dA = 2\pi r \, dr$. Then the integral becomes

$$2\pi \int_0^\infty e^{-ar^2} r \, dr = \frac{\pi}{a} \int_0^\infty e^{-u} \, du$$

$$= \frac{\pi}{a}. \tag{C.4}$$

Equation C.2 follows.

A Gaussian probability density has the general form

$$\mathcal{P}(x) = \sqrt{\frac{a}{\pi}} e^{-a(x-b)^2}. \tag{C.5}$$

The constant ensures that the probability density is normalized.

To determine more about the Gaussian probability density, it helps to have two further integrals:

$$\int_{-\infty}^{+\infty} x e^{-ax^2} = 0, \tag{C.6}$$

$$\int_{-\infty}^{+\infty} x^2 e^{-ax^2} = \frac{1}{2a}\sqrt{\frac{\pi}{a}}. \tag{C.7}$$

From these we can evaluate $\langle x \rangle$ and $\langle x^2 \rangle$ for the Gaussian $\mathcal{P}(x)$.

Exercise C.1 Show that for the Gaussian distribution in Eq. C.5, $\langle x \rangle = b$ and $\langle x^2 \rangle = \frac{1}{2a} + b^2$.

Let us denote the mean of x by μ and the variance of x by σ^2. Then it follows that $b = \mu$ and $a = (2\sigma^2)^{-1}$. We can therefore write the Gaussian probability distribution as

$$\mathcal{P}(x) = \frac{1}{\sqrt{2\pi\sigma^2}} \exp\left(-\frac{(x-\mu)^2}{2\sigma^2}\right). \tag{C.8}$$

Notice that a Gaussian distribution is completely determined by its mean μ and variance σ^2. The form in Eq. C.8 is particularly useful, since these parameters appear explicitly.

Let $g(x) = e^{-ax^2}$ be a Gaussian function. Then the Fourier transform of $g(x)$ is

$$\widetilde{g}(k) = \frac{1}{\sqrt{2a}} e^{-k^2/4a}. \tag{C.9}$$

Note that the Fourier transform of a Gaussian function is also a Gaussian function. If the parameter a is increased, the width of the bell curve $g(x)$ decreases – but the width of the Gaussian $\widetilde{g}(k)$ increases by exactly the same factor. In general, a narrowly peaked function $g(x)$ has a spread-out transform $\widetilde{g}(k)$.

The Dirac delta function can be viewed as a limiting case of a normalized Gaussian, in the limit as the curve becomes narrower and higher. That is, as $a \to \infty$,

$$\delta_a(x) = \sqrt{\frac{a}{\pi}} e^{-ax^2} \longrightarrow \delta(x). \tag{C.10}$$

The Fourier transform of $\delta_a(x)$ is

$$\widetilde{\delta}_a(k) = \frac{1}{\sqrt{2\pi}} e^{-k^2/4a} \longrightarrow \frac{1}{\sqrt{2\pi}}, \tag{C.11}$$

which is the (constant) Fourier transform of the delta function.

Appendix D Generalized evolution

D.1 Completely positive maps

In Chapter 9 we described the dynamics of a system Q that interacted with an external "environment" system E, which was initially in a standard state $|0\rangle$. This open system evolution was described by the map $\mathcal{E}$, so that the initial density operator ρ evolved by

$$\rho \rightarrow \mathcal{E}(\rho) = \text{Tr}_{(E)} U\left(\rho \otimes |0\rangle\langle 0|\right) U^{\dagger}, \tag{D.1}$$

for a unitary operator U on the composite system QE. From this, we derived a second way of expressing the map $\mathcal{E}$, the Kraus or operator–sum representation:

$$\mathcal{E}(\rho) = \sum_k A_k \rho A_k^{\dagger}, \tag{Re 9.10}$$

where the Kraus operators A_k are normalized according to Eq. 9.12.

We then made the claim that any "physically reasonable" time evolution of density operators could be expressed in this way. Our business here is to explain what we mean by this, and to prove the theorem.

What do we mean by a "physically reasonable" map $\mathcal{E}$? We begin by noting that $\mathcal{E}$ must be linear in the input density operator.[2]

Exercise D.1 Use an ensemble argument to show this. (See the argument leading up to Eq. 9.51.)

Since the map is linear, it is a *superoperator*, an element of $\mathcal{B}(\mathcal{B}(\mathcal{H}))$.

The necessary properties of the evolution superoperator $\mathcal{E}$ are determined by the requirement that the output of $\mathcal{E}$ must be a valid density operator for any valid input state. This leads us at once to two properties:

- $\mathcal{E}$ must be a *positive* map, in the sense that it maps positive operators to positive operators.
- $\mathcal{E}$ must be a *trace-preserving* map, so that $\text{Tr}\,\mathcal{E}(A) = \text{Tr}\,A$ for all operators A.

The first property is easy to prove because every positive operator is a scalar multiple of a possible density operator. The second is not much harder.

---

[2] Throughout, we are assuming that the later state of affairs depends on the initial system density operator ρ only, and not on any more elaborate description of the system.

Exercise D.2 Prove that a superoperator $\mathcal{E}$ is trace-preserving for all operators if and only if it is trace-preserving for density operators. To do this, you will need to write an arbitrary operator as a linear combination of density operators. See Exercise 3.40 and Problem 3.4 for ideas.

Exercise D.3 Suppose $\mathcal{E}$ is a linear map on operators, and we know that $\mathcal{E}(|\psi\rangle\langle\psi|)$ is a positive, trace-1 operator for input pure states $|\psi\rangle\langle\psi|$. Show that $\mathcal{E}$ must be a positive, trace-preserving map – that is, show that $\mathcal{E}(\rho)$ must be a positive, trace-1 operator for any input state ρ.

We might hope that we are finished, that any positive, trace-preserving map $\mathcal{E}$ is a possible time evolution for an open quantum system. However, things are not so simple! There are positive, trace-preserving maps that cannot correspond to the time evolution of a quantum system.

We will show this by considering a qubit example. We saw in Section 8.4 that any operator A on the qubit Hilbert space can be written in terms of basis operators $\mathbf{1}$, X, Y, and Z:

$$A = \frac{1}{2}(a_0\mathbf{1} + a_X X + a_Y Y + a_Z Z). \tag{D.2}$$

(In Eq. 8.37, we took advantage of the fact that $a_0 = \operatorname{Tr} A = 1$ for density operators.) Our proposed map $\mathcal{T}$ acts on the basis operators as follows:

$$\begin{aligned}
\mathcal{T}(\mathbf{1}) &= \mathbf{1}, & \mathcal{T}(X) &= X, \\
\mathcal{T}(Y) &= Y, & \mathcal{T}(Z) &= -Z.
\end{aligned} \tag{D.3}$$

Thus, $\mathcal{T}$ maps an operator given by components (a_0, a_X, a_Y, a_Z) to one given by components $(a_0, a_X, a_Y, -a_Z)$.

Exercise D.4 Use the geometry of the Bloch sphere to describe how $\mathcal{T}$ acts on qubit density operators.

Exercise D.5 Show that $\mathcal{T}$ is a positive, trace-preserving map. (Hint: For positivity, first show that an operator is positive if and only if $\vec{a} \cdot \vec{a} \le a_0^2$.)

Why does $\mathcal{T}$ not describe a possible evolution of the state of a qubit? The reason is that our qubit is not necessarily alone in the Universe. Consider a second, independent qubit whose state evolves according to the identity map $\mathcal{I}$. If the first qubit evolved according to $\mathcal{T}$, then the composite system would evolve according to $\mathcal{T} \otimes \mathcal{I}$, defined by its action on product operators:

$$\mathcal{T} \otimes \mathcal{I}(A \otimes B) = \mathcal{T}(A) \otimes B. \tag{D.4}$$

Amazingly, even though both $\mathcal{T}$ and $\mathcal{I}$ are positive superoperators, the combination $\mathcal{T} \otimes \mathcal{I}$ is not. There is an entangled input state of the two qubits which is mapped by $\mathcal{T} \otimes \mathcal{I}$ to an operator that is not positive, and thus not a legitimate output density operator. The following pair of exercises construct such an example.

Exercise D.6 Show that

$$T(|0\rangle\langle 0|) = |1\rangle\langle 1|, \qquad T(|1\rangle\langle 1|) = |0\rangle\langle 0|,$$
$$T(|0\rangle\langle 1|) = |0\rangle\langle 1|, \qquad T(|1\rangle\langle 0|) = |1\rangle\langle 0|. \qquad (D.5)$$

Exercise D.7 Calculate $T \otimes \mathcal{I}(|\Phi_+\rangle\langle\Phi_+|)$. Show that this operator is not positive by showing that

$$\langle\Phi_-| T \otimes \mathcal{I}(|\Phi_+\rangle\langle\Phi_+|)|\Phi_-\rangle < 0. \qquad (D.6)$$

Here $|\Phi_\pm\rangle$ are the Bell states from Eq. 7.13.

What does all this mean? If we wish to confine our attention to *possible* dynamical evolutions of an open system, we will need to be more choosy. We need a stronger property for $\mathcal{E}$ that will exclude non-physical cases like the T map. Luckily, the structure of our example provides a clue to what this stronger property looks like.

The map $\mathcal{E}$ for a system is said to be *completely positive* if the map $\mathcal{E} \otimes \mathcal{I}$ is positive whenever we include an independent quantum system in our consideration. This means that, for any initial pure state $|\Psi\rangle$ of the composite system, the operator $\mathcal{E} \otimes \mathcal{I}(|\Psi\rangle\langle\Psi|)$ is positive. We conclude that any map describing the evolution of the state of a quantum system must be a linear, trace-preserving completely positive map. These are the "physically reasonable" maps describing the generalized evolution of a quantum system.

Exercise D.8 For a qubit system, suppose that $\mathcal{R}(\rho) = |0\rangle\langle 0|$ for every input system ρ. This map simply "resets" the state of the system to $|0\rangle\langle 0|$. Show that $\mathcal{R}$ is completely positive.

Now we ask the question of whether every generalized evolution has to be regarded as a physically reasonable time evolution of a quantum system. Our answer is contained in the following important theorem:

Representation theorem for generalized dynamics. The following conditions are equivalent for a map $\mathcal{E}$:

(a) $\mathcal{E}$ is a linear, trace-preserving, completely positive map on Q-operators.

(b) $\mathcal{E}$ has a "unitary representation;" that is, we can introduce an environment system E, an initial environment state $|0\rangle$ and a joint unitary evolution U on QE so that

$$\mathcal{E}(G) = \text{Tr}_{(E)} U \left(G \otimes |0\rangle\langle 0| \right) U^\dagger. \qquad (Re\ 9.2)$$

(c) $\mathcal{E}$ has a Kraus representation; that is, we can find operators A_k such that

$$\mathcal{E}(\rho) = \sum_k A_k \rho A_k^\dagger. \qquad (Re\ 9.10)$$

The Kraus operators satisfy the normalization condition $\sum_k A_k^\dagger A_k = \mathbf{1}$ (Eq. 9.12).

Any generalized evolution map $\mathcal{E}$ that is linear, trace-preserving, and completely positive can be "realized" by unitary dynamics on a larger system. We thus should regard any such map as a "physically reasonable" time evolution for the system.

The representation theorem is a fundamental and powerful result. Its proof, which is somewhat technical, is presented in the next section.

D.2 The representation theorem

We now prove the representation theorem stated above. The structure of our proof is to show that (a) implies (b), (b) implies (c), and (c) implies (a). Some of these implications are quite easy to see. Our discussion of generalized evolution in Section 9.1 *defined* $\mathcal{E}$ via its unitary representation, and then derived its Kraus representation. We have thus already proven that (b) implies (c) in our theorem. Another implication is equally straightforward.

Exercise D.9 Show that any map described by a normalized Kraus representation is linear, trace-preserving, and completely positive. Thus, (c) implies (a).

It only remains to prove that (a) implies (b). There are several systems to keep track of in that proof, together with their various states and operators, so for the remainder of this section we will be scrupulous to indicate the quantum system associated with each mathematical object.

Let $\mathcal{E}^{(Q)}$ be a linear, trace-preserving completely positive map for a quantum system Q, which has a Hilbert space of finite dimension $\dim \mathcal{H}^{(Q)} = d$. As in Section 7.5, we append an identical quantum system R and consider the maximally entangled state

$$\left| \Phi^{(RQ)} \right\rangle = \frac{1}{\sqrt{d}} \sum_k \left| k^{(R)}, k^{(Q)} \right\rangle. \tag{Re 7.28}$$

Any initial state $|\psi^{(Q)}\rangle$ of Q could arise in the following way. The system RQ is initially in the entangled state $|\Phi^{(RQ)}\rangle$, and then a measurement is performed on R. The resulting state of Q, conditional on the particular measurement outcome for R, happens to be $|\psi^{(Q)}\rangle$.

Let us do this a bit more mathematically. Given $|\psi^{(Q)}\rangle = \sum_k c_k |k^{(Q)}\rangle$, we can construct the R-state

$$\left| \tilde{\psi}^{(R)} \right\rangle = \sum_k c_k^* \left| k^{(R)} \right\rangle. \tag{D.7}$$

The significance of this state is seen here:

Exercise D.10 With the definitions given, show that

$$\left\langle \tilde{\psi}^{(R)} \middle| \Phi^{(RQ)} \right\rangle = \frac{1}{\sqrt{d}} \left| \psi^{(Q)} \right\rangle. \tag{D.8}$$

If $\left| \tilde{\psi}^{(R)} \right\rangle$ corresponds to one outcome of a basic measurement on R, then the probability of that outcome is $1/d$ and the conditional state of Q is $|\psi^{(Q)}\rangle$.

In terms of operators, we can let $\pi_\psi^{(R)} = \left| \tilde{\psi}^{(R)} \right\rangle\left\langle \tilde{\psi}^{(R)} \right|$ and write

$$
\left| \psi^{(Q)} \right\rangle\left\langle \psi^{(Q)} \right| = d \operatorname{Tr}_{(R)} \left(\left| \Phi^{(RQ)} \right\rangle\left\langle \Phi^{(RQ)} \right| \pi_\psi^{(R)} \right)
$$

$$
= \mathcal{M}_\psi^{(R)} \left(\left| \Phi^{(RQ)} \right\rangle\left\langle \Phi^{(RQ)} \right| \right). \tag{D.9}
$$

The operation $\mathcal{M}_\psi^{(R)}$ can be loosely translated, "Make the appropriate measurement on R and obtain result $\tilde{\psi}$."[3] This map takes the maximally entangled state $\left| \Phi^{(RQ)} \right\rangle\left\langle \Phi^{(RQ)} \right|$ to the particular Q-state $\left| \psi^{(Q)} \right\rangle\left\langle \psi^{(Q)} \right|$. As both a physical and mathematical fact, we can see that $\mathcal{M}_\psi^{(R)}$ only acts on R and therefore commutes with operations on Q and any other systems.

Note: To simplify our notation, we often omit the identity operator when the context is clear. Thus, in Eq. D.9, we wrote $\pi_\psi^{(R)}$ rather than $\pi_\psi^{(R)} \otimes \mathbf{1}^{(Q)}$. We will do the same for superoperators. The map $\mathcal{D}^{(A)}$ acting on the state $\rho^{(AB)}$ is automatically taken to be $\mathcal{D}^{(A)} \otimes \mathcal{I}^{(B)}$, where $\mathcal{I}^{(B)}$ is the identity map for B.

For instance, suppose we wish to evolve the initial state $\left| \psi^{(Q)} \right\rangle\left\langle \psi^{(Q)} \right|$ by the map $\mathcal{E}^{(Q)}$. We can begin with $\left| \Phi^{(RQ)} \right\rangle\left\langle \Phi^{(RQ)} \right|$, apply $\mathcal{M}_\psi^{(R)}$ to obtain $\left| \psi^{(Q)} \right\rangle\left\langle \psi^{(Q)} \right|$, and then act with $\mathcal{E}^{(Q)}$. But since $\mathcal{M}_\psi^{(R)}$ and $\mathcal{E}^{(Q)}$ commute, we could do things in the reverse order: act with $\mathcal{E}^{(Q)}$ on $\left| \Phi^{(RQ)} \right\rangle\left\langle \Phi^{(RQ)} \right|$ and then apply $\mathcal{M}_\psi^{(R)}$. Either way, we must wind up with $\mathcal{E}^{(Q)}(\left| \psi^{(Q)} \right\rangle\left\langle \psi^{(Q)} \right|)$. This can be represented diagrammatically:

$$
\begin{array}{ccc}
\left| \Phi^{(RQ)} \right\rangle\left\langle \Phi^{(RQ)} \right| & \xrightarrow{\ \mathcal{M}_\psi^{(R)}\ } & \left| \psi^{(Q)} \right\rangle\left\langle \psi^{(Q)} \right| \\[2mm]
\mathcal{E}^{(Q)} \downarrow & & \mathcal{E}^{(Q)} \downarrow \\[2mm]
\omega^{(RQ)} & \xrightarrow{\ \mathcal{M}_\psi^{(R)}\ } & \mathcal{E}^{(Q)}(\left| \psi^{(Q)} \right\rangle\left\langle \psi^{(Q)} \right|),
\end{array} \tag{D.10}
$$

where $\omega^{(RQ)} = \mathcal{E}^{(Q)} \left(\left| \Phi^{(RQ)} \right\rangle\left\langle \Phi^{(RQ)} \right| \right)$. We have shown that

$$
\mathcal{E}^{(Q)}(\left| \psi^{(Q)} \right\rangle\left\langle \psi^{(Q)} \right|) = \mathcal{M}_\psi^{(R)} \left(\mathcal{E}^{(Q)}(\left| \Phi^{(RQ)} \right\rangle\left\langle \Phi^{(RQ)} \right|) \right). \tag{D.11}
$$

This gives us a rather curious way of interpreting the diagram in Eq. D.10. We can either select the initial Q-state $\left| \psi^{(Q)} \right\rangle\left\langle \psi^{(Q)} \right|$ (using $\mathcal{M}_\psi^{(R)}$) and then evolve it (according to $\mathcal{E}^{(Q)}$); or we can *first* evolve the system Q *and then* select the input Q-state $\left| \psi^{(Q)} \right\rangle\left\langle \psi^{(Q)} \right|$. To put it another way, applying the generalized evolution map $\mathcal{E}^{(Q)}$ to the entangled state $\left| \Phi^{(RQ)} \right\rangle\left\langle \Phi^{(RQ)} \right|$ of RQ is formally like applying it to *every pure input Q-state at once*. The operator $\omega^{(RQ)}$ somehow includes a complete description of how $\mathcal{E}^{(Q)}$ acts on those inputs.

---

[3] The reader may have some misgivings about the operation $\mathcal{M}_\psi^{(R)}$. After all, we cannot *force* the R-measurement to yield the result ψ. Two points may provide some reassurance. (1) $\mathcal{M}_\psi^{(R)}$ can be regarded as a purely mathematical construction, defined by Eq. D.9, whose properties do not rely on our physical interpretation. (2) If we insist on the physical interpretation, we have to think of $\mathcal{M}_\psi^{(R)}$ as a chancy sort of operation – one which may or may not work in a given instance, but which has the specified effect when it does.

Operator $\omega^{(RQ)}$ is a density operator for RQ. (This operator is positive and has unit trace because the map $\mathcal{E}^{(Q)}$ is completely positive and trace-preserving.) We can append a system E and find a global pure state $|\Omega^{(RQE)}\rangle$ such that

$$\omega^{(RQ)} = \text{Tr}_{(E)} |\Omega^{(RQE)}\rangle\langle\Omega^{(RQE)}|. \tag{D.12}$$

Exercise D.11 Explain why the system E need have a Hilbert space dimension no larger than d^2 to do the job. Under what circumstances might it be smaller?

Since we obtain $\omega^{(RQ)}$ from $|\Phi^{(RQ)}\rangle\langle\Phi^{(RQ)}|$ by $\mathcal{E}^{(Q)}$, which acts only on Q, it follows that the subsystem state of R is the same in each case. Appending E does not change this fact,

$$\text{Tr}_{(Q)} |\Phi^{(RQ)}\rangle\langle\Phi^{(RQ)}| = \text{Tr}_{(QE)} |\Omega^{(RQE)}\rangle = \frac{1}{d} \mathbf{1}^{(R)}. \tag{D.13}$$

Choose an E-state $|0^{(E)}\rangle$ and consider the two pure state vectors $|\Phi^{(RQ)}\rangle \otimes |0^{(E)}\rangle$ and $|\Omega^{(RQE)}\rangle$. In the terminology of Exercise 8.16, both of these are "purifications" of the same R-state in the system QE. Therefore (by the same exercise) they are related by a unitary operator that acts only on QE:

$$|\Omega^{(RQE)}\rangle = U^{(QE)} |\Phi^{(RQ)}\rangle \otimes |0^{(E)}\rangle. \tag{D.14}$$

This unitary transformation on vectors defines a map $\mathcal{U}^{(QE)}$ on operators:

$$\mathcal{U}^{(QE)}(\sigma^{(QE)}) = U^{(QE)}\sigma^{(QE)}U^{(QE)\dagger}. \tag{D.15}$$

At this point it is worthwhile to remind ourselves that the RQE pure state $|\Omega^{(RQE)}\rangle\langle\Omega^{(RQE)}|$, the E-state $|0^{(E)}\rangle\langle0^{(E)}|$ and the unitary map $\mathcal{U}^{(QE)}$ are constructed on the basis of the maximally entangled state $|\Phi^{(RQ)}\rangle\langle\Phi^{(RQ)}|$ and not on any particular choice $|\psi^{(Q)}\rangle\langle\psi^{(Q)}|$ of input state for Q. The "input selection" operation $\mathcal{M}_{\psi}^{(R)}$ commutes with both $\mathcal{U}^{(QE)}$ and the partial trace operation $\text{Tr}_{(E)}$. We can therefore construct the following diagram:

$$
\begin{array}{ccc}
|\Phi^{(RQ)}\rangle\langle\Phi^{(RQ)}| \otimes |0^{(E)}\rangle\langle0^{(E)}| & \xrightarrow[\text{(Eq. D.9)}]{\mathcal{M}_{\psi}^{(R)}} & |\psi^{(Q)}\rangle\langle\psi^{(Q)}| \otimes |0^{(E)}\rangle\langle0^{(E)}| \\
\\
\mathcal{U}^{(QE)} \downarrow \text{ (Eq. D.14)} & & \mathcal{U}^{(QE)} \downarrow \text{ (Eq. D.15)} \\
\\
|\Omega^{(RQE)}\rangle\langle\Omega^{(RQE)}| & & U^{(QE)}(|\psi^{(Q)}\rangle\langle\psi^{(Q)}| \\
& & \otimes |0^{(E)}\rangle\langle0^{(E)}|)U^{(QE)\dagger} \\
\\
\text{Tr}_{(E)} \downarrow \text{ (Eq. D.12)} & & \text{Tr}_{(E)} \downarrow \\
\\
\omega^{(RQ)} & \xrightarrow[\text{(Eq. D.11)}]{\mathcal{M}_{\psi}^{(R)}} & \mathcal{E}^{(Q)}(|\psi^{(Q)}\rangle\langle\psi^{(Q)}|).
\end{array}
\tag{D.16}
$$

Equation D.16 is the "road map" of our argument. The whole diagram must lead to the same lower right-hand result whichever way it is traversed. The various arrows are justified by the referenced equations. This leaves one last arrow, which must also express a valid relation. For any pure Q-state $|\psi^{(Q)}\rangle$,

$$\mathcal{E}^{(Q)}(|\psi^{(Q)}\rangle\langle\psi^{(Q)}|) = \mathrm{Tr}_{(E)} U^{(QE)}\left(|\psi^{(Q)}\rangle\langle\psi^{(Q)}| \otimes |0^{(E)}\rangle\langle0^{(E)}|\right) U^{(QE)\dagger}. \tag{D.17}$$

Since this is true for all pure states, it must by linearity also hold for any mixture of pure states. Thus, we have established the existence of a unitary representation for the map $\mathcal{E}^{(Q)}$. The representation theorem is thus proved.

D.3 Measurement and conditional dynamics

The projection rule of Section 4.3 tells us how the state of a quantum system is changed in a measurement process, at least in some situations. An outcome x of the measurement is associated with a projection operator Π_x. If the system starts out in the state $|\psi\rangle$, and if a measurement is made yielding the outcome x, then the subsequent state of the system is $K\Pi_x|\psi\rangle$, where the coefficient K is determined by normalization.

Unfortunately, the projection rule applies only to highly idealized measurement processes. Now that we have developed the mathematical machinery of generalized measurements and evolution maps, it is time to formulate a more universal rule.

Our measurement is a generalized one characterized by positive operators E_μ whose sum is the identity $\mathbf{1}$. If the system is in the state ρ, the probability that the outcome μ appears is $P(\mu) = \mathrm{Tr}\,\rho E_\mu$. When the measurement yields μ, the system ends up in a state that we denote ρ_μ. Our goal is to find the rule by which this state ρ_μ arises from ρ and the measurement.

If we average the final state over all measurement outcomes, then the resulting map is a generalized evolution:

$$\rho \to \mathcal{E}(\rho) = \sum_\mu P(\mu)\rho_\mu. \tag{D.18}$$

We can call this the *unconditional evolution* of the state in the measurement process, since it describes a situation in which the measurement is made, but its result is not specified. For the individual measurement outcomes, we break the unconditional $\mathcal{E}$ into a set of partial maps:

$$\mathcal{E} = \sum_\mu \mathcal{E}_\mu, \tag{D.19}$$

where $\mathcal{E}_\mu(\rho) = P(\mu)\rho_\mu$. The partial maps, of course, are not trace-preserving, since $\mathrm{Tr}\,\mathcal{E}_\mu(\rho) = P(\mu)$ rather than 1. However, the maps make up for this deficiency by being both linear and completely positive.

Exercise D.12 By an ensemble argument, show that $\mathcal{E}_\mu$ is linear. Also explain, by means of a plausible counter-example, why the map $\rho \to \rho_\mu$ (without the $P(\mu)$ factor) might not be linear in ρ.

Exercise D.13 By considering a measurement made on one subsystem of a composite system, show that $\mathcal{E}_\mu$ must be completely positive.

Since each $\mathcal{E}_\mu$ is linear and completely positive, it has an operator–sum representation:

$$\mathcal{E}_\mu(\rho) = \sum_k \mathsf{A}_{k\mu} \rho \mathsf{A}_{k\mu}^\dagger, \tag{D.20}$$

(see Problem D.3).

What are these Kraus operators $\mathsf{A}_{k\mu}$? If we sum over both k and μ, then we get the unconditional evolution map:

$$\mathcal{E}(\rho) = \sum_{k\mu} \mathsf{A}_{k\mu} \rho \mathsf{A}_{k\mu}^\dagger, \tag{D.21}$$

with $\sum_{k\mu} \mathsf{A}_{k\mu}^\dagger \mathsf{A}_{k\mu} = \mathbf{1}$. The probability of the measurement outcome μ is

$$P(\mu) = \operatorname{Tr} \mathcal{E}_\mu(\rho) = \operatorname{Tr} \rho \left(\sum_k \mathsf{A}_{k\mu}^\dagger \mathsf{A}_{k\mu} \right). \tag{D.22}$$

Since this is true for all input states ρ, we can conclude that the outcome operator E_μ is related to the $\mathsf{A}_{k\mu}$ operators via

$$\mathsf{E}_\mu = \sum_k \mathsf{A}_{k\mu}^\dagger \mathsf{A}_{k\mu}. \tag{D.23}$$

Equation D.20 gives us the conditional evolution of the quantum state in the measurement process, while Eq. D.23 relates the generalized measurement (the E_μ operators) to that evolution.

Sometimes things are a bit simpler. We say that a measurement process is *ideal* if the operator sum in Eq. D.20 has only one term for each outcome μ. Then

$$\mathcal{E}_\mu(\rho) = \mathsf{A}_\mu \rho \mathsf{A}_\mu^\dagger, \tag{D.24}$$

where $\mathsf{E}_\mu = \mathsf{A}_\mu^\dagger \mathsf{A}_\mu$. As suggested in Problem D.4, we can regard any measurement process as an ideal measurement process in which the outcome is incompletely recorded. An ideal measurement is called *minimally disturbing* if $\mathsf{A}_\mu = \mathsf{E}_\mu^{1/2}$ for all μ.

Exercise D.14 Show that the projection rule holds for minimally disturbing (and thus ideal) projection measurements, and only for those.

(See also Problem D.5.)

For a specific example, let us once again recall the discussion of the projection rule in Section 4.3. There, we considered a two-beam interferometer which had either one or zero photons in it. The Hilbert space for the system had basis states $|0\rangle$ (no photon), $|u\rangle$ (one photon in the upper beam), and $|l\rangle$ (one photon in the lower beam). We make a measurement by putting an absorbing photodetector in the lower beam and registering whether the detector "clicks" either once or not at all. (This is the situation described in Exercise 4.10.) A detector "click" registers that a photon was in the lower beam, but it also destroys the photon, leaving the system in the final state $|0\rangle$.

Exercise D.15 Show that the measurement process described is an ideal one with Kraus operators

$$A_0 = |0\rangle\langle 0| + |u\rangle\langle u|, \qquad A_1 = |0\rangle\langle l|. \qquad (D.25)$$

What are the corresponding measurement outcome operators E_0 and E_1? Verify that these add up to $\mathbf{1}$.

Problems

Problem D.1 Consider a generalized evolution map on a qubit system as a transformation of the Bloch sphere of qubit states.

(a) Suppose the linear map $\mathcal{F}$ leaves both $|0\rangle\langle 0|$ and $|1\rangle\langle 1|$ unaffected, but that $\mathcal{F}(|0\rangle\langle 1|) = \mathcal{F}(|1\rangle\langle 0|) = 0$. Describe how $\mathcal{F}$ acts on the Bloch sphere and show that it is trace-preserving and completely positive.
(b) Suppose the linear map $\mathcal{G}$ leaves both $|0\rangle\langle 1|$ and $|1\rangle\langle 0|$ unaffected, but that $\mathcal{G}(|0\rangle\langle 0|) = \mathcal{G}(|1\rangle\langle 1|) = \frac{1}{2}\mathbf{1}$. Describe how $\mathcal{G}$ acts on the Bloch sphere and show that it is trace-preserving but *not* completely positive.

Problem D.2 Suppose $\mathcal{E}$ is a generalized evolution map for a quantum system whose Hilbert space has dim $\mathcal{H} = d$. Show that there is an operator–sum representation for $\mathcal{E}$ that has no more than d^2 operators A_μ (see Exercise D.11).

Problem D.3 Review the arguments of Section D.2. Adapt them to prove that a linear and completely positive map has an operator–sum representation of the form in Eq. 9.10. If the map is not trace-preserving, however, the representation need not be normalized, so that Equation 9.12 need not hold.

Problem D.4

(a) Suppose we make a generalized measurement described by the outcome operators $E_{k\mu}$. The outcome of the measurement is designated by the compound index $k\mu$. However, only part of this outcome is recorded, so that we find out the value of μ but not of k. Show that this procedure can be effectively described by outcome operators

$$E_\mu = \sum_k E_{k\mu}. \qquad (D.26)$$

(b) The conditional evolution map for a measurement process is given in Eq. D.20. Show that we can view this as the result of an ideal measurement followed by a partial recording of the result, as in part (a).

Problem D.5

(a) First, prove the *polar decomposition theorem* for operators. This states that any operator A can be written $A = UP$, where U is unitary and P is positive. (This is analogous to the polar decomposition of complex numbers: $z = e^{i\phi}r$, where ϕ and r are real and $r \geq 0$.)

The operator P is given by

$$P = \left(A^\dagger A\right)^{1/2}.$$

(D.27)

For a positive operator G, $G^{1/2}$ is the unique positive operator whose square is G. We know that $A^\dagger A$ is positive, a fact first proven in Exercise 3.39.

(b) Show that any ideal measurement operation can be viewed as a minimally disturbing measurement operation followed by *conditional* unitary evolution. (That is, if the outcome μ arises in the minimally disturbing measurement, then the unitary operator U_μ is applied to the conditional state ρ_μ.)

Index

CPSIA information can be obtained
at www.ICGtesting.com
Printed in the USA
LVHW061035190122
708888LV00005B/358